T0141973

Advances in Intelligent Systems and Computing

Volume 1433

The series "Advances in Intelligent Systems and Computing" contains publications on theory, applications, and design methods of Intelligent Systems and Intelligent Computing. Virtually all disciplines such as engineering, natural sciences, computer and information science, ICT, economics, business, e-commerce, environment, healthcare, life science are covered. The list of topics spans all the areas of modern intelligent systems and computing such as: computational intelligence, soft computing including neural networks, fuzzy systems, evolutionary computing and the fusion of these paradigms, social intelligence, ambient intelligence, computational neuroscience, artificial life, virtual worlds and society, cognitive science and systems, Perception and Vision, DNA and immune based systems, self-organizing and adaptive systems, e-Learning and teaching, human-centered and human-centric computing, recommender systems, intelligent control, robotics and mechatronics including human-machine teaming, knowledge-based paradigms, learning paradigms, machine ethics, intelligent data analysis, knowledge management, intelligent agents, intelligent decision making and support, intelligent network security, trust management, interactive entertainment, Web intelligence and multimedia.

The publications within "Advances in Intelligent Systems and Computing" are primarily proceedings of important conferences, symposia and congresses. They cover significant recent developments in the field, both of a foundational and applicable character. An important characteristic feature of the series is the short publication time and world-wide distribution. This permits a rapid and broad dissemination of research results.

Indexed by DBLP, INSPEC, WTI Frankfurt eG, zbMATH, Japanese Science and Technology Agency (JST).

All books published in the series are submitted for consideration in Web of Science.

For proposals from Asia please contact Aninda Bose (aninda.bose@springer.com).

More information about this series at https://link.springer.com/bookseries/11156

Luis A. García-Escudero ·
Alfonso Gordaliza · Agustín Mayo ·
María Asunción Lubiano Gomez ·
Maria Angeles Gil · Przemyslaw Grzegorzewski ·
Olgierd Hryniewicz
Editors

Building Bridges between Soft and Statistical Methodologies for Data Science

Springer

Editors
Luis A. García-Escudero
Departamento de Estadística e I.O.
University of Valladolid
Valladolid, Spain

Alfonso Gordaliza
Departamento de Estadística e I.O.
University of Valladolid
Valladolid, Spain

Agustín Mayo
Departamento de Estadística e I.O.
University of Valladolid
Valladolid, Spain

María Asunción Lubiano Gomez
Departamento de Estadística, I.O. y D.M.
University of Oviedo
Oviedo, Spain

Maria Angeles Gil
Departamento de Estadística, I.O. y D.M.
University of Oviedo
Oviedo, Spain

Przemyslaw Grzegorzewski
Department of Computational Statistics and Data Analysis
Warsaw University of Technology
Warsaw, Poland

Olgierd Hryniewicz
Department of Stochastic Methods, Systems Research Institute
Polish Academy of Sciences
Warsaw, Poland

ISSN 2194-5357 ISSN 2194-5365 (electronic)
Advances in Intelligent Systems and Computing
ISBN 978-3-031-15508-6 ISBN 978-3-031-15509-3 (eBook)
https://doi.org/10.1007/978-3-031-15509-3

This Springer imprint is published by the registered company Springer Nature Switzerland AG
The registered company address is: Gewerbestrasse 11, 6330 Cham, Switzerland

Preface

This volume is a selection of peer-reviewed papers presented at the 10th International Conference on Soft Methods in Probability and Statistics, SMPS 2022, held in Valladolid (Spain) during September 14–16, 2022. The series of biannual international conferences on Soft Methods in Probability and Statistics (SMPS) started in Warsaw in 2002 and continued in Oviedo (2004), Bristol (2006), Toulouse (2008), Oviedo/Mieres (2010), Konstanz (2012), Warsaw (2014), Rome (2016) and Compiègne (2018), progressively consolidating itself in the international agenda of events in Probability, Statistics and Soft Computing.

The 10th edition of the SMPS conference was initially scheduled for 2020, however, the SARS-COV-2 pandemic obligated to postpone twice the event. Although it could have been held on the initially scheduled date through an online modality, organizers have decided to wait for having a face-to-face activity. For sure, the warm atmosphere associated with the close exchange of scientific discussions is one of the best assets of this conference.

SMPS 2022 has been organized by the Departamento de Estadística e Investigación Operativa and the Instituto de Investigación en Matemáticas (IMUVA) of the University of Valladolid, Spain. Valladolid has a long tradition in research in the field of Probability and Statistics and has successfully welcomed important national and international events in this area, such as the Spanish Conference on Statistics an Operations Research (SEIO 2007), the International Conference on Robust Statistics (ICORS 2011) or the recent conference New Bridges between Mathematics and Data Science (NBMDS 2021). We are convinced that this event will also be a success, thanks to the effort and good work of the entire local committee. We have done our best in order to guarantee the participants in the conference have a comfortable stay in Valladolid. Thus, this will meet their expectations both scientifically and personally and will allow them to enjoy our rich culture and history.

We are grateful to the Executive Board of this conference for allowing us to have the opportunity to hold the event in Valladolid. We are also grateful to the Program Committee members for their support on the scientific aspects of the conference, specifically to all the session organizers. We are grateful to the keynote speakers of the conference, Christophe Croux (EDHEC Business School, France), Francisco Herrera (University of Granada, Spain) and Frank Klawonn (Helmholtz Centre for Infection Research, Germany) for accepting our invitation. We would also like to thank the members of the Local Committee for their valuable contribution to the logistics of the event.

In recent years, we are experiencing a revolution around data analysis motivated by the emergence of new data collection sources due to great technical developments. A by-product of this enormous availability of data is the appearance of a myriad of new data typologies that need to be analyzed. The complexity of these data sets requires the development of new probabilistic and statistical approaches capable of dealing with the difficulties associated with them. Different communities of experts, with very different origins, including mathematicians, statisticians, engineers, computer scientists, biotechnologists, econometricians and psychologists try to respond to these challenges with tools based on their own background. All these varied origins and different approaches motivate the importance of building bridges between all these fields for Data Science.

Soft methods are designed either to address, among others, difficulties related to imprecise or other complex data, or to create/combine alternatives to deal with traditional data. Consequently, they will certainly play an important role in the near future to cope with these current challenges. Furthermore, contaminated data are ubiquitous, and therefore, robust methodologies are required when certain degree of inaccuracy and noise are present in data.

The volume contains more than fifty selected contributions that are clearly useful in establishing those important bridges between soft and statistical methodologies for Data Science. These contributions cover very different and relevant aspects such as imprecise probabilities, information theory, random sets and random fuzzy sets, belief functions, possibility theory, dependence modeling and copulas, clustering, depth concepts, dimensionality reduction of complex data and robustness. The editors are grateful to all the contributing authors, Program Committee members and additional referees who made it possible to put together this interesting volume and preparing such attractive program for the SMPS 2022 conference. The priceless effort and good work of M. Asunción Lubiano as Publication Chair deserve our sincere recognition. Without her excellent work, this publication would not have been possible.

We would like to thank Publishing Editor of the Springer Series of Advances in Intelligent Systems and Computing, Dr. Thomas Ditzinger, as well as Series Editor, Professor Janusz Kacprzyk, and Springer for their dedication to the production of this volume.

May 2022

Luis A. García-Escudero
Alfonso Gordaliza
Agustín Mayo
María Asunción Lubiano Gomez
Maria Angeles Gil
Przemyslaw Grzegorzewski
Olgierd Hryniewicz

Organization

General Chairs

Luis A. García-Escudero	University of Valladolid, Spain
Alfonso Gordaliza	University of Valladolid, Spain
Agustín Mayo-Iscar	University of Valladolid, Spain

Executive Board (Core SMPS Group)

María Ángeles Gil	University of Oviedo, Spain
Przemyslaw Grzegorzewski	Polish Academy of Sciences, Warsaw, Poland
Olgierg Hryniewicz	Polish Academy of Sciences, Warsaw, Poland

Program Committee

Christian Borgelt, Germany, Austria
Paula Brito, Portugal
Giulianella Coletti, Italy
Ana Colubi, Spain, Germany
Inés Couso, Spain
Bernard De Baets, Belgium
Gert De Cooman, Belgium
Thierry Denoeux, France
Sébastien Destercke, France
Francesco Dotto, Italy
Didier Dubois, France
Fabrizio Durante, Italy
Pierpaolo D'Urso, Italy
Erol Eğrioğlu, Turkey
Alessio Farcomeni, Italy
Maria Brigida Ferraro, Italy
Peter Filzmoser, Austria
Alfonso García-Pérez, Spain
Jonathan Garibaldi, UK
Paolo Giordani, Italy
Gil González-Rodríguez, Spain
Francesca Greselin, Italy
Salvatore Ingrassia, Italy
Janusz Kacprzyk, Poland
Frank Klawonn, Germany
Vladik Kreinovich, USA
Rudolf Kruse, Germany
Mark Last, Israel
Jonathan Lawry, UK
Jacek M. Lęski, Poland
Miguel López-Díaz, Spain
M. Asunción Lubiano, Spain
Enrique Miranda, Spain
Susana Montes, Spain

Witold Pedrycz, Canada
Henri Prade, France
Madan L. Puri, USA
Dan A. Ralescu, USA
Marco Riani, Italy
Beatriz Sinova, Spain
Martin Štěpnička, Czech Republic
Pedro Terán, Spain
Valentin Todorov, Austria
Wolfgang Trutschnig, Austria
Stefan Van Aelst, Belgium
Barbara Vantaggi, Italy
Maurizio Vichi, Italy
José Antonio Vilar, Spain
Christian Wagner, UK
Jin Hee Yoon, South Korea
Slawomir Zadrożny, Poland

Additional Referees

Sebastian Fuchs, Austria
Pedro Miranda, Spain
Ignacio Montes, Spain
Arthur Van Camp, UK

Organizing Local Committee

Pedro C. Álvarez-Esteban, Spain
Eustasio del Barrio, Spain
Miguel A. Fernández, Spain
José Luis García-Lapresta, Spain
Raquel González del Pozo, Spain
Paula Gordaliza, Spain
Hristo Inouzhe, Spain
Carlos Matrán, Spain
Julio Pastor (Conference Website Chair), Spain
Azucena Prieto (Administrative Staff), Spain

Publication Chair

M. Asunción Lubiano University of Oviedo, Spain

Contents

Multi-dimensional Maximal Coherent Subsets Made Easy: Illustration on an Estimation Problem

Loïc Adam and Sébastien Destercke(✉)

UMR CNRS 7253 Heudiasyc, Sorbonne Université, Université de Technologie de Compiègne, CS 60319, 60203 Compiègne Cedex, France
{loic.adam,sebastien.destercke}@hds.utc.fr

Abstract. Fusing uncertain pieces of information to obtain a synthetic estimation when those are inconsistent is a difficult task. A particularly appealing solution to solve such conflict or inconsistency is to look at maximal coherent subsets of sources (MCS), and to concentrate on those. Yet, enumerating MCS is a difficult combinatorial task in general, making the use of MCS limited in practice. In this paper, we are interested in the case where the pieces of information are multi-dimensional sets or polytopes. While the problem remains difficult for general polytopes, we show that it can be solved more efficiently for hyperrectangles. We then illustrate how such an approach could be used to estimate linear models in the presence of outliers or in the presence of misspecified model.

1 Introduction

This paper deals with the problem of fusing multiple pieces of information (Dubois et al. 2016). In this problem, handling conflict between contradicting sources of information is one of the most difficult tasks. This is often a mandatory task, including in situations where we want a synthetic estimation from all the sources. Moreover, analyzing the reasons of the contradiction and trying to explain its appearance can be of equal importance, as it can give important insights about the situation.

Dubois et al. (1999) reviewed different methods for aggregating conflict, some requiring additional data like the reliability of the different sources. Yet, having additional data is sometimes difficult or even impossible. A quite appealing way for dealing with contradiction, requiring no additional information, is based on maximal coherent subsets (MCS), which are groups of consistent sources that are as big as possible. MCS have been used in the past both in logic (Manor and Rescher 1970) and in numerical settings (Destercke et al. 2008). Here, we illustrate their application to estimation problems.

Detecting and enumerating MCS are NP-hard problems, with intervals being a well-known exception (Dubois et al. 2000). In this paper, we show that we

L. A. García-Escudero et al. (Eds.): SMPS 2022, AISC 1433, pp. 1–8, 2023.
https://doi.org/10.1007/978-3-031-15509-3_1

can extend this exception to hyperrectangles, which can in turn be used as approximations of polytopes.

In Sect. 2 we further introduce maximal coherent subset and explain the difficulty of enumerating them. We then show in Sect. 3 that it is easy to list the MCSs of a set of axis-aligned hyperrectangles. Lastly, Sect. 4 illustrates our approach on linear regression problems.

2 Maximal Coherent Subsets

As mentioned previously, maximal coherent subsets are in theory a nice solution to manage conflict between information sources. Moreover, they can be used with different structures of information, like polytopes. However, we will show that listing the different MCSs is usually a difficult combinatorial problem.

General Definition: Let us suppose we have a set $\mathcal{S} = \{S_1, ..., S_N\}$ of sources of information providing a subset $S_i \subseteq \mathcal{X}$ of some space $\mathcal{X}$ of information, for which intersection $\cap$ is well-defined. A maximal coherent subset $c \subseteq \{1, \ldots, N\}$ is a list of source indices such that $\cap_{i \in c} S_i \neq \emptyset$, and for any $j \notin c$, $\cap_{i \in c} S_i \cap S_j = \emptyset$, i.e., the subset c of sources is consistent and is maximal with this property.

Example 1. Let us suppose we have a set of sources $\mathcal{S} = \{S_1, ..., S_4\}$ as shown on Fig. 1. As we can see, $\{1, 2\}$ is a coherent subset, but not a maximal coherent subset, as it is possible to add S_3 and have a non-empty intersection. $\{1, 2, 3\}$ is a maximal coherent subset, as S_4 is contradicting S_3 (empty intersection).

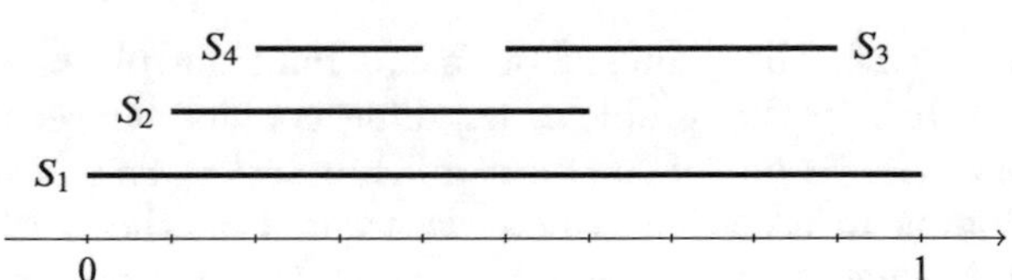

Fig. 1. Visualization of not fully consistent sources $S_i \in [0, ..., 1]$

MCSs of Polytopes: When considering sources of information in the d-dimensional Euclidean space R^d, polytopes are a quite versatile tool to model set-valued information. Such shapes can either be defined through their vertices or extreme points (V-representation) or through a system of linear constraints defining intersections of half-planes (H-representation).

While finding whether two polytopes given by their V-representation intersect is a NP-hard problem (Tiwary 2008), the same problem can be solved easily in H-representation through linear programming. Switching between representations is a NP-complete problem,[1] with some efficient algorithms for specific H-polytopes (Khachiyan et al. 2009).

Finding a single MCS in H-representation is thus easy: we add polytopes one by one, and we have a MCS when it is not possible to add another polytope without having an empty intersection. Checking if a set is a MCS is also easy: we check that the intersection of the corresponding H-polytopes is not empty and maximal. However, listing all the MCSs of a set of polytopes $\mathscr{P} = \{P_1, ..., P_N\}$ requires in the worst case to consider all the subsets of $\mathscr{P}$, thus at most 2^N sets for which we need to check if the intersection is not empty. When the number of polytopes is important, it becomes impossible to list all the MCSs. In Sect. 3, we propose an efficient algorithm to list all the MCSs through an approximation of the polytopes with minimum bounding axis-aligned hyperrectangles.

3 Enumerating the MCSs of Axis-Aligned Hyperrectangles

Enumerating all the MCS of a set of polytopes is a difficult problem in the general case. However, polynomial algorithms (Dubois et al. 2000) exist in the case of intervals. We will show in this section that such results can also be used in the case where we consider a set $\mathscr{H}$ of axis-aligned hyperrectangles, in order to efficiently determine the set of MCS $\mathscr{C}_{\mathscr{H}}$.

We denote by $I^d_{H_i} \in \mathbb{R}$ the projection of H_i onto the dth dimension of the space $\mathbb{R}^D$. We have an important equivalence between the intersection of hyperrectangles and the intersection of their projections:

Proposition 1. *Given a set* $\mathscr{H} = \{H_1, ..., H_N\}$ *of axis-aligned hyperrectangles in the space* $\mathbb{R}^D$*, and their projections* $I^d_{H_i} \in \mathbb{R}$ *onto the different dimensions* $d \in \{1, ..., D\}$*, we have:*

$$\bigcap_{H_i \in \mathscr{H}} H_i \neq \emptyset \Longleftrightarrow \bigcap_{\mathscr{H}} I^d_{H_i} \neq \emptyset \; \forall d \in \{1, ..., D\}. \tag{1}$$

Proof. To see the equivalence, it is sufficient to observe that $\times_{d=1}^{D} \cap_{i=1}^{N} I^d_{H_i} = \bigcap_{H_i \in \mathscr{H}} H_i$. This means in particular that any point $x \in \mathbb{R}^D$ such that its projection $x^d \in \cap_{i=1}^{N} I^d_{H_i}$ for all $d \in \{1, \ldots, D\}$ will also be in $\bigcap_{H_i \in \mathscr{H}} H_i$. Note that this is only true for axis-aligned hyperrectangles. □

The following corollary, which is merely the negation of Proposition 1, will be useful in further demonstrations.

Corollary 1

$$\bigcap_{\mathscr{H}} H_i = \emptyset \Longleftrightarrow \exists d \in \{1, ..., D\} \; s.t. \; \bigcap_{\mathscr{H}} I^d_{H_i} = \emptyset. \tag{2}$$

[1] Otherwise the two problems would have the same complexity.

In the next proof, we show that the MCS of axis-aligned hyperrectangles can be found exactly by combining the MCS of their projections, which we recall can be found in polynomial time.

Proposition 2. *Given a set* $\mathscr{H} = \{H_1, ..., H_N\}$ *of axis-aligned hyperrectangles in the space* $\mathbb{R}^D$*, its set of MCSs* $\mathscr{C}_{\mathscr{H}}$ *and the sets of MCSs* $\mathscr{C}_d$ *on their projection on the d-dimension, we have:*

$$\mathscr{C}_{\mathscr{H}} = \left\{\cap_{d=1}^{D} c_d \mid c_d \in \mathscr{C}_d \ \forall d \in \{1, ..., D\}, \cap_{d=1}^{D} c_d \neq \emptyset\right\}. \tag{3}$$

Proof. We proceed by showing a double inclusion for a given MCS $c_h \in \mathscr{C}_{\mathscr{H}}$:

- Let us first show that there exists $c_i \in \left\{\cap_{d=1}^{D} c_d \mid c_d \in \mathscr{C}_d\right\}$ s.t. $c_h \subseteq c_i$, i.e. c_i is an outer approximation of c_h. Because $\cap_{i \in c_h} H_i \neq \emptyset$ (it being a MCS), Proposition 1 tells us that for any dimension d, $\cap_{i \in c_h} I^d_{H_i} \neq \emptyset$, meaning that there will be a MCS $c_d \in \mathscr{C}_d$ such that $c_h \subseteq c_d$. Since this is true for all $d \in \{1, \dots, D\}$, this means that $c_h \subseteq \cap_{d=1}^{D} c_d$ for some collection of c_d, showing the inclusion.
- To show the other inclusion, we consider a set $c_d, d \in \{1, \dots, D\}$ of MCSs on dimension d which outer-approximate $c_h \subseteq c_d$, that we know exists from the first part of the proof. We will then demonstrate that $j \notin c_h$ implies $j \notin \cap_{d=1}^{D} c_d$, therefore $\cap_{d=1}^{D} c_d \subseteq c_h$. To see this, simply consider the set of hyperrectangles $H_k, k \in c_h \cup \{j\}$, then by Corollary 1, there will be a dimension d such that $\cap_{k \in c_h \cup \{j\}} I^d_{H_k} = \emptyset$, yet $\cap_{k \in c_h} I^d_{H_k} \neq \emptyset$ (c_h being a MCS). This shows that $j \notin c_h$ implies $j \notin \cap_{d=1}^{D} c_d$. □

Proposition 2 provides us with an easy approach to get MCS: we start by projecting the hyperrectangles onto the different dimensions, in order to obtain intervals. Then we enumerate the MCS on each dimension, which is polynomial. Lastly we determine $\mathscr{C}_{\mathscr{H}}$ as the set of common sources among all the enumerations, i.e., the conjunctive combination (i.e., intersection) of the different sources such that for each element c_h of $\mathscr{C}_{\mathscr{H}}$. By Proposition 2 and Eq. (3), this gives us exactly the set of MCS.

4 Illustration on Linear Estimation

In this section, we illustrate our method on small imprecise linear regression problems with only two dimensions. Note that the purpose of this section is purely illustrative, so as to show the potential usefulness of our result when performing estimation from a logical, set-theoretic standpoint.

Given a data set $\{y_i, x_i\}_{i=1}^{N}$ with a single input and a single output, a linear model assumes that the relationship between the response variable y_i and the input variable x_i is linear:

$$y_i = \beta_0 + \beta_1 x_i \tag{4}$$

Usually, the y_i are considered to be observed with a (normal) noise ϵ, and a statistical regression is performed. In our case, we will adopt a more logical, version space point of view (Mitchell 1982): we assume that data is set-valued, and will consider the linear models consistent with it.

4.1 Estimating Possible Linear Models with MCS and Rectangles

In our setting, we assume that we observe imprecise data points $R_i = ([\underline{y_i}, \overline{y_i}], [\underline{x_i}, \overline{x_i}])$. Given the Cartesian equation of a line L $= \{(x, y) \mid ax + by = c\}$ and two imprecise points R_i and R_j, finding all the lines that intersect both rectangles is formalized as:

$$\mathscr{L}_{ij} = \{(a, b) \mid (L \cap H_i \neq \emptyset) \wedge (L \cap H_j \neq \emptyset)\} \tag{5}$$

n points can be intersected by a single line if and only if all the corresponding $\mathscr{L}_{ij}$ ($\binom{n}{2}$) in total) have a non-empty intersection, i.e., their indices belong to a single coherent subset. It is maximal if no other points can be intersected by the same line. Listing all the MCSs is hard, as mentioned in Sect. 2, but our results tell us that if we approximate the different $\mathscr{L}_{ij}$ with minimum bounding axis-aligned rectangles, minimal outer approximations of the $\mathscr{L}_{ij}$ with axis-aligned rectangles, then enumerating their MCSs is easily done as shown in Sect. 3.

Determining the minimum axis-aligned bounding rectangle $H_{ij} = [\underline{y}_i, \overline{y}_i] \times [\underline{x}_i, \overline{x}_i]$ of $\mathscr{L}_{ij}$, i.e., the minimal volume that is fully enclosing $\mathscr{L}_{ij}$, is equivalent to finding the minimum and maximum values of the parameters a and b. It can be done quite easily, as maximizing (respectively minimizing) a is equivalent to minimizing (respectively maximizing) b.

4.2 Application

We first start with the case where the model is indeed linear, but where one data point is an outlier (box 2), as pictured on Fig. 2. As we can see, the statistical regression model, due to the outlier, does not capture the true model. In contrast, Fig. 3 shows the 5 different MCS (and their intersection) obtained using axis-aligned rectangle approximations. We can see that the biggest MCS is c_3, and it includes the true parameters, while the other MCSs are smaller and point out possible model outliers (i.e., observation 2 is in all the remaining MCSs, but not in the biggest one). Note that such an approach is quite different from standard imprecise regressions, that still uses least-square inspired approaches (Ferraro et al. 2010).

The second illustration considers the case where the observations are not faulty (the model goes through all observations), but where the model assumption is wrong, as we have a piecewise linear regression as shown on Fig. 4, but not a linear one. We have two partitions over x: $[0, 0.5]$ and $[0.5, 1]$. The two ground truths are very different, and a single statistical linear regression (in dotted blue) over the whole domain of x fits poorly to the data.

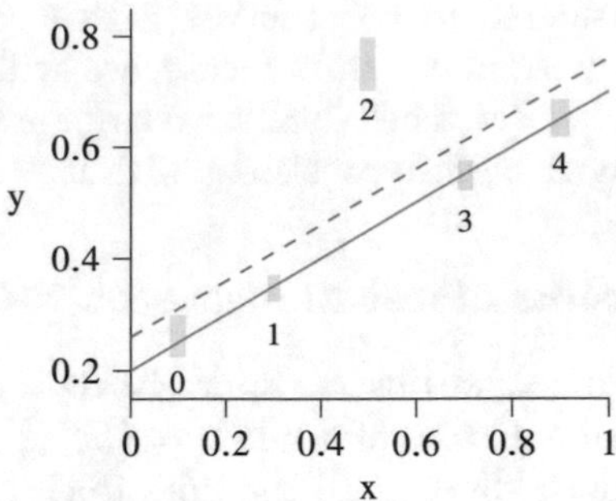

Fig. 2. Example of an imprecise linear regression problem. The dotted line corresponds to the linear regression including the outlier. The continuous line corresponds to the ground truth

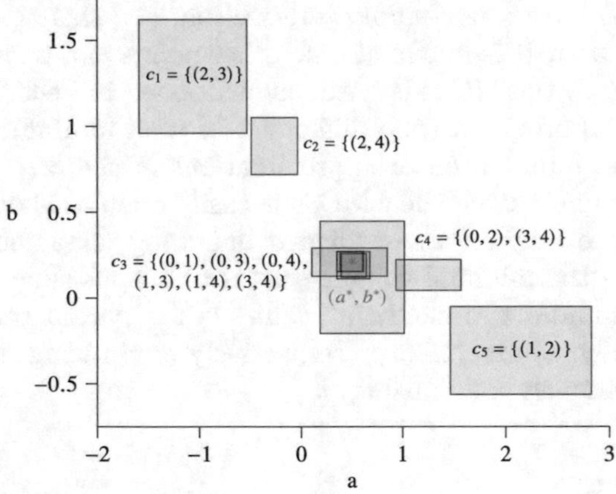

Fig. 3. Approximation with rectangles of the different $\mathscr{L}_{ij}$ from the example shown on Fig. 2

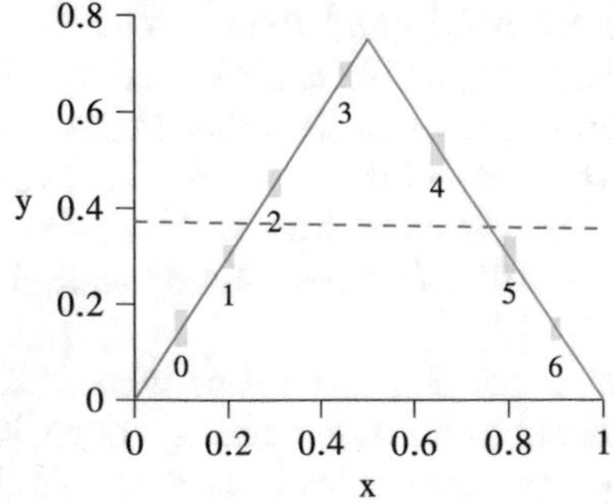

Fig. 4. Example of an imprecise piecewise linear regression problem. The dotted line corresponds to the full linear regression. The continuous lines correspond to the piecewise ground truths

Figure 5 shows what happens when we consider a MCS approach in this case. Instead of having one large MCS and quite smaller ones, we have now two quite large MCSs c_1 and c_2, along with other isolated ones. Those two distinct MCSs indicates that the error is likely in the model, that may only be locally valid. It also shows that it is not clear whether observation 3 belongs to one line or the other, being in both MCS. This ambiguity also cause c_1 to not include the true parameter (a_2^*, b_2^*).

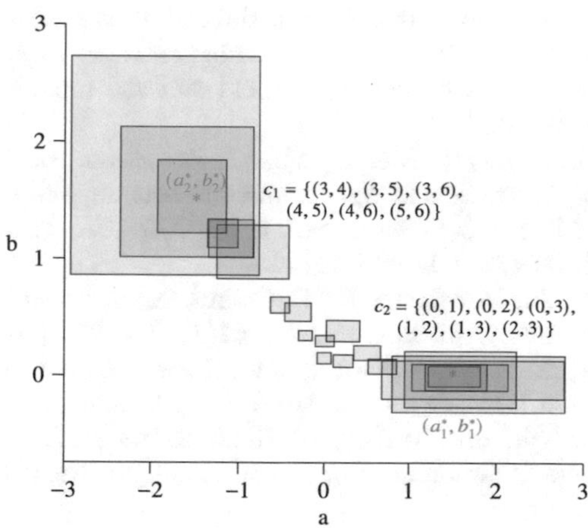

Fig. 5. Approximation with rectangles of the different $\mathscr{L}_{ij}$ from the example shown on Fig. 4

5 Conclusion

MCS is a theoretically very interesting notion for handling conflicting observations or pieces of information, but hard to enumerate in practice. We showed that such an enumeration was easier with hyperrectangles, as we can use known polynomial algorithms on their interval-valued projection to perform it. We illustrated their possible use on estimation problems, showing that their different behaviors could provide useful information (erroneous observations vs erroneous models).

In the future, we plan to perform some comparisons of such estimation approaches with Bayesian approaches in case of model misspecification, similarly to previous works intending to solve inverse problems (Shinde et al. 2021). Another case where our results may be useful is in the repair of inconsistent preference information (Adam and Destercke 2021).

Acknowledgements. This work was done within the PreServe Project, funded by grant ANR-18-CE23-0008 of the national research agency (ANR).

References

Adam, L., Destercke, S.: Possibilistic preference elicitation by minimax regret. In: Uncertainty in Artificial Intelligence, pp. 718–727. PMLR (2021)

Destercke, S., Dubois, D., Chojnacki, E.: Possibilistic information fusion using maximal coherent subsets. IEEE Trans. Fuzzy Syst. **17**(1), 79–92 (2008)

Dubois, D., Fargier, H., Prade, H.: Multiple-sources informations fusion-a practical inconsistency-tolerant approach. In: 8th International Conference on Information Processing and Management of Uncertainty in Knowledge-based Systems (IPMU 2000, Madrid, Spain) (2000). https://hal.archives-ouvertes.fr/hal-03405306

Dubois, D., Liu, W., Ma, J., Prade, H.: The basic principles of uncertain information fusion. An organised review of merging rules in different representation frameworks. Inf. Fusion **32**, 12–39 (2016)

Dubois, D., Prade, H., Yager, R.: Merging fuzzy information. In: Bezdek, J.C., Dubois, D., Prade, H. (eds.) Fuzzy Sets in Approximate Reasoning and Information Systems. The Handbooks of Fuzzy Sets Series, vol. 5, pp. 335–401. Springer, Boston (1999). https://doi.org/10.1007/978-1-4615-5243-7_7

Ferraro, M.B., Coppi, R., Rodríguez, G.G., Colubi, A.: A linear regression model for imprecise response. Int. J. Approx. Reason. **51**(7), 759–770 (2010)

Khachiyan, L., Boros, E., Borys, K., Gurvich, V., Elbassioni, K.: Generating all vertices of a polyhedron is hard. In: Twentieth Anniversary Volume, pp. 1–17. Springer, New York (2009). https://doi.org/10.1007/978-0-387-87363-3_17

Manor, R., Rescher, N.: On inference from inconsistent premises. Theor. Decis. **1**, 179–219 (1970)

Mitchell, T.M.: Generalization as search. Artif. Intell. **18**(2), 203–226 (1982)

Shinde, K., Feissel, P., Destercke, S.: Dealing with inconsistent measurements in inverse problems: set-based approach. Int. J. Uncertain. Quantif. **11**(3), 59–84 (2021)

Tiwary, H.R.: On the hardness of computing intersection, union and Minkowski sum of polytopes. Discrete Comput. Geom. **40**(3), 469–479 (2008)

On Convergence in Distribution of Fuzzy Random Variables

Miriam Alonso de la Fuente(✉) and Pedro Terán

Escuela Politécnica de Ingeniería, Departamento de Estadística e I.O. y D.M., Universidad de Oviedo, Oviedo, Spain
{alonsofmiriam,teranpedro}@uniovi.es

Abstract. We study whether convergence in distribution of fuzzy random variables, defined as the weak convergence of their probability distributions, is consistent with the additional structure of spaces of fuzzy sets. Positive results are obtained which reinforce the viability of that definition.

1 Introduction

Convergence in distribution is one of the most useful notions of convergence for random variables, notably because of its role in the central limit theorem. It is typically defined as follows: $\zeta_n \to \xi$ in distribution when $F_{\xi_n}(x) \to F_\xi(x)$ (where F_{ξ_n} and F_ξ are the respective cumulative distribution functions) for each point of continuity x of F_ξ.

Throughout the years, a number of proposals trying to extend the notion of a cumulative distribution function to fuzzy random variables have been made. Without judging their usefulness for specific problems, it is fair to say (Terán 2012) that they fail to have the theoretical properties that make the cumulative distribution function important in the case of random variables and random vectors. Specifically, they do not determine the probability distribution of the fuzzy random variable. Thus they are not useful to study convergence in distribution of fuzzy random variables.

Since a fuzzy random variable can be equivalently described (Krätschmer 2001) as a random element of a metric space of fuzzy sets (endowed with any of the d_p-metrics), it is possible to study probability distributions of fuzzy random variables using the general theory of probability distributions in metric spaces (e.g., Billingsley 1968). This approach was taken by the authors in recent papers (Alonso de la Fuente and Terán 2022a,b). In particular, it provides a way to define convergence in distribution as being tantamount to *weak convergence* of the probability distributions (since the Helly–Bray theorem and its converse establish the equivalence of those two notions for ordinary random variables).

The theoretical properties of weak convergence are well understood (see Billingsley 1968). But spaces of fuzzy sets have more structure than a generic

L. A. García-Escudero et al. (Eds.): SMPS 2022, AISC 1433, pp. 9–15, 2023.
https://doi.org/10.1007/978-3-031-15509-3_2

metric space, which raises the question whether defining convergence in distribution via weak convergence works well with that additional structure.

In this contribution, we show that convergence in distribution of fuzzy random variables with convex values can be studied using the support function embedding into an L^p-type space (i.e., convergence in distribution of the fuzzy random variables and of their support functions are equivalent). We also show that this type of convergence is consistent with some known structures in the space of fuzzy sets. A sequence of trapezoidal fuzzy random variables converges in distribution if and only if the vertices of the trapezoid converge jointly as a 4-dimensional random vector. A sequence of random vectors converges in distribution if and only if their indicator functions converge as fuzzy random variables. Finally, we show a consistency result between convergence and the sum and product by scalars which parallels the corresponding property of ordinary random variables.

2 Preliminaries

Let $\mathcal{F}_c(\mathbb{R}^d)$ be the space of fuzzy sets $U : \mathbb{R}^d \to [0, 1]$ whose α-cuts U_α are non-empty compact convex subsets of $\mathbb{R}^d$. Every fuzzy set $U \in \mathcal{F}_c(\mathbb{R}^d)$ is uniquely determined by its support function

$$\begin{aligned} s_U : [0,1] \times \mathbb{S}^{d-1} &\to \mathbb{R} \\ (r, \alpha) &\mapsto s_U(r, \alpha) = \sup_{x \in U_\alpha} \langle r, x \rangle \end{aligned}$$

where $\mathbb{S}^{d-1}$ denotes the unit sphere in $\mathbb{R}^d$.

For each $p \in [1, \infty)$, the metric d_p in $\mathcal{F}_c(\mathbb{R}^d)$, introduced by Klement et al. (1986) and Puri and Ralescu (1986), is defined by

$$d_p(U, V) = \left[\int_{[0,1]} (d_H(U_\alpha, V_\alpha))^p \, d\alpha \right]^{1/p}.$$

The metric ρ_p is defined by

$$\rho_p(U, V) = \left[\int_{[0,1]} \int_{S^{d-1}} |s_U(r, \alpha) - s_V(r, \alpha)|^p \, dr \, d\alpha \right]^{1/p}.$$

Denote by $\mathcal{K}_c(\mathbb{R}^d)$ the space of non-empty compact convex subsets of $\mathbb{R}^d$. Given a probability space $(\Omega, \mathcal{A}, P)$, a mapping $X : \Omega \to \mathcal{K}_c(\mathbb{R}^d)$ is called *random set* (also a *random compact convex set* in the literature) if X is measurable with respect to the Borel σ-algebra $\mathcal{B}_{\mathcal{K}_c(\mathbb{R}^d)}$ generated by the topology of the Hausdorff metric.

Definition 1. A mapping $X : (\Omega, \mathcal{A}, P) \to \mathcal{F}_c(\mathbb{R}^p)$ is called a *fuzzy random variable* if $X_\alpha : \omega \mapsto X(\omega)_\alpha$ is a random compact set for each $\alpha \in [0, 1]$.

Denote by σ_L the smallest σ-algebra that makes the mappings $U \in \mathscr{F}_c(\mathbb{R}^d) \mapsto U_\alpha \in \mathscr{K}_c(\mathbb{R}^d)$ measurable. Thus a fuzzy random variable is the same thing as a $(\mathscr{A}, \sigma_L)$-measurable mapping. A sequence of probability measures $\{P_n\}_n$ on σ_L is said to *converge weakly in d_p* to a probability measure P if

$$\int f dP_n \to \int f dP$$

for every $f : \mathscr{F}_c(\mathbb{R}^d) \to \mathbb{R}$ which is d_p-continuous and bounded. A sequence $\{X_n\}_n$ of fuzzy random variables converges *weakly* or *in distribution in d_p* to a fuzzy random variable X if their distributions P_{X_n} converge weakly to P_X, namely

$$E[f(X_n)] \to E[f(X)]$$

for each bounded d_p-continuous function $f : \mathscr{F}_c(\mathbb{R}^d) \to \mathbb{R}$.

The Lebesgue measure in $[0, 1]$ will be denoted by ℓ. The following results will be used in the sequel.

Lemma 1 (Billingsley 1968, **Theorem 2.1).** *Let $\mathbb{E}$ be a metric space, P a probability measure and $\{P_n\}_n$ a sequence of probabilities in $(\mathbb{E}, \mathscr{B}_{\mathbb{E}})$. Then $P_n \to P$ weakly if and only if for every open set G we have $\liminf_{n\to\infty} P_n(G) \geq P(G)$.*

Lemma 2 (Alonso de la Fuente and Terán 2022a, **Theorem 3.5).** *Let $p \in [1, \infty)$. Let P_n, P be probability measures on σ_L, such that $P_n \to P$ weakly. Then there exist fuzzy random variables $X_n, X : ([0,1], \mathscr{B}_{[0,1]}, \mathbb{P}) \to (\mathscr{F}_c(\mathbb{R}^d), d_p)$, such that*

(a) The distributions of X_n and X are P_n and P, respectively.
(b) $X_n(t) \to X(t)$ in d_p for every $t \in [0, 1]$.

Lemma 3 (Alonso de la Fuente and Terán 2022a, **Theorem 5.1).** *Let X_n and X be fuzzy random variables such that $X_n \to X$ in distribution in d_p. If $f : \mathscr{F}_c(\mathbb{R}^d) \to \mathscr{F}_c(\mathbb{R}^d)$ is a P_X-almost surely continuous function, then $f(X_n) \to f(X)$ weakly in d_p.*

Lemma 4 (Parthasarathy 1967, **Corollary 3.3, p. 22).** *If $\mathbb{E}$ is a Borel subset of a complete separable metric space X and φ is a one-one measurable map of $\mathbb{E}$ into a separable metric space Y, then $\varphi(\mathbb{E})$ is a Borel subset of Y, $\mathbb{E}$ and $\varphi(\mathbb{E})$ are isomorphic as measurable spaces and φ is an isomorphism.*

Recall that a trapezoidal fuzzy number $Tra(a, b, c, d)$ has the following expression:

$$U(x) = \begin{cases} 0 & \text{if } x < a \\ \frac{x-a}{b-a} & \text{if } a \leq x < b \\ 1 & \text{if } b \leq x \leq c \\ \frac{d-x}{d-c} & \text{if } c < x \leq d \\ 0 & \text{if } x > d \end{cases}$$

We will denote the space of trapezoidal fuzzy numbers by $\mathscr{F}_c^{tra}(\mathbb{R})$.

3 Main Results

Our first result states that d_p-convergence in distribution of fuzzy random variables is equivalent with the convergence obtained by embedding them into an L^p-type space. Note that this is not an immediate consequence of the embedding.

Theorem 1. *Let $p \in [1, \infty)$. Let X_n, X be fuzzy random variables. Then the following conditions are equivalent.*

1. $X_n \to X$ *in distribution in* $(\mathscr{F}_c(\mathbb{R}^d), d_p)$.
2. $s_{X_n} \to s_X$ *in distribution in* $L^p(\mathbb{S}^{d-1} \times [0,1], \lambda \otimes \ell)$,

where λ denotes the uniform measure in $\mathbb{S}^{d-1}$.

Proof. Denote by φ the mapping given by

$$\begin{aligned} \varphi : (\mathscr{F}_c(\mathbb{R}^d), \rho_p) &\to L^p(\mathbb{S}^{d-1} \times [0,1], \lambda \otimes \ell) \\ U &\mapsto s_U. \end{aligned}$$

By, e.g., Krätschmer (2006, p. 444), φ is an isometry.

Let $Y_n, Y : ([0,1], \mathscr{B}_{[0,1]}, \ell) \to (\mathscr{F}_c(\mathbb{R}^d), d_p)$ be the fuzzy random variables given by Lemma 2. We have $Y_n(t) \to Y(t)$ in d_p for all $t \in [0,1]$, $P_{X_n} = \ell_{Y_n}$ and $P_X = \ell_Y$. Since d_p and ρ_p are topologically equivalent (Diamond and Kloeden 1994, Proposition 7.4.5, p. 65), $\rho_p(Y_n(t), Y(t)) \to 0$ for every $t \in [0,1]$.

Set $s_{Y_n} = \varphi \circ Y_n$ and $s_Y = \varphi \circ Y$. Since φ is an isometry, we have $s_{Y_n}(t) \to s_Y(t)$. By Alonso de la Fuente and Terán (2022a, Proposition 5.4), $(\mathscr{F}_c(\mathbb{R}^d), d_p)$ is a Lusin space, hence it is Borel measurable in every metric space it embeds into (see Frolík 1970, Proposition 7.11). There follows that φ is Borel measurable and thus s_{Y_n} and s_Y are random elements of $L^p(\mathbb{S}^{d-1} \times [0,1], \lambda \otimes \ell)$.

We need to check $\ell_{s_{Y_n}} = P_{s_{X_n}}$. For any measurable subset A of $L^p(\mathbb{S}^{d-1} \times [0,1], \lambda \otimes \ell)$,

$$\begin{aligned} P_{s_{X_n}}(A) &= P(\{\omega \in \Omega : s_{X_n}(\omega) \in A\}) = P(\{\omega \in \Omega : (\varphi \circ X_n)(\omega) \in A\}) \\ &= P(\{\omega \in \Omega : X_n(\omega) \in \varphi^{-1}(A)\}) = \ell(\{t \in [0,1] : Y_n(t) \in \varphi^{-1}(A)\}) \\ &= \ell(\{t \in [0,1] : (\varphi \circ Y_n)(t) \in A\}) = \ell(\{t \in [0,1] : s_{Y_n}(t) \in A\}) = \ell_{s_{Y_n}}(A). \end{aligned}$$

Analogously, $\ell_{s_Y} = P_{s_X}$. Since $s_{Y_n} \to s_Y$ almost surely, by Kallenberg (2002, Lemma 4.2) almost sure convergence implies convergence in probability and by Kallenberg (2002, Lemma 4.7) convergence in probability implies weak convergence $\ell_{s_{Y_n}} \to \ell_{s_Y}$. In conclusion, $P_{s_{X_n}} \to P_{s_X}$ weakly, that is, $s_{X_n} \to s_X$ in distribution.

For the converse, notice that $P_{X_n} = P_{s_{X_n}} \circ \varphi$ and $P_X = P_{s_X} \circ \varphi$. Since φ is an isometry, for any open set G of $\mathscr{F}_c(\mathbb{R}^d)$ there exists an open set $\mathbf{G}$ of $L^p(\mathbb{S}^{d-1} \times [0,1], \lambda \otimes \ell)$ such that $\varphi(G) = \mathbf{G} \cap \varphi(\mathscr{F}_c(\mathbb{R}^d))$. Then

$$\liminf_{n\to\infty} P_{s_{X_n}} \circ \varphi(G) = \liminf_{n\to\infty} P_{s_{X_n}}(\varphi(G)) = \liminf_{n\to\infty} P_{s_{X_n}}(\mathbf{G} \cap \varphi(\mathscr{F}_c(\mathbb{R}^d)))$$

$$= \liminf_{n\to\infty} P_{s_{X_n}}(\mathbf{G}) \geq P_{s_X}(\mathbf{G}) = P_{s_X}(\mathbf{G} \cap \varphi(\mathscr{F}_c(\mathbb{R}^d))) = P_{s_X}(\varphi(G)) = P_{s_X} \circ \varphi(G)$$

by Lemma 1 and knowing that s_{X_n} and s_X take on values in $\varphi(\mathscr{F}_c(\mathbb{R}^d))$. Again by Lemma 1, $\liminf_{n\to\infty} P_{X_n} \circ \varphi(G) \geq P_X \circ \varphi(G)$ yields $X_n \to X$ in distribution in d_p. □

Therefore this type of convergence can indeed be studied using support functions. Another question concerns the relationship between convergence of fuzzy random variables taking on values in parametric families of fuzzy sets (in this case, trapezoidal fuzzy sets but the study could be extended to other families) and convergence in distribution of their defining parameters. The content of the following lemma is intuitively clear although its proof is not trivial. For space reasons we skip the proof, which may appear elsewhere.

Lemma 5. *Let $p \in [1, \infty)$. Let $U_n, U \in \mathscr{F}_c^{tra}(\mathbb{R})$. If $U_n \to U$ in d_p, then the sequence $\{\|(U_n)_0\|\}_n$ is bounded.*

Theorem 2. *Let $p \in [1, \infty)$. Let X_n be $Tra(X_{n,1}, X_{n,2}, X_{n,3}, X_{n,4})$ where $X_{n,1} \leq X_{n,2} \leq X_{n,3} \leq X_{n,4}$ are random variables, and analogously $X = Tra(X_1, X_2, X_3, X_4)$. Then $X_n \to X$ in distribution in d_p if and only if, as random vectors in $\mathbb{R}^4$, $(X_{n,1}, X_{n,2}, X_{n,3}, X_{n,4}) \to (X_1, X_2, X_3, X_4)$ in distribution.*

Proof. Set $A = \{(u_1, u_2, u_3, u_4) \in \mathbb{R}^4 : u_1 \leq u_2 \leq u_3 \leq u_4\}$. The mapping

$$\varphi : A \to (\mathscr{F}_c(\mathbb{R}), d_p)$$
$$(u_1, u_2, u_3, u_4) \mapsto Tra(u_1, u_2, u_3, u_4)$$

is injective. Let us show that φ is continuous. Let $(u_{n,1}, u_{n,2}, u_{n,3}, u_{n,4}) \to (u_1, u_2, u_3, u_4)$ in $\mathbb{R}^4$. Denote by U_n the fuzzy set $Tra(u_{n,1}, u_{n,2}, u_{n,3}, u_{n,4})$ and by U the fuzzy set $Tra(u_1, u_2, u_3, u_4)$.

Now let $\alpha \in [0, 1]$,

$$d_H(U_{n_\alpha}, U_\alpha) = \max\{|\inf(U_n)_\alpha - \inf U_\alpha|, |\sup(U_n)_\alpha - \sup U_\alpha|\}$$
$$= \max\{|(1-\alpha)\inf(U_n)_0 + \alpha \inf(U_n)_1 - (1-\alpha)\inf U_0 - \alpha \inf U_1|,$$
$$|\alpha \sup(U_n)_1 + (1-\alpha)\sup(U_n)_0 - \alpha \sup U_1 - (1-\alpha)\sup U_0|\}$$
$$= \max\{|(1-\alpha)(u_{n,1} - u_1) + \alpha(u_{n,2} - u_2)|, |\alpha(u_{n,3} - u_3) + (1-\alpha)(u_{n,4} - u_4)|\}$$
$$\leq \max\{|u_{n,1} - u_1|, |u_{n,2} - u_2|, |u_{n,3} - u_3|, |u_{n,4} - u_4|\}.$$

Since the last term is the max distance between both vectors in $\mathbb{R}^4$ and is independent of α, indeed it bounds $d_p(U_n, U)$, making φ be d_p-continuous.

We will establish now two further facts which will be used in the proof. Firstly, since A is closed in $\mathbb{R}^4$, it is complete and separable. Moreover $(\mathscr{F}_c(\mathbb{R}), d_p)$ is separable, hence by Lemma 4 the image $\varphi(A) = \mathscr{F}_c^{tra}(\mathbb{R})$ is Borel measurable.

Secondly, set

$$A_{a,b} = \{(u_1, u_2, u_3, u_4) \in \mathbb{R}^4 : a \leq u_1 \leq u_2 \leq u_3 \leq u_4 \leq b\} \subseteq A$$

for each $a, b \in \mathbb{R}$. Since $A_{a,b}$ is compact, φ is continuous and $\varphi(A) = \mathscr{F}_c^{tra}(\mathbb{R})$ is a Hausdorff space, the restriction $\varphi|_{A_{a,b}}$ is a homeomorphism (Joshi 1983, Corollary 2.4, p. 169).

($\Rightarrow$) By Lemma 2, there exist fuzzy random variables Y_n, Y such that $Y_n(t) \to Y(t)$ in d_p for each $t \in [0,1]$, $\ell_{Y_n} = P_{X_n}$ and $\ell_Y = P_X$. Since

$$\ell_{Y_n}(\mathscr{F}_c^{tra}(\mathbb{R})) = P_{X_n}(\mathscr{F}_c^{tra}(\mathbb{R})) = 1,$$

Y_n and Y are almost surely trapezoidal fuzzy sets. For clarity, we assume without loss of generality that all $Y_n(t), Y(t)$ are trapezoidal fuzzy sets (otherwise it would suffice to modify the value of those variables in a null set, which would not change their probability distributions). Set $Tra(Y_{n,1}, Y_{n,2}, Y_{n,3}, Y_{n,4}) = Y_n$, $Tra(Y_1, Y_2, Y_3, Y_4) = Y$ and let us show that $(Y_{n,1}, Y_{n,2}, Y_{n,3}, Y_{n,4})$ converges in distribution to (Y_1, Y_2, Y_3, Y_4).

By Lemma 5, each sequence $\{\|(Y_n)_0(t)\|\}_n$ is bounded by some constant M_t. Therefore $(Y_{n,1}(t), Y_{n,2}(t), Y_{n,3}(t), Y_{n,4}(t)) \in A_{-M_t,M_t}$ for all $t \in [0,1]$. By the homeomorphism between A_{-M_t,M_t} and $\varphi(A_{-M_t,M_t})$, the 4-dimensional vector converges to $(Y_1(t), Y_2(t), Y_3(t), Y_4(t))$. Almost sure convergence of those vectors implies their convergence in distribution. To finish the proof, we just need to check $\ell_{(Y_{1n},\dots,Y_{4n})} = P_{(X_{1n},\dots,X_{4n})}$. For any Borel subset $B \subseteq \mathbb{R}^4$,

$$\ell_{Y_{1,n},\dots,Y_{4,n}}(B) = \ell(\{t \in [0,1] : (Y_{1,n}, \dots, Y_{4,n})(t) \in A \cap B\})$$

$$= \ell(\{t \in [0,1] : Y_n(t) \in \varphi(A \cap B)\}) = P(\{\omega \in \Omega : X_n(\omega) \in \varphi(A \cap B)\})$$

$$= P(\{\omega \in \Omega : (X_{1,n}, \dots, X_{4,n})(\omega) \in A \cap B\}) = P_{X_{1,n},\dots,X_{4,n}}(B).$$

Analogously, $\ell_{Y_1,\dots,Y_4} = P_{X_1,\dots,X_4}$.

($\Leftarrow$) By the Skorokhod representation theorem in $\mathbb{R}^4$, there exist random vectors $(Y_{n,1}, Y_{n,2}, Y_{n,3}, Y_{n,4})$, (Y_1, Y_2, Y_3, Y_4) such that $\ell_{(Y_{n,1},Y_{n,2},Y_{n,3},Y_{n,4})} = P_{(X_{n,1},X_{n,2},X_{n,3},X_{n,4})}$, $\ell_{(Y_1,Y_2,Y_3,Y_4)} = P_{(X_1,X_2,X_3,X_4)}$ and $(Y_{n,1}, Y_{n,2}, Y_{n,3}, Y_{n,4})$ converges to (Y_1, Y_2, Y_3, Y_4) pointwise. Set

$$Y_n = Tra(Y_{n,1}, Y_{n,2}, Y_{n,3}, Y_{n,4}), Y = Tra(Y_1, Y_2, Y_3, Y_4).$$

By the continuity of φ, $Y_n(t) \to Y(t)$ in d_p for each $t \in [0,1]$. By Lemmas 4.2 and 4.7 in Kallenberg (2002), almost sure convergence implies convergence in distribution. Finally, one shows like before $\ell_{Y_n} = P_{X_n}$ and $\ell_Y = P_X$, whence $X_n \to X$ in distribution in d_p. □

Since a random variable ξ can be identified with the trapezoidal fuzzy set $Tra(\xi, \xi, \xi, \xi)$, which is the indicator function $I_{\{\xi\}}$, the following corollary holds.

Corollary 1. *Let ξ_n, ξ be random variables. Then $\xi_n \to \xi$ in distribution if and only if $I_{\{\xi_n\}} \to I_{\{\xi\}}$ in distribution in d_p.*

The following proposition is analogous to an important property of convergence in distribution for random variables. It states that convergence is compatible with the operations in $\mathscr{F}_c(\mathbb{R}^d)$.

Proposition 1. *Let X_n, X be fuzzy random variables such that $X_n \to X$ in distribution in d_p. Then*

1. *For every $U \in \mathscr{F}_c(\mathbb{R}^d)$, we have $X_n + U \to X + U$ in distribution in d_p.*
2. *For every $a \in \mathbb{R}$, we have $aX_n \to aX$ in distribution in d_p.*

Proof. Since the mappings $V \in \mathscr{F}_c(\mathbb{R}^d) \mapsto V + U$ and $V \in \mathscr{F}_c(\mathbb{R}^d) \mapsto aV$ are d_p-continuous, we obtain the result with an application of the continuous mapping theorem (Lemma 3). □

Remark 1. It is not true, in general, that $X_n + Y \to X + Y$ in distribution in d_p provided $X_n \to X$ in distribution. That fails even for random variables.

We close the paper by pointing out another parallel with ordinary random variables: if the limit is a degenerate fuzzy random variable U, then $X_n \to U$ in distribution in d_p if and only if $X_n \to U$ in probability in d_p (by an application of Kallenberg 2002, Lemma 4.7).

Acknowledgements. Research in this paper was partially funded by grants and fellowships from Spain (PID2019-104486GB-I00), the Principality of Asturias (SV-PA-21-AYUD/2021/50897 and PA-21-PF-BP20-112), and the University of Oviedo (PAPI-20-PF-21). Their contribution is gratefully acknowledged.

References

Alonso de la Fuente, M., Terán, P.: Some results on convergence and distributions of fuzzy random variables. Fuzzy Sets Syst. **435**, 149–163 (2022)

Alonso de la Fuente, M., Terán, P.: Convergence theorems for random elements in convex combination spaces (2022b, submitted for publication)

Billingsley, P.: Convergence of Probability Measures. Wiley, New York (1968)

Diamond, P., Kloeden, P.: Metric Spaces of Fuzzy Sets. World Scientific, Singapore (1994)

Frolík, Z.: A survey of separable descriptive theory of sets and spaces. Czechoslov. Math. J. **20**, 406–467 (1970)

Joshi, K.D.: Introduction to General Topology. New Age, New Delhi (1983)

Klement, E.P., Puri, M.L., Ralescu, D.A.: Limit theorems for fuzzy random variables. Proc. R. Soc. Lond. Ser. A **407**, 171–182 (1986)

Kallenberg, O.: Foundations of Modern Probability, 2nd edn. Springer, New York (2002). https://doi.org/10.1007/978-3-030-61871-1

Krätschmer, V.: A unified approach to fuzzy random variables. Fuzzy Sets Syst. **123**, 1–9 (2001)

Krätschmer, V.: Integrals of random fuzzy sets. TEST **15**, 433–469 (2006). https://doi.org/10.1007/BF02607061

Parthasarathy, K.R.: Probability Measures on Metric Spaces. Academic Press, New York (1967)

Puri, M.L., Ralescu, D.A.: Fuzzy random variables. J. Math. Anal. Appl. **114**, 409–422 (1986)

Terán, P.: Characterizing the distribution of fuzzy random variables. In: Book of Abstracts of the 5th International Conference ERCIM Working Group on Computing and Statistics, CFE-ERCIM, Oviedo, Spain, p. 22 (2012)

A Fuzzy Survival Tree (FST)

Jorge Luis Andrade(✉) and José Luis Valencia

Universidad Complutense, 28040 Madrid, Spain
jorandra@ucm.es, joseval@estad.ucm.es

Abstract. A fuzzy survival tree (FST) is introduced as an alternative proposal for survival tree learning. Fuzzy logic theory and Harrell's index (c-index) are combined as a new rule in node splitting. The introduction of fuzzy sets in tree learning improves FST performance and provides robustness to the algorithm when data are missing. FST performance improves significantly over other tree-based machine learning algorithms as demonstrated in public clinical datasets.

1 Introduction

Right-censored data have been commonly analysed using the Cox regression model (Cox 1972) under the proportional hazards assumption. Recently, studies have used algorithms such as the random survival forest (RSF). It's proposed by Ishwaran et al. (2008) as an extension of the random forest algorithm (Breiman 2001) to the right-censored survival problem. In the learning process at each tree, the maximization of the log-rank statistical test is used for nodes splitting. Ishwaran et al. (2008) demonstrates that the RSF performance is higher than the Cox model and its use is recommended when the assumptions of proportional risks are not met (Omurlu et al. 2009) or when the effect of the explanatory variables is nonlinear.

In decision trees for classification and regression problems, Ferri et al. (2002) and Lee (2019) proposed including the performance metric, the area under the curve (AUC), in the base learner. In trees for right-censored data analysis, the inclusion of the c-index (the AUC equivalent) is proposed by Schmid et al. (2016) with good results.

Recently, fuzzy set theory has been included in learning decision trees (Zhai et al. 2018; Mitra et al. 2002; Olaru and Wehenkel 2003, etc.) because datasets can contain missing values, noise in the class or outlier elements, etc. Fuzzy logic provides the flexibility to deal with these types of datasets without affecting the performance of the algorithm.

This article introduces fuzzy survival tree (FST). This new algorithm presents as a novelty the inclusion of fuzzy logic in combination with the c-index in the learning process of survival trees for splitting at each node.

L. A. García-Escudero et al. (Eds.): SMPS 2022, AISC 1433, pp. 16–23, 2023.
https://doi.org/10.1007/978-3-031-15509-3_3

2 Fuzzy Survival Tree Algorithm

Here the variables used for the survival tree building are explained. The datasets (T_e, δ_e, A_e) are composed of a set of data organized in rows and columns, each row represents an individual or element e, who has a series of attributes $A_e = (a_{e1}, ..., a_{ep})$ (explanatory variables); in addition, the response vector $[T_e, \delta_e]$ which for this case is composed of the variable time in the study, and a binary variable (0, 1) for censoring status.

Let $|E|$ denote the universe of all elements in a dataset. A fuzzy set $S \subset |E|$ is characterized by a membership function $\mu_S : |E| \rightarrow [0,1]$ that associates each element e of $|E|$ with a number $\mu_S(e)$ in the interval $[0,1]$ representing the degree of membership of element e to S.

2.1 FST Learning Process

FST is a supervised learning algorithm, and the base learner is a binary survival tree grown by recursive splitting of tree nodes by analogy with RSF and random forest methodology (Breiman 2001). A tree is grown starting at the root node which comprises all the data. After the split of the root node using the new criterion (fuzzy logic & c-index), two daughter nodes are created, left and right. The process is repeated in a recursive fashion at each node until the stopping criteria are met. Finally, a label is attached to every terminal node.

At each new node, considering membership of the elements to the node, the c-index method is maximized; in this splitting in a second step the fuzzy sets are integrated using a genetic algorithm to determine a fuzzy split.

Unlike crisp trees where an element activates a unique leaf, in fuzzy or soft trees if the element is within the node overlap region, it goes to both daughter nodes following multiple paths in parallel as a consequence an element can activate multiple terminal nodes.

FST grows recursively; at the root node are all the training elements $|E|$ with a membership degree equal to 1; $\mu_{nroot}(e) = 1$.

A discriminator function is attached to each node, and it determines the split of the test node and its fuzziness by $(a,\ a(e), \theta_n, \beta_n) \rightarrow [0,1]$; the parameters defining it are: a the selected explanatory variable, $a(e)$ the value that element e has for that explanatory variable, the value of the split is θ_n, and β_n is the parameter that determines the overlap region.

2.2 FST Notation

The degree of membership μ of an element (individual) e to the set of elements of the daughter nodes $\{n_L, n_R\}$ is given by the linear membership function:

$$\mu_{nL}(e) = \begin{cases} 1 & a(e) \leq \theta_n - \beta_n/2 \\ \dfrac{(\theta_n + \beta_n/2) - a(e)}{\beta_n}; & \theta_n - \beta_n/2 < a(e) \leq \theta_n + \beta_n/2 \, ; \mu_{nR} = 1 - \mu_{nL} \\ 0 & a(e) > \theta_n + \beta_n/2 \end{cases} \quad (1)$$

When an element has a missing value in the explanatory variable selected for the split $a\left(e\right)$, element e continues to both daughter nodes and the degree of membership to each daughter node is: $\mu_{n_L} = \mu_{n_R}\left(e\right) = \frac{1}{2}$.

At each node, a class is represented by the cumulative hazard function (CHF) value, H. The membership degree of an element e to a defined class $\mu_H\left(e\right)$ is computed taking into account the CHF value in both daughter nodes:

$$\mu_H\left(e\right) = \sum_{i\in\{n_L,n_R\}} \mu_{n_i}\left(e\right) H_i \tag{2}$$

where $H_i = \sum_k \frac{\delta_{i,k}}{|E_{i.k}|}$, and are the events produced and the exposed elements at time tk. The times are discretized and can only take the values $t\epsilon t_1, t_2, t_3, \ldots, t_m$. The calculated CHF is common for all elements of node n; then, $H\left(t \mid A_e\right) = H_n\left(t\right)$, if $e \;\epsilon\; n$.

2.3 FST Algorithm Steps

The following subsections present the steps of the FST algorithm for growing and estimation. For growing, splitting at each node is decomposed into two parts, Steps 1 and 2 (Olaru and Wehenkel 2003). Once the tree finishes learning, the estimation starts, and the CHF for a new element with explanatory variables $H\left(t \mid A_e\right)$ is estimated by defuzzification, Step 3.

Step 1 - Searching for the Explanatory Variable and Splitting
The cut-off value θ_n is chosen among explanatory variables A and its values which maximize the c-index. The RSF iterative algorithm for trees construction is adapted to consider the fuzzy sets dynamics. In this first step the parameter β_n is equal to zero (crisp split).

For the search of θ_n, the RSF algorithm is modified to incorporate the degree of membership of every element to the node set as follows, at each node the value of every element $a\left(e\right)$ is modified to consider the cumulative membership degree and is renamed as $a\left(e\right)_n$. The cumulative membership degree is calculated recursively from the root to the parent node $\mu_p\left(e\right)$. The renamed element $a\left(e\right)_n$ is computed as: $a(e)_{nL} = a(e)\mu_{nL}(e)\mu_p(e)$, and $a(e)_{nR} = a(e)\mu_{nR}(e)\mu_p(e)$, for the left and right daughter nodes, respectively.

The c-index is designed to estimate $P\left(H_j > H_i \mid T_i < T_j\right)$ the concordance probability, it's implemented as a coefficient that accounts the concordant pairs over permissible for the whole range of the observed survival times. Its inclusion as splitting criterion is to maximize the $P\left(j \in n_r\ ,\ i \in n_l \mid T_i\ <\ T_j\right)$ in the division of elements into daughter nodes by the cut-off of some candidate variable; and, also to overcome the discrepancy between splitting and evaluation criteria in survival trees. Schmid et al. (2016) justifies and recommends c-index as splitting criterion because it improves RSF in smaller scale clinical studies.

The parameter θ_n as cut-off value to determine daughter nodes is temporary, because the fuzzification step and its overlap region has not been executed.

Step 2 - Fuzzification and Labelling
In the FST the fuzzification degree is determined within the internal learning process of the algorithm, fuzzification is represented by overlap region and is computed using a genetic algorithm; as the overlap region is necessary for each node, the genetic algorithm runs at each node splitting.

At each new node, the best overlap region for the most discriminant explanatory variable is determined by the genetic algorithm proposed in Bozorg-Haddad et al. (2017) and the cut-off point θ_n obtained in the previous step.

The β_n width defines the overlap region; it is obtained by minimizing the fitness function:

$$\sum_{e\ \in n} \mu_p\,(e)\,[\mu_H\,(e) - \mu\prime_H\,(e)]^2 \tag{3}$$

where, $\mu\prime_H\,(e) = \mu\prime_{nL}\,(e)\,H_L + \mu'_{nR}(e)\,H_R$ is the membership function to the left and right daughter nodes defined in the previous subsection, but in this case, it is updated with every candidate value of β_n that comes from every iteration of the genetic algorithm.

The range of possible values for the β_n parameter is delimited by the range of variation of the explanatory variable a at the node given by the interval: $\max\,(0, \min\,[(\max\,(a(\cdot)_n) - \theta_n)\,, (\theta_n - \min\,(a(\cdot)_n))])$.

The genetic algorithm was run with the following recommended configuration, it provided satisfactory solutions in a feasible amount of time:

- number of iterations: 10, population size: 10, mutation probability: 0.01,
- elit ratio: 0.20, crossover probability: 0.7, crossover type: 'uniform'.

Step 3 - Defuzzification
Once the growing process finished, the defuzzification process starts, which consists of predicting the unique CHF of an unknown element. For this, the dataset of the test data is taken, and the new element is dropped from the root to the terminal nodes. At each test node, element e is evaluated according to the parameters obtained in the learning process (a, θ_n, β_n).

As mentioned before, it is possible for an element to reach multiple terminal nodes; for the prediction of the single CHF of an element $H\,(t \mid A_e)$ you can follow multiple strategies (Bonissone et al. 2010), some basic aggregation operators are the maximum, minimum or average.

In the following sections, to calculate the c-index results the H is calculated using the average aggregation operator where the multiple terminal nodes activated. The average operator for defuzzification provides the best results.

3 Results

In this section, we describe the results that indicate the performance of the proposed FST algorithm in 3 experiments.

For comparison, the algorithms have the same parametrization. In addition, the algorithms were grown to their maximum size, without pruning.

The RSF algorithms of Ishwaran et al. (2008) and Schmid et al. (2016) using parameter ntree = 1 (number of trees) and mtry = all (number of explanatory variables for node splitting) in this section are called survival trees (ST).

Prior to the algorithms training, the numerical explanatory variables were normalized [0–1]. Other variables are not transformed or entered into the fuzzification process.

The following tables present the mean or median results of the 5-fold cross validation technique performed independently 10 times.

These experiments used publicly available datasets in: http://web.archive.org/web/20170114043458/http:/www.umass.edu/statdata/statdata/data/ and contain information related to clinical studies. The experiments results are validated using nonparametric hypothesis tests.

3.1 Behaviour and Stability of the FST with Missing Data

A percentage d% of missing data is introduced in a dataset of $|E|$ elements, which have A attributes. We randomly select d%·E·A elements of the dataset that will be uniformly distributed; at each $a(e)$ selected a missing value is introduced. The decrease in the percentage of the c-index is computed, following Bonissone et al. (2010), as:

$$\%\text{decrease c-index} = 100 \times \frac{\text{c-index(original)} - \text{c-index(imperfect)}}{\text{c-index(original)}}.$$

Table 1 shows the FST stability and good results in the presence of missing data.

Table 1. Testing c-index of the FST for different percentages of missing values

Dataset	Without	Introducing missing values		
		1%	3%	5%
		% decrease in average c-index		
Breast	65.52	0.05	1.11	1.51
AIDS	70.20	−0.31	−2.67	3.19

3.2 Comparison of FST with Other Survival Algorithms

The FST is compared to other survival tree algorithms found in the literature. These algorithms are based on growing a single survival tree and are built with the ST technique using as splitting rules: the logrank test (ST_logrank), and the c-index (ST_cindex).

To show the FST behaviour in distinct cases, datasets analysed in this section have incidence rates ($event/|E|$), and number of elements $|E|$ quite differentiated. Breast and AIDS datasets are used, see its description in Table 3.

Table 2. Breast and AIDS dataset results for three survival trees

Data		ST(log_rank)a	ST(c_index)b	FST	FST/a -1	FST/b -1
Breast	c-index median	56.075	61.487	64.451	14.94	4.82
	c-index mean	55.885	61.717	64.352	15.15	4.27
	CPU time	23.15 s	2.91 min	13.45 min	3385	362
AIDS	c-index median	56.147	64.779	67.294	19.85	3.88
	c-index mean	56.421	64.232	68.578	21.55	6.77
	CPU time	24.03 s	2.37 min	23.28 min	5712	882

Table 2 shows the better performance of the FST algorithm than the ST(log_rank) and ST(c_index) in c-index mean and median.

To compare the performance of the three algorithms in this experiment, the nonparametric Friedman aligned test was used, the null-hypothesis is rejected in both datasets with confidence levels of 99.99% which indicates that there are significant differences between the algorithms. Finally, a post-hoc Holm test is performed to find statistically significant differences between the control algorithm-FST and the other algorithms. It is confirmed that there are significant differences in both datasets between the FST and the other two techniques, with 99.9% and 97.1% in the Breast dataset; and 99.9% and 97.3% in the AIDS dataset.

Table 2 shows the CPU times of ST and FST. They come from an average of 50 runs of each algorithm, we have used a computer Intel(R) Core(TM) i7-1065G7 1.50 GHz, 16 GB RAM. There is a big difference between STs and FST, and indicates the difficulty to implement a random forest methodology based in multiple FSTs.

3.3 FST Performance Compared to ST(c_index)

The results of the previous section validate that the performance of ST(c_index) is higher than ST(log_rank) for this type of datasets, it is also concluded in Schmid et al. (2016). In this section, we compare the results between the FST and the ST (c_index) in four clinical datasets.

Table 3 shows the results to compare results between both algorithms. Clearly, the performance of the FST is higher than the ST(c_index).

The Wilcoxon signed-rank test is used to compare the two algorithms' results. The tests applied to the results presented in Table 3, showed significant differences with a confidence level of 96%.

Table 3. C-index mean of the FST vs. ST(c_index)

Description	Dataset	$\|E\|$	$\delta = 1$	$\|E\|$	ST(c_index)	FST	FST/ST-1
German breast cancer	Breast	686	299	6	61.72	64.35	4.27
AIDS clinical trial	AIDS	1121	26	11	64.23	68.58	6.77
Veteran lung cancer	Lung	137	128	6	59.76	60.73	1.62
Primary biliary cirrhosis	Pbc	418	161	17	76.42	77.01	0.78

4 Conclusions

In this article we introduced FST as a new algorithm to solve the right-censored data problem. In the learning process, at each node a new rule is considered in which fuzzy sets theory was applied by using a genetic algorithm.

The obtained results demonstrates that the FST has a higher generalization capacity than the other two state-of-the-art methods.

Regarding the computational time, the FST demands more time than the crisp ST algorithms. It's the price paid for having a better performance.

The FST technique allows individuals to work with missing data without a significant reduction in the capacity of generalization.

The FST technique is implemented in Python code, which facilitates its manipulation and future extension; the code is available in https://github.com/Jorandra/FST-surv.

Future work on this topic may focus on research how to incorporate the FST results for the construction and ensemble the multiple survival trees.

References

Bonissone, P., Cadenas, J.M., Garrido, M.C., Díaz-Valladares, R.A.: A fuzzy random forest. Int. J. Approx. Reason. **51**(7), 729–747 (2010)

Bozorg-Haddad, O., Solgi, M., Loáiciga, H.A.: Meta-heuristic and Evolutionary Algorithms for Engineering Optimization. Wiley, Hoboken (2017)

Breiman, L.: Random forests. Mach. Learn. **45**, 5–32 (2001)

Cox, D.: Regression models and life-tables. J. Roy. Stat. Soc.: Ser. B (Methodol.) **34**(2), 187–202 (1972)

Ferri, C., Flach, P., Hernández-Orallo, J.: Learning decision trees using the area under the ROC curve. In: Sammut, C., Hoffmann, A. (eds.) Proceedings of the 19th International Conference on Machine Learning, pp. 139–146 (2002)

Ishwaran, H., Kogalur, U.B., Blackstone, E.H., Lauer, M.S.: Random survival forests. Ann. Appl. Stat. **2**, 841–860 (2008)

Lee, J.S.: AUC4.5: AUC-based C4.5 decision tree algorithm for imbalanced data classification. IEEE Access **7**, 106034–106042 (2019)

Mitra, S., Konwar, K., Pal, S.: Fuzzy decision tree, linguistic rules and fuzzy knowledge-based network: generation and evaluation. IEEE Trans. Syst. Man Cybern. Part C **32**(4), 328–339 (2002)

Olaru, C., Wehenkel, L.: A complete fuzzy decision tree technique. Fuzzy Sets Syst. **138**, 221–254 (2003)

Omurlu, I.K., Ture, M., Tokatli, F.: The comparisons of random survival forests and cox regression analysis with simulation and an application related to breast cancer. Expert Syst. Appl. **36**, 8582–8588 (2009)

Schmid, M., Wright, M.N., Ziegler, A.: On the use of Harrell's C for clinical risk prediction via random survival forests. arXiv preprint arXiv:1507.03092 (2016). https://doi.org/10.1016/j.eswa.2016.07.018

Zhai, J., Wang, X., Zhang, S., Hou, S.: Tolerance rough fuzzy decision tree. Inf. Sci. **465**, 425–438 (2018)

Robust Rao-Type Tests for Non-destructive One-Shot Device Testing Under Step-Stress Model with Exponential Lifetimes

Narayanaswamy Balakrishnan[1], María Jaenada[2](✉), and Leandro Pardo[2]

[1] McMaster University, Hamilton, ON L8S 4K1, Canada
bala@mcmaster.ca

[2] Complutense University of Madrid, Plaza Ciencias 3, 28040 Madrid, Spain
mjaenada@ucm.es, lpardo@mat.ucm.es

Abstract. One-shot devices analysis involves an extreme case of interval censoring, wherein one can only know whether the failure time is before the test time. Some kind of one-shot units do not get destroyed when tested, and then survival units can continue within the test providing extra information for inference. This not-destructiveness is a great advantage when the number of units under test are few. On the other hand, one-shot devices may last for long times under normal operating conditions and so accelerated life tests (ALTs), which increases the stress levels at which units are tested, may be needed. ALTs relate the lifetime distribution of an unit with the stress level at which it is tested via log-linear relationship, so inference results can be easily extrapolated to normal operating conditions. In particular, the step-stress model, which allows the experimenter to increase the stress level at pre-fixed times gradually during the life-testing experiment is specially advantageous for non-destructive one-shot devices. In this paper, we develop robust Rao-type test statistics based on the density power divergence (DPD) for testing linear null hypothesis for non-destructive one-shot devices under the step-stress ALTs with exponential lifetime distributions. We theoretically study their asymptotic and robustness properties, and empirically illustrates such properties through a simulation study.

1 Introduction

One shot devices, also known as current status data in survival analysis, is an extreme case of interval censoring. One-shot devices are tested at pre-specified inspections times, when we can only know if a test unit have failed or not. In this paper we focus on non-destructive one-shot devices, which do not get destroyed when tested and, therefore, all units that did not failed before an inspection

L. A. García-Escudero et al. (Eds.): SMPS 2022, AISC 1433, pp. 24–31, 2023.
https://doi.org/10.1007/978-3-031-15509-3_4

time can continue within the experiment. The non-destructiveness assumption is reasonable in many practical applications and makes best use of all units under test. For example, the proposed techniques can be applied for analyzing the effect of temperature in electronic components when instantaneous status data is not available (Gouno 2001, among others).

On the other hand, many real one-shot devices have large mean lifetimes under normal operating conditions, and so accelerated life tests (ALTs) plans, which accelerate the time to failure by increasing the stress level at which units are tested, are inevitable to infer on their reliability. This acceleration process will shorten the life of devices as well as reduce the costs associated with the experiment. In particular, we assume the lifetime of one-shot devices follows an exponential distribution, which is widely used as a lifetime model in engineering and physical sciences. Step-stress ALTs apply stress to devices progressively at pre-specified times. The step-stress are specially suitable for testing non-destructive one devices, as several stress levels can applied to the same unit until the failure occurs.

Further, we assume that the lifetime distribution of the one-shot devices follows the cumulative exposure model, which relates the lifetime distribution of a device at one stress level to the distribution at preceding stress levels by assuming the residual life of that device depends only on the cumulative exposure it had experienced, with no memory of how this exposure was accumulated. In particular, if we consider a multiple step-stress ALT with k ordered stress levels, $x_1 < x_2 < \cdots < x_k$ and their corresponding times of stress change $\tau_1 < \tau_2 \cdots < \tau_k$, the cumulative distribution function is given by:

$$G_T(t) = \begin{cases} G_1(t) = 1 - e^{-\lambda_1 t}, & 0 < t < \tau_1 \\ G_2\left(t + a_1 - \tau_1\right) = 1 - e^{-\lambda_2 (t + a_1 - \tau_1)}, & \tau_1 \leq t < \tau_2 \\ \vdots & \vdots \\ G_k\left(t + a_{k-1} - \tau_{k-1}\right) = 1 - e^{-\lambda_k (t + a_{k-1} - \tau_{k-1})}, & \tau_{k-1} \leq t < \infty, \end{cases} \tag{1}$$

with

$$a_{i-1} = \frac{\sum_{l=1}^{i-1} \left(\tau_l - \tau_{l-1}\right) \lambda_l}{\lambda_i}, \quad i = 1, ..., k-1. \tag{2}$$

and

$$\lambda_i(\boldsymbol{\theta}) = \theta_0 \exp(\theta_1 x_i), \quad i = 1, .., k, \tag{3}$$

where $\boldsymbol{\theta} = (\theta_0, \theta_1) \in \mathbb{R}^+ \times \mathbb{R} = \Theta$ is the unknown parameter vector of the model. This log-linear relation in (3) is frequently assumed in accelerated life test models.

Now, let consider a grid of inspection times, $t_1 < t_2 < \cdots < t_L$, including the times of stress change. The probability of a failure within the interval $(t_{j-1}, t_j]$ is given by

$$\pi_j(\boldsymbol{\theta}) = G_T(t_j) - G_T(t_{j-1}), \quad j = 1, .., L, \tag{4}$$

and the probability of survival at the end of the experiment is $\pi_{L+1}(\boldsymbol{\theta}) = 1 - G_T(t_L)$. Further, given a sample of one-shot data, $(n_1, ..., n_{L+1})$, the empirical probability vector can be defined as $\widehat{p} = (n_1/N, ..., n_{L+1}/N)$.

Classical inferential methods for one-shot are based on the maximum likelihood estimator (MLE), which is very efficient by it lacks of robustness. To overcome the robustness drawback, Balakrishnan et al. (2022a) proposed robust estimators for one-shot devices based on the popular density power divergence (DPD) under exponential lifetimes. They developed minimum DPD estimators (MDPDE) as well as Wald-type test based on them, and studied their asymptotic properties. Later, Balakrishnan et al. (2022b) extended the method and developed the restricted MDPDE for the same model, and they examined its robustness and asymptotic properties as well.

The DPD between the theoretical and empirical probability vectors is given by

$$d_\beta\left(\widehat{\boldsymbol{p}}, \boldsymbol{\pi}\left(\boldsymbol{\theta}\right)\right) = \sum_{j=1}^{L+1}\left(\pi_j(\boldsymbol{\theta})^{1+\beta} - \left(1+\frac{1}{\beta}\right)\widehat{p}_j\pi_j(\boldsymbol{\theta})^\beta + \frac{1}{\beta}\widehat{p}_j^{\beta+1}\right). \tag{5}$$

If we consider the restricted parameter space given by

$$\Theta_0 = \{\theta |\ g(\boldsymbol{\theta}) = \boldsymbol{m}^T\boldsymbol{\theta} - d = 0\},$$

with $\boldsymbol{m} = (m_0, m_1)^T \in \mathbb{R}^2$ and $d \in \mathbb{R}$, the restricted MDPDE, $\widetilde{\boldsymbol{\theta}}^\beta$, is naturally defined by

$$\widetilde{\boldsymbol{\theta}}^\beta = \arg\min\nolimits_{\boldsymbol{\theta}\in\boldsymbol{\Theta}_0} d_\beta(\widehat{\boldsymbol{p}}, \boldsymbol{\pi}(\boldsymbol{\theta})). \tag{6}$$

As Wald-type tests, Rao-type tests play a fundamental role in hypothesis testing. Indeed, each of these tests have their own important positions in the statistical literature, which are not eclipsed by the other. Classical Wald and Rao tests are based on the MLE and the restricted MLE. However, the non-robust nature of the procedures based on the MLE has motivated several researchers to look for robust generalizations of such tests. Basu et al. (2021) developed a robust generalization of the Rao test based on the DPD for general statistical models, and Jaenada et al. (2015) extended the method using the Rényi pseudistance.

In this paper, we develop Rao-type test statistics for non-destructive one-shot devices tested under step-stress ALT for testing linear null hypothesis. In Sect. 2 we define the Rao-type test statistics based on the restricted MDPDEs and we derive their asymptotic distribution. Section 3 theoretically analyzes the robustness properties of the tests through its IF. Finally, in Sect. 4 a simulation study is carried out to evaluate the performance of the proposed statistics.

2 Robust Rao-Type Test

Let us consider the score of the DPD loss function for the step-stress ALT model

$$\boldsymbol{U}_{\beta,N}(\boldsymbol{\theta}) = \boldsymbol{W}^T\boldsymbol{D}_{\boldsymbol{\pi}(\boldsymbol{\theta})}^{\beta-1}(\widehat{\boldsymbol{p}} - \boldsymbol{\pi}(\boldsymbol{\theta})) \tag{7}$$

where $\boldsymbol{D}_{\boldsymbol{\pi}(\boldsymbol{\theta})}$ denotes a $(L+1)\times(L+1)$ diagonal matrix with diagonal entries $\pi_j(\boldsymbol{\theta})$, $j = 1, ..., L+1$, and $\boldsymbol{W}$ is a $(L+1)\times 2$ matrix with rows $\boldsymbol{w}_j = \boldsymbol{z}_j - \boldsymbol{z}_{j-1}$, with

$$z_j = g_T(t_j)\left(\frac{\frac{t_j+a_{i-1}-\tau_{i-1}}{\theta_0}}{(t_j+a_{i-1}-\tau_{i-1})x_i+a^*_{i-1}}\right), \quad j=1,...,L, \tag{8}$$

$$a^*_{i-1} = \frac{1}{\lambda_i}\sum_{l=1}^{i-1}\lambda_l\left(\tau_l-\tau_{l-1}\right)\left(-x_i+x_l\right), \quad i=2,..,k, \tag{9}$$

$\boldsymbol{z}_{-1} = \boldsymbol{z}_{L+1} = \mathbf{0}$ and i is the stress level at which the units are tested after the $j-$th inspection time. Then, the MDPDE verifies the estimating equations given by $\boldsymbol{U}_{\beta,N}(\widehat{\boldsymbol{\theta}}^{\beta}) = \mathbf{0}$ (see Balakrishnan et al. 2022a, for more details).

We define Rao-type test statistics for testing linear null hypothesis

$$\mathrm{H}_0 : \boldsymbol{m}^T\boldsymbol{\theta} = d, \tag{10}$$

as

Definition 1. The Rao-type statistics, based on the restricted to the linear null hypothesis (10) MDPDE, $\widetilde{\boldsymbol{\theta}}^{\beta}$, for testing (10) is given by

$$\boldsymbol{R}_{\beta,N}(\widetilde{\boldsymbol{\theta}}^{\beta}) = N\boldsymbol{U}_{\beta,N}(\widetilde{\boldsymbol{\theta}}^{\beta})^T\boldsymbol{Q}_{\beta}(\widetilde{\boldsymbol{\theta}}^{\beta})\left[\boldsymbol{Q}_{\beta}(\widetilde{\boldsymbol{\theta}}^{\beta})^T\boldsymbol{K}_{\beta}(\widetilde{\boldsymbol{\theta}}^{\beta})\boldsymbol{Q}_{\beta}(\widetilde{\boldsymbol{\theta}}^{\beta})\right]^{-1}\boldsymbol{Q}_{\beta}(\widetilde{\boldsymbol{\theta}}^{\beta})^T\boldsymbol{U}_{\beta,N}(\widetilde{\boldsymbol{\theta}}^{\beta}), \tag{11}$$

where

$$\begin{aligned}\boldsymbol{K}_{\beta}(\boldsymbol{\theta}_0) &= \boldsymbol{W}^T\left(D^{2\beta-1}_{\boldsymbol{\pi}(\boldsymbol{\theta}_0)} - \boldsymbol{\pi}(\boldsymbol{\theta}_0)^{\beta}\boldsymbol{\pi}(\boldsymbol{\theta}_0)^{\beta T}\right)\boldsymbol{W}\\ \boldsymbol{Q}_{\beta}(\boldsymbol{\theta}_0) &= \boldsymbol{J}_{\beta}(\boldsymbol{\theta}_0)^{-1}\boldsymbol{m}(\boldsymbol{m}^T\boldsymbol{J}_{\beta}(\boldsymbol{\theta}_0)^{-1}\boldsymbol{m})^{-1}, \text{with } \boldsymbol{J}_{\beta}(\boldsymbol{\theta}_0) = \boldsymbol{W}^T D^{\beta-1}_{\boldsymbol{\pi}(\boldsymbol{\theta}_0)}\boldsymbol{W}.\end{aligned} \tag{12}$$

and $\boldsymbol{U}_{\beta,N}(\boldsymbol{\theta})$, is defined in (7).

Here, the matrix $\boldsymbol{Q}_{\beta}(\boldsymbol{\theta})$ depends on the null hypothesis trough $\boldsymbol{m}$, and the term d is only used to obtain the restricted MDPDE. The proposed testing procedure can be easily extended to more general composite null hypothesis defined by a set of restrictions.

Before presenting the asymptotic distribution of the Rao-type test statistics, $\boldsymbol{R}_{\beta,N}(\widetilde{\boldsymbol{\theta}}^{\beta})$, we shall establish the asymptotic distribution of the score $\boldsymbol{U}_{\beta,N}(\widetilde{\boldsymbol{\theta}}^{\beta})$.

Theorem 1. *The asymptotic distribution of the score $\boldsymbol{U}_{\beta,N}(\widetilde{\boldsymbol{\theta}}^{\beta})$ for the step-stress ALT model under exponential lifetimes, is given by*

$$\sqrt{N}\boldsymbol{U}_{\beta,N}(\widetilde{\boldsymbol{\theta}}^{\beta}) \xrightarrow[N\to\infty]{L} \mathcal{N}\left(\mathbf{0}, \boldsymbol{K}_{\beta}(\boldsymbol{\theta})\right)$$

where the variance-covariance matrix $\boldsymbol{K}_{\beta}(\boldsymbol{\theta})$ is defined in (12).

Proof. See proof in Balakrishnan et al. (2022b). □

Now, the following results states the asymptotic distribution of the Rao-type test statistics

Theorem 2. *The asymptotic distribution of the Rao-type test statistics defined in (11) under the linear null hypothesis (10) is a chi-square with 1 degree of freedom.*

Proof. The proof can be found in Balakrishnan et al. (2022b). □

Based on Theorem 2, for any $\beta \geq 0$ and $\boldsymbol{m} \in \mathbb{R}^2$, the critical region with significance level α for the hypothesis test with null hypothesis (10) is given by

$$\mathscr{R}_\alpha = \{(n_1, ..., n_{L+1}) \text{ s.t. } \boldsymbol{R}_{\beta,N}(\boldsymbol{\theta}) > \chi^2_{1,\alpha}\} \tag{13}$$

where $\chi^2_{1,\alpha}$ denotes the lower α-quantile of a chi-square with 1 degree of freedom.

3 Influence Function Analysis

We evaluate the robustness of the Rao-type test statistics through its IF. The robustness of an estimator or test statistic is widely analyzed using the concept of Influence Function (IF), which intuitively describes the effect of an infinitesimal contamination of the model on the estimate. The IF of the restricted MDPDE for the step-stress model with one-shot devices was established in Balakrishnan et al. (2022b), and the boundedness of the function was discussing there, concluding that the IF is always bounded for positive values of the tuning parameter.

The IF is computed as the Gateaux derivative of the functional defining the Rao-type test statistics (14) at a direction $\Delta_{\boldsymbol{x}}$. The functional associated to the Rao-type test statistic in terms of the statistical functional associated to the restricted MDPDE, $\widetilde{\boldsymbol{T}}_\beta$, is given by

$$\begin{aligned}\boldsymbol{R}_{\beta,N}(\widetilde{\boldsymbol{T}}_\beta(G)) =& N\boldsymbol{U}_{\beta,N}(\widetilde{\boldsymbol{T}}_\beta(G))^T\boldsymbol{Q}_\beta(\boldsymbol{\theta}_0)\left[\boldsymbol{Q}_\beta(\boldsymbol{\theta}_0)^T\boldsymbol{K}_\beta(\boldsymbol{\theta}_0)\boldsymbol{Q}_\beta(\boldsymbol{\theta}_0)\right]^{-1} \\ & \boldsymbol{Q}_\beta(\boldsymbol{\theta}_0)^T\boldsymbol{U}_{\beta,N}(\widetilde{\boldsymbol{T}}_\beta(G)).\end{aligned} \tag{14}$$

We define $G_\varepsilon = (1-\varepsilon)G + \varepsilon\Delta_{\boldsymbol{x}}$ the contaminated version of the distribution G. Taking derivatives in (14) and evaluating at $\varepsilon = 0$ we obtain, by the consistency of the restricted MDPDE under the null hypothesis, that the first order IF vanishes so it is inadequate to evaluate the robustness properties of the proposed Rao-type tests.

The second-order influence function is computed as the second-order derivative of the functional associated to the Rao-type test statistic, $\widetilde{\boldsymbol{T}}_\beta(G_\varepsilon)$, evaluated at $\varepsilon = 0$. Therefore, simple calculations yield the expression of the IF of the Rao-type test statistics,

$$\begin{aligned}\text{IF}^{(2)}\left(\boldsymbol{x}, \boldsymbol{R}_{\beta,N}(\widetilde{\boldsymbol{T}}_\beta), G\right) =& 2\text{IF}\left(\boldsymbol{x}, \widetilde{\boldsymbol{T}}_\beta, G\right)^T \boldsymbol{Q}_\beta(\boldsymbol{\theta}_0)\left[\boldsymbol{Q}_\beta(\boldsymbol{\theta}_0)^T\boldsymbol{K}_\beta(\boldsymbol{\theta}_0)\boldsymbol{Q}_\beta(\boldsymbol{\theta}_0)\right]^{-1} \\ & \boldsymbol{Q}_\beta(\boldsymbol{\theta}_0)^T\text{IF}\left(\boldsymbol{x}, \widetilde{\boldsymbol{T}}_\beta, G\right).\end{aligned}$$

Since $\boldsymbol{Q}_\beta(\boldsymbol{\theta}_0)\left[\boldsymbol{Q}_\beta(\boldsymbol{\theta}_0)^T\boldsymbol{K}_\beta(\boldsymbol{\theta}_0)\boldsymbol{Q}_\beta(\boldsymbol{\theta}_0)\right]^{-1}\boldsymbol{Q}_\beta(\boldsymbol{\theta}_0)^T$ are typically assumed to be bounded, the boundedness of the second order influence function of the Rao-type test statistics under the null hypothesis is determined by the boundedness of the IF of the restricted MDPDEs. Then, Rao-type statistics based on the restricted MDPDEs are robust for positives values of β, but lack of robustness for $\beta = 0$, corresponding to the MLE.

4 Simulation Study

In this section we empirically examine the performance of the Rao-type test statistics based on the restricted MDPDE for the step-stress model with exponential lifetime distributions, $\widetilde{\boldsymbol{\theta}}^{\beta}$, under different contamination scenarios. We consider a 2-step stress ALT experiment with $L = 11$ inspection times and a total of $N = 180$ one-shot devices. We consider two stress levels $x_1 = 35$, switched at $\tau_1 = 25$ to $x_2 = 45$. The experiment ends at $\tau_2 = 70$. During the experiment, inspection is performed at a grid of inspection times containing the times of stress change, $\mathrm{IT} = (10, 15, 20, 25, 30, 35, 40, 45, 50, 60, 70)$.

In our context, we must consider "outlying intervals" rather than "outlying devices", so we introduce contamination by increasing (or decreasing) the probability of failure in (4) for (at least) one interval. In our simulations the probability of failure in the third interval is switched as

$$\tilde{\pi}_3(\boldsymbol{\theta}) = G_{\boldsymbol{\theta}}(IT_3) - G_{\tilde{\boldsymbol{\theta}}}(IT_2) \tag{15}$$

where $\tilde{\boldsymbol{\theta}} = (\widetilde{\theta}_0, \widetilde{\theta}_1)$ is a contaminated parameter with $\tilde{\theta}_0 \leq \theta_0$ and $\tilde{\theta}_1 \leq \theta_1$. Of course, the probability vector is normalized after introducing contamination.

We introduce contamination in two different ways: decreasing the value of the first parameter θ_0, (first scenario of contamination) and decreasing the value of the second parameter θ_1 (second scenario of contamination). In both scenarios, the mean lifetime is decreased for the outlying cell. Moreover, we consider the linear hypothesis test defined by

$$\mathrm{H}_0 : \theta_1 = 0.03 \quad \text{vs} \quad \mathrm{H}_1 : \theta_1 \neq 0.03.$$

We evaluate the empirical significance level and power of the Rao-type test statistics under increasing contamination rate on the third cell. We compute the empirical level generating data with true parameter value $\boldsymbol{\theta} = (0.003, 0.03)^T$ (satisfying the null hypothesis) and the empirical power generating data with true parameter value $\boldsymbol{\theta} = (0.003, 0.06)^T$ (violating H_0) over $R = 1000$ repetitions. The contamination rate on the third cell is introduced in one of the model parameters, either θ_0 or θ_1, and correspondingly the contamination rate is calculated as $\varepsilon = 1 - \widetilde{\theta}_i/\theta_i$, $i = 0, 1$ where $\widetilde{\theta}_i$ denotes the contaminated parameter. It is interesting to note that, for $\beta = 0$, the associated Rao-type test statistic is not based on the MLE but on the restricted MLE under the constrain defined by the null hypothesis.

Figures 1 and 2 show the empirical level and power under increasing contamination rates for a $\alpha = 5\%$ significance level. As expected, the Rao test based on the restricted MLE is the most efficient in the absence of contamination, but it performance gets worse when there is data contamination. However, the Rao-type test statistics based on restricted MDPDEs with positive values of β keep competitive in the absence of contamination, and outperform the Rao-type statistics based on the restricted MLE when increasing the contamination rate. Furthermore, greater values of β produce more robust statistics.

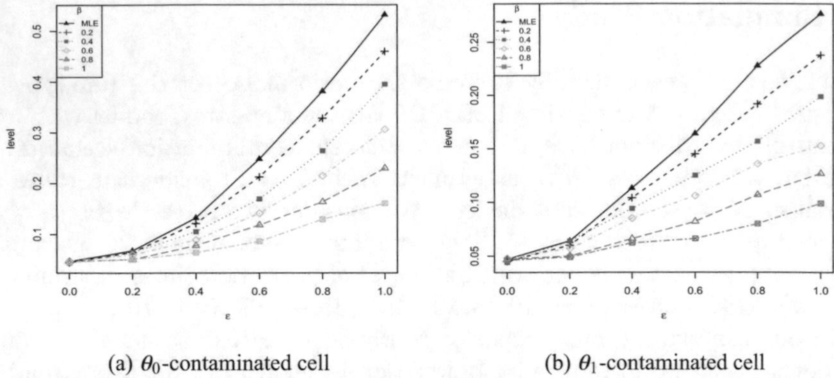

(a) θ_0-contaminated cell

(b) θ_1-contaminated cell

Fig. 1. Empirical significance level against contamination cell proportion in $R = 1000$ replications

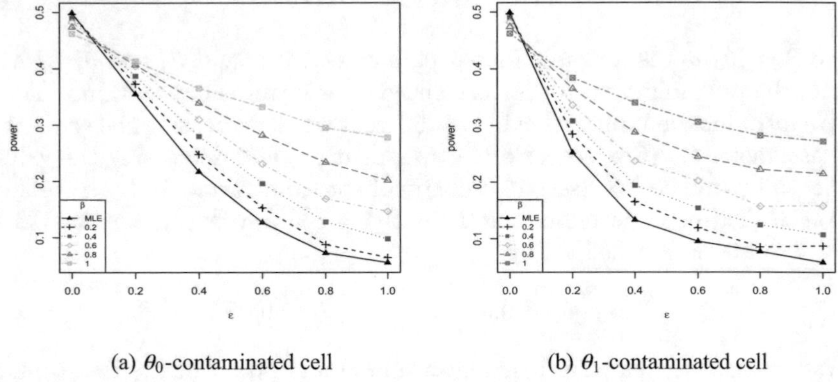

(a) θ_0-contaminated cell

(b) θ_1-contaminated cell

Fig. 2. Empirical power against contamination cell proportion in $R = 1000$ replications

Funding. This research is supported by the Spanish Grants PGC2018-095 194-B-100 and FPU 19/01824. M. Jaenada and L. Pardo are members of the Interdisciplinary Mathematics Institute.

References

Balakrishnan, N., Castilla, E., Jaenada, M., Pardo, L.: Robust inference for non-destructive one-shot devicetesting under step-stress model with exponential lifetimes. arXiv preprint, arXiv: 2204.11560 (2022)

Balakrishnan, N., Jaenada, M., Pardo, L.: The restricted minimum density power divergence estimator for non-destructive one-shot device testing the under step-stress model with exponential lifetimes. arXiv preprint, arXiv: 2205.07103 (2022)

Basu, A., Ghosh, A., Martin, N., Pardo, L.: A robust generalization of the Rao Test. J. Bus. Economic Stat. **40**(2), 868–879 (2021)

Gouno, E.: An inference method for temperature step-stress accelerated life testing. Qual. Reliab. Eng. Int. **17**(1), 11–18 (2001)

Jaenada, M., Miranda, P., Pardo, L.: Robust test statistics based on restricted minimum Rényi's pseudodistance estimators. Entropy **24**(5), 616 (2022)

An Imprecise Label Ranking Method for Heterogeneous Data

Tathagata Basu(✉), Sébastien Destercke, and Benjamin Quost

UMR CNRS 7253 Heudiasyc, Université de Technologie de Compiègne, Compiègne, France
{tathagata.basu,sebastien.destercke,benjamin.quost}@hds.utc.fr

Abstract. Learning to rank is an important problem in many sectors ranging from social sciences to artificial intelligence. However, it remains a rather difficult task to perform. Therefore, in some cases, it is preferable to perform cautious inference. For this purpose, we look into the possibility of an imprecise probabilistic approach for the Plackett-Luce model, a popular probabilistic model for label ranking. We aim at extending current Bayesian inference techniques for the Plackett-Luce model to an imprecise probabilistic setting so that we can deal with heterogeneous data by means of cautious mixture modelling. To achieve this, we perform a robust Bayesian analysis over a set of imprecise Dirichlet priors, which allows us to perform cautious label ranking. Finally, we use a synthetic dataset to illustrate our imprecise estimation method.

1 Introduction

Ranking objects is an important problem in many areas, such as social sciences, stock markets, e-commerce, etc. Sometimes, such rankings proceed from pairwise comparisons between the objects: one such treatment can be found in the model defined by Bradley and Terry (1952). This model naturally extends to the Plackett-Luce model in the case of multiple comparisons, as suggested by Plackett (1975) and Luce (1959). Several frequentist and Bayesian estimation methods have been developed based on these models.

Once the ranking model has been constructed, it can be used to estimate an optimal ranking between the objects. However, estimation of this ranking requires the data to be homogeneous, i.e. the objects are compared by a subpopulation of rankers which are assumed to be consistent with each other. In reality this might not be the case as we may gather this ranking data from different sources making the data heterogeneous. In such cases, mixtures of ranking models allow us to capture the sample information efficiently and also opens up the possibility of predicting a ranking after observing the ranker.

One of the first works on heterogeneous data was done by Gormley and Murphy (2006) where they suggested a mixture model using the Plackett-Luce model. Later, Caron et al. (2014) suggested a Bayesian alternative using a Dirichlet process model, where infinitely many models are assumed to be present in the

L. A. García-Escudero et al. (Eds.): SMPS 2022, AISC 1433, pp. 32–39, 2023.
https://doi.org/10.1007/978-3-031-15509-3_5

mixture. A similar idea involving a finite mixture was proposed by Mollica and Tardella (2016) for partially ranked data. Recently, Adam et al. (2020) proposed an imprecise probabilistic approach for the Plackett-Luce model where imprecise estimation was carried out using likelihood cuts.

In this paper, we discuss the notion of robust Bayesian analysis for the Plackett-Luce model. Section 2 presents our approach, very similar to that of Mollica and Tardella (2016). In our approach we overcome the difficulty of obtaining closed forms for the imprecise estimates by using a non-linear optimiser for certain parameter estimates. Section 3 illustrates our approach on a synthetic dataset. Section 4 concludes the paper along with a discussion on future works.

2 The Plackett-Luce Model

The Plackett-Luce model (Plackett 1975) is a simple and intuitive probabilistic model which gives us a probability for any observed ranking of p objects. Each object is associated with a strength parameter λ, which determines its probability of being preferred over others when drawing a sequence of objects. This model gives the probability of n independent rankings as

$$P(X \mid \lambda) = \prod_{i=1}^{n} \prod_{j=1}^{p_i - 1} \frac{\lambda_{x_{ij}}}{\sum_{m=j}^{p_i} \lambda_{x_{im}}}. \tag{1}$$

where $p_i \leq p$ is the number of objects in the i-th ranking, $\lambda := (\lambda_1, \cdots, \lambda_p)$ is the vector of strength parameters, and $X := [x_{ij}]$ is the $n \times p$ matrix containing the rankings (i.e., x_{ij} is the rank of the jth object or participant in the ith observed ranking, with $i = 1, \ldots, n$ and $j = 1, \ldots, p_i$). Note that Eq. (1) may be called the Plackett-Luce distribution because of its probabilistic formulation.

Example 1. Table 1 displays $n = 2$ rankings observed over $p = 4$ different objects 'A', 'B', 'C' and 'D'; where A and B have rank 4 and 1 in the first observed ranking ($x_{12} = 1$ and $x_{11} = 4$). The probability of these data is given by Eq. (2):

Table 1. Toy example with $n = 2$ rankings of $p = 4$ objects

	Obj. 1	Obj. 2	Obj. 3	Obj. 4
Ranking 1	B	D	C	A
Ranking 2	B	A	C	–

$$P(X \mid \lambda) = \left[\frac{\lambda_B}{\lambda_B + \lambda_D + \lambda_C + \lambda_A} \cdot \frac{\lambda_D}{\lambda_D + \lambda_C + \lambda_A} \cdot \frac{\lambda_C}{\lambda_C + \lambda_A}\right] \cdot \left[\frac{\lambda_B}{\lambda_B + \lambda_A + \lambda_C} \cdot \frac{\lambda_A}{\lambda_A + \lambda_C}\right] \tag{2}$$

We aim to estimate the strength parameters which maximise this probability.

2.1 Hierarchical Model

We follow Mollica and Tardella (2016) to construct our hierarchical mixture model. Though, we interpret a partial ordering involving $K \leq p$ objects as a top-K ordering, assuming the remaining objects to be absent from the model. This also simplifies the expression of the PL distribution, which is beneficial for faster computation.

Assuming a total of G components in the mixture, the model can be written as

$$X_i \mid \lambda, \omega \sim \sum_{g=1}^{G} \omega_g \mathrm{PL}(X_i \mid \lambda_g), \quad \mathrm{PL}(X_i \mid \lambda_g) = \prod_{j=1}^{p_i - 1} \frac{\lambda_{g,x_{ij}}}{\sum_{m=j}^{p_i} \lambda_{g,x_{im}}}.$$

We associate each observation with a unique latent membership indicator z_i, which follows a categorical distribution

$$z_i \mid \omega \sim \mathrm{Cat}(\omega_1, \cdots, \omega_G),$$

with ω_g being the weight of the g-th mixture component. For a full Bayesian treatment, we assign a set of imprecise Dirichlet priors on these weights:

$$\omega \mid s, \alpha \sim \mathrm{Dir}(s; \alpha_1, \cdots, \alpha_G),$$

where $\underline{\alpha}_g \leq \alpha_g \leq \overline{\alpha}_g$ for $g = 1, \cdots, G$, with in addition $\sum_g \alpha_g = 1$ and $s > 0$.

For the data augmentation process, we follow Mollica and Tardella (2016) and use exponentially distributed variables y_{ij} so that

$$y_{ij} \overset{\text{ind}}{\sim} \mathrm{Exp}\left(\prod_{g=1}^{G} \left(\sum_{m=j}^{p_i} \lambda_{g,x_{im}} \right)^{z_{ig}} \right).$$

Finally, we specify the strength parameters λ using gamma priors:

$$\lambda_{gk} \sim \mathrm{Gamma}(a_{gk}, b_k),$$

with $\underline{a}_{gk} \leq a_{gk} \leq \overline{a}_{gk}$, for $1 \leq g \leq G$ and $1 \leq k \leq p$.

2.2 Parameter Estimation

To discuss parameter estimation, we first look at the following iterative steps. These formulas can be obtained by adding log-prior components to the complete data log-likelihood given by Gormley and Murphy (2006). Since details are provided in this latter reference, we omit them here.

$$\widehat{z}_{ig}^{(t+1)} = \frac{\widehat{\omega}_g^{(t)} \mathrm{PL}(X_i \mid \widehat{\lambda}_g^{(t)})}{\sum_g \widehat{\omega}_g^{(t)} \mathrm{PL}(X_i \mid \widehat{\lambda}_g^{(t)})}, \quad \widehat{\omega}_g^{(t+1)} = \frac{s\,\alpha_g - 1 + \sum_i \widehat{z}_{ig}^{(t+1)}}{s - G + N},$$

$$\widehat{\lambda}_{gk}^{(t+1)} = \frac{a_{gk} - 1 + \sum_i \widehat{z}_{ig}^{(t+1)} u_{ik}}{b_g + \sum_i \widehat{z}_{ig}^{(t+1)} \sum_j \frac{\delta_{ijk}}{\sum_{m=j}^{p_i} \widehat{\lambda}_{g,x_{im}}^{(t)}}}, \quad (3)$$

where $u_{ik} = \mathbb{I}_{k\in\{x_{i1},\cdots,x_{i(p_i-1)}\}}$ and $\delta_{ijk} = \mathbb{I}_{k\in\{x_{ij},\cdots,x_{ip_i}\}}$. Note that, to start this iterative process we need suitable initial guesses for λ_g and ω_g.

2.2.1 Imprecise Estimates

In order to compute imprecise estimates, we need to calculate the bounds of the parameters in Eq. (3) over the sets of all possible values of $\alpha := (\alpha_1, \cdots, \alpha_G)$. From the iterative formula of the posterior membership probability ($\widehat{z}_{ig}$), we have

$$\widehat{z}_{ig}^{(t+1)} = \frac{\widehat{\omega}_g^{(t)}\mathrm{PL}(X_i \mid \widehat{\lambda}_g^{(t)})}{\sum_g \widehat{\omega}_g^{(t)}\mathrm{PL}(X_i \mid \widehat{\lambda}_g^{(t)})} = \frac{1}{1+\frac{\sum_{g'\neq g}\widehat{\omega}_{g'}^{(t)}\mathrm{PL}(X_i|\widehat{\lambda}_{g'}^{(t)})}{\widehat{\omega}_g^{(t)}\mathrm{PL}(X_i|\widehat{\lambda}_g^{(t)})}}$$
$$\geq \frac{1}{1+\frac{\sum_{g'\neq g}\max_{\widehat{\omega}_{g'}^{(t)},\widehat{\lambda}_{g'}^{(t)}}\left\{\widehat{\omega}_{g'}^{(t)}\mathrm{PL}(X_i \mid \widehat{\lambda}_{g'}^{(t)})\right\}}{\min_{\widehat{\omega}_g^{(t)},\widehat{\lambda}_g^{(t)}}\left\{\widehat{\omega}_g^{(t)}\mathrm{PL}(X_i \mid \widehat{\lambda}_g^{(t)})\right\}}}.$$

Now, let

$$\underline{\mathrm{PL}}(X_i \mid \widehat{\lambda}_g^{(t)}) := \min_{\widehat{\lambda}_g^{(t)}}\left\{\mathrm{PL}(X_i \mid \widehat{\lambda}_g^{(t)})\right\},\quad \overline{\mathrm{PL}}(X_i \mid \widehat{\lambda}_g^{(t)}) := \max_{\widehat{\lambda}_g^{(t)}}\left\{\mathrm{PL}(X_i \mid \widehat{\lambda}_g^{(t)})\right\}$$

such that $\underline{\widehat{\lambda}}_{gk}^{(t)} \leq \widehat{\lambda}_{gk}^{(t)} \leq \overline{\widehat{\lambda}}_{gk}^{(t)}$ for $1 \leq k \leq p$. Then, the lower bound of the posterior membership probability is given by:

$$\underline{\hat{z}}_{ig} = \frac{\underline{\widehat{\omega}}_g^{(t)}\underline{\mathrm{PL}}(X_i \mid \widehat{\lambda}_g^{(t)})}{\underline{\widehat{\omega}}_g^{(t)}\underline{\mathrm{PL}}(X_i \mid \widehat{\lambda}_g^{(t)}) + \sum_{g'\neq g}\overline{\widehat{\omega}}_{g'}^{(t)}\overline{\mathrm{PL}}(X_i \mid \widehat{\lambda}_{g'}^{(t)})}, \tag{4}$$

and similarly the upper bound is given by:

$$\overline{\hat{z}}_{ig} = \frac{\overline{\widehat{\omega}}_g^{(t)}\overline{\mathrm{PL}}(X_i \mid \widehat{\lambda}_g^{(t)})}{\overline{\widehat{\omega}}_g^{(t)}\overline{\mathrm{PL}}(X_i \mid \widehat{\lambda}_g^{(t)}) + \sum_{g'\neq g}\underline{\widehat{\omega}}_{g'}^{(t)}\underline{\mathrm{PL}}(X_i \mid \widehat{\lambda}_{g'}^{(t)})}, \tag{5}$$

where the lower and upper bounds $\underline{\widehat{\omega}}_g^{(t)}$ and $\overline{\widehat{\omega}}_g^{(t)}$ on the mixture weights $\widehat{\omega}_g^{(t)}$ is obtained from Eq. (3), so that

$$\underline{\widehat{\omega}}_g^{(t)} = \frac{s\underline{\alpha}_g - 1 + \sum_i \underline{\widehat{z}}_{ig}^{(t)}}{s - G + N} \quad \text{and} \quad \overline{\widehat{\omega}}_g^{(t)} = \frac{s\overline{\alpha}_g - 1 + \sum_i \overline{\widehat{z}}_{ig}^{(t)}}{s - G + N}. \tag{6}$$

Even though the update equations for these parameter bounds can be derived easily, computing these bounds is difficult and we need to employ an optimiser to compute $\underline{\mathrm{PL}}(X_i \mid \widehat{\lambda}_g^{(t)})$ and $\overline{\mathrm{PL}}(X_i \mid \widehat{\lambda}_g^{(t)})$

We face similar issues for the strength parameters as well, as $\widehat{\lambda}_{gk}^{(t+1)}$ are not monotone with respect to $\widehat{z}_{ig}^{(t+1)}$'s. As a result, in order to compute $\underline{\widehat{\lambda}}_{gk}^{(t+1)}$ and $\overline{\widehat{\lambda}}_{gk}^{(t+1)}$; we need to solve the following multiobjective optimisation problems:

$$\min\left(\widehat{\lambda}_{1k}^{(t+1)}, \widehat{\lambda}_{2k}^{(t+1)}, \cdots, \widehat{\lambda}_{Gk}^{(t+1)}\right) \quad \text{and} \quad \max\left(\widehat{\lambda}_{1k}^{(t+1)}, \widehat{\lambda}_{2k}^{(t+1)}, \cdots, \widehat{\lambda}_{Gk}^{(t+1)}\right), \quad (7)$$

such that $\underline{\widehat{z}}_{ig} \leq \widehat{z}_{ig} \leq \overline{\widehat{z}}_{ig}$, for $1 \leq g \leq G$ and $1 \leq i \leq N$.

3 Illustration

We illustrate our approach using a synthetic dataset, which allows us to assess the performance of our method in estimating the strength parameters. We consider a set of $p = 8$ objects, together with two different sets of strengths λ: in one case, $(\lambda_{1,1}, \lambda_{1,2}, \cdots \lambda_{1,8}) = (8, 7, \cdots, 1)$, and in the other case $(\lambda_{2,1}, \lambda_{2,2}, \cdots \lambda_{2,8}) = (1, 2, \cdots, 8)$. These strengths correspond to two ranking processes which order the objects in an exactly opposite way. We use these parameters to randomly generate our ranking dataset using the generative model suggested by Caron and Doucet (2012). Then, we pick 60% samples from the first set, and the remaining 40% from the second one. Note that, this is a randomised generative process and in the estimation the orderings of the components may not be estimated based on this generation.

Results

To show our results, we consider two different settings. In the first case, we consider the weights of the Dirichlet model to be precise and equal. In the second case, we define imprecise initial weights, such that $0.5 \leq \alpha_1 \leq 0.7$ and $0.3 \leq \alpha_2 \leq 0.5$. In both cases, we set our initial estimates for a_{gk}, so that they satisfy the orderings of λ_{gk} estimated from a precise estimation process. This initialisation step is crucial to avoid $\widehat{\lambda}_g^{(t)}$ taking extreme values, which would be problematic for the convergence of the iterative algorithm. Moreover, in each iteration, we need to enforce orderings of these $\widehat{\lambda}_{gk}^{(t)}$ estimates whilst computing $\underline{\text{PL}}(X_i \mid \widehat{\lambda}_g^{(t)})$ and $\overline{\text{PL}}(X_i \mid \widehat{\lambda}_g^{(t)})$ using the optimiser. Nevertheless, to ensure robustness, we take a large interval for each a_{gk} for capturing the imprecision in the data.

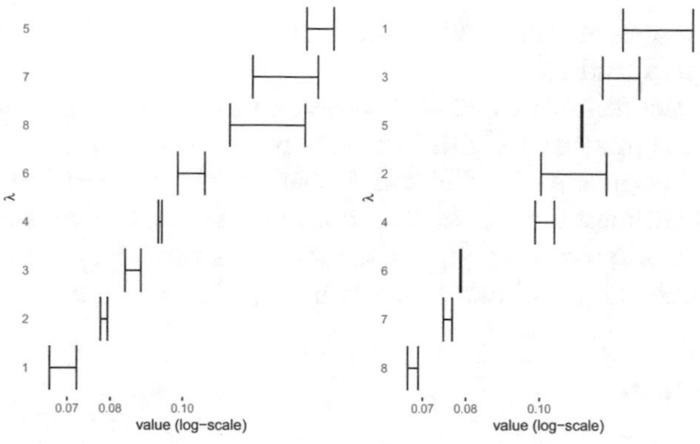

Fig. 1. Strength parameter estimates for $\alpha_1 = \alpha_2 = 0.5$ and $s = 5$

For precise and equal initial weights, we get estimated mixture weights $\omega_1 \in [0.36; 0.63]$ and $\omega_2 \in [0.37; 0.64]$. The corresponding strengths are displayed in Fig. 1 (first component on left and the second one on right). We notice that the strength parameters follow a strict ordering for the first component, which is not the case for the second one[1]: in this latter case, we notice that '2' and '5' are not comparable, which leads to a partial ordering.

We notice a similar outcome for imprecise initial weights (Fig. 2). In this case, we obtain mixture weight estimates $\omega_1 \in [0.21; 0.62]$ and $\omega_2 \in [0.38; 0.78]$. We can observe from the figure that the intervals tend to be larger than the intervals in the first case, confirming the initial imprecise prior. Furthermore, we see that the estimated ranking from the first component is the same as in the precise initialisation case (see Fig. 1). However, the estimates are slightly different for

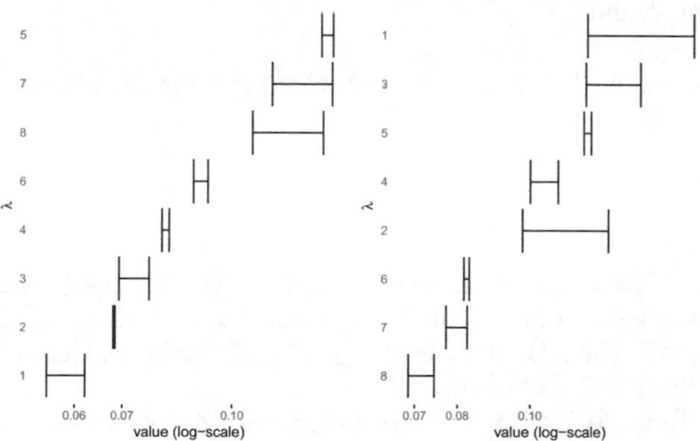

Fig. 2. Strength parameter estimates for $0.3 \leq \alpha_1 \leq 0.5; 0.5 \leq \alpha_2 \leq 0.7$ and $s = 5$

[1] We consider, object i is preferred to object j if $\underline{\lambda_i} > \underline{\lambda_j}$ and $\overline{\lambda_i} > \overline{\lambda_j}$.

the second component: then, '2' is not comparable to both '4' and '5', though '5' is strictly preferred to '4'.

We also notice that our choice of ordered imprecise weights gives us better estimates for the mixture weights, which is not the case for starting with equal weights. This happens as for the equal weights, the imprecision in the data is only captured through our ignorance about the size of the mixture components whereas for the ordered weights, our model tends to assign membership probabilities more efficiently and can also capture the imprecision in rankings.

4 Conclusion

In this article, we investigate inferring a probabilistic ranking model from heterogeneous data: we assume the data to come from several sub-populations, each of which can be associated with a Plackett-Luce ranking model. We propose a robust Bayesian approach to estimate the strength parameters of the resulting mixture of PL models. We notice that estimating the model is computationally expensive, since it requires to repeatedly calculate the bounds of each PL distribution—which involves solving nonlinear optimisation problems.

We illustrate our method using a synthetic dataset, in order to study its efficiency in inferring a mixture of rankings. The experiments show that imprecise initial mixture weights tend to produce wider intervals for the strength estimates, which reflects our methods ability to showcase the inherent imprecision in scarce data. This is extremely important for cautious ranking as we may want to abstain from ordering in certain scenarios.

A major issue remaining to be addressed is the high computation time of the estimation procedure. This motivates us to find tight approximate bounds to ensure monotonicity. Hopefully, this will help us to reduce the computation cost significantly, while allowing us to investigate problems with a high number of mixture components.

Acknowledgements. This research was funded by the project PreServe (ANR Grant ANR-18-CE23-0008).

References

Adam, L., Van Camp, A., Destercke, S., Quost, B.: Inferring from an imprecise Plackett–Luce model: application to label ranking. In: Davis, J., Tabia, K. (eds.) SUM 2020. LNCS (LNAI), vol. 12322, pp. 98–112. Springer, Cham (2020). https://doi.org/10.1007/978-3-030-58449-8_7

Bradley, R.A., Terry, M.E.: Rank analysis of incomplete block designs: I. the method of paired comparisons. Biometrika **39**(3/4), 324–345 (1952)

Caron, F., Doucet, A.: Efficient bayesian inference for generalized Bradley-Terry models. J. Comput. Graph. Stat. **21**(1), 174–196 (2012)

Caron, F., Teh, Y.W., Murphy, T.B.: Bayesian nonparametric Plackett-Luce models for the analysis of preferences for college degree programmes. Ann. Appli. Stat. **8**(2), 1145–1181 (2014)

Gormley, I.C., Murphy, T.B.: Analysis of Irish third-level college applications data. J. R. Stat. Soc. A. Stat. Soc. **169**(2), 361–379 (2006)
Luce, R.D.: Individual Choice Behavior: A Theoretical Analysis. Wiley, New York (1959)
Mollica, C., Tardella, L.: Bayesian Plackett-Luce mixture models for partially ranked data. Psychometrika **82**(2), 442–458 (2016)
Plackett, R.L.: The analysis of permutations. J. Roy. Stat. Soc.: Ser. C (Appl. Stat.) **24**(2), 193–202 (1975)

The Choice of an Appropriate Stochastic Order to Aggregate Random Variables

Juan Baz[1(✉)], Irene Díaz[2], and Susana Montes[1]

[1] Department of Statistics and I.O. and Didactics of Mathematics, University of Oviedo, Calle Federico García Lorca 18, 33007 Oviedo, Spain
{bazjuan,montes}@uniovi.es

[2] Department of Computer Science, University of Oviedo, Campus de Viesques, 33004 Gijón, Spain
sirene@uniovi.es

Abstract. Aggregation functions have been widely used as a method to fuse data in a large number of applications. In most of them, the data can be modeled as a simple random sample. Thus, it is reasonable to treat the aggregated values as random variables. In this paper, the concept of aggregation functions of random variables with respect to a stochastic order is presented. Additionally, four alternatives for the choice of the adequate order are considered and their benefits and drawbacks are studied.

1 Introduction

An aggregation function takes some values on the cartesian product of n identical intervals and returns a new value in the considered interval, satisfying monotony in every component and some boundary conditions. Aggregation functions are widely used in fields such as decision theory (Mohd and Abdullah 2017), fusion of predictions (Nungesser et al. 1999; Shanmugam et al. 2021) and are the basis of mathematical areas such us fuzzy set theory (Zimmermann 2010).

In many applications, the aggregated values are data obtained as a measurement process over a population. In this direction, as it is classically done in statistics, it is reasonable to model these measurements as random variables. With this assumption the aggregation function, if measurable, can be seen as a function that takes a random vector and returns a random variable.

Specifically, given an interval I (bounded or not), we can define the concept of aggregation function of random variables as a function that, given a random vector with support I^n, returns a random variable with support I. In addition, the conditions of monotony and the boundary conditions must be extended from vectors to random vectors.

This paper is devoted to define the aforementioned concept and study the adequate settlement of the boundary and monotony conditions, considering 4 different alternatives as stochastic orders between random vectors. In Sect. 2,

L. A. García-Escudero et al. (Eds.): SMPS 2022, AISC 1433, pp. 40–47, 2023.
https://doi.org/10.1007/978-3-031-15509-3_6

the basic concepts about aggregation functions and stochastic orders that are needed for the development of the study are provided. A definition of aggregation function of random variables with respect to a stochastic order is proposed in Sect. 3. The choice of the adequate stochastic order is studied in Sect. 3.1. Finally, in Sect. 3.2, the definition of aggregation function of random variables induced by a classical (measurable) aggregation function is provided and the conclusions are discussed in Sect. 4.

2 Basic Concepts

This section is devoted to introduce the main concepts we will work with, the definition of aggregation function and the stochastic orders.

2.1 Aggregation Functions

An aggregation function is typically referred as a function which summarize the information of a number of values by a single number. The formal definition is provided below:

Definition 1 (Grabisch et al. 2009). Let I be an interval in the real line $\mathbb{R}$. An aggregation function is a function $A : I^n \to I$ satisfying:

- Is non decreasing (in each variable).
- The following boundary conditions are fulfilled:

$$\inf_{\mathbf{x}\in I^n} A(\mathbf{x}) = \inf I, \ \sup_{\mathbf{x}\in I^n} A(\mathbf{x}) = \sup I$$

Let us remark that the considered infimum and supremum are elements of extended real line $\bar{R} = [-\infty, \infty]$. Notice that the main properties of an aggregation function are defined with respect to the usual order of real numbers and the component-wise order of real vectors. When extending this concept to random vectors, the choice of an order will be especially relevant.

2.2 Stochastic Orders

When ordering random variables, the most used alternative is the usual stochastic order, also known as Stochastic Dominance, which compares the cumulative distribution functions of the considered random variables. In particular, it is defined as follows:

Definition 2 (Shaked and Shanthikumar 2007). Let X and Y be two random variables with cumulative distribution functions F_X and F_Y. Then, it is said that X is smaller than X in the usual stochastic order if $F_Y(t) \leq F_X(t)$ for any $t \in \mathbb{R}$ and it is denoted as $X \leq_{st} Y$.

Qualitatively, X is smaller than Y if Y takes greater values with greater probability. Although its widespread use in literature, the usual stochastic order, even is reflexive and transitive, it is not total relation. For instance, a Gaussian random variable only can be smaller than another Gaussian random variable if both have the same variance. In addition, it does not consider the possible dependence between the variables. In order to fix some of these lacks, the Statistical Preference can be considered.

Definition 3 (De Schuymer et al. 2003**).** Let X and Y be two random variables defined in the same probability space. Then, it is said that **Y** is statistical preferred over X if $P(X \leq Y) + \frac{1}{2}P(X = Y) \leq 0.5$ and is denoted as $X \leq_{SP} Y$.

Let us clarify the usual abuse of notation when working with random variables in the same probability space (Ω, Σ, P). When denoting $P(X \leq Y)$ in the latter definition, we refer to $P(\{w \in \Omega : X(w) \leq Y(w)\})$. Notice that the Statistical Preference is a total relation. On the other hand, this relation is neither transitive nor antisymmetric.

We need to use orders not only defined for random variables but also for random vectors. Let us first introduce the concept of upper set.

Definition 4 (Davey and Priestley 2002**).** A subset $M \subset \mathbb{R}^n$ is said to be an upper set if for any pair of vectors $\mathbf{t}, \mathbf{v} \in \mathbb{R}^n$, if $t_i \leq v_i$ for any $i \in \{1, \ldots, n\}$ and $\mathbf{t} \in M$, then $\mathbf{v} \in M$.

In this paper, we are going to consider four different alternatives to order random vectors. The first three are generalizations of the usual stochastic order for random vectors, see Shaked and Shanthikumar (2007); Kopa and Petrová (2018), while for the latter one we have extended the concept of Statistical Preference to random vectors.

Definition 5. Let **X** and **Y** be two random vectors of dimension n. Then:

- If $X_i \geq_{st} Y_i$ for any $i \in \{1, \ldots, n\}$, it is said that **X** is smaller than **Y** in the component-wise stochastic order and it is denoted as $\mathbf{X} \geq_{cst} \mathbf{Y}$.
- If for any set of n positive real numbers $\alpha_1, \ldots, \alpha_n \in \mathbb{R}^+$, $\sum_{i=1}^n \alpha_i X_i \leq_{LSD} \sum_{i=1}^n \alpha_i Y_i$, it is said that **Y** has Linear Stochastic Dominance over **X** and it is denoted by $\mathbf{X} \leq_{LSD} \mathbf{Y}$.
- If for any (Borel) upper set $M \subset \mathbb{R}^n$, $P(\mathbf{X} \in M) \leq P(\mathbf{Y} \in M)$, it is said that **X** is smaller than **Y** in the usual stochastic order and it is denoted by $\mathbf{X} \leq_{st} \mathbf{Y}$.
- When **X** and **Y** are defined in the same probability space, if $P(\mathbf{X} \geq \mathbf{Y}) + \frac{1}{2}P(\mathbf{X} = \mathbf{Y}) \leq 0.5$ (component-wise), it is said that **Y** is statistical preferred over **X** and it is denoted as $\mathbf{X} \leq_{SP} \mathbf{Y}$.

For simplicity, in the following we will refer to the latter relations as the considered stochastic orders, although the Statistical Preference is neither a transitive nor antisymmetric relation. In addition, we will use the notation $\mathbf{X} <_R \mathbf{Y}$ for the case when $\mathbf{X} \leq_R \mathbf{Y}$ but $\mathbf{Y} \nleq_R \mathbf{X}$, regardless of the considered relation R.

3 Aggregation of Random Variables

As discussed before, the main objective of this paper is to define aggregation functions of random variables. Since the aggregation functions are defined over intervals, we need to establish a similar concept for random vectors. The most intuitive and straightforward way to extend the intervals for random vectors is to consider the random vectors with support over Cartesian products of intervals.

Definition 6. Let (Ω, Σ, P) be a probability space and let I be a real interval. Then, we define $L_I^n(\Omega)$ as the set of random vectors of (Ω, Σ, P) with support in I^n, i.e. $L_I^n(\Omega) = \{\mathbf{X} : \Omega \to I^n \mid X \text{ is measurable}\}$.

In addition, it must be introduced an order between random vectors. Notice that the main properties of aggregation functions, the monotony and the boundary conditions are related to the component-wise order of real vectors. As introduced in Sect. 2.2, several alternatives can be considered when ordering random vectors. Let us, at least for the moment, consider a general stochastic order $\leq_{SO}$, considering also as an alternative the Statistical Preference. The choice of the most adequate stochastic order will be discussed in Sect. 3.1. With this consideration, a definition of aggregation of random variables can be stated as follows:

Definition 7. Let (Ω, Σ, P) a probability space, a stochastic order $\leq_{SO}$ and I a real non empty interval. An aggregation function of random variables is a function $A : L_I^n(\Omega) \to L_I(\Omega)$ which satisfies:

- For any $\mathbf{X}, \mathbf{Y} \in L_I^n(\Omega)$ such that $\mathbf{X} \leq_{SO} \mathbf{Y}$, $A(\mathbf{X}) \leq_{SO} A(\mathbf{Y})$(non decreasing).
- The following boundary conditions with respect to $\leq_{SO}$ are fulfilled:

$$\inf_{\mathbf{X} \in L_I^n(\Omega)} A(\mathbf{X}) = \inf L_I(\Omega), \quad \sup_{\mathbf{X} \in L_I^n(\Omega)} A(\mathbf{X}) = \sup L_I(\Omega)$$

Considering the latter definition, which is defined as a function between random vectors and random variables, we are specially interested in using classical aggregation functions for aggregating random variables. More precisely, we wonder if composing an aggregation function with random vectors is, for a properly determined stochastic order, an aggregation function of random variables.

In this direction, two main question arise. Firstly, we need the composition of an aggregation function and a random vector to be a random variable, that is, measurable. Secondly, a stochastic order that fits the properties of classical aggregation functions must be stated.

Starting from the first one, we need to obtain a random variable when applying an aggregation function to a random vector. Trivially, if we consider a measurable aggregation function, this property holds noticing that the composition of measurable functions are measurable (Bauer 2011). Typically, usual aggregation functions are always measurable. In fact, examples of non-measurable aggregation functions are pathological and uncommon.

3.1 The Choice of the Stochastic Order

The second issue to be addressed is the choice of a order for random vectors. Since we are defining the concept, actually, any choice is valid, but we are looking for a definition of aggregation of random variables that fits to the case of using usual aggregation functions over random vectors. In particular, the 4 alternatives introduced in Definition 5 has been studied.

3.1.1 Component-Wise Stochastic Order

Component-wise stochastic order regards on the usual stochastic order of the components of the random vectors. Unfortunately, this way to compare random vectors does not consider the possible dependence among the random variables, thus it is not suitable for our purpose. Let us illustrate this with an example:

Example 1. Consider two bivariate random vectors $\mathbf{X} = (X_1, X_2)$ and $\mathbf{Y} = (Y_1, Y_2)$, such that $P(X_1 = 1, X_2 = 0) = P(X_0 = 1, X_2 = 1) = 0.5$ and $P(X_1 = 0, X_2 = 0) = 0.16$, $P(X_1 = 0, X_2 = 1) = P(X_1 = 1, X_2 = 0) = 0.24$ and $P(X_1 = 1, X_2 = 1) = 0.36$.

It is immediate that $X_1 <_{st} Y_1$ and $X_2 <_{st} Y_2$, so $\mathbf{X} <_{cst} \mathbf{Y}$. However, when aggregating the random vectors by using the minimum function, it holds that $P(\min \mathbf{X} = 1) = 0.5$ and $P(\min \mathbf{Y} = 1) = 0.36$. Thus, $\min \mathbf{Y} <_{cst} \min \mathbf{X}$.

It is concluded that the component-wise stochastic order is not an adequate order to consider, since the monotony property does not have to follow when using an usual aggregation function, in the case of the latter example the minimum, to aggregate the random variables.

3.1.2 Linear Stochastic Dominance

Since the component-wise stochastic order is not a suitable order to consider, one may consider the Linear Stochastic Dominance, which impose a stronger condition over the random vectors. Now, to have one random vector dominating the another, it is not only required to study the usual stochastic order for the components but also for all the positive linear combinations of them. However, even with this addition, the behaviour is not the desired one.

Example 2. Consider two bivariate random vectors $\mathbf{X} = (X_1, X_2)$ and $\mathbf{Y} = (Y_1, Y_2)$, such that $P(X_1 = 0, X_2 = 1) = P(X_1 = 1, X_2 = 0) = 0.5$ and $P(X_1 = 0, X_2 = 0) = P(X_1 = 0, X_2 = 0.5) = P(X_1 = 0.5, X_2 = 0) = P(X_1 = 0.5, X_2 = 0.5) = 0.25$.

Now, let us consider the positive linear combinations of the components $\alpha X_1 + \beta X_2$ and $\alpha Y_1 + \beta Y_2$, with $\alpha, \beta \in \mathbb{R}^+$. Notice that $\alpha X_1 + \beta X_2$ takes the value α and β with the same probability 0.5 and $\alpha Y_1 + \beta Y_2$ takes 4 equiprobable values, 0, $\frac{\alpha}{2}$, $\frac{\beta}{2}$ or $\frac{\alpha+\beta}{2}$. For any case of $\alpha < \beta$, $\alpha > \beta$ or $\alpha = \beta$, we always have $\alpha Y_1 + \beta Y_2 \leq_{st} \alpha X_1 + \beta X_2$. Moreover, is immediate to check that $\mathbf{Y} <_{LSD} \mathbf{X}$.

On the other hand, when applying the minimum to both random vectors, we get $P(\min \mathbf{X} = 0) = 1$ and $P(\min \mathbf{Y} > 0) = 0.25$. Thus, we have that $\min \mathbf{X} <_{LSD} \min \mathbf{Y}$ holds.

3.1.3 Usual Stochastic Order

Since the two latter alternatives are not adequate for our purpose, let us consider a more restrictive concept, the usual stochastic order. In this case, the use of upper sets on its definition fits completely with the monotony of usual aggregation functions.

Proposition 1. *Let* $\mathbf{X}$ *and* $\mathbf{Y}$ *be two random vectors of dimension* $n \in \mathbb{N}$ *with support* I^n *such us* $\mathbf{X} \leq_{st} \mathbf{Y}$ *and* $A : I^n \to I$ *a measurable aggregation function. Then,* $A(\mathbf{X}) \leq_{st} A(\mathbf{Y})$.

Proof. Consider the cumulative distribution functions of $A(\mathbf{X})$ and $A(\mathbf{Y})$, F_X and F_Y respectively. Let $a \in \mathbb{R}$ an arbitrary real number. Now, take the pre-image of (a, ∞) by A, $A^{-1}((a, \infty)) = \{\mathbf{t} \in \mathbb{R}^n : A(\mathbf{t}) \in (a, \infty)\}$. Notice that, since A is monotone, if $A(\mathbf{t}) \geq a$, then for any $\mathbf{v} \geq \mathbf{t}$ it holds $A(\mathbf{v}) \geq a$. Thus, $A^{-1}((a, \infty))$ is an upper set. If $\mathbf{X} \leq_{st} \mathbf{Y}$, then $P(A(\mathbf{Y}) \leq a) \leq P(A(\mathbf{Y}) \leq a)$, which is equivalent to $F_Y(a) \leq F_X(a)$ and it is concluded that $A(\mathbf{X}) \leq_{st} A(\mathbf{Y})$. □

3.1.4 Statistical Preference

Moving on from Stochastic Dominance relations, we end by considering Statistical Preference. Likely as the latter case, its definition is convenient for our objectives. We remark that, for dimension greater than one, the considered concept of Statistical Preference for random vectors is not a total relation.

Proposition 2. *Let* $\mathbf{X}$ *and* $\mathbf{Y}$ *be two random vectors of dimension* $n \in \mathbb{N}$ *with support* I^n *such us* $\mathbf{X} \leq_{SP} \mathbf{Y}$ *and* $A : I^n \to I$ *a measurable aggregation function. Then,* $A(\mathbf{X}) \leq_{SP} A(\mathbf{Y})$.

Proof. Consider the probability space (Ω, Σ, P). Consider the probabilities $P(A(\mathbf{X}) \leq A(\mathbf{Y}))$ and $P(A(\mathbf{X}) = A(\mathbf{Y}))$. Then, applying the monotony of A, have that if $\mathbf{X} \leq \mathbf{Y}$ (component-wise), then $A(\mathbf{X}) \leq A(\mathbf{Y})$ as well as if $\mathbf{X} = \mathbf{Y}$, then $A(\mathbf{X}) = A(\mathbf{Y})$. Thus, $P(A(\mathbf{X}) \leq A(\mathbf{Y})) + \frac{1}{2}P(A(\mathbf{X}) = A(\mathbf{Y})) \geq P(\mathbf{X} \leq \mathbf{Y}) + \frac{1}{2}P(\mathbf{X} = \mathbf{Y})$ and using that $\mathbf{X} \leq_{SP} \mathbf{Y}$ it is concluded that $A(\mathbf{X}) \leq_{SP} A(\mathbf{Y})$. □

3.2 Induced Aggregation of Random Variables

We end the paper with the concept of induced aggregation function of random variables. The main idea is to compose an usual aggregation function with the random vector in order to obtain an aggregation of random variables.

Proposition 3. *Let* I *be a real interval and let* $A : I^n \to I$ *be a measurable aggregation function. Consider the function* $\hat{A} : L_I^n \to L_I$ *such that for any random vector* $\mathbf{X} \in L_I^n$*, it holds* $\hat{A}(\mathbf{X}) = A \circ \mathbf{X}$*. Then,* $\hat{A}$ *is an aggregation function of random variables with respect to the usual stochastic order and with respect to the Statistical Preference.*

The function $\hat{A}$ *is referred to as the aggregation function of random variables induced by the aggregation function* A*.*

Proof. Firstly, we recall that the composition of measurable functions is measurable (Bauer 2011). Secondly, notice that result of aggregation must be a random variable with support in the considered interval, as a consequence of the associated property of the aggregation function. Then, the monotony is fulfilled as a consequence of Proposition 1 and Proposition 2, respectively for usual stochastic order and Statistical Preference. Finally, for the boundary conditions, consider a degenerate distribution for the bounded ends or a sequence of degenerate distributions for the unbounded ends of the interval. □

4 Conclusions

The concept of aggregation function of random variables with respect to a stochastic order has been defined. The adequate choice of the stochastic order, with the objective of use compositions of usual aggregation functions as aggregation functions of random variables, has been studied. As a conclusion, it has been shown that the component-wise stochastic order and the Linear Stochastic Dominance are not suitable in this regard. On the other hand, the usual stochastic order and the Statistical Preference allow any measurable aggregation function to induce an aggregation function of random variables.

The main future topic of research as a continuation of this proposal is the definition of the extension of properties of aggregation functions, such us idempotence, shift-invariance or associativity. Since this properties are defined using inequalities, we may extend this inequalities using the considered stochastic order.

Acknowledgements. This research has been partially supported by the Spanish Ministry of Science and Technology (TIN-2017-87600-P and PGC2018- 098623-B-I00).

References

Bauer, H.: Measure and Integration Theory. De Gruyter, Berlin (2011)

Davey, B.A., Priestley, H.A.: Introduction to Lattices and Order. Cambridge University Press, Cambridge (2002)

De Schuymer, B., De Meyer, H., De Baets, B., Jenei, S.: On the cycle-transitivity of the dice model. Theor. Decis. **54**(3), 261–285 (2003)

Grabisch, M., Marichal, J.L., Mesiar, R., Pap, E.: Aggregation Functions, vol. 127. Cambridge University Press, Cambridge (2009)

Kopa, M., Petrová, B.: Strong and weak multivariate first-order stochastic dominance (2018). https://ssrn.com/abstract=3144058

Mohd, W.R.W., Abdullah, L.: Aggregation methods in group decision making: a decade survey. Informatica **41**(1), 71–86 (2017)

Nungesser, M.K., Joyce, L.A., McGuire, A.D.: Effects of spatial aggregation on predictions of forest climate change response. Climate Res. **11**(2), 109–124 (1999)

Shaked, M., Shanthikumar, J.G.: Stochastic Orders. Springer, New York (2007). https://doi.org/10.1007/978-0-387-34675-5

Shanmugam, D., Blalock, D., Balakrishnan, G., Guttag, J.: Better aggregation in test-time augmentation. In: Proceedings of the IEEE/CVF International Conference on Computer Vision, pp. 1214–1223 (2021)
Zimmermann, H.J.: Fuzzy set theory. WIREs Comput. Stat. **2**(3), 317–332 (2010)

A Framework for Probabilistic Reasoning on Knowledge Graphs

Luigi Bellomarini[2], Davide Benedetto[1(✉)], Eleonora Laurenza[2], and Emanuel Sallinger[3]

[1] Universitá degli Studi di Roma Tre, Rome, Italy
davide.benedetto@uniroma3.it
[2] Banca d'Italia, Rome, Italy
luigi.bellomarini@bancaditalia.it, eleonora.laurenza@bancaditalia.it
[3] TU Wien, Vienna, Austria
sallinger@dbal.tuwien.ac.at

Abstract. In this paper we introduce a framework for probabilistic reasoning on knowledge graphs. The framework leverages the notion of probabilistic knowledge graphs (PKGs), a dedicated probabilistic graphical model, as well as Soft Vadalog, a specific language for knowledge representation and reasoning on such model. We illustrate PKGs, the language and the general problem of probabilistic reasoning, providing approximate algorithmic tools to make it feasible and efficient. This work—a short version of our recent contribution to the International Joint Conference on Rules and Reasoning 2020—aims at making our results available to the broader statistical community.

1 Introduction

The common trait of the large set of different notions of Knowledge Graphs (KGs) offered in the literature (Hogan et al. 2021) is the presence of a (hyper)graph-based structure used to represent a domain of interest. The recent soar of KGs has been favoured by the development of new mature languages for knowledge representation and reasoning (KRR), as well as *Knowledge Graph Management Systems* (KGMS) leveraging such languages to implement reasoning at scale (Bellomarini et al. 2020b).

KRR and Reasoning. Recent work has recalled how the KRR languages adopted in KGMSs should support a number of desiderata such as simple syntax, high expressive power, ontological reasoning, low complexity, and probabilistic reasoning (Bellomarini et al. 2017).

VADALOG is a state-of-the-art logic-based KRR language of the Datalog$^{\pm}$ family (Calì et al. 2012), and the VADALOG system (Bellomarini et al. 2018) is a KGMS with VADALOG at its core. VADALOG supports recursion and existential quantification, with syntactic restrictions to limit complexity and enable scalable logical reasoning on KGs.

L. A. García-Escudero et al. (Eds.): SMPS 2022, AISC 1433, pp. 48–56, 2023.
https://doi.org/10.1007/978-3-031-15509-3_7

Ontological Reasoning on Knowledge Graphs. VADALOG models a KG with facts and with existential rules describing the domain.

Example 1. Consider a Knowledge Graph G, with facts describing semantics relationships between constants a, b, c, l, m, n:

$$\{\text{Triple}(a, b, c), \text{Inverse}(b, l), \text{Restriction}(m, l), \text{Subclass}(m, n)\}$$

Let us extend G with the following existential rules, encoding the membership part of the OWL 2 semantics entailment regime for OWL 2 QL (see (Bellomarini et al. 2017; Gottlob and Pieris 2015)):

$$0.9 :: \text{Type}(x, y), \text{Restriction}(y, z) \rightarrow \exists v\, \text{Triple}(x, z, v) \quad (1)$$

$$0.8 :: \text{Type}(x, y), \text{SubClass}(y, z) \rightarrow \text{Type}(x, z) \quad (2)$$

$$0.7 :: \text{Triple}(x, y, z), \text{Inverse}(y, w) \rightarrow \text{Triple}(z, w, x) \quad (3)$$

$$\text{Triple}(x, y, z), \text{Restriction}(w, y) \rightarrow \text{Type}(x, w). \quad (4)$$

Ignoring what precedes the :: symbols, Rule (1) encodes that if x is of type y (as expressed by the atom Type) and is involved in a binary relation z (the atom Restriction), then there exists some value v s.t. the tuple (x, v) occurs in some instance of z (as specified by the atom Triple). Similarly, Rules (2–4) encode usual notions of subclass, inverse, and type restriction.

An example of (ontological) reasoning task over G is the query: *"What are all the entailed* Triples?". We see such triples are Triple(c, l, a) and Triple(c, l, v_0), where v_0 is a fresh arbitrary value (a *labeled null*). Let us now consider a modified version Example 1, where Rules (1–3) are not definitive but hold with a certain probability. We prefix them with a weight proportional to such bias (indicated by the number before the :: symbol). A probabilistic reasoning task would then consist in answering, over such uncertain logic programs, queries like: *"What is the probability for each* Triple *to be entailed?"*. We wish to compute the marginal probability of entailed facts, so, e.g., of Triple(c, l, a) and Triple(c, l, v_0).

Related Work. To enable such scenarios, we need KRR languages able to perform probabilistic reasoning and, at the same time, satisfy the requirements for ontological reasoning, which are: (i) adoption of *well-founded semantics* (Lee and Wang 2016), (ii) powerful existential quantification, supporting the quantification of SPARQL and OWL 2 QL, (iii) full recursion, (iv) ability to express non-ground inductive definitions (Lee and Wang 2016; Fierens et al. 2015). While probabilistic reasoning is of central interest of three research areas, *probabilistic logic programming* (PLP) (e.g., ProbLog) (Alberti et al. 2017; De Raedt and Kimmig 2015; Poole 2008; Riguzzi 2007; Sato 1995; Sato and Kameya 1997; Vennekens et al. 2004), *probabilistic programming languages* (PPL) (e.g., BLOG) (Goodman et al. 2008; Kersting and Raedt 2008; Milch et al. 2005; Pfeffer and River Analytics 2009), and *statistical relational learning* (SRL) (e.g., Markov Logic Networks) (Jaeger 2018; Lee and Wang 2016; Richardson and

Domingos 2006), none of the approaches is fully satisfactory for our requirements: they either do not exhibit high expressive power not supporting full recursion and existential quantification, or do not achieve low computational complexity.

Contribution. In this work, a short version of our recent contribution (Bellomarini et al. 2020a), we present a framework for probabilistic reasoning on knowledge graphs, organized as follows:

- We introduce the notion of Probabilistic Knowledge Graph, a probabilistic graphical model conceived for probabilistic reasoning and discuss SOFT VADALOG, a probabilistic extension of VADALOG (Sect. 2).
- We study the general problem of probabilistic reasoning on KGs and introduce algorithmic tools for approximate solutions, discussing their properties (Sect. 3).

2 Probabilistic Knowledge Graphs

Our approach consists of the following steps: i) first we define a Probabilistic Knowledge Graph as the combination of an input database D, which can be considered as the evidence set, and a set Σ of uncertain first-order rules; ii) then, given a query Q, i.e., an n-ary predicate appearing in Σ, we construct a structure, called *chase network*, that reflects all possible databases that can be obtained from D and Σ. This structure is already enough to compute marginal probabilities. However, at this point there are two issues we need to mitigate: first, logical inference in the presence of general first-order rules is undecidable or intractable even under heavy restrictions; second, computing exact marginal probabilities is intractable as well (in fact, #P-hard). For the first issue, we leverage the language VADALOG, for which logical inference can be done in polynomial time in the size of input data. To address the second issue, we compute *approximate* marginal probability instead. Thus, iii) we introduce an MCMC method that simultaneously performs logical and marginal inference and allows to efficiently answer queries over large PKGs.

Datalog$^\pm$ and Vadalog. SOFT VADALOG is an extension of VADALOG, a language in the Datalog family (Calì 2011; Gottlob et al. 2014). We consider pairwise disjoint countably infinite sets of *constants*, *labeled nulls* and *variables*. An expression $R(\bar{v})$, where R is a predicate symbol of arity $n \geq 0$ and $\bar{v}$ is a tuple of length n, is called an *atom* if $\bar{v}$ consists of variables or constants, and a *fact* if $\bar{v}$ consists of constants or labeled nulls. A *database instance* (or simply *instance*) is a set of facts. By $dom(D)$ we denote the constants and labeled nulls of D.

Datalog$^\pm$ generalizes Datalog, a popular language in the database community, with existential quantification in the rule conclusion. An *existential rule* (or a *rule*) is a first-order sentence of the form $\forall\bar{x}\forall\bar{y}(\varphi(\bar{x},\bar{y}) \rightarrow \exists\bar{z}\,\psi(\bar{x},\bar{z}))$, where φ (the *body*) and ψ (the *head*) are conjunctions of atoms. For brevity, we write such a rule as $\varphi(\bar{x},\bar{y}) \rightarrow \exists\bar{z}\,\psi(\bar{x},\bar{z})$ and use comma to denote conjunction. The semantics of a set of existential rules Σ over an instance D, denoted $\Sigma(D)$, is defined via the *chase procedure*. This procedure adds new facts to D (possibly involving generation of new labeled nulls used to satisfy the existentially quantified variables) until the final result $\Sigma(D)$ satisfies all the existential rules of Σ. More formally, initially $\Sigma(D) = D$. By a *unifier* we mean a mapping from variables to constants or labeled nulls. We say $\rho = \varphi(\bar{x}, \bar{y}) \rightarrow \exists \bar{z}\, \psi(\bar{x}, \bar{z})$ is *applicable* to $\Sigma(D)$ if there is a unifier θ_ρ such that $\varphi(\bar{x}\theta_\rho, \bar{y}\theta_\rho) \subseteq \Sigma(D)$ and θ_ρ has not been used to generate new facts in $\Sigma(D)$ via ρ. If ρ is applicable to $\Sigma(D)$ with a unifier θ_ρ, then it performs a *chase step*, i.e., it *generates* new facts $\psi(\bar{x}\theta'_\rho, \bar{z}\theta'_\rho)$ that are added to $\Sigma(D)$, where $\bar{x}\theta_\rho = \bar{x}\theta'_\rho$ and $z_i\theta'_\rho$, for each $z_i \in \bar{z}$, is a fresh labeled null that does not occur in $\Sigma(D)$. The chase step easily generalizes to a set of rules. The chase procedure performs chase steps until no rule in Σ is applicable.

Now, given a pair $Q = (\Sigma, \text{Ans})$ called a *query* and an instance D, called an *extensional database* (EDB), where Σ is a set of rules and Ans an n-ary predicate, a tuple $\bar{t} \in dom(D)^n$ is an *answer* to Q over D if $\text{Ans}(\bar{t}) \in \Sigma(D)$. Since $\Sigma(D)$ is potentially infinite, the number of answers to a query could be infinite as well. For this, we are interested in finding a representative set of answers, called *universal answer set*, that can be embedded into any other answer set by means of renaming labeled nulls. In our setting, a *logical reasoning task* is computing a universal answer set.

Note that in general in the presence of existential rules, an algorithm that solves the above reasoning task can even not exist (Calì et al. 2012). Because of this, certain restrictions are imposed on the syntax of the rules in order not only for an algorithm to exist but also to be scalable. A VADALOG program is a set of facts and rules that obey such syntactic restrictions, namely *wardedness* (Gottlob and Pieris 2015), which ensure that there is a finite subset $\Sigma'(D) \subset \Sigma(D)$ such that the universal answer sets for a query Q over D calculated via $\Sigma(D)$ and $\Sigma'(D)$ are isomorphic. Thanks to this property, in VADALOG, a solution to the logical reasoning task can be obtained by executing just a finite number of chase steps using a so-called *termination strategy*(Bellomarini et al. 2018). In particular, given two isomorphic facts h and h' (i.e., having same terms up to renaming of the labeled nulls), one needs to explore only h and so never perform chase steps starting from h'. By a *warded chase step* we mean a chase step limited to those unifiers allowed by the termination strategy described in (Bellomarini et al. 2018).

Soft Vadalog and PKGs. Towards defining PKGs, we extend VADALOG to SOFT VADALOG with soft rules. A SOFT VADALOG rule is a pair (ρ, w), where ρ is a (usual) VADALOG rule and $w \in R \cup \{+\infty, -\infty\}$, a real number, is a *weight.* For brevity we assume that weights are positive as the results of this paper can be extended to include negative weights as well. A soft rule $(\rho, +\infty)$ is called a *hard* rule. By abuse of notation, ρ denotes a soft rule and $w(\rho)$ its weight. A SOFT VADALOG program is a set of SOFT VADALOG rules.

We define a Probabilistic Knowledge Graph *as the pair* $\langle D, \Sigma \rangle$, *where* D *is a database instance and* Σ *is a soft vadalog program.*

A PKG can be viewed as a template for constructing chase networks, the ground networks required for marginal inference in reasoning tasks.

Given a PKG $\mathscr{G} = \langle D, \Sigma \rangle$ and a set of database instances $\mathbf{D}$, closed under the hard rules of Σ, a *chase network* $\Gamma(\mathscr{G})$ is a n-uple $\langle \mathbf{W}, \mathbf{T}, \lambda, W_0 \rangle$, where:

1. $\mathbf{W}$ is a set of nodes and $\mathbf{T}$ is a set of edges.
2. $\lambda : \mathbf{W} \rightarrow \mathbf{D}$ *is a total injective labeling function associating nodes* W *of* $\mathbf{W}$ *to database instances* $\lambda(W)$.
3. $W_0 \in \mathbf{W}$ *is a source node, s.t.* $\lambda(W_0) = cl_\Sigma(D)$, *i.e.,* W_0 *is associated to the closure of* D *w.r.t. the hard rules of* Σ.
4. *There is an edge* $t \in \mathbf{T}$ *from* W *to* W' *iff* $\lambda(W')$ *can be obtained from* $\lambda(W)$ *by one transition step. A* transition step *from* $\lambda(W)$ *to* $\lambda(W')$ *consists of a warded chase step of at least one applicable soft rule with one unifier followed by the closure w.r.t. the hard rules of* Σ. *Edge* t *is then labeled by* $\sum_{\rho \in \sigma} w(\rho)$, *where* σ *is the set of soft rules applied.*

Note that since a transition step always adds new facts, there are no (directed) cycles in the chase network; also, as, λ is injective, all the paths in the chase network leading to the same database instance will converge into the same terminal node; moreover, the chase network is finite for the finiteness of the chase algorithm in VADALOG (see the Datalog$^\pm$ and VADALOG Section). We define the *weight* $w(W)$ of a node W in $\Gamma(\mathscr{G})$ as the sum of the edge labels on all the paths from W_0 to W. The chase network induces the following probability distribution over its nodes: $P(W) = \frac{1}{Z} \exp w(W)$, where Z is a normalization constant (a *partition function*), to make $P(W)$ a proper distribution, defined as $Z = \sum_W \exp w(W)$. For a given fact f, its *marginal probability* $P(f)$ can be calculated as $\sum_{W_i : f \in \lambda(W_i)} P(W_i)$. Figure 1 summarizes the chase network for Example 1. Nodes are facts f in database instances $\lambda(W_i)$, where W_i is a node of the chase network. Facts f are annotated with a set $\{W_0, \ldots, W_n\}$ of nodes of the chase network s.t. for each W_i in the set, $f \in \lambda(W_i)$. Solid edges are warded chase steps applying hard rules; dashed edges are for soft rules, with weight γ.

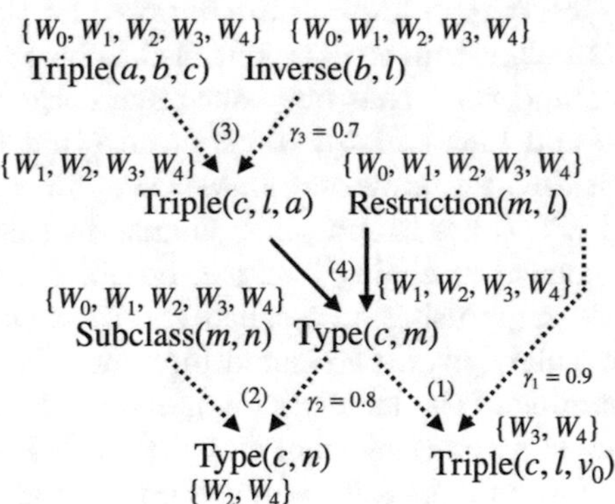

Fig. 1. Chase network for Example 1

Let us now compute the marginal probability of Triples. We have $w(W_0) = 0$. Then it follows $w(W_1) = 0.7$ and $w(W_2) = 0.7 + 0.8 = 1.5$, $w(W_3) = 0.7 + 0.9 = 1.6$, and $w(W_4) = 0.7 + (0.8 + 0.9) \times 2 = 4.1$. So we can calculate marginal probability for Triple(c, l, a). This fact appears for W_1, W_2, W_3, W_4, so we have: $(e^{0.7} + e^{1.5} + e^{1.6} + e^{4.1})/Z = 0.99$, with $Z = 1 + e^{0.7} + e^{1.5} + e^{1.6} + e^{4.1}$. Similarly, for Triple$(c, l, v_0)$, we have: $(e^{1.6} + e^{4.1})/Z = 0.9$.

Probabilistic Reasoning. Given an instance D and a query $Q = (\Sigma, \text{Ans})$, the *probabilistic reasoning task* consists in computing the set $\{\langle \bar{t}, P(\bar{t})\rangle\}$, where $\text{Ans}(\bar{t}) \in \mathbf{D}$, with $\mathbf{D}$ being the instances associated to nodes in $\Gamma(\mathcal{G})$, defined on $\mathcal{G} = \langle D, \Sigma\rangle$, and $P(\bar{t})$ is the marginal probability of $\text{Ans}(\bar{t})$.

Reasoning on PKGs is a computationally hard problem. For marginal inference, an exploration of the full chase network is needed and thus an exponential number of chase executions, each with polynomial complexity in the size of D. This makes it NP-hard. By adapting the proof for #P-hardness of query answering over probabilistic databases (Suciu et al. 2011), it can be shown that marginal inference in SOFT VADALOG is $\#P$-hard, where the program is assumed to be fixed.

3 The MCMC-Chase Algorithm

MCMC-chase is an independence sampling MCMC where the chase is seen as a Markov process (Gilks et al. 1995) over the nodes of chase network. Given a PKG $\mathcal{G} = \langle D, \Sigma\rangle$, the MCMC-chase starts from the facts in D and incrementally satisfies soft rules of Σ with a probability that is proportional to the rule weight and generates nodes of $\Gamma(\mathcal{G})$. Meanwhile, the algorithm keeps track of the weight of the current node. After a number of steps, a node is accepted or rolled back according to an *acceptance probability*, in a Metropolis-Hastings style.

Algorithm 1 gives pseudo-code for the MCMC-chase. It takes as input a PKG $\mathcal{G}$ and returns samples from the distribution $P(W) = \frac{1}{Z} \exp w(W)$ over the nodes $\mathbf{W}$ of the chase network $\Gamma(\mathcal{G})$. The algorithm performs N iterations, each consisting of S steps. The number of steps per iteration is extracted from a Poisson (*jump*) distribution; each step can be forward or backward, depending on a value δ uniformly chosen. In a step, the algorithm selects subsets $\mathbf{R_a}$ and $\mathbf{R_u}$ of rules from Σ with a probability proportional to $w(\rho)$ (lines 10–11) of applicable or undoable (that generated leaf facts) rules. *Forward steps* (line 12) try to apply a transition step with the selected rules $\mathbf{R_a}$ to the current node $\mathcal{T}$ of the chase network. *Backward steps* (line 14) try to undo a transition step with rules in $\mathbf{R_u}$.

In forward (resp. backward) steps, applicable (resp. undoable) hard rules are used to add (resp. remove) facts from the closure of $\mathcal{T}$; applicable (resp. undoable) soft rules are applied and the respective weight is summed to (resp. subtracted from) the total weight of $\mathcal{T}$. Note that, when applicable, hard rules are in $\mathbf{R_a}$ and $\mathbf{R_u}$ and therefore applied or undone, but they do not contribute to

Algorithm 1. MCMC-chase

function MCMC-CHASE($\mathscr{G} = \langle D, \Sigma \rangle, N$)
  $\mathbf{W}_S = \emptyset$ ▷ samples from the distribution over the nodes $\mathbf{W}$ of $\Gamma(\mathscr{G})$
  $\mathscr{D}^0 = D$
  for $n \leftarrow 1$ to N **do** ▷ N: # of iterations
    Sample $S \sim \mathscr{P}(\lambda)$ ▷ # of steps, from a Poisson distr.
    $\mathscr{T} \leftarrow \mathscr{D}^{n-1}$
    $w(\mathscr{T}) \leftarrow w(\mathscr{D}^{n-1})$
    for $s \leftarrow 1$ to S **do**
      Sample $\delta \sim \mathscr{U}(0,1)$; Sample $\mu \sim \mathscr{U}(0,1)$
      $\mathbf{R_f} \leftarrow$ all applicable ρ in Σ s.t. $\mu < 1 - e^{-w(\rho)}$
      $\mathbf{R_u} \leftarrow$ all undoable ρ in Σ s.t. $\mu < 1 - e^{-w(\rho)}$
      if $\delta < 0.5$ **then** ▷ forward step
        TRANSITION_STEP($\mathscr{T}, \mathbf{R_a}$)
      else ▷ backward step
        UNDO_TRANSITION_STEP($\mathscr{T}, \mathbf{R_u}$)
    $\alpha \leftarrow f(\mathscr{T})/f(\mathscr{D}^{n-1})$ ▷ acceptance probability
    With prob. $\min(1, \alpha)$, accept and add $\langle \mathscr{T}, w(\mathscr{T}) \rangle$ to $\mathbf{W}_{\mathscr{S}}$
    if accepted **then**
      $\mathscr{D}^n \leftarrow \mathscr{T}$
    else
      $\mathscr{D}^n \leftarrow \mathscr{D}^{n-1}$ ▷ rollback to prev. possible world
  return $\mathbf{W}_S$

the total weight. After S steps, the acceptability of the current node is evaluated with an *acceptance function* $f(\mathbf{Y}) = \exp w(\mathbf{Y})$. All accepted nodes and their weights are finally returned.

The Markov chain generated by the MCMC-chase respects the *Markov property* since at each iteration a candidate node inherits all the facts from the previous iteration node. The MCMC-chase samples from the required distribution. In fact, it can be proven that our distribution is stationary with respect to the defined Markov chain and that our sampling process is ergodic.

4 Conclusion

In this paper, we considered the reasoning desiderata for KGs and introduced the syntax and semantics of SOFT VADALOG. Within a new probabilistic reasoning framework, SOFT VADALOG allows to induce a probability distribution over the facts defined through the warded chase. We introduced the notion of Probabilistic Knowledge Graphs, a template for chase networks, a new probabilistic graphical model where marginal inference can be performed. To cope with intractability of marginal inference, we introduced the MCMC-chase, whose core idea is performing logical and probabilistic inference at the same time, while sampling the chase space with a Monte Carlo technique.

References

Alberti, M., Bellodi, E., Cota, G., Riguzzi, F., Zese, R.: Cplint on SWISH: probabilistic logical inference with a web browser. Intelligenza Artificiale **11**(1), 47–64 (2017)

Bellomarini, L., Gottlob, G., Pieris, A., Sallinger, E.: Swift logic for big data and knowledge graphs. In: Proceedings of the Twenty-Sixth International Joint Conference on Artificial Intelligence (IJCAI-2017), pp. 2–10 (2017)

Bellomarini, L., Sallinger, E., Gottlob, G.: The Vadalog system: Datalog-based reasoning for knowledge graphs. Proc. VLDB Endow. **11**(9), 975–987 (2018)

Bellomarini, L., Laurenza, E., Sallinger, E., Sherkhonov, E.: Reasoning under uncertainty in knowledge graphs. In: Gutiérrez-Basulto, V., Kliegr, T., Soylu, A., Giese, M., Roman, D. (eds.) RuleML+RR 2020. LNCS, vol. 12173, pp. 131–139. Springer, Cham (2020). https://doi.org/10.1007/978-3-030-57977-7_9

Bellomarini, L., Sallinger, E., Vahdati, S.: Knowledge graphs: the layered perspective. In: Janev, V., Graux, D., Jabeen, H., Sallinger, E. (eds.) Knowledge Graphs and Big Data Processing. LNCS, vol. 12072, pp. 20–34. Springer, Cham (2020). https://doi.org/10.1007/978-3-030-53199-7_2

Calì, A., Gottlob, G., Lukasiewicz, T., Pieris, A.: Datalog$^{\pm}$: a family of languages for ontology querying. In: de Moor, O., Gottlob, G., Furche, T., Sellers, A. (eds.) Datalog 2.0 2010. LNCS, vol. 6702, pp. 351–368. Springer, Heidelberg (2011). https://doi.org/10.1007/978-3-642-24206-9_20

Calì, A., Gottlob, G., Pieris, A.: Towards more expressive ontology languages: the query answering problem. Artif. Intell. **193**, 87–128 (2012)

De Raedt, L., Kimmig, A.: Probabilistic (logic) programming concepts. Mach. Learn. **100**(1), 5–47 (2015). https://doi.org/10.1007/s10994-015-5494-z

Fierens, D., et al.: Inference and learning in probabilistic logic programs using weighted Boolean formulas. Theory Pract. Logic Program. **15**(3), 358–401 (2015)

Gilks, W., Richardson, S., Spiegelhalter, D. (eds.): Markov Chain Monte Carlo in Practice. Chapman & Hall/CRC Interdisciplinary Statistics, London (1995)

Goodman, N.D., Mansinghka, V.K., Roy, D.M., Bonawitz, K., Tenenbaum, J.B.: Church: a language for generative models. In: Proceedings of the Twenty-Fourth Conference on Uncertainty in Artificial Intelligence (UAI 2008), pp. 220–229. AUAI Press (2008)

Gottlob, G., Lukasiewicz, T., Pieris, A.: Datalog$^{\pm}$: questions and answers. In: Proceedings of the Fourteenth International Conference on Principles of Knowledge Representation and Reasoning, pp. 682–685 (2014)

Gottlob, G., Pieris, A.: Beyond SPARQL under OWL 2 QL entailment regime: rules to the rescue. In: Proceedings of the Twenty-Fourth International Joint Conference on Artificial Intelligence (IJCAI 2015), pp. 2999–3007 (2015)

Hogan, A., et al.: Knowledge graphs. ACM Comput. Surv. **54**(4), 71 (2021). https://doi.org/10.1145/3447772

Jaeger, M.: Probabilistic logic and relational models. In: Alhajj, R., Rokne, J. (eds.) Encyclopedia of Social Network Analysis and Mining, 2nd edn., pp. 1907–1921. Springer, New York (2018). https://doi.org/10.1007/978-1-4939-7131-2_157

Kersting, K., De Raedt, L.: Basic principles of learning Bayesian logic programs. In: De Raedt, L., Frasconi, P., Kersting, K., Muggleton, S. (eds.) Probabilistic Inductive Logic Programming. LNCS (LNAI), vol. 4911, pp. 189–221. Springer, Heidelberg (2008). https://doi.org/10.1007/978-3-540-78652-8_7

Lee, J., Wang, Y.: Weighted rules under the stable model semantics. In: Proceedings of the 15th International Conference on Principles of Knowledge Representation and Reasoning (KR 2016), pp. 145–154. AAAI Press (2016)

Milch, B., Marthi, B., Russell, S.J., Sontag, D., Ong, D.L., Kolobov, A.: BLOG: probabilistic models with unknown objects. In: Proceedings of 19th International Joint Conference on Artificial Intelligence (IJCAI 2005), pp. 1352–1359 (2005)

Pfeffer, A.: Figaro: An Object-Oriented Probabilistic Programming Language (2009). www.cs.tufts.edu/~nr/cs257/archive/avi-pfeffer/figaro.pdf

Poole, D.: The independent choice logic and beyond. In: De Raedt, L., Frasconi, P., Kersting, K., Muggleton, S. (eds.) Probabilistic Inductive Logic Programming. LNCS (LNAI), vol. 4911, pp. 222–243. Springer, Heidelberg (2008). https://doi.org/10.1007/978-3-540-78652-8_8

Richardson, M., Domingos, P.M.: Markov logic networks. Mach. Learn. **62**(1–2), 107–136 (2006)

Veloso, M.: Learning to select team strategies in finite-timed zero-sum games. In: Basili, R., Pazienza, M.T. (eds.) AI*IA 2007. LNCS (LNAI), vol. 4733, p. 1. Springer, Heidelberg (2007). https://doi.org/10.1007/978-3-540-74782-6_1

Sato, T.: A statistical learning method for logic programs with distribution semantics. In: Proceedings of the 12th International Conference on Logic Programming (ICLP 1995), pp. 715–729. MIT Press (1995)

Sato, T., Kameya, Y.: PRISM: a language for symbolic-statistical modeling. In: Proceedings of the 15th International Joint Conference on Artificial Intelligence (IJCAI 1997), pp. 1330–1339 (1997)

Suciu, D., Olteanu, D., Ré, C., Koch, C.: Probabilistic Databases. Synthesis Lectures on Data Management, Morgan & Claypool Publishers (2011)

Vennekens, J., Verbaeten, S., Bruynooghe, M.: Logic programs with annotated disjunctions. In: Demoen, B., Lifschitz, V. (eds.) ICLP 2004. LNCS, vol. 3132, pp. 431–445. Springer, Heidelberg (2004). https://doi.org/10.1007/978-3-540-27775-0_30

Biological Age Imputation by Data Depth

A Proposal and Some Preliminary Results

Stefano Cabras[1], Ignacio Cascos[1(✉)], Bernardo D'Auria[2], María Durbán[1], Vanesa Guerrero[1], and Maicol Ochoa[1]

[1] Department of Statistics, Universidad Carlos III de Madrid, Getafe-Leganés, Spain
{stefano.cabras,ignacio.cascos,marialuz.durban,vanesa.guerrero}@uc3m.es
[2] Department of Economics, Università degli Studi "G. d'Annunzio", Chieti-Pescara, Italy
bernardo.dauria@unich.it

Abstract. The biological age is an indicator of the functional condition of an individual's body. Unlike the chronological age, which just measures the time from birth, the biological age of a human is also affected by its medical condition, life habits, some sociodemographic variables, as well as biomarkers. Taking advantage of the statistical concept of depth, which serves as a measurement of the degree of centrality of a multivariate observation with respect to a dataset, we assess the biological age of an individual as the chronological age that would make her selected records as deep as possible when compared with those of other individuals with a chronological age similar to hers. Some direct conclusions of this imputation technique are presented.

1 Introduction

Human ageing is today a global phenomena and a major issue. One remarkable fact to support this claim is that while in 1990 only 6% of the world population aged 65 or over, the proportion increased to 9% in 2019, and it is expected to rise further to 16% by 2050, see UN DESA (2019). Actually Sustainable Development Goal #3 of the 2030 Agenda for Sustainability, see UN General Assembly (2015), reads: "Ensure healthy lives and promote well-being for all at *all ages*", while the Stakeholder Group on Ageing (also named Older Persons) is one of the major groups that actively worked in the implementation and is now monitoring the 2030 Agenda, see UN DESA and UNITAR (2020).

All living organisms gradually undergo an ageing process due to the physiological changes that take place over time. According to the World Health Organization, see WHO (2015), from a biological viewpoint, ageing is the result of molecular and cellular damage accumulated in the course of time decreasing the physical and mental capacity that leads to higher disease rates, and finally to death. Further, these changes are "only loosely related to a person's chronological age in years". This *vague* relation between chronological age and ageing gives rise to the introduction of the concept of biological age, see Rodríguez-Pardo and López-Farré (2017).

L. A. García-Escudero et al. (Eds.): SMPS 2022, AISC 1433, pp. 57–64, 2023.
https://doi.org/10.1007/978-3-031-15509-3_8

The biological age is an indicator of the functional condition of an individual's body. It is related with the chronological age, but also with the individual's medical condition, life habits, sociodemographic variables, as well as biomarkers, including some genetical ones. An accurate assessment of the biological age is relevant for care resources allocation of both public and private funds. Furthermore, a good understanding of the biological age can be used to promote healthy life habits.

For ease of interpretation, the biological age is commonly given in the same unit as the chronological one (years) and an individual whose biological age is x years can be considered as someone whose body, from a functional viewpoint, and habits resemble the ones of a standard individual who is chronologically x years old. As a consequence, statistical estimation techniques can be used to assess the biological age of an individual. As a matter of fact, other procedures already introduced in the literature for the biological age assessment are based on multiple linear regression (Voitenko and Tokar 1983), Principal Component Analysis (Zhang et al. 2014), machine learning techniques, see Zhand and Chen (2017) for a literatuve review, or deep learning (Rahman et al. 2021).

In multivariate Statistics, the term data depth refers to the degree of centrality of an observation with respect to a dataset or a probability distribution. The most central observations, which are highly representative of the whole of the dataset, assume the highest depth values, while depth decreases as we move away from the centre towards outwarding observations.

Our proposal is to assess the biological age of an individual as the chronological age that would make her selected records as deep as possible when compared with those of other individuals with a chronological age similar to hers. This methodology can handle the situation when there is missing data in any specific variable different from the chronological age.

In Sect. 2 we review some classical concepts of data depth. Section 3 is devoted to the introduction of a biological age imputation technique based on data depth, while in Sect. 4 we introduce the used dataset and present our results. Finally, in Sect. 5 we present some conclusions of our work and discuss future work lines.

2 Preliminaries, Data Depth

Individual biological characteristics are represented by a random vector X in the d-dimensional Euclidean space. For any $x \in \mathbb{R}^d$, the depth of x with respect to X quantifies its degree of centrality, see Zuo and Serfling (2000). If x is central with respect to the distribution of X, its depth is high, while peripheral points assume low depth values. We introduce next four data depth functions:

- Mahalanobis depth: $\mathrm{MhD}(x;X) = \left(1 + (x - \mathbb{E}X)^\top \Sigma_X^{-1}(x - \mathbb{E}X)\right)^{-1}$, where $\mathbb{E}X$ stands for the expectation of X and Σ_X for its covariance matrix.
- L_2 depth: $\mathrm{L_2D}(x;X) = (1 + \mathbb{E}\|x - X\|)^{-1}$, where $\|\cdot\|$ is the Euclidean norm in $\mathbb{R}^d$, see Mosler (2013).
- Halfspace depth: $\mathrm{HD}(x;X) = \inf\{\Pr(X \in H) : x \in H \text{ halfspace}\}$, see Tukey (1975).

- Zonoid depth: $\mathrm{ZD}(x;X) = \sup\{\alpha \in (0,1] : x = \int y g(y) \mathrm{d}F_X(y),$ for some $g : \mathbb{R}^d \mapsto [0, \alpha^{-1}]$ measurable with $\int g(y)\mathrm{d}F_X(y) = 1\}$, see Koshevoy and Mosler (1997).

The classical properties of a data depth function are that it attains its maximal value at the centre (or point of symmetry, whenever there is one) of the distribution, it decreases on rays from the deepest point, it vanishes on points of arbitrarily large norm, and it is independent of the system or coordinates (at least invariant under rigid Euclidean motions).

The Mahalanobis and L_2 depths never assume value 0, which is a very useful property for applications, since they allow us to compare points that lie out of the convex hull of the support of a distribution in terms of their centrality. Nevertheless, there are extensions of the halfspace and zonoid depths that overcome this difficulty.

3 Biologial Age Imputation by Data Depth

Based on Mozharovskyi et al. (2020), who described imputation techniques based on data depth, we assess the biological age of an individual as the chronological age that make her records as deep as possible when compared with those of other individuals whose chronological age is similar to hers. Since it is commonly assumed that the biological age is within a range of 15 years the chronological one, we have set a gap of ±15 years for the similarity.

If A represents the chronological age, BA the biological age, and the remaining variables are given by X, the biological age of the i-th individual (chronological age a_i and remaining records x_i) is computed as

$$ba_i = \arg\max_b \mathrm{D}((b, x_i); (a, x)_{(i)}),$$

where by $(a, x)_{(i)}$ we represent the chronological age and remaining variables evaluated over the subsample consisting of all but the i-th individual.

This procedure is illustrated in Fig. 1. An individual who is 45 years old (chronological age, axis X) and has some given value in other characteristic (axis Y) is represented by means of a black bullet. The biological age of this individual is imputed through a sample of 500 individuals sampled uniformly at random from those with a chronological age between $30 = 45-15$ and $60 = 45+15$ years. Each black cross corresponds to one of the 500 individuals, while the darker a region is, the deeper it is with respect to the zonoid depth. The horizontal dashed line is set at the value of the other characteristic of the individual whose biological age is imputed. Finally, the dashed vertical line is set at the value that maximizes the depth over the horizontal line and corresponds to the imputed depth, 42.6 years.

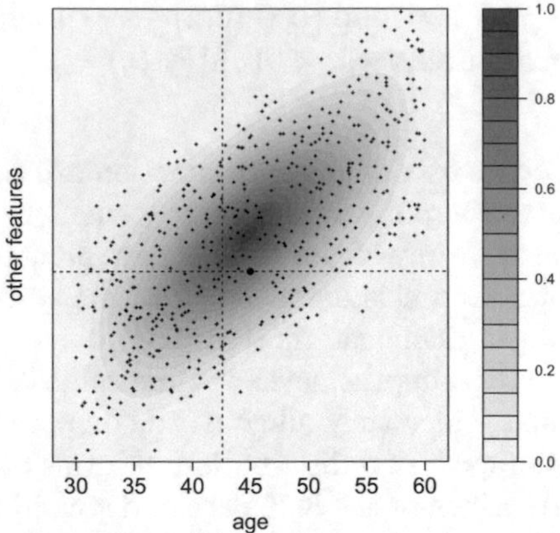

Fig. 1. Biologial age imputation by the zonoid data depth

3.1 Sample Balancing

The main problem that we face on a real dataset with information on the chronological age and some other variables is that, for any given chronological age a, the sample might not be correctly balanced about it. That is, it migth be skewed consisting mainly of either younger (chronological age less than a) or older (chronological age greater than a) individuals. In order to overcome this issue, when assessing the biological age of an individual with chronological age a, we sample uniformly at random 5000 ages in the interval $[a-15, a+15]$. For each sampled age, we select, again at random, an individual whose chronological age is precisely that one (or as close as possible in case there are no individuals of that age). This way we obtain a resample of 5000 observations and, finally, compute the age that maximizes the depth of an artifical observation whose variables different from the chronological age match those of the considered individual with respect to it. If some variable is missing on a give individual, it is simply ignored in the depth computation.

4 Data and Results

The Spanish *Mutualidad de la Abogacía* provided us a data set with 107 variables (some are transformations of others) on 911 individuals. These variables are the chronological age, biomarkers, indicators of some medical condition, life habits, sociodemographic variables, and data on six genes. Twenty five of these variables were used in the assessment of the biological age, specifically the (chronological) age, biometric variables (sex, Boby Mass Index, Diastolic Blood Pressure, and Systolic Blood Pressure), twelve binary variables about medical conditions, and eigth variables about life habits (stress, which is binary, and seven numerical variables about the physical activity).

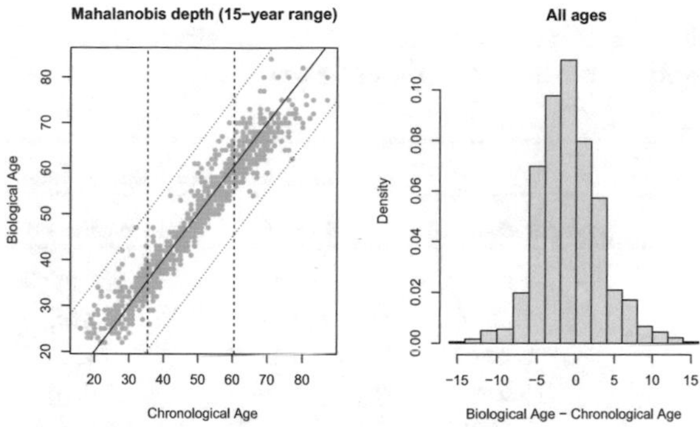

Fig. 2. Biological age vs. chronological age scatterplot (left); histogram of their difference (right)

In Fig. 2, we present the biological age assessed using the Mahalanobis depth together with the chronological one in a scatterplot. The dashed vertical lines split the sample in three age groups: 18 to 35, 36 to 30, and 61 to 88, while the dotted lines represent the maximum allowed separation between the two ages (15 years).

The effect of each of the variables used in the imputation process together with the one of other variables dealing mainly with the diet of the individuals is analyzed in Table 1 in four multiple linear regression models, one with all the individuals and the other three with the individuals in each age group.

The best two models are those built for all individuals, and the age group 36 to 60. At both of them the coefficient of the chronological age is close to 0.8. In all cases, women have a younger biological age than men (2 years on average), while the Body Mass Index has a positive coefficient, except for the older than 60 group. Some stress is good, except for people younger than 35 and the sign of the coefficients associated to the binary variables that assume value 1 on individuals that suffer a particular medical condition have all positive signs, except for the

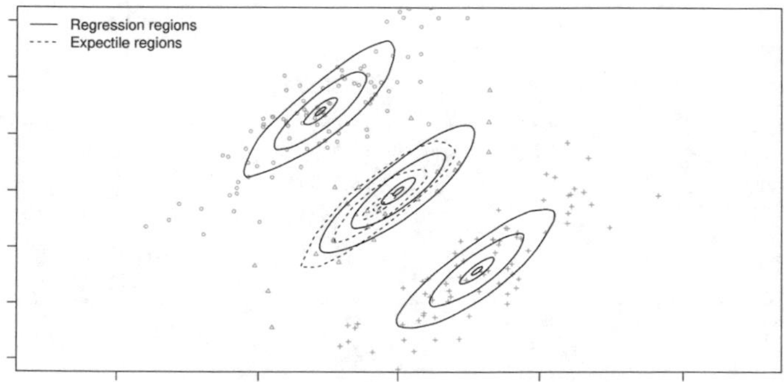

Fig. 3. Data depth in a regression model

Table 1. Multiple regression models to analyze the effect of several variables on the biological age. Standard errors are in parentheses

	Dependent variable: biological.age			
	younger than 36	36 to 60	older than 60	ALL
Constant	−0.677	−0.379	42.54***	−0.741
	(1.601)	(0.997)	(2.579)	(0.991)
chronological.age	0.483***	0.866***	0.410***	0.799***
	(0.033)	(0.012)	(0.029)	(0.008)
sex[F]	−1.682***	−1.956***	−2.225***	−2.092***
	(0.380)	(0.181)	(0.331)	(0.215)
BMI	0.408***	0.144***	−0.213***	0.153***
	(0.037)	(0.022)	(0.042)	(0.025)
DBP	0.115***	−0.051***	−0.221***	−0.055***
	(0.016)	(0.010)	(0.021)	(0.012)
SBP		0.070***	0.140***	0.072***
		(0.007)	(0.013)	(0.008)
stress[Y]	1.333***	−1.685***	−2.762***	−1.197***
	(0.309)	(0.168)	(0.343)	(0.199)
hypertension[Y]	7.241***	5.351***	2.330***	3.674***
	(2.164)	(0.307)	(0.365)	(0.319)
angina[Y]			−4.091***	−4.594***
			(1.081)	(1.280)
COPD[Y]		7.299***	3.213***	2.858***
		(1.032)	(0.769)	(0.800)
cancer[Y]		5.474***	3.414***	4.665***
		(0.405)	(0.435)	(0.406)
asthma[Y]	2.023***		−1.789**	
	(0.666)		(0.696)	
heart.attack[Y]			4.754***	
			(1.094)	
heart.failure[Y]			4.099***	
			(0.885)	
metabolic.syndrome[Y]			2.788**	
			(1.086)	
brain.stroke[Y]		1.573*		
		(0.802)		
soft.drinks	3.733***	1.753**		2.027**
0 to 1	(1.375)	(0.778)		(0.898)
physical.activity		−2.728***		
0 to 1		(0.575)		
legumes			2.720**	
0 to 1			(1.131)	
Observations	197	444	224	849
R^2	0.797	0.955	0.832	0.959
Adjusted R^2	0.788	0.954	0.819	0.959

angina pectoris, which can be due to the medical treatment prescripted to such patients, and the asthma on eldery people. Finally, soft drinks are bad at all ages, despite they don't appear in the model for elderly people (probably because they drink less of them), the physical activity is particularly good for people in the age group 36 to 60, while eating legumes does not seem good for people who are older than 60.

5 Conclusions and Future Work

We have presented a technique for biological age imputation based on data depth and checked, by means of multiple linear regression, that the imputed biological ages assume meaningful values. We must adapt the depth notions in order to use the available gene data, properly incorporate the binary variables and also ordered categorial variables.

In order to incorporate information about individuals that do not lie in the age range of the one whose biological age is to be assessed, we are considering using the halfspace and expectile depths (Cascos and Ochoa 2021) in a multiple-output regression model in the manner of Hallin et al. (2010), where the explanatory variable has three values: younger, same age group, and older. Figure 3 shows a scatterplot of these three age groups, with axis X for the age and Y for the other variables, as in Fig. 1.

Acknowledgement. The statistical team of the project "Desarrollo de un algoritmo predictivo de edad biológica basado en hábitos de vida y biomarcadores genéticos y no genéticos" (Fundación Mutua Abogacía 2020–2021) would like to dedicate this work to the memory of the deceased Prof. Antonio López Farré.

References

Cascos, I., Ochoa, M.: Expectile depth: theory and computation for bivariate datasets. J. Multivar. Anal. **184**, 104757 (2021)

Hallin, M., Paindaveine, D., Šiman, M.: Multivariate quantiles and multiple-output regression: from L_1 optimization to halfspace depth. Ann. Stat. **38**, 635–669 (2010)

Koshevoy, G., Mosler, K.: Zonoid trimming for multivariate distributions. Ann. Stat. **25**, 1998–2017 (1997)

Mosler, K.: Depth statistics. In: Becker, C., Fried, R., Kuhnt, S. (eds.) Robustness and Complex Data Structures: Festschrift in Honour of Ursula Gather, pp. 17–34. Springer, Berlin (2013). https://doi.org/10.1007/978-3-642-35494-6_2

Mozharovskyi, P., Josse, J., Husson, F.: Nonparametric imputation by data depth. J. Am. Stat. Assoc. **115**, 241–253 (2020)

Rahman, S.A., Giacobbi, P., Pyles, L., Mullett, C., Doretto, G., Adjeroh, D.A.: Deep learning for biological age estimation. Brief. Bioinform. **22**, 1767–1781 (2021)

Rodríguez-Pardo, J.M., López-Farré, A.: Longevity and aging in the third millennium: new perspectives. Fundación MAPFRE, Madrid (2017)

Tukey, J.W.: Mathematics and the picturing of data. In: Proceedings of the International Congress of Mathematicians, Vancouver, B.C., 1974, vol. 2, pp 523–531. Canadian Mathematical Congress, Montreal (1975)

United Nations General Assembly, Transforming our world: the 2030 Agenda for Sustainable Development, A/RES/70/1, 21 October 2015
United Nations, Department of Economic and Social Affairs, Population Division. World Population Ageing 2019: Highlights, ST/ESA/SER.A/430 (2019)
United Nations, Department of Economic and Social Affairs and Institute for Training and Research. Stakeholder engagement & the 2030 Agenda. A practical guide (2020)
Voitenko, V.P., Tokar, A.V.: The assessment of biological age and sex differences of human aging. Exp. Aging Res. **9**, 239–244 (1983)
World Health Organization: World report on ageing and health. WHO Library, Luxembourg (2015)
Zhang, W.-G.: Construction of an integral formula of biological age for a healthy Chinese population using principle component analysis. J. Nutrition Health Aging **18**(2), 137–142 (2014). https://doi.org/10.1007/s12603-013-0345-8
Zhang, J.L., Chen, X.: Common methods of biological age estimation. Clin. Interv. Aging **12**, 759–772 (2017)
Zuo, Y., Serfling, R.: General notions of statistical depth function. Ann. Stat. **28**, 461–482 (2000)

Monitoring Tools in Robust CWM for the Analysis of Crime Data

Andrea Cappozzo[1], Luis Angel García-Escudero[2], Francesca Greselin[3](✉), and Agustín Mayo-Iscar[2]

[1] Department of Mathematics, Politecnico di Milano, Milan, Italy
andrea.cappozzo@polimi.it

[2] Departamento de Estadística e Investigación Operativa, Universidad de Valladolid, Valladolid, Spain
{lagarcia,agustin.mayo.iscar}@uva.es

[3] Department of Statistics and Quantitative Methods, University of Milano-Bicocca, Milan, Italy
francesca.greselin@unimib.it

Abstract. Robust inference for the Cluster Weighted Model requires the specification of a few hyper-parameters. Their role is crucial for increasing the quality of the estimators, while arbitrary decisions about their value could severely hamper inferential results. To guide the user in the delicate choice of such parameters, a monitoring approach has been introduced in the recent literature, yielding an adaptive method. The approach is here exemplified, via the analysis of a dataset on the effect of punishment regimes on crime rates.

1 Introduction and Notation

The purpose of the present paper is to demonstrate how to perform hyper-parameters selection when applying Cluster Weighted Modelling (CWM) to a real dataset. In detail, through the employment of graphical tools, we propose a two-stage monitoring procedure to sequentially detect the number of potential outliers and the thresholds used in constrained estimation.

The contribution advances the studies on the semi-automation of clustering techniques, which is a relevant topic in statistics and in real statistical applications. Our proposal has been developed along the lines of Cerioli et al. (2018) and Torti et al. (2021).

We begin by introducing very briefly the notation, then provide the main ideas of the monitoring methodology, and in Sect. 2 we present and discuss an application to Crime data. Final remarks end the paper in Sect. 3.

Let $\mathbf{X}$ be a vector of covariates with values in $\mathbb{R}^d$, and let Y be a dependent (or response) variable with values in $\mathbb{R}$. Assume that the regression of Y on $\mathbf{X}$ varies across G levels, say groups or clusters, of a categorical latent variable Z. The linear gaussian CWM, introduced by Gershenfeld (1997), decomposes the

L. A. García-Escudero et al. (Eds.): SMPS 2022, AISC 1433, pp. 65–72, 2023.
https://doi.org/10.1007/978-3-031-15509-3_9

joint distribution of $(\mathbf{X}, Y)$, in each mixture component, as the product of the marginal and the conditional distributions, as follows:

$$p(\mathbf{x}, y; \mathbf{\Theta}) = \sum_{g=1}^{G} \pi_g \phi_1(y; \boldsymbol{b}_g'\mathbf{x} + b_g^0, \sigma_g)\phi_d(\mathbf{x}; \boldsymbol{\mu}_g, \boldsymbol{\Sigma}_g), \tag{1}$$

where $\phi_d(\cdot; \boldsymbol{\mu}_g, \boldsymbol{\Sigma}_g)$ denotes the density of the d-variate Gaussian distribution with mean vector $\boldsymbol{\mu}_g$ and covariance matrix $\boldsymbol{\Sigma}_g$. Y is related to $\mathbf{X}$ by a linear model in (1), that is, $Y = \boldsymbol{b}_g'\mathbf{x} + b_g^0 + \varepsilon_g$ with $\varepsilon_g \sim N(0, \sigma_g^2)$, $\boldsymbol{b}_g \in \mathbb{R}^d$, $b_g^0 \in \mathbb{R}$, $\sigma_g^2 \in \mathbb{R}^+$, $\forall g = 1, \ldots, G$. Unfortunately, Maximum Likelihood inference on models based on normal assumptions suffers from two major drawbacks: (i) the likelihood is unbounded over the parameter space, hence its maximization is in an ill-posed mathematical problem; (ii) the resulting inference is strongly affected by outliers (see, e.g., Huber and Ronchetti 2009). To overcome both issues, García-Escudero et al. (2017) introduced the Cluster Weighted Robust Model (CWRM), where a fixed fraction α of the less plausible observations is trimmed out and the estimation of the scatter matrices and the regression errors is constrained to ensure robust inference. The first constraint is applied to the set of eigenvalues $\{\lambda_l(\boldsymbol{\Sigma}_g)\}_{l=1,\ldots,d}$ of the scatter matrices $\boldsymbol{\Sigma}_g$ by requiring

$$\lambda_{l_1}(\boldsymbol{\Sigma}_{g_1}) \leq c_X \lambda_{l_2}(\boldsymbol{\Sigma}_{g_2}) \qquad \text{for every } 1 \leq l_1 \neq l_2 \leq d \text{ and } 1 \leq g_1 \neq g_2 \leq G. \tag{2}$$

The second bound is enforced to the variances σ_g^2 of the regression error terms as follows

$$\sigma_{g_1}^2 \leq c_y \sigma_{g_2}^2 \qquad \text{for every } 1 \leq g_1 \neq g_2 \leq G. \tag{3}$$

The constants $c_X, c_y \geq 1$ prevent degenerate cases with $|\boldsymbol{\Sigma}_g| \to 0$ and $\sigma_g^2 \to 0$ leading to an unbounded likelihood or non interesting spurious solutions. Therefore, the percentage of trimmed data and the threshold for eigenvalues ratio play an important role and should be carefully set.

We will exemplify how the monitoring tools introduced in Cappozzo et al. (2021) can be applied to real data. A first monitoring step is devoted to screen the space of solutions for CWRM, in view of making an informed choice for the trimming level α, which is the most crucial parameter. Metrics such as the group proportion, the total sum of squares decomposition, the regression slopes and standard deviations, the cluster volumes, and the Adjusted Rand Index (ARI) between consecutive cluster allocations are monitored when varying α, to uncover the most sensible trimming level to be employed.

Afterwards, the second monitoring step screens the space of solutions $\mathcal{E}_0$ generated by varying the number of clusters G, and the pair of hyper-parameters c_X and c_y over a grid, conditioned on a fixed trimming level. We aim at collecting a reduced list $\mathcal{O}$ of "optimal" solutions, qualified by two features: their optimality in terms of a CWRM-specific BIC criterion (Cappozzo et al. 2021), and their stability across hyper-parameter values. Stability of solution A means that, varying the values of the constraints, the estimation yields a partition B pretty close to A (the ARI between A and B is greater than a threshold η).

Finally, to explore the quality of the clustering obtained in the optimal solutions and to uncover the nature of the outliers, silhouette plots (Rousseeuw 1987) can be employed. Silhouette plots have been introduced for representing the quality of the clustering solution, and may be defined in the spirit of discriminant factors introduced in García-Escudero et al. (2011). Specific discriminant factors, tailored for the CWM characterization in (1) should be considered here. The first discriminant factor $DF(i)$ assesses the strength of the assignment, or the strength of the trimming decision for unit i in the joint modeling expressed by the CWM. On the other side, $DF_{Y|X}(i)$ and $DF_X(i)$ break down the overall mixture density in the contribution of the G regression hyperlanes and the component-wise random covariates, respectively.

2 Crime Dataset

The dataset originates from a study on the effect of punishment regimes on crime rates. We analyse aggregate data on 47 US states taken place in 1960 illustrated by Ehrlich (1973), available in the `MASS` `R` package. The crime rate, measured as the number of offenses per 100.000 population, is the response variable Y.

The goal is to infer whether the structure of dependence among covariates differently affects the crime rate Y depending on the geographical area.

Available predictors $\mathbf{X}$ for each of the 47 states are the following: *percentage of males aged 14–24*, *mean years of schooling*, *police expenditure*, *labour force participation rate*, *number of males per 1000 females*, *state population*, *number of non-whites per 1000 people*, *unemployment rate of urban males 14–24*, *unemployment rate of urban males 35–39*, *gross domestic product per head*, *income inequality*, *probability of imprisonment*, and *average time served in state prisons*. Finally, the indicator variable denoting the *16 Southern states* will be considered as the grouping variable hereafter. To this extent, robust CWM is applied on $(\mathbf{X}, Y)$ to uncover geographical grouping. The high number of variables involved in the estimation and the small sample size represent a challenge for the discriminating task.

The first step of our monitoring tools provides the outcome displayed in Fig. 1, where α takes values on a grid from 0 to 0.255. We opted for setting in advance $G = 2$ the number of geographical areas we would like to uncover. This choice is in line with the monitoring philosophy, for which any domain-related knowledge that may guide hyper-parameter selection shall be included in the analysis.

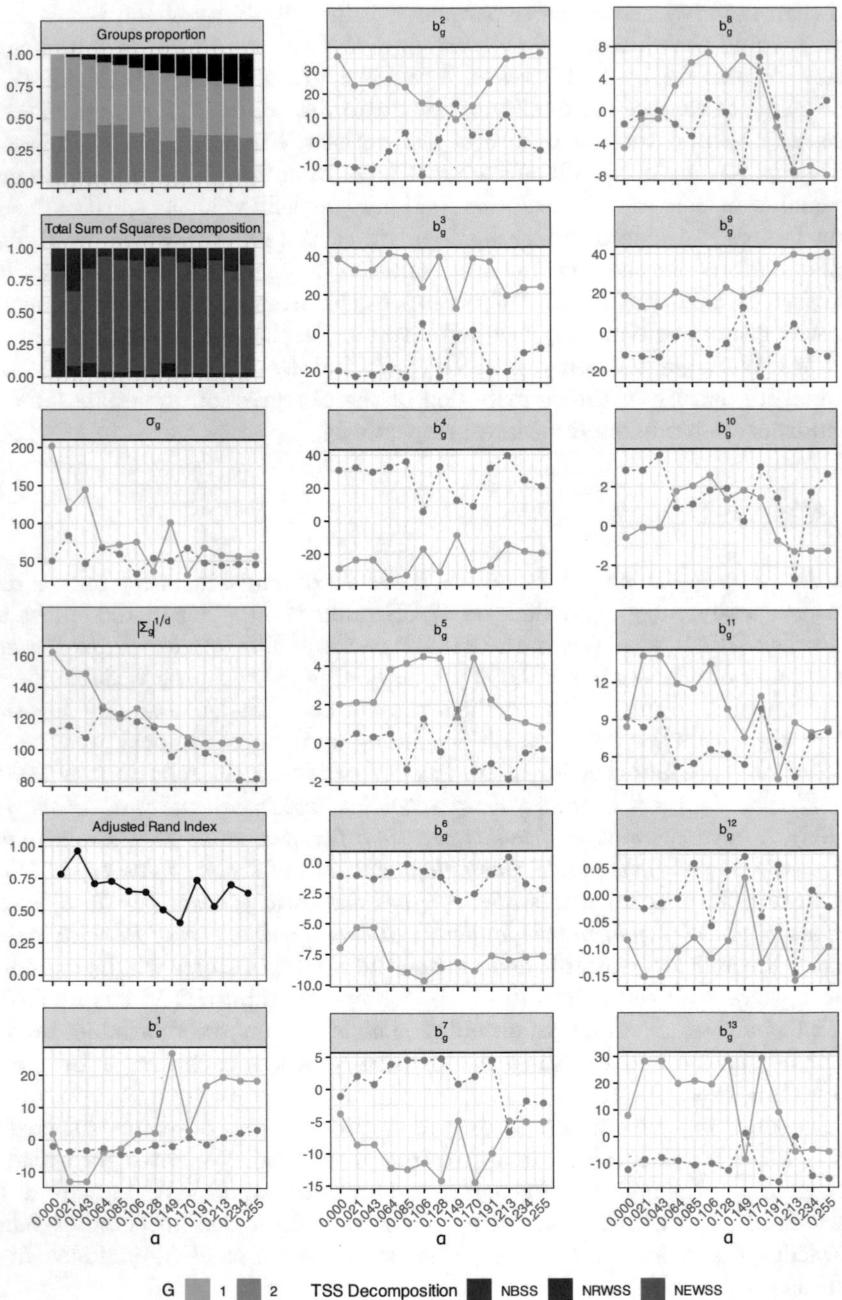

Fig. 1. Crime data: monitoring tools obtained in Step 1. Groups proportion (black bars denote the trimmed units), total sum of squares decomposition, regression coefficients, standard deviations, cluster volumes, ARI between consecutive cluster allocations are shown as a function of the trimming level α. G is kept fixed and equal to 2

We see that the choice of $\alpha = 0.064$ stabilizes the variance of the regression errors σ_1 and de-inflate the determinant of the scatter matrix $|\Sigma_1|^{1/d}$ in the first cluster (represented in green in Fig. 1), aligning them to the order of magnitude of the analogous quantities in the second cluster. Therefore, an estimation based on 44 observations seems sufficient to assure robustness without sacrificing efficiency.

In the second step, the monitoring procedure assesses the validity and stability of the solutions. Figure 2 reports the results, while the pairs plots for the first and second optimal solutions, based on a subset of explanatory variables, are displayed in Figs. 3 and 4, respectively. The first optimal solution remains best for $c_X = 10$ and for c_Y ranging from 1 to 8, while when $c_Y = 10$ the second optimal solution appears. On the one hand, the first solution shows a wide stability varying the values of the constraints; on the other hand, the second solution offers the highest classification accuracy. Indeed, the latter partition possesses only 2 misclassified units, maintained even after assigning the three trimmed units using the MAP rule: the proposed monitoring procedure provides a highly accurate classification for this dataset. In this regard, the best result present in the literature has been achieved by means of a competing method, introduced in Subedi et al. (2015), in which a variant of the CWM was developed considering a mixture of t-factor analyzers for modeling the marginal distribution of $\mathbf{X}$ and a t-distribution for the regression error. Such a model certainly has nice features in terms of explainability and parsimony, it nonetheless showcases 4 misclassified units when applied to the US crime dataset. The t distribution is known to be a very good option in presence of mild outliers, but hard trimming seems to achieve slightly better performance in this context. Note that, as already reported in Subedi et al. (2015), mixture of regression (DeSarbo and Cron 1988; De Veaux 1989) and mixtures of regression with concomitants (Dayton and Macready 1988) do not show good clustering performance whenever the distribution of the covariates plays a role on the cluster structure of the data (see the pairs plot in Figs. 3 and 4): such approaches are, by construction, unable to capture it.

Lastly, from the right panel in Fig. 3, the silhouette plots tell us that observations 4 and 29 are bad leverage points, having low values of the discriminant factors $DF_{Y|X}$ and DF_X respectively assessing the strength of the assignment/trimming for each unit in relation to the regression lines and the covariates. Without trimming such observations, inferential results on the regression parameters would have been biased. Observation 11 is instead a non-outlying point in the covariates and with a fitting regression line for one of the $G = 2$ components, but with an outlying pattern according to the joint CWM density (revealed by the high negative value for DF).

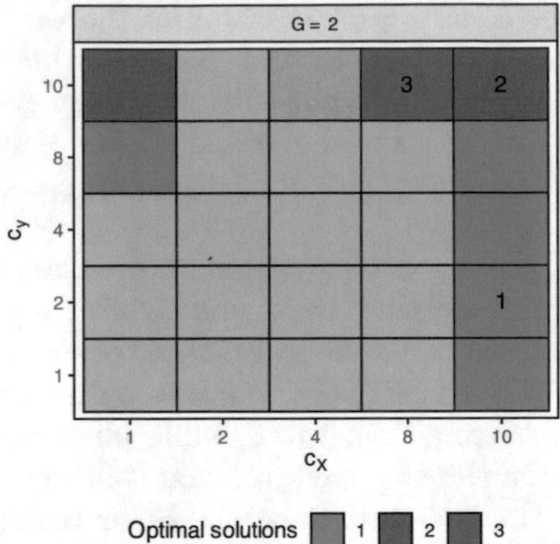

Fig. 2. Crime data. Step 2: monitoring the optimal solutions, indicated by the cells with ordinal numbers 1, 2, 3 and 4 ($\alpha = 0.064$). Each solution is featured by one color, showing the range of cases in which it is best (darker opacity cells), and stable (lighter opacity cells), varying c_X (horizontal axis) and c_Y (vertical axis) in $\mathscr{E}_0$. G is fixed and equal to 2

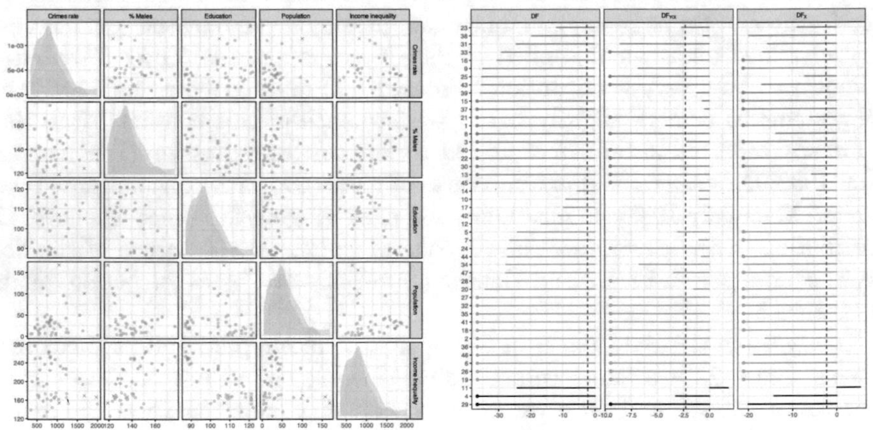

Fig. 3. Crime data. Pairs plot of the first optimal solution obtained in Step 2, different colors denote the partition induced by the CWRM, trimmed units are denoted by $\times$ (left panel) and silhouette plots displaying $DF(i)$, $DF_{Y|X}(i)$ and $DF_X(i)$ for observation i in the dataset (right panel)

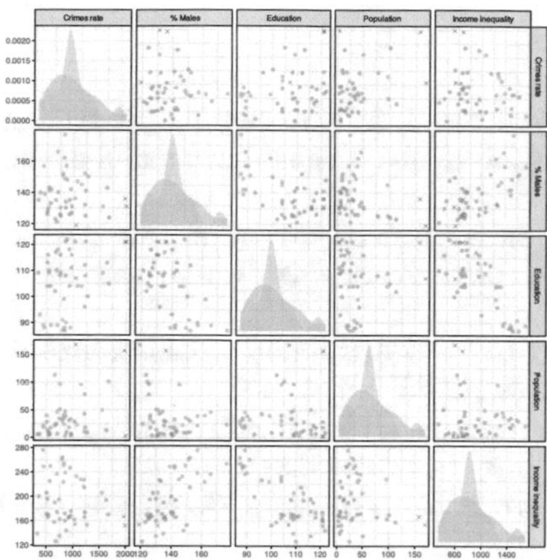

Fig. 4. Crime data. Pairs plot of the second optimal solution obtained in Step 2, different colors denote the partition induced by the CWRM. Trimmed units are denoted by ×

3 Conclusions

Robustness is the property of statistical methods to infer the generating process that originates the main body of the data in presence of contamination. A wide class of robust procedures for clustering is featured by constrained estimation and impartial trimming, for which specific hyper-parameters are introduced. In this paper, we presented an application of a recent contribution to the literature for the case of cluster-wise regression, where a monitoring approach is proposed. We have shown how the graphical and computational tools are able to assist the practitioner in the delicate task of setting hyper-parameters in the estimation of robust cluster weighted models, to analyze the effect of punishment regimes on crime rates.

The method relies on the combination of two exploratory steps. Sensible options for the trimming proportion α are identified in the first monitoring step. Afterwards, in the second monitoring step, the whole space of solutions is explored, varying the hyper-parameters governing the heterogeneity in the covariates and the regression error terms, as well as the number of groups.

In the analysis of the crime dataset, our semiautomatic procedure yields two final solutions, qualified by the interval of hyper-parameters values in which their optimality, stability and validity hold true. The first solution shows a wide stability; on the other hand, the second solution offers the highest classification accuracy for this dataset. New silhouette plots reveal the nature and the extent of the outlying observations, distinguishing between outliers with respect to the clustering of the covariate $\mathbf{X}$, and the local regression lines Y, following the nature of the Cluster Weighted model.

Further research can be devoted to reducing the computational burden of the proposed methodology, and extending it to other robust clustering models.

References

Cappozzo, A., García-Escudero, L.A., Greselin, F., Mayo-Iscar, A.: Parameter choice, stability and validity for robust cluster weighted modeling. Stats **4**(3), 602–615 (2021)

Cerioli, A., Riani, M., Atkinson, A.C., Corbellini, A.: The power of monitoring: how to make the most of a contaminated multivariate sample. Stat. Meth. Appl. **27**(4), 661–666 (2018)

Dayton, C.M., Macready, G.B.: Concomitant-variable latent-class models. J. Am. Stat. Assoc. **83**(401), 173–178 (1988)

DeSarbo, W.S., Cron, W.L.: A maximum likelihood methodology for clusterwise linear regression. J. Classif. **5**(2), 249–282 (1988)

De Veaux, R.D.: Mixtures of linear regressions. Comput. Stat. Data Anal. **8**(3), 227–245 (1989)

Ehrlich, I.: Participation in illegitimate activities: a theoretical and empirical investigation. J. Polit. Econ. **81**(3), 521–565 (1973)

García-Escudero, L.A., Gordaliza, A., Greselin, F., Ingrassia, S., Mayo-Iscar, A.: Robust estimation of mixtures of regressions with random covariates, via trimming and constraints. Stat. Comput. **27**(2), 377–402 (2016). https://doi.org/10.1007/s11222-016-9628-3

García-Escudero, L.A., Gordaliza, A., Matrán, C., Mayo-Iscar, A.: Exploring the number of groups in robust model-based clustering. Stat. Comput. **21**(4), 585–599 (2011)

Gershenfeld, N.: Nonlinear inference and cluster-weighted modeling. Ann. N. Y. Acad. Sci. **808**(1 Nonlinear Signal and Image Analysis), 18–24 (1997)

Huber, P.J., Ronchetti, E.M.: Robust Statistics. Wiley Series in Probability and Statistics. Wiley, Hoboken, NJ, USA (2009)

Rousseeuw, P.J.: Silhouettes: a graphical aid to the interpretation and validation of cluster analysis. J. Comput. Appl. Math. **20**(C), 53–65 (1987)

Subedi, S., Punzo, A., Ingrassia, S., McNicholas, P.D.: Cluster-weighted t-factor analyzers for robust model-based clustering and dimension reduction. Stat. Meth. Appl. **24**(4), 623–649 (2015)

Torti, F., Riani, M., Morelli, G.: Semiautomatic robust regression clustering of international trade data. Stat. Meth. Appl. **30**(3), 863–894 (2021). https://doi.org/10.1007/s10260-021-00569-3

Penalized Model-Based Clustering with Group-Dependent Shrinkage Estimation

Alessandro Casa[1(✉)], Andrea Cappozzo[2], and Michael Fop[3]

[1] Faculty of Economics and Management, Free University of Bozen-Bolzano, Bolzano, Italy
alessandro.casa@unibz.it

[2] MOX - Laboratory for Modeling and Scientific Computing, Politecnico di Milano, Milan, Italy
andrea.cappozzo@polimi.it

[3] School of Mathematics and Statistics, University College Dublin, Dublin, Ireland
michael.fop@ucd.ie

Abstract. Gaussian mixture models (GMM) are the most-widely employed approach to perform model-based clustering of continuous features. Grievously, with the increasing availability of high-dimensional datasets, their direct applicability is put at stake: GMMs suffer from the curse of dimensionality issue, as the number of parameters grows quadratically with the number of variables. To this extent, a methodological link between Gaussian mixtures and Gaussian graphical models has recently been established in order to provide a framework for performing penalized model-based clustering in presence of large precision matrices. Notwithstanding, current methodologies do not account for the fact that groups may be under or over-connected, thus implicitly assuming similar levels of sparsity across clusters. We overcome this limitation by defining data-driven and component specific penalty factors, automatically accounting for different degrees of connections within groups. A real data experiment on handwritten digits recognition showcases the validity of our proposal.

1 Introduction and Motivation

In model-based clustering finite mixture models are employed to delineate a one-to-one correspondence between mixture components and sought clusters, with the Gaussian distribution being the conventional choice to group multivariate continuous samples (Bouveyron et al. 2019). Unfortunately, in the big data era the applicability of this well-established procedure is jeopardized as Gaussian mixture models (GMM) tend to be over-parameterized in high-dimensional settings (Bouveyron and Brunet-Saumard 2014). To mitigate this issue, several solutions have been proposed that include constrained modelling, variable selection and sparse estimation (Fop and Murphy 2018). Particularly, within the latter

L. A. García-Escudero et al. (Eds.): SMPS 2022, AISC 1433, pp. 73–78, 2023.
https://doi.org/10.1007/978-3-031-15509-3_10

family, Zhou et al. (2009) proposed a penalized approach in which the number of parameters to be estimated is drastically reduced by enforcing a graphical lasso penalty in the objective function (Friedman et al. 2008). The resulting penalized likelihood allows to detect different sparsity patterns in the estimated precision matrices, but it falls short when these matrices have a substantially different number of non-zero entries, as the method explicitly assumes a common shrinkage factor for each and every component of the mixture. Such a behavior may hinder the resulting clustering in applications where sparse intensity is cluster-wise different. To overcome this limitation, the present paper extends the methodology of Zhou et al. (2009) by devising group-wise penalty factors which automatically enforce under or over-connectivity in the precision matrices. The approach is entirely data-driven and does not require any additional hyper-parameter specification.

The remainder of the paper is structured as follows. In Sect. 2 we introduce our new proposal and we discuss two strategies to compute cluster-specific penalty factors. Section 3 presents a digits recognition application, in which dependence structures between pixels differ across digits. Section 4 summarizes the novel contributions and highlights future research directions.

2 Proposed Solution

Consider a set of n observed data $\mathbf{X} = \{\mathbf{x}_1, \ldots, \mathbf{x}_n\}$, with $\mathbf{x}_i \in \mathbb{R}^p$ for $i = 1, \ldots, n$. With the aim of partitioning $\mathbf{X}$ in K subpopulations or clusters, the present work proposes to carry out parameter estimation by maximizing the following penalized log-likelihood function:

$$\sum_{i=1}^{n} \log \sum_{k=1}^{K} \pi_k \phi(\mathbf{x}_i; \boldsymbol{\mu}_k, \boldsymbol{\Omega}_k) - \lambda \sum_{k=1}^{K} \|\mathbf{P}_k * \boldsymbol{\Omega}_k\|_1 . \tag{1}$$

The first term in (1) is the log-likelihood of a GMM, with K the number of mixture components, π_ks the mixing proportions ($\pi_k > 0$, $\sum_k \pi_k = 1$), and $\phi(\cdot; \boldsymbol{\mu}_k, \boldsymbol{\Omega}_k)$ the density of a multivariate Gaussian distribution with mean vector $\boldsymbol{\mu}_k = (\mu_{1k}, \ldots, \mu_{pk})$ and precision matrix $\boldsymbol{\Omega}_k$, $k = 1, \ldots, K$. The second term in (1) identifies a graphical lasso penalty with shrinkage factor λ that is applied to the K precision matrices. In details, $\|\cdot\|_1$ is the L_1 norm taken element-wise ($\|A\|_1 = \sum_{ij} |A_{ij}|$), with $*$ we denote the Hadamard product, and $\mathbf{P}_k$s are weighting matrices that scale the effect of the common penalty λ depending on the component-specific sparsity underlying cluster k, $k = 1, \ldots, K$. Such a penalty forces some entries in the precision matrices to be shrunk to 0, uncovering group-wise conditional independence among the variables.

The original proposal by Zhou et al. (2009) implicitly assumed $\mathbf{P}_k$ to be an all-one matrix $\forall k$, our specification of $\mathbf{P}_1, \ldots, \mathbf{P}_K$ instead allows to encode information about class specific sparsity patterns, accounting for under or over-connectivity scenarios. We rely on carefully initialized sample precision matrices $\hat{\boldsymbol{\Omega}}_1^{(0)}, \ldots, \hat{\boldsymbol{\Omega}}_K^{(0)}$ (based on model-based and/or ensemble initialization strategies) to define $\mathbf{P}_k = f(\hat{\boldsymbol{\Omega}}_k^{(0)})$, with $f : \mathbb{S}_+^p \to \mathbb{S}^p$ a function from the space of positive semi-definite matrices to the space of symmetric matrices of dimension p. Two viable options for defining $f(\cdot)$ are briefly described hereafter.

Option 1: $f(\cdot)$ via inversely weighted sample precision matrices

The first proposal for defining $\mathbf{P}_k$ is as follows:

$$P_{k,ij} = 1/\left(|\hat{\Omega}_{k,ij}^{(0)}|\right), \tag{2}$$

where $P_{k,ij}$, $\hat{\Omega}_{k,ij}^{(0)}$ are respectively the (i,j)-th elements of the matrices $\mathbf{P}_k$ and $\hat{\boldsymbol{\Omega}}_k^{(0)}$. Intuitively, an high $|\hat{\Omega}_{k,ij}^{(0)}|$ value induces a deflation on the penalty enforced on the (i,j)-th element of $\boldsymbol{\Omega}_k$, whereas when $|\hat{\Omega}_{k,ij}^{(0)}|$ is close to 0 we are imposing an extra shrinkage on $\Omega_{k,ij}$. This strategy can be seen as a multiclass extension of the approach proposed in Fan et al. (2009).

Option 2: $f(\cdot)$ via distance measures in the $\mathbb{S}_+^p$ space

A second data-driven alternative involves setting $\mathbf{P}_k$ entries proportional to the distance between $\hat{\boldsymbol{\Omega}}_k^{(0)}$ and $\operatorname{diag}\left(\hat{\boldsymbol{\Omega}}_k^{(0)}\right)$, where $\operatorname{diag}\left(\hat{\boldsymbol{\Omega}}_k^{(0)}\right)$ is a diagonal matrix whose diagonal elements are equal to the ones in $\hat{\boldsymbol{\Omega}}_k^{(0)}$. Such a strategy mathematically reads as follows:

$$P_{k,ij} = \frac{1}{\mathscr{D}\left(\hat{\boldsymbol{\Omega}}_k^{(0)}, \operatorname{diag}\left(\hat{\boldsymbol{\Omega}}_k^{(0)}\right)\right)}, \quad \forall i,j = 1,\ldots,p \quad \text{and} \quad i \neq j, \tag{3}$$

where with $\mathscr{D}(\cdot,\cdot)$ we identify a distance measure in the space of positive semi-definite matrices. Given the non-Euclidean nature of the $\mathbb{S}_+^p$ space several $\mathscr{D}(\cdot,\cdot)$ may be considered when defining (3): we subsequently employ Frobenius and Riemannian distances, but other options are at our disposal (see, e.g., Dryden et al. 2009).

The two above-described strategies for defining $f(\cdot)$ force entries corresponding to weaker sample conditional dependencies to be more strongly penalized. Once the definition of P_k has been established, coherently to Zhou et al. (2009), the model is estimated employing an EM algorithm where, in the M step, a graphical lasso strategy is adopted to compute $\Omega_1, \ldots, \Omega_K$ with $\lambda_k = 2\lambda P_k/n_k^{(t)}$ with $n_k^{(t)}$ denoting the estimated sample size of the k-th cluster at the t-th iteration of the algorithm.

3 Application to Handwritten Digits Recognition

The methodology presented in the previous section is employed to perform automatic handwritten digits recognition. The considered dataset is publicly available in the University of California Irvine Machine Learning data repository (http://archive.ics.uci.edu/ml/datasets/optical+recognition+of+handwritten+digits) and it contains $n = 5620$ handwritten samples of $K = 10$ digits. After having performed a preprocessing step to eliminate the near-zero variance pixels, we are left with $p = 47$ features onto which perform model based clustering. This translates to a challenging modeling task due to the narrow separation between classes and the high dimensionality of the parameter space. Indeed, a standard GMM with full precision matrices would require the estimation of $(K - 1) + Kp + Kp(p - 1)/2 = 11759$ parameters. We fit the penalized GMM methodology in (1) to the handwritten digits recognition dataset with different specification of P_ks: results are reported in Table 1. In details, $\mathbf{P}_k = \boldsymbol{J}$ identifies the original procedure of Zhou et al. (2009), with $\boldsymbol{J}$ the all-one matrix, while the remaining models describe the novel proposals of Sect. 2.

Table 1. BIC, Adjusted Rand Index (ARI), number of estimated parameters in the precision matrices for different penalized model-based clustering methods and for digits 0, 5 and 9. Handwritten digits dataset

	BIC	ARI	d_Ω	d_0	d_5	d_9
$\mathbf{P}_k = \boldsymbol{J}$	−388862	0.6837	4914	701	989	1271
$\mathbf{P}_k$ as in (2)	−368604	0.6820	3436	535	651	721
$\mathbf{P}_k$ as in (3), Frobenius distance	−391359	0.6827	6066	771	1003	1041
$\mathbf{P}_k$ as in (3), Riemannian distance	−388902	0.6841	5206	723	1059	1295

The penalized methods are able to shrink the estimates in a group-wise manner, recovering fairly well the underlying data partition. This is especially true in our proposals for which, even though the resulting Adjusted Rand Index Rand (1971) is not dramatically affected, the number of covariance parameters shrunk to 0 is digits-wise different thanks to the $\mathbf{P}_k$ specification. In Fig. 1 we report the averaged images for digits 0, 5, and 9 and the estimated graphs in the precision matrices for the $\mathbf{P}_k$ via Riemannian distance approach. This method showcases the highest ARI and we can appreciate how the number of estimated non-zero entries appreciably differ between the selected digits.

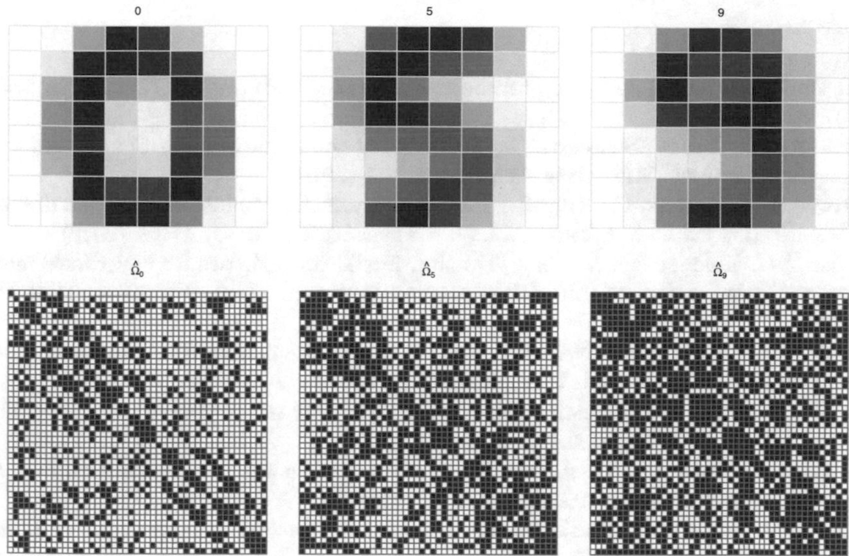

Fig. 1. Averaged images for digits 0, 5, and 9 and estimated graphs in the precision matrices for the $\mathbf{P}_k$ via Riemannian distance approach. Dark blue squares denote the presence of an edge between the two variables. Handwritten digits dataset

4 Conclusion and Discussion

In this work we have proposed an extension to the approach outlined in Zhou et al. (2009). Two different procedures have been suggested to account for under or over-connected sparsity patterns in the precision matrices within a model-based clustering framework. The first solution provides an entry-wise inflation/deflation on the common penalty factor, while the second relies on distance metrics in the space of positive semi-definite matrices to determine group-wise adjustments to the overall shrinkage term. An experiment on handwritten digits recognition has demonstrated the promising applicability of the devised procedure.

A direction for future research involves the development of a flexible mix-and-match methodology in which the penalization could interchangeably be applied to sparse precision and/or covariance matrices (Bien and Tibshirani 2011). Such a framework, coupled with a penalty in the component means, can ultimately be employed to discard variables irrelevant for the clustering: ideas are being explored and they will be the object of future work.

References

Bien, J., Tibshirani, R.J.: Sparse estimation of a covariance matrix. Biometrika **98**(4), 807–820 (2011)

Bouveyron, C., Brunet-Saumard, C.: Model-based clustering of high-dimensional data: a review. Comput. Stat. Data Anal. **71**, 52–78 (2014)

Bouveyron, C., Celeux, G., Murphy, T.B., Raftery, A.E.: Model-Based Clustering and Classification for Data Science, vol. 50. Cambridge University Press (2019)

Dryden, I.L., Koloydenko, A., Zhou, D.: Non-Euclidean statistics for covariance matrices, with applications to diffusion tensor imaging. Ann. Appl. Stat. **3**(3), 1102–1123 (2009)

Fan, J., Feng, Y., Wu, Y.: Network exploration via the adaptive LASSO and SCAD penalties. Ann. Appl. Stat. **3**(2), 521–541 (2009)

Fop, M., Murphy, T.B.: Variable selection methods for model-based clustering. Stat. Surv. **12**, 18–65 (2018)

Friedman, J., Hastie, T., Tibshirani, R.: Sparse inverse covariance estimation with the graphical lasso. Biostatistics **9**(3), 432–441 (2008)

Rand, W.M.: Objective criteria for the evaluation of clustering methods. J. Am. Stat. Assoc. **66**(336), 846 (1971)

Zhou, H., Pan, W., Shen, X.: Penalized model-based clustering with unconstrained covariance matrices. Electron. J. Stat. **3**, 1473–1496 (2009)

Robust Diagnostics for Linear Mixed Models with the Forward Search

Aldo Corbellini[1], Luigi Grossi[2], and Fabrizio Laurini[1(✉)]

[1] Ro.S.A. and Department of Economics, University of Parma, Parma, Italy
{aldo.corbellini,fabrizio.laurini}@unipr.it

[2] Ro.S.A. and Department Statistical Sciences, University of Padova, Padova, Italy
luigi.grossi@unipd.it

Abstract. Robustness of Linear Mixed Models (LMM) with random effects is investigated with the forward search (FS). Extending the FS to LMM offers new computational challenges, as some restrictions, imposed by the model and their estimates, are required. The method is illustrated by an application to real data where exports of coffee to European Union are analyzed to identify outliers that might be linked to potential frauds. An additional short simulation is presented to strengthen the usefulness of the proposed method.

1 Introduction and Background

In this paper a robust approach to the study of random effect models is suggested. Random effect models are a special case of Linear Mixed Models (LMM) which are particularly attractive when repeated measures of variables are collected on a sample of individuals.

For these models, the estimate of parameters are badly affected by the presence of outliers and influential observations, the effect of which can be monitored via robust estimators. Some existing diagnostics are mostly based on the leave-k-out approach, but they suffer from "masking" when the real number of outliers is greater than k.

An alternative "monitoring", called the forward search (FS) method, has revealed to be particularly effective to cope with the masking effect in several multivariate settings due to its great flexibility, being efficient and robust, with fairly expensive computational costs.

In this paper we present a forward search approach to LMM to monitor the influence of outliers on estimated coefficients and to suggest a way to obtain a robust estimator. The correlation structure imposed by LMM models is more complex than that of regression models for cross-sectional data and this opens new challenging issues that we try to address in this paper.

The paper presents a real data problem, which is used as an illustration, because there are many interesting features. The extension of the forward search to LMM and the main related issues are discussed in Sect. 4. Then the application

L. A. García-Escudero et al. (Eds.): SMPS 2022, AISC 1433, pp. 79–86, 2023.
https://doi.org/10.1007/978-3-031-15509-3_11

of the forward search to a simulated data and to real trade data is presented, with some discussion of results and possible further extensions.

2 Data, Their Features and Economic Characteristics

Coffee production takes place mainly in developing countries. One of the biggest producers is Brazil. Coffee's consumption, instead, is well spread all over European countries, none of which produces it. The coffee price is characterized by relatively low price elasticity of supply because new plants require more than two years to become productive. Likewise demand is characterized by low price elasticity, because it is quite stable and changes only if big movements in price arise independently from any income increase. As a consequence, coffee price time series show a very high variability and outliers are very likely to appear.

In this work we consider the coffee traded from the origin Brazil to destination of 18 countries in Europe. For each destination country the monthly value (in Euro) and the monthly quantity (in tons) of coffee sold from Brazil, to the specific country, are recorded. Monthly figures of sales are recorded for approximately 5 years. We stick on the 18 countries that traded with Brazil every month in these 5 years span. Therefore, our dataset represents a case of a balanced design in the framework of longitudinal data analysis. However, our results and methods are not sensitive to deviations from that balanced structure, although minor adjustments might be required to handle some missing data.

To make notation specific, we have that for each destination country (group) it is available the value of coffee imported from Brazil, denoted as $V_{i,t}$ with $i = 1, \dots, g$ and $t = 1, \dots, ng$. Additionally, the quantity imported by each country is denoted by $W_{i,t}$. From the above notation, it is already considered the option to have a different number of replicates for each country as the number of replicates ng might be different for each country.

A snapshot of the whole set of data is given in Fig. 1 where each destination country is plotted with a different symbol.

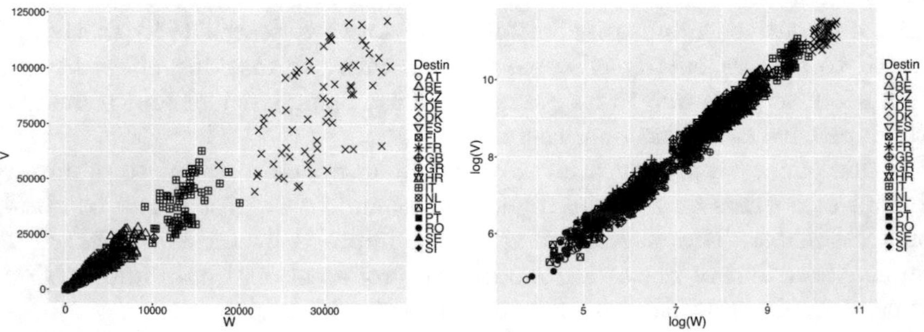

Fig. 1. Left panel: Scatterplot of the data with symbol for each destination. Largest importers are Germany and Italy. Right panel: scatterplot of the log-log data

To highlight the monthly time evolution of the trades, we also show, in Fig. 2, the logged time series of V and W for Brazil versus Italy. A clear relationship is visible between the two series, but the presence of a long-term trend is questionable. The seasonality is quite clear in June and December, when the minimum and maximum yearly values happen, while it does not look particularly strong in other months.

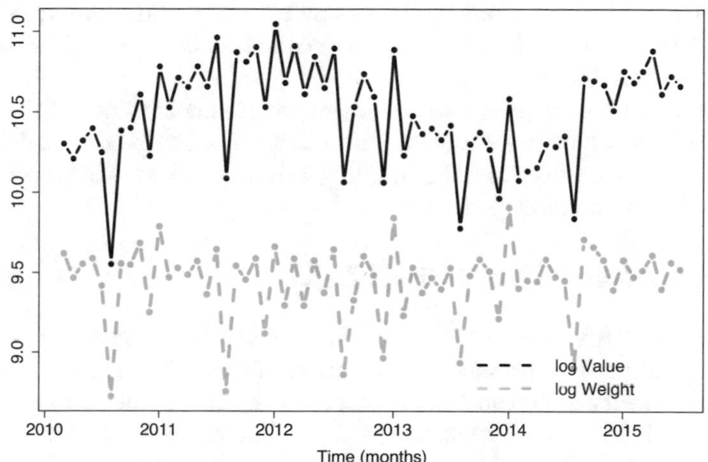

Fig. 2. Time series of relevant quantities, in log scale, for destination Italy

3 LMM for Repeated Measures

We consider the unified theory of LMM with fixed component $X\beta$ and the random effect u written in the form

$$y = X\beta + Zu + \epsilon, \tag{1}$$

where Z is a kind of indicator matrix. The model is well specified under a number of subsequent assumptions. Denoting with $\mathbf{I}$ the identity matrix and with $\mathbf{0}$ either a vector or a matrix with zeros (clear from the context), the random component u and the error term ϵ have the following features:

$$E\begin{bmatrix} u \\ \epsilon \end{bmatrix} = \begin{bmatrix} \mathbf{0} \\ \mathbf{0} \end{bmatrix}, \quad \text{Cov}\begin{bmatrix} u \\ \epsilon \end{bmatrix} = \begin{bmatrix} G & \mathbf{0} \\ \mathbf{0} & R \end{bmatrix} \quad \text{with } G = \sigma_u^2\mathbf{I} \text{ and } R = \sigma_\epsilon^2\mathbf{I}.$$

Model (1) is very flexible, despite its simplicity. For instance it offers a very convenient framework when repeated measures, which are generally dependent, have to be analyzed. For repeated measures it can be used as a "simple" random intercept model (the constant correlation model) where all components of β, but the intercept, are common among groups. Representation (1) is appropriate even if regression coefficients are assumed to vary among groups. Model (1) can also be used when groups are correlated to each others, by considering a richer structure for either R or G, or both. Once the vector β is estimated we obtain the fitted values of the model given by

$$\hat{y} = X\hat{\beta} + Z\hat{u}. \tag{2}$$

Estimated elements (which are BLUPs) have two sources of variability, namely estimation of the fixed and random effects pair β and u and estimation of covariance matrices G and V. This is somehow critical as both should be taken into account when making inference.

All statistical results discussed so far are key elements to derive the residuals from model (1), which are computed by taking the observed y and the fitted values from (2). Residuals are the building bricks to set up the forward search, whose details will be discussed next.

4 The Forward Search for LMM

The forward search is a sequential procedure which is based on a set of algorithms such that we start from an outlier-free subset of size $m^\star < n$ (the Basic Subset, BSB) and, at every step, the BSB is increased by including units closer to the selected model. In general, inclusion is such that at every step we move from a subset of size m to a subset of size $m + 1$, with iterations until all n data are included. When outliers or other influential observations enter the subset, through the monitoring of relevant statistics, sharp movements are recorded. For regression models, this procedure is illustrated in Atkinson and Riani (2000) and made precise in Riani et al. (2009).

There are several differences from regression to LMM. We start by highlighting that it is common to have time series even for simple random effects models, thus the selection of units belonging to $m^\star$ is made accordingly. To be specific, a sensible approach is to build the initial subset by taking contiguous observations in each group (here represented by all destination countries). Using proper notation, tailored to the set of available regressors (one explanatory variable, one time index for the trend and eleven dummies for the monthly seasonality), we have that, for each destination country i, with $i = 1, \ldots, g$, a coherent set of observations is given by fixing a time index $t^{(i)}$, and then selecting contiguous observations, the number of which is driven by k, the number of columns of the X matrix. Specifically, we select an arbitrary set of observations for each group $m_{t^{(i)},t^{(i)}+1,\ldots,t^{(i)}+k+1}$ which ensures that the model is identifiable. A similar choice is made for all groups but, in general, $t^{(i)} \neq t^{(j)}$, with $j = 1, \ldots, g$ and $i \neq j$. An initial subset M1 given by

$$\mathrm{M1} = \bigcup_{i=1}^{g} \{m_{t^{(i)},t^{(i)}+1,\ldots,t^{(i)}+k+1}\},$$

is then used to fit the random effect model. After fitting the model the following algorithm is performed.

1. Squared residuals $r_{i,t}^2 = (y_{i,t} - \hat{y}_{i,t})^2$, for all units are computed (even for those that did not contribute to the fitting). With this notation $y_{i,t}$ and $\hat{y}_{i,t}$ denote a single element from the vector expressed in Eqs. (1) and (2) respectively.
2. Squared residuals are sorted yielding to $r_{i,t}^2[\star_i]$, with the argument $[\star_i]$ denoting that the sorting is kept separated for each group.
3. The median of $r_{i,t}^2[\star_i]$ is computed for each i and then stored.

Steps 1 to 3 above are then repeated by changing, for each group, the time index $t^{(i)}$ and leading to sets M2, M3, ..., all having the same size. This procedure is repeated 10000 times, as suggested by Riani et al. (2009), since choosing among all possible subsets is unfeasible for almost all practical applications.

For each group the observations that lead to the smallest median of sorted residuals are those contributing to the BSB $m^\star$. Hence, not necessarily the time index to build $m^\star$ is identical for all groups, but inside each group, the contiguity of observation is at this stage guaranteed. This last restriction, however, might be relaxed when having an X matrix with a time index as explanatory variable, since the sorting of the data is irrelevant in the fitting when the time dependence is explicitly modelled with a time trend variable.

As stated above, once fitted the model, computing residuals during the forward search is quite similar to standard regression. The fitted values (2) are obtained after getting $\hat{\beta}$, $\hat{\sigma}_u$ and $\hat{\sigma}_\epsilon$. Plugging such estimates into BLUP gives estimates of $\hat{u}$ and $\hat{y}$.

Some discussion is needed when fitted values (2) and residuals are computed for all units, i.e. also for units that did not contribute to the estimation step. Fitted values and residuals are obtained by considering all entries of X (similarly to the regression model) but using only units that contributed to the fit to extract entries of Z (unlike the regression model). This last restriction is needed to ensure the conformability of product of matrices.

The key difference with regression model, however, comes when moving from m to $m+1$ because of the presence of groups in the data (here represented by each destination). The move from m to $m+1$ requires, once again, the sorting of squared residuals.

Our forward search algorithm is quite flexible and relatively general, as it does not require any group membership balancing during the procedure. In other words, at every step of the forward search, all groups must have only the minimum number of observations, say $m_{t^{(i)},t^{(i)}+1,\ldots,t^{(i)}+k+1}$, such that the full model is identifiable.

As stated above, the inclusion of new units, i.e. the move from size m to $m+1$, is based on the squared residuals computed for all data. As for the choice of the BSB, once residuals have been computed, the sorting is made separately for every group. Therefore, the difference compared to the forward search in regression, is that the sorting is not made on the whole dataset, but split by group. In practice, there are squared residuals sorted for the first group, then squared residuals sorted for the second group, and so forth.

The $(m+1)$-th observation joining the dataset is the one for which the squared sorted residuals not belonging to the m-th step is smallest. As a consequence, if observations belonging to the same group are well described by the model, and consequently have all small residuals, then the units forming the subset at steps $m + 1$, $m + 2$, ..., will belong to the same group. Therefore, at each step of the forward search, the size of each group belonging to the subset of size m can be, potentially, very different. As stated above, when discussing the choice of the BSB, the time index in the inclusion is not considered, as there is an explanatory variable taking the time trend into account, so that the fit does not require, necessarily, observations to be contiguous.

The peculiar feature of the forward search is that the inclusion of outliers or the inclusion of influential observations is highlighted by sharp peaks moving from m to $m + 1$. This is also true for LMM. Influential observations, hidden structures, or outliers display similar pattern in many diagnostics summaries, like plot of residuals and standardized estimates of random effects.

5 Illustration for Simulated and Trade Coffee Data

An example with simulated data, where two outliers were added, is reported in Fig. 3. The two outliers are the two central lowest extreme observations in the log-log transform (see the right panel of Fig. 3) and, for the sake of the illustration, they have been assigned to Germany (DE). Their effect is visible in the $\log V$ variable, but less visible, in this scale, for the $\log W$.

The structure is quite different from our example coffee data, but a similar pattern with some heteroschedasticity can be seen.

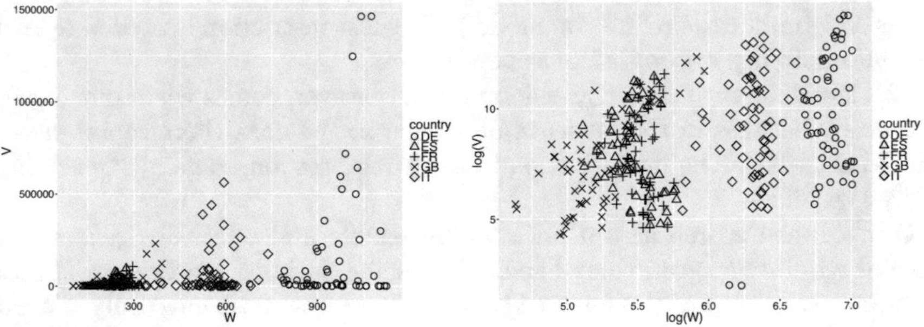

Fig. 3. Scatterplot of the simulated data with five groups (left panel). Simulated data have also two outliers. The right panel is the associated log-log transform, with the two outliers located in the lowest part of the plot

After running the forward search we monitor several graphical diagnostics. One of the most important plot in this framework is the monitor of estimates of standardized random effects, which is sketched in the left panel of Fig. 4. There is evidence, from findings in Fig. 4, that toward the last steps of the forward search, influential observations were included into the procedure, showing sharp peaks. Not all groups, in the simulated data, were affected in the same way by the inclusion of outliers. In general, the random components is absorbing influential observations, so outliers are likely to appear in the estimated random components. There are peaks also in the residuals, but since the random effects are fewer, the diagnostic plot is simpler to interpret.

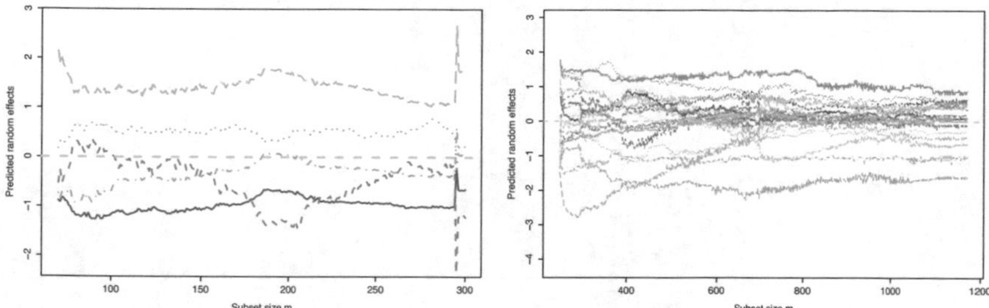

Fig. 4. Estimates of random effects on the simulated and contaminated data during the forward search (left panel). Peaks toward the end of procedure shows the steps when outliers are included. In the right panel estimates of random effects on the coffee trade data during the forward search. A quite smooth behaviour is visible, denoting lack of evidence of suspicious trades

For the coffee data, introduced and discussed in Sect. 2, running the forward search leads to results which do not show any influential observations in the data. A snapshot of such a summary is visible by exploring the estimated standardized random effects during the forward search. Since there are 18 destination countries, it is hard to understand whether or not outliers, or hidden structures, are present into the dataset. These are illustrated in the right panel of Fig. 4.

Robust procedures, often, tend to spot outliers even when they are not present (false signal). The forward search does not suffer from that unappealing feature. However, in the future, we will perform more simulations, with cleaned and contaminated data, for better understanding the false discovery rate of the forward search, and compare it with other robust techniques.

6 Final Remarks

We have introduced the forward search for LMM where correlation is induced by repeated measures. The benefits of the introduced robust and efficient technique are, essentially, two-folds: when outliers are present, they are properly highlighted and estimates are unaffected by their presence. When data are outlier-free or when no hidden structure is inside the data, the forward search does not flag observations as outliers (false signals).

Future research is needed as more general models should be considered, with the additional problem to have a robust procedure for proper model selection.

References

Atkinson, A., Riani, M.: Robust Diagnostic Regression Analysis. SSS, Springer, New York (2000). https://doi.org/10.1007/978-1-4612-1160-0

Riani, M., Atkinson, A.C., Cerioli, A.: Finding an unknown number of multivariate outliers. J. Roy. Stat. Soc. B **71**, 447–466 (2009)

K-Partitioning with Imprecise Probabilistic Edges

Tom Davot(✉), Sébastien Destercke, and David Savourey

Université de Technologie de Compiègne, CNRS, Heudiasyc (Heuristics and Diagnosis of Complex Systems), CS 60319, 60203 Compiègne Cedex, France
{tom.davot,sebastien.destercke,david.savourey}@hds.utc.fr

Abstract. Partitioning a set of elements into disjoint subsets is a common problem in unsupervised learning (clustering) as well as in networks (e.g., social, ecological) where one wants to find heterogeneous subgroups such that the elements within each subgroup are homogeneous. In this paper, we are concerned with the case where we imprecisely know the probability that two elements should belong to the same partition, and where we want to search the set of most probable partitions. We study the corresponding algorithmic problem on graphs, showing that it is difficult, and propose heuristic procedures that we test on data sets.

1 Introduction

Partitioning a set of elements into heterogeneous groups such that elements within each group are as homogeneous as possible is a common task. It is at the very core of unsupervised learning and clustering problems, as well as when one considers networks of different kinds (e.g., social, voting, ...).

A natural way to encode the relations existing between elements is through graphs, where the presence of an edge indicates that elements should be grouped together. However, the existence of such a link may be subject to various uncertainties. For instance, if one thinks of grouping persons (e.g., in parliament) voting in the same way, it may be that we rarely observe two persons voting at the same time, or that two persons do not always have the same behaviour (sometimes voting in the same way, sometimes not). Imprecise probabilities offer a rich and natural model to describe this uncertainty.

However, once one has modelled link uncertainty by imprecise probabilities, it remains to infer what are the more likely clusters. In this paper, we study the problem of extracting possibly optimal clusters from imprecise probabilistic graphs. We show that solving this problem exactly is NP-hard, and propose some heuristics. While there are existing approaches trying to extract partial clusters from imprecise probabilistic knowledge (Denoeux and Kanjanatarakul 2016; Masson et al. 2020), to our knowledge this is the first paper to view the problem as a robust decision-making one.

L. A. García-Escudero et al. (Eds.): SMPS 2022, AISC 1433, pp. 87–95, 2023.
https://doi.org/10.1007/978-3-031-15509-3_12

Organisation of the Paper[1]. Next section defines some notation and introduces the problem. Section 3 presents some theoretical results on the complexity of the problem. In Sect. 4 introduces an heuristic method. Finally, Sect. 5 is devoted to the presentation of some numerical experiments.

2 Notations and Problem Definition

Let G be a simple loopless graph. We denote $V(G)$ and $E(G)$ the set of vertices and edges of G, respectively (or simply V or E if no ambiguity occurs). The *complement graph* of G, denoted $\bar{G}$ is the graph defined by $V(\bar{G}) = V(G)$ and $E(\bar{G}) = \{uv \mid uv \notin E(G)\}$. A *cluster graph* is a disjoint union of complete graphs [2], called *cliques*. A k-cluster graph is a cluster graph that contains k non-empty connected components. Let G be a complete graph, a *k-partition* of G is a k-cluster subgraph G' of G such that $V(G) = V(G')$.

In this paper, an *imprecise probability* p is an interval $[\underline{p}, \overline{p}] \subseteq [0, 1]$ of probabilities, and p is called *precise* if $\overline{p} = \underline{p}$. An *imprecise probabilistic graph* $(G, \mathscr{P})$ is a graph with a function $\mathscr{P}$ that associates to each edge in the graph an imprecise probability. If uv is an edge, we denote $\underline{p}_{uv}$ and $\overline{p}_{uv}$ the lower and upper bounds of $\mathscr{P}(uv)$, respectively. The probability bounds of an absence of an edge can be deduced by duality (*i.e.* $[1 - \overline{p}_{uv}, 1 - \underline{p}_{uv}]$). p_{uv} being the marginal probability uv is an edge (and $1 - p_{uv}$ that it is not), we only assume $[\underline{p}_{uv}, \overline{p}_{uv}] \subseteq [0, 1]$. A *Probability Realisation* $R : E(G) \mapsto [0, 1]$ of $\mathscr{P}$ is a function that associates to each edge uv a probability within $[\underline{p}_{uv}, \overline{p}_{uv}]$. We denote $\mathscr{R}_{\mathscr{P}}$ the set of probability realisations of $\mathscr{P}$. Let G' be a subgraph of G and $R \in \mathscr{R}_{\mathscr{P}}$ be a probability realisation. The probability of G' under R, denoted $R(G')$ corresponds to

$$R(G') = \prod_{uv \in E(G')} R(uv) \prod_{uv \notin E(G')} 1 - R(uv).$$

Let G_1 and G_2 be two vertices k-partitions of G. We say that G_1 is *certainly more probable* than G_2, denoted by $G_1 \succ_p G_2$, if

$$\forall R \in \mathscr{R}_{\mathscr{P}}, R(G_1) - R(G_2) > 0.$$

Let ${}^{G_1}/_{G_2}$ denote the following value

$${}^{G_1}/_{G_2} = \prod_{uv \in E(G_1) \setminus E(G_2)} \frac{\underline{p}_{uv}}{1 - \overline{p}_{uv}} \prod_{uv \in E(G_2) \setminus E(G_1)} \frac{1 - \overline{p}_{uv}}{\underline{p}_{uv}}.$$

Notice that if an edge uv belongs (resp. does not belong) to both G_1 and G_2, then the factor $R(uv)$ (resp. $1 - R(uv)$) is present on both sides of the substraction. Thus, to verify if G_1 is certainly more probable than G_2, we only need to consider edges in $E(G) \setminus (E(G1) \cap E(G_2))$. Moreover, by duality $R(G_1) - R(G_2)$ is minimum if for every edge $uv \in E(G_1) \setminus E(G_2)$ (resp. $uv \in E(G_2) \setminus E(G_1)$), we have $R(uv) = \underline{p_{uv}}$ (resp. $R(uv) = \overline{p_{uv}}$). Hence, we have the following property.

[1] Due to space restriction, the proofs has been ommited. The full version is available here: https://hal.utc.fr/hal-03665950.

[2] A complete graph is a simple graph where every pair of vertices is connected.

Property 1. Given two k-partitions G_1 and G_2 of the imprecise probabilistic graph $(G, \mathscr{F})$, we have

$$G_1 \succ_p G_2 \Leftrightarrow {}^{G_1}/_{G_2} > 1.$$

Notice that the order given by $\succ_p$ is partial since we may have $G_1 \not\succ_p G_2$ and $G_2 \not\succ_p G_1$. Given a constant k, we are then interested in finding the most probable k-partitions of G. Let $P_k(V(G))$ be the set of k-partitions of G. We define $\mathscr{M}_{G,k} = \{G \in P_k(V(G)) \mid \nexists G' \in P_k(V(G)), G' \succ_p G\}$ the set of *non-dominated* k-partitions under $\succ_p$. In the following, we are interested in enumerating every partition of $\mathscr{M}_{G,k}$. Hence, we define the following problem.

Most probable k-partitions (k-MPP)
Input A complete graph G and an integer k.
Output Enumeration of $\mathscr{M}_{G,k}$.

3 Analysis

3.1 Computational Complexity

We first show that finding one element of $\mathscr{M}_{G,k}$ is NP-hard, even if $k = 2$ and $\mathscr{R}_{\mathscr{P}}$ has one element. To do so, we construct in the following way a reduction from the Max Cut (Karp 1972) problem (that aims at finding a spanning bipartite subgraph with a maximum number of edges in a graph H).

Construction 1. *Given H an instance of* Max Cut, *we construct an imprecise probabilistic graph $(G, \mathscr{P})$ such that:*

- *$V(G) = V(H)$,*
- *for each pair of vertices u and v, $\underline{p}_{uv} = \overline{p}_{uv} = 0.1$ (*red edges*) if $uv \in E(H)$ and, $\underline{p}_{uv} = \overline{p}_{uv} = 0.5$ (*blue edges*), otherwise.*

The proof idea is that a 2-partition is non-dominated if and only if it contains a minimum number of red edges, and thus its complement graph is a bipartite graph with a maximum number of edges. Hence, we can show the following.

Theorem 1. *Let $(G, \mathscr{P})$ be an imprecise probabilitstic graph. Computing any element of $\mathscr{M}_{G,k}$ is NP-hard, even if $k = 2$ and $\mathscr{P}$ is a singleton.*

3.2 Easy Cases

In this section we present three easy cases in which some element of k-MPP can be polynomially computed in the size of graph. These easy cases appear when one value appears in every probabilistic interval of $\mathscr{P}$. The first case is when 0.5 is contained in every probabilistic interval which implies that any k-partition is non-dominated.

Theorem 2. *Let $(G, \mathscr{P})$ be an imprecise probabilistic graph such that $\forall uv \in E(G), 0.5 \in \mathscr{P}(uv)$. We have $\mathscr{M}_{G,k} = P_k(V)$.*

The second case is when a value inferior to 0.5 is contained in every probabilistic interval. In that case, every k-partition that contains a minimum number of edges (i.e., is balanced) is non-dominated.

Theorem 3. *Let $x < 0.5$ and $(G, \mathscr{P})$ be an imprecise probabilistic graph such that $\forall uv \in E(G), x \in \mathscr{P}(uv)$. Let G' be a k-partition of G with connected components $\{V_1, \ldots, V_k\}$ of respective orders $n_1, \ldots, n_k$. If we have*

$$\forall i, j, |n_i - n_j| \leq 1$$

then, $G' \in \mathscr{M}_{G,k}$.

Finally, the last case is when a value greater to 0.5 is contained in every probabilistic interval. In that case, every k-partition that contains a maximum number of edges (i.e., is unbalanced) is non-dominated.

Theorem 4. *Let $x > 0.5$ and $(G, \mathscr{P})$ be an imprecise probabilistic graph such that $\forall uv \in E(G), x \in \mathscr{P}(uv)$. Let G' be a k-partition of G with connected components $\{V_1, \ldots, V_k\}$ such that $\forall i < j, |V_i| \leq |V_j|$. If we have*

1. *$|V_i| = 1, \forall i < k$, and*
2. *$|V_k| = |V(G)| - k + 1$.*

then, $G' \in \mathscr{M}_{G,k}$.

4 Heuristic

In this section, we describe some heuristic method used to approach $\mathscr{M}_{G,k}$ or to improve the computation time. This method relies on the use of a pattern and some associated reductions rules.

4.1 Pattern and Reduction Rules

Let $(G, \mathscr{F})$ be an imprecise probabilistic graph. A *pattern* X of G is a subset of edges. We say that a k-partition G' respects a pattern X if G' contains every edge of X (*i.e.* $X \subset E(G')$). We denote $P_k(V(G), X)$ the set of k-partitions that respects X. Let $\mathscr{M}_{G,k}(X) = \{G \in P_k(V(G), X) \mid \nexists G' \in P_k(V(G), X), G' \succ_p G\}$ the set of non-dominated k-partitions respecting X. We give some reduction rules that reduce the size of G without altering the computation of $\mathscr{M}_{G,k}(X)$.

Let uv be an edge of X and x be any vertex. Note that for any k-partition $G' \in P_k(V(G), X)$, u and v are contained in the same clique and G' contains either both xu and xv or none of them. Hence, u and v are acting like a single vertex and thus, we can contract uv into a single vertex and merge xu and xv together. Formally, the *contraction* of uv, denoted f_{uv}, is the application which given any graph G, constructs the graph H where:

- $V(H) = (V(G) \setminus \{uv\}) \cup \{w\}$,
- $E(H) = (E(G) \setminus \{xy \mid \forall xy \in E(G), xy \cap uv \neq \emptyset\}) \cup \{xw \mid \forall x \in V(G) \setminus \{uv\}\}$.

The contraction rule uses f_{uv} and adapts the imprecise probability set and the pattern so that the sets of non-dominated k-partitions respecting the pattern are equivalent in the original graph and the newly created graph.

Rule 1 (Contraction rule)
Let $(G, \mathscr{F})$ be an imprecise probabilistic graph and let $X = (C, A)$ be a pattern. Let uv be an edge of C. We reduce G to the following imprecise probabilistic graph $(H, \mathscr{G})$.

- $H = f_{uv}(G)$,
- *for any edge xy of $E(G)$ such that $xy \cap uv = \emptyset$, we set $\mathscr{G}(xy) = \mathscr{F}(xy)$, and*
- *for any vertex $x \notin uv$ of $V(G)$, we set*

$$\mathscr{G}(xw) = [\frac{\underline{p}_{xu} \cdot \underline{p}_{xv}}{\underline{p}_{xu} \cdot \underline{p}_{xv} + (1 - \underline{p}_{xu}) \cdot (1 - \underline{p}_{xv})}, \frac{\bar{p}_{xu} \cdot \bar{p}_{xv}}{\bar{p}_{xu} \cdot \bar{p}_{xv} + (1 - \bar{p}_{xu}) \cdot (1 - \bar{p}_{xv})}].$$

This corresponds to compute bounds over the conditional probability $P(xu \wedge xv | (xu \wedge xv) \vee (\neg xu \wedge \neg xv))$. We construct a new pattern X' for H as follows.

- *Let $T_X = \{xy \mid xy \in X, xy \cap uv \neq \emptyset\}$ and $R_X = \{xw \mid \exists xy \in X, y \in uv\}$. We set $X' = (X \setminus T_X) \cup R_X$.*

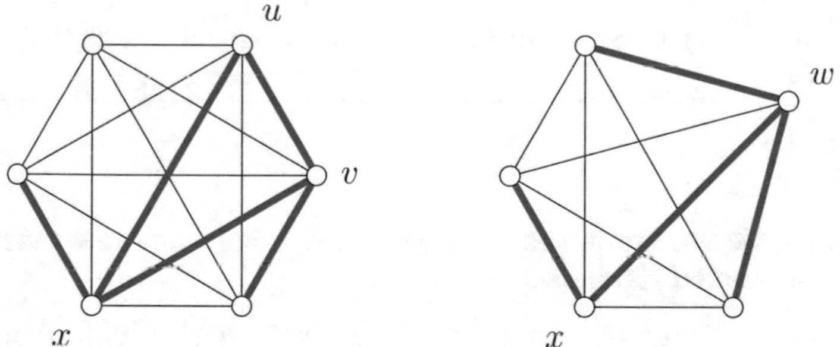

Fig. 1. Example of an application of Rule 1 on the edge uv in an imprecise probabilistic graph with a motif X (blue edges)

An example of the application of Rule 1 is depicted in Fig. 1. The two next properties show that Rule 1 is safe, that is, the computation of $\mathscr{M}_{G,k}(X)$ is equivalent to the computation of $\mathscr{M}_{H,k}(Y)$.

Property 2. f_{uv} is a bijection from $P_k(V(G), X)$ to $P_k(V(H), Y)$.

Property 3. Given two k-partitions G_1 and G_2 in $P_k(V(G), X)$ we have

$$G_1 \succ_p G_2 \Leftrightarrow f_{uv}(G_1) \succ_g f_{uv}(G_2).$$

4.2 Algorithm Description

Our method to approach $\mathscr{M}_{G,k}$ is described by Algorithm 1. The basic idea is to reduce the size of the graph by computing a pattern X and applying the reduction rules described above. Once the size of the reduced graph is small enough, it seems possible to enumerate every k-partitions and thus compute $\mathscr{M}_{G,k}(X)$ in a relatively small amount of time. The difficulty of this method is to find a pattern X such that $|\mathscr{M}_{G,k}(X)\Delta\mathscr{M}_{G,k}|$ is minimum. In the next section, we take the pattern $X_i = \{uv \mid \underline{p}_{uv} \geq 0.9\}$.

5 Numerical Experiments

In the following, we provide some tests for the special case where $k = 2$.

Algorithm 1 Heuristic method

Data: An imprecise probabilistic graph $(G, \mathscr{F})$ and two integers t and k.
Result: A set of k-partitions for G.

```
for all i ≤ t do
    Compute a feasible pattern X_i for (G, F);
    Apply the contraction rule until X_i is empty;
end for
X ← ⋃_{i≤k} X_i;
Enumerate every possible k-partitions of G to compute M ← M_{G,k}(X);
return M;
```

5.1 Dataset

For our tests, we use two types of dataset, one using some real data and one using some generated instances.

Real Dataset: We use the dataset used by Arinik et al. (2017) that contains vote information for French and Italian members of the european parliament. Each member is represented by a vertex in the graph. To generate the imprecise probability sets, we use the following formula. Let u and v be two members. Let k be the number of sessions in which both u and v participated and let t be the number of sessions in which u and v voted the same. We set $\mathscr{P}(uv) = [\frac{t}{k+s}, \frac{t+s}{k+s}]$ where s is a parameter settling the speed at which the intervals converge to a precise value. In the following, we set $s = 5$. Since the generated graph contains 870 vertices, it is not possible to compute exactly the set of non-dominated 2-partitions, we create twenty subinstances by randomly drawing 15 vertices of the original graph.

Randomly Generated Instances: We proceed as follows. First, we define k groups of vertices with a given size and an application $f : \{1, \ldots, k\}^2 \mapsto$

$[0,1]^4$ which associate to each pair of integers $\{i,j\}$ a tuple $\{\underline{m}_{i,j}, \bar{m}_{i,j}, \underline{\ell}_{i,j}, \bar{\ell}_{i,j}\}$. Then for each pair of vertices u and v such that u belongs to the group i and v belongs to the group j (i can be equal to j), we draw two real numbers $m \in [\underline{m}_{i,j}, \bar{m}_i, j]$ and $\ell \in [\underline{\ell}_{i,j}, \bar{\ell}_{i,j}]$. Finally, we introduce the edge uv with the imprecise probabilistic interval $\mathscr{F}(uv) = [max(0, m - \ell), min(1, m + \ell)]$.

We test two different groups configurations.

- *Configuration A.* The graph contains two groups of 7 vertices, and we set $f(1,1) = f(2,2) = \{0.9, 0.95, 0, 0.3\}$ and $f(1,2) = \{0.1, 0.2, 0, 0.3\}$.
- *Configuration B.* The graph contains three groups of 6 vertices, and we set $f(1,1) = f(2,2) = f(3,3) = \{0.9, 0.95, 0, 0.1\}$ and $f(1,2) = \{0.1, 0.2, 0, 0.1\}$. For the values of $f(1,3)$ and $f(2,3)$, we test three different variations.
  - B1: $f(1,3) = f(2,3) = \{0.45, 0.55, 0, 0.1\}$,
  - B2: $f(1,3) = f(2,3) = \{0.45, 0.55, 0.2, 0.0.35\}$,
  - B3: $f(1,3) = f(2,3) = \{0.45, 0.55, 0.3, 0.7\}$.

For each configuration and each variation, we generate twenty instances.

5.2 Results

The tests were run on a personal laptop with 16 GB of RAM and with an Intel Core 7 processor 2.5 GHz. Results are displayed in Table 1. In our tests, we compare two exacts algorithms with our heuristic with two different values of t. The "brute" version enumerate every 2-partitions to construct the solution set S. For each enumerated 2-partition G', if G' is not dominated by another 2-partition of S, then G' is added to S and every 2-partition of S that is dominated by G' is removed from S. The "init" version does the same thing but S is initialized by a set of 2-partitions. To initialize S, we take the value returned by Algorithm 1 with $t = 1$. The idea behind the "init" version is that initializing S with a set of possibly non-dominated solutions can reduce computational time (similar ideas

Table 1. Results on real and generated data for the 2-MPP problem. The columns "Brute" and "Init" correspond to exact algorithms. The columns "t = 1" and "t = 2" correspond to Algorithm 1 with two different values of t. "Sol" is the size of the enumerated 2-partitions, "n" is the size of the graph after the application of the reduction rules. "ER1" is the percentage of enumerated 2-partitions that do not belong to $\mathscr{M}_{G,k}$. "ER2" is the percentage of 2-partitions that belong to $\mathscr{M}_{G,k}$ and that are not enumerated. Time is given in seconds

Config	Brute		Init	t = 1					t = 2				
	Time	Sol	Time	Time	n	Sol	ER1	ER2	Time	n	Sol	ER1	ER2
A	15.5	1	0.36	10^{-3}	9.55	1	0%	0%	10^{-5}	4.3	2	0%	0%
B1	1.8	2	3	10^{-3}	6.25	2	0%	0%	10^{-5}	3.05	2	0%	0%
B2	296	2.7	3.05	10^{-3}	6.4	2	0%	17.2%	10^{-5}	3	2	0%	17.2%
B3	5017	71.35	34.9	10^{-4}	6.55	2.7	0%	97%	10^{-5}	3	2	0%	97%
Real	29	827.35	28.5	10^{-3}	6.9	9.15	30%	99.25%	10^{-5}	3.7	2.15	37.5%	99.5%

can be found in Nakharutai et al. 2019). For every configuration, the results correspond to the average of the twenty instances.

We can see for the generated instances every 2-partition enumerated by the heuristic belongs to the exact solution. However, the number of 2-partitions returned by the heuristic can be relatively small compared to the size of $\mathscr{M}_{G,k}$. For example, for the B3 configuration, 97% of $\mathscr{M}_{G,k}$ is not enumerated by the heuristic. Nevertheless, the results of 1 help to drastically reduce the computation time of the exact algorithm. For instances from real data, the results are more mixed: almost all of $\mathscr{M}_{G,k}$ is not enumerated by the heuristic and at least 30% of the 2-partitions returned by the heuristic does not belong to $\mathscr{M}_{G,k}$. Moreover, since the results of the heuristic are not good enough, the computation time is not significantly reduced for the "init" version. We can explain this bad performance by the fact that drawing randomly 15 vertices in a real instance can lead to a subinstance that is not really representative since the 15 vertices can belong to the same group.

6 Conclusion

In this paper, we addressed the problem of the most probable k-partition with imprecise probabilistic edges. After some theoretical results, we developed a heuristic to tackle this problem. We show that this heuristic can have good results in practice but becomes less performant if the probability intervals are to large. A natural perspective of our work can be to find another way to compute some k-partitions to make the initialisation for the exact version, since we show that it can significantly reduce the computation time. It can be interesting to find another method since our heuristic can not perform well for some instances.

Acknowledgements. This work was funded in the framework of the Labex MS2T. It was supported by the French Government, through the program "Investments for the future" managed by the ANR (Reference ANR-11-IDEX-0004-02).

References

Arinik, N., Figueiredo, R., Labatut, V.: Signed graph analysis for the interpretation of voting behavior. In: International Conference on Knowledge Technologies and Data-driven Business (i-KNOW) (2017). https://doi.org/10.48550/arXiv.1712.10157

Denœux, T., Kanjanatarakul, O.: Evidential clustering: a review. In: Huynh, V.-N., Inuiguchi, M., Le, B., Le, B.N., Denoeux, T. (eds.) IUKM 2016. LNCS (LNAI), vol. 9978, pp. 24–35. Springer, Cham (2016). https://doi.org/10.1007/978-3-319-49046-5_3

Karp, R.M.: Reducibility among combinatorial problems. In: Miller, R.E., Thatcher, J.W., Bohlinger, J.D. (eds.) Complexity of Computer Computations. The IBM Research Symposia Series, pp. 85–103. Springer, Boston (1972). https://doi.org/10.1007/978-1-4684-2001-2_9

Masson, M.H., Quost, B., Destercke, S.: Cautious relational clustering: a thresholding approach. Exp. Syst. Appl. **139**, 112837 (2020)

Nakharutai, N., Troffaes, M.C., Caiado, C.C.: Improving and benchmarking of algorithms for decision making with lower previsions. Int. J. Approx. Reason. **113**, 91–105 (2019)

Decision-Making with E-Admissibility Given a Finite Assessment of Choices

Arne Decadt(✉), Alexander Erreygers, Jasper De Bock, and Gert de Cooman

Foundations Lab, Ghent University, Ghent, Belgium
{arne.decadt,alexander.erreygers,gert.decooman}@ugent.be,
Jasper.DeBock@UGent.be

Abstract. Given information about which options a decision-maker definitely rejects from given finite sets of options, we study the implications for decision-making with E-admissibility. This means that from any finite set of options, we reject those options that no probability mass function compatible with the given information gives the highest expected utility. We use the mathematical framework of choice functions to specify choices and rejections, and specify the available information in the form of conditions on such functions. We characterise the most conservative extension of the given information to a choice function that makes choices based on E-admissibility, and provide an algorithm that computes this extension by solving linear feasibility problems.

1 Introduction

A decision-maker's uncertainty is typically modelled by a probability measure, and it is often argued that her rational decisions maximise expected utility with respect to this probability measure. However, she may not always have sufficient knowledge to come up with a unique and completely specified probability measure. It is then often assumed, as a work-around, that there is some set of probability measures that describes her uncertainty. In this setting, *E-admissibility* is among the more popular criteria for making choices, as indicated by Troffaes (2007).

In this paper, we study and propose an algorithm for decision-making based on this criterion, starting from a finite uncertainty assessment. As E-admissibility is popular, we are not the first to try and deal with this. Utkin and Augustin (2005) and Kikuti et al. (2005) have gone before us, but their assessments essentially only deal with pairwise comparison of options, while we can handle comparisons between sets of options. Decadt et al. (2020) have also studied more general assessments, but for other decision criteria than E-admissibility. In order to achieve this generality, we will use choice functions as tools to model the decision-making process, because they lead to a very general framework, as argued elsewhere by, for instance, Seidenfeld et al. (2010), De Bock and De Cooman (2019) and De Bock (2020).

L. A. García-Escudero et al. (Eds.): SMPS 2022, AISC 1433, pp. 96–103, 2023.
https://doi.org/10.1007/978-3-031-15509-3_13

2 Setting and Choice Functions

A choice function is a function that, for any given set of options, selects some subset of them. The set $\mathscr{V}$ collects all *options* and $\mathscr{Q}$ is the set of all non-empty finite subsets of $\mathscr{V}$. Formally, a *choice function* C is then a map from $\mathscr{Q}$ to itself such that $C(A) \subseteq A$ for all $A \in \mathscr{Q}$. If $C(A)$ is a singleton consisting of a single option u, this means that u is chosen from A. If $C(A)$ has more than one element, however, we don't take this to mean that all the options in $C(A)$ are chosen, but rather that the options in $A \setminus C(A)$ are rejected and that the model does not contain sufficient information to warrant making a choice between the remaining options in $C(A)$. Depending on the desired behaviour, various axioms can be imposed, leading to different types of choice functions; see for example Seidenfeld et al. (2010), De Bock and De Cooman (2019) and De Bock (2020). In this contribution we consider choice functions under E-admissibility, as introduced in Sect. 3.

We furthermore assume that we have an uncertain experiment with n possible outcomes, and we order the set of all outcomes $\mathscr{X}$ as $\{x_1, \ldots, x_n\}$. We interpret an option u as a function that maps each outcome x in $\mathscr{X}$ to the real-valued utility $u(x)$ that we get when the outcome of the uncertain experiment turns out to be x. So we take the set of all options $\mathscr{V}$ to be the real vector space of all real-valued maps on $\mathscr{X}$.

3 E-Admissibility

A decision-maker's uncertainty about an experiment is typically modelled by means of a probability mass function $p\colon \mathscr{X} \to [0, 1]$, which represents the probability of each outcome in $\mathscr{X}$; we will use Σ to denote the set of all such probability mass functions on $\mathscr{X}$. The standard way—see for example (Savage 1972, Chap. 5)—to choose between options u proceeds by maximising expected utility with respect to p, where the expected utility of an option $u \in \mathscr{V}$ is given by $\mathrm{E}_p(u) := \sum_{x \in \mathscr{X}} u(x)p(x)$.

For every probability mass function $p \in \Sigma$, the resulting choice function C_p that maximises expected utility is defined by

$$C_p(A) := \{u \in A\colon (\forall a \in A)\mathrm{E}_p(u) \geq \mathrm{E}_p(a)\} \text{ for all } A \in \mathscr{Q}. \tag{1}$$

It is, however, not always possible to pin down exact probabilities for the outcomes (Walley 1991, Chap. 1). Yet, the decision-maker might have some knowledge about these probabilities, for example in terms of bounds on the probabilities of some events. Such knowledge gives rise to a set of probability mass functions $\mathscr{P} \subseteq \Sigma$, called a *credal set* (Levi 1978, Sect. 1.6.2). In this context, there need no longer be a unique expected utility and so the decision-maker cannot simply maximise it. Several other decision criteria can then be used instead; Troffaes (2007) gives an overview. One criterion that is often favoured is *E-admissibility*: choose those options that maximise expected utility with respect to at least one of the probability mass functions p in $\mathscr{P}$ (Levi 1978).[1] If $\mathscr{P}$ is non-empty, the corresponding choice function $C^{\mathrm{E}}_{\mathscr{P}}$ is defined by

$$C^{\mathrm{E}}_{\mathscr{P}}(A) := \bigcup_{p\in\mathscr{P}} C_p(A) \text{ for all } A \in \mathcal{Q}. \tag{2}$$

It will prove useful to extend this definition to the case that $\mathscr{P} = \emptyset$. Equation (2) then yields that $C^{\mathrm{E}}_{\emptyset}(A) = \emptyset$ for all $A \in \mathcal{Q}$, so $C^{\mathrm{E}}_{\emptyset}$ is no longer a choice function. In either case, it follows immediately from Eqs. (1) and (2) that

$$C^{\mathrm{E}}_{\mathscr{P}}(A) = \{u \in A\colon (\exists p \in \mathscr{P})(\forall a \in A)\mathrm{E}_p(u) \geq \mathrm{E}_p(a)\} \text{ for all } A \in \mathcal{Q}. \tag{3}$$

The behaviour of choice functions under E-admissibility was first studied for horse lotteries by Seidenfeld et al. (2010), characterised in a very general context by De Cooman (2021), and captured in axioms by De Bock (2020).

4 Assessments and Extensions

We assume that there is some choice function C that represents the decision-maker's preferences, but we may not fully know this function. Our partial information about C comes in the form of preferences regarding some—so not necessarily all—option sets. More exactly, for some option sets $A \in \mathcal{Q}$, we know that the decision-maker rejects all options in $W \subseteq A$, meaning that $C(A) \subseteq A \setminus W$; this can be also be stated as $C(V \cup W) \subseteq V$, with $V := A \setminus W$. We will represent such information by an *assessment*: a set $\mathscr{A} \subseteq \mathcal{Q}^2$ of pairs (V, W) of disjoint option sets with the interpretation that, for all $(V, W) \in \mathscr{A}$, the options in W are definitely rejected from $V \cup W$.

Given such an assessment, it is natural to ask whether there is some choice function $C^{\mathrm{E}}_{\mathscr{P}}$ under E-admissibility that agrees with it, in the sense that $C^{\mathrm{E}}_{\mathscr{P}}(V \cup W) \subseteq V$ for all $(V, W) \in \mathscr{A}$. Whenever this is the case, we call the assessment $\mathscr{A}$ *consistent* with E-admissibility. It follows from Eq. (2) that $C^{\mathrm{E}}_{\mathscr{P}}$ agrees with the assessment $\mathscr{A}$ if and only if

$$\mathscr{P} \subseteq \mathscr{P}(\mathscr{A}) := \{p \in \Sigma\colon (\forall (V, W) \in \mathscr{A}) C_p(V \cup W) \subseteq V\}.$$

Hence, $\mathscr{A}$ is consistent if and only if $\mathscr{P}(\mathscr{A}) \neq \emptyset$. To check if $\mathscr{A}$ is consistent, the following alternative characterisation will also be useful: for any $A \in \mathcal{Q}$,

$$\mathscr{P}(\mathscr{A}) \neq \emptyset \Leftrightarrow C^{\mathrm{E}}_{\mathscr{P}(\mathscr{A})}(A) \neq \emptyset. \tag{4}$$

If an assessment $\mathscr{A}$ is consistent and there is more than one choice function that agrees with it, the question remains which one we should use. A careful decision-maker would only want to reject options if this is implied by the

[1] Levi's original definition considered credal sets that are convex, whereas we do not require this. In fact one of the strengths of our approach is that an assessment can lead to non-convex credal sets; see the example in Sect. 5 further on.

assessment. So she wants a most conservative agreeing choice function under E-admissibility, one that rejects the fewest number of options. Since larger credal sets lead to more conservative choice functions, this most conservative agreeing choice function under E-admissibility clearly exists, and is equal to $C^{\mathrm{E}}_{\mathscr{P}(\mathscr{A})}$. For this reason, we call $C^{\mathrm{E}}_{\mathscr{P}(\mathscr{A})}$ the *E-admissible extension* of the assessment $\mathscr{A}$.

So we conclude that checking the consistency of an assessment $\mathscr{A}$, as well as finding the E-admissible extension of a consistent assessment $\mathscr{A}$, amounts to evaluating $C^{\mathrm{E}}_{\mathscr{P}(\mathscr{A})}$. In the following sections we provide a method for doing this, which makes use of the following more practical expression for $\mathscr{P}(\mathscr{A})$.

Proposition 1. *Consider an assessment $\mathscr{A}$. Then*

$$\mathscr{P}(\mathscr{A}) = \{p \in \Sigma\colon (\forall (V,W) \in \mathscr{A})(\forall w \in W)(\exists v \in V)\mathrm{E}_p(v) > \mathrm{E}_p(w)\}\,.$$

5 A Characterisation of the E-Admissible Extension

Having defined the E-admissible extension $C^{\mathrm{E}}_{\mathscr{P}(\mathscr{A})}$ of an assessment $\mathscr{A}$, it is only natural to wonder whether we can easily compute it. We now turn to a method for doing so, albeit only for finite assessments. As a first step, we derive a convenient characterisation of $C^{\mathrm{E}}_{\mathscr{P}(\mathscr{A})}$. For any positive integer m, it uses the notations $[1{:}m] := \{1,\dots,m\}$ and $d_{1:m} := (d_1,\dots,d_m)$.

Theorem 1. *Consider an option set A, an option $u \in A$ and a non-empty, finite assessment $\mathscr{A}$. Enumerate the set $\{v - wv \in V\colon (V,W) \in \mathscr{A}, w \in W\}$ as $\{D_1,\dots,D_m\}$ and the set $\{u - a\colon a \in A \setminus \{u\}\}$ as $\{u_1,\dots,u_\ell\}$. Then $\mathscr{P}(\mathscr{A}) = \bigcup_{d_{1:m} \in \times_{j=1}^m D_j} \mathscr{P}(d_{1:m})$, where, for each $d_{1:m} \in \times_{j=1}^m D_j$, we let*

$$\mathscr{P}(d_{1:m}) := \{p \in \Sigma\colon (\forall j \in [1{:}m])\mathrm{E}_p(d_j) > 0\}\,. \tag{5}$$

Furthermore, $u \in C^{\mathrm{E}}_{\mathscr{P}(\mathscr{A})}(A)$ if and only if there is some $p \in \mathscr{P}(\mathscr{A})$—or equivalently some $d_{1:m} \in \times_{j=1}^m D_j$ and $p \in \mathscr{P}(d_{1:m})$—such that $\mathrm{E}_p(u_i) \geq 0$ for all $i \in [1{:}\ell]$.

Let us illustrate the use of Theorem 1 in determining, for a given option set, the resulting choices under the conservative E-admissible extension of a given assessment. Let $\mathscr{X} := \{1,2,3\}$. In order to allow for a graphical representation, we identify options and probability mass functions with vectors in $\mathbb{R}^3$, where for any $x \in \mathscr{X}$, the x-th component corresponds to the value of the option or probability mass function in x; so for example the option $w_1 := (1,-3,1)$ corresponds to the option that maps 1 to 1, 2 to -3 and 3 to 1. We will choose from the option set $A := \{w_1,w_2,w_3\}$, where we also let $w_2 := (1,1,-2)$ and $w_3 := (0,0,0)$.

For the assessment, we will consider $v_1 := (-1,2,-2)$, $v_2 := (-2,2,-1)$, $v_3 := (0,3,-11)$, $v_4 := (0,-7,-1)$, $v_5 := (2,5,-9)$ and $v_6 := (0,-2,-1)$. Suppose that we are given the information that v_2, v_3 and v_4 are rejected from $\{v_1,v_2,v_3,v_4\}$ and that v_1 is rejected from $\{v_1,v_5,v_6\}$. This corresponds to the assessment $\mathscr{A} = \{(\{v_1\},\{v_2,v_3,v_4\}),(\{v_5,v_6\},\{v_1\})\}$.

Now we will check for every option in A whether it is in $C^{\mathrm{E}}_{\mathscr{P}(\mathscr{A})}(A)$, by applying Theorem 1. For the sake of efficiency, we note that for all options, the assessment $\mathscr{A}$ is the same, so they all have

$$\begin{aligned}\{D_1,\dots,D_4\} \\ &= \{\{v_1-v_2\},\{v_1-v_3\},\{v_1-v_4\},\{v_5-v_1,v_6-v_1\}\} \\ &= \{\{(1,0,-1)\},\{(-1,-1,9)\},\{(-1,9,-1)\},\{(3,3,-7),(1,-4,1)\}\}\,.\end{aligned}$$

In Fig. 1, we have drawn the credal set $\mathscr{P}(\mathscr{A})$ in blue in a ternary plot, using the characterisation in Theorem 1.

For w_1, the probability mass function $p_1 := (^{12}/_{20}, ^3/_{20}, ^5/_{20})$ is consistent with the assessment, and we have $\mathrm{E}_{p_1}(u_1) = {}^3/_{20} \geq 0$ and $\mathrm{E}_{p_1}(u_2) = {}^2/_5 \geq 0$, with $u_1 := w_1 - w_2 = (0,-4,3)$ and $u_2 := w_1 - w_3 = (1,-3,1)$. Therefore, it follows from Theorem 1 that w_1 is not rejected from A by $C^{\mathrm{E}}_{\mathscr{P}(\mathscr{A})}$. That w_2 is not rejected either can be inferred similarly, for example using $p_2 := (^3/_5, ^1/_5, ^1/_5)$. For w_3, we have $u_1 := w_3 - w_1 = (-1,3,-1)$ and $u_2 := w_3 - w_2 = (-1,-1,2)$. The set of probability mass functions for which $\mathrm{E}_p(u_1) \geq 0$ and $\mathrm{E}_p(u_2) \geq 0$ corresponds to the green region in Fig. 1, which has no overlap with the blue region. Therefore, w_3 is rejected from A by $C^{\mathrm{E}}_{\mathscr{P}(\mathscr{A})}$. So we conclude that $C^{\mathrm{E}}_{\mathscr{P}(\mathscr{A})}(A) = \{w_1, w_3\}$.

6 An Algorithmic Approach

For larger problems, when the graphical approach in the example above is no longer feasible, we can translate Theorem 1 into an algorithm. A first way is to

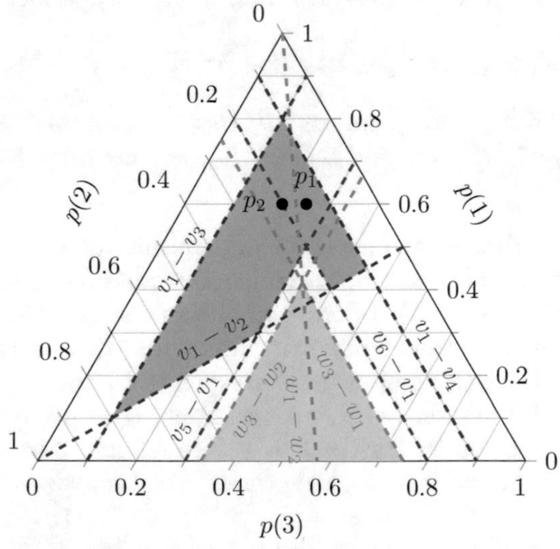

Fig. 1. Ternary plot where the credal set $\mathscr{P}(\mathscr{A})$ consists of those probability mass functions $p\colon \{1,2,3\} \to [0,1]$ that correspond to the blue region. A line labelled with an option v means that $\mathrm{E}_p(v) = 0$ for all p on the line. The green region corresponds to the probability mass functions p for which $\mathrm{E}_p(w_3 - w_1) \geq 0$ and $\mathrm{E}_p(w_3 - w_2) \geq 0$

directly search, for each $d_{1:m}$, for a probability mass function p in $\mathscr{P}(d_{1:m})$ such that $u \in C_p(A)$. To this end, we consider the function $\text{Primal}\colon \mathscr{V}^\ell \times \mathscr{V}^m \to \{\text{True}, \text{False}\}$ that returns True if the following feasibility problem has a solution for input $(u_{1:\ell}, d_{1:m}) \in \mathscr{V}^\ell \times \mathscr{V}^m$ and False otherwise, where λ_k can be seen as a scaled version of $p(x_k)$:

$$\begin{aligned} \text{find}\quad & \lambda_1, \ldots, \lambda_n \in \mathbb{R}, & \\ \text{subject to}\quad & \textstyle\sum_{k=1}^n \lambda_k u_i(x_k) \geq 0 & \text{for all } i \in [1\colon\ell], \\ & \textstyle\sum_{k=1}^n \lambda_k d_j(x_k) \geq 1 & \text{for all } j \in [1\colon m], \\ & \lambda_k \geq 0 & \text{for all } k \in [1\colon n]. \end{aligned}$$

For another way to translate the condition in Theorem 1 into an algorithm, we use duality. That is, we consider the function $\text{Dual}\colon \mathscr{V}^\ell \times \mathscr{V}^m \to \{\text{True}, \text{False}\}$ that returns True if the following feasibility problem has a solution for input $(u_{1:\ell}, d_{1:m}) \in \mathscr{V}^\ell \times \mathscr{V}^m$ and False otherwise:

$$\begin{aligned} \text{find}\quad & \lambda_1, \ldots, \lambda_{\ell+m} \in \mathbb{R}, & \\ \text{subject to}\quad & \textstyle\sum_{i=1}^\ell \lambda_i u_i(x_k) + \sum_{j=1}^m \lambda_{\ell+j} d_j(x_k) \leq 0 & \text{for all } k \in [1\colon n], \\ & \textstyle\sum_{j=\ell+1}^{\ell+m} \lambda_j \geq 1, & \\ & \lambda_i \geq 0 & \text{for all } i \in [1\colon \ell+m]. \end{aligned}$$

In practice, either of these feasibility problems can be solved by linear programming.[2] Our next result relates these feasibility problems to the conditions in Theorem 1.

Theorem 2. *Consider option sequences $u_{1:\ell} \in \mathscr{V}^\ell$ and $d_{1:m} \in \mathscr{V}^m$ and let $\mathscr{P}(d_{1:m})$ be as defined in Eq. (5). Then the following statements are equivalent:*

(i) There is some $p \in \mathscr{P}(d_{1:m})$ such that $\mathrm{E}_p(u_i) \geq 0$ for all $i \in [1\colon\ell]$.
(ii) $\text{Primal}(u_{1:\ell}, d_{1:m}) = \text{True}$.
(iii) $\text{Dual}(u_{1:\ell}, d_{1:m}) = \text{False}$.

Theorems 1 and 2 guarantee that we can determine $C^{\mathrm{E}}_{\mathscr{P}(\mathscr{A})}(A)$ for any set $A \in \mathscr{Q}$ using Algorithm 1, by checking for each option $u \in A$ whether $u \in C^{\mathrm{E}}_{\mathscr{P}(\mathscr{A})}(A)$. For any single u, this amounts to solving a linear feasibility program for each $d_{1:m}$, using Primal or Dual, as preferred. Interestingly, consistency is equivalent to $C^{\mathrm{E}}_{\mathscr{P}(\mathscr{A})}(A) \neq \emptyset$, by Eq. (4). In practice, consistency can also be easily verified beforehand, by checking if $0 \in C^{\mathrm{E}}_{\mathscr{P}(\mathscr{A})}(\{0\})$, where ‘0’ is the constant option that is 0 everywhere.

[2] It can for instance be considered as a linear programming problem, by adding the trivial objective function that is zero everywhere. Feeding this into a linear programming software package, the software will announce whether the problem is feasible. For a deeper understanding of how software solves such feasibility problems, we refer to the explanation of initial feasible solutions in Matoušek and Gärtner (2006, Sect. 5.6).

Algorithm 1. Check for an option set $A \in \mathscr{Q}$ and a finite assessment $\mathscr{A}$ if an option $u \in A$ is in $C^{\mathrm{E}}_{\mathscr{P}(\mathscr{A})}(A)$.

Precondition: Let $\{D_1, \dots, D_m\}$ and $\{u_1, \dots, u_\ell\}$ be as in Theorem 1.

```
for all d_{1:m} ∈ ×_{j=1}^m D_j do
    if PRIMAL(u_{1:ℓ}, d_{1:m}) then                    ▷ Or ¬DUAL(u_{1:ℓ}, d_{1:m}).
        return True                 ▷ For one of the d_{1:m} the condition is fulfilled.
return False                ▷ When all elements of ×_{j=1}^m D_j have been checked.
```

If the assessment $\mathscr{A}$ consists solely of pairs (V, W) where V is a singleton, then the corresponding set $\times_{j=1}^{m} D_j$ is a singleton, and the for-loop in Algorithm 1 vanishes. Our algorithm can therefore be seen as repeatedly solving problems that have assessments of that form.

7 Conclusion

We have shown how to make choices using the E-admissibility criterion, starting from a finite assessment, using choice functions. Our main conclusion is that calculating the most conservative E-admissible extension of this assessment reduces to checking linear feasibility multiple times. Our setup is similar to the one previously studied by Utkin and Augustin (2005) and Kikuti et al. (2005), the essential difference being that they have pairwise comparisons in the form of non-strict inequalities on expected utilities, whereas our assessments consider comparisons between sets of options, which leads to strict inequalities and allows for non-convex credal sets. Future work could look into also implementing assessments that incorporate non-strict inequalities. One way to do so would be through infinite assessments, so it might pay to look at which types of infinite assessments can still be handled finitely.

References

De Bock, J.: Archimedean choice functions. In: Lesot, M.J., et al. (eds.) IPMU 2020. CCIS, vol. 1238, pp. 195–209. Springer, Cham (2020). https://doi.org/10.1007/978-3-030-50143-3_15

De Bock, J., De Cooman, G.: Interpreting, axiomatising and representing coherent choice functions in terms of desirability. In: Proceedings of the 11th International Symposium on Imprecise Probabilities: Theories and Applications, ISIPTA 2019, vol. 103, pp. 125–134. Proceedings of Machine Learning Research (2019)

De Cooman, G.: Coherent and Archimedean choice in general Banach spaces. Int. J. Approx. Reason. **140**, 255–281 (2021)

Decadt, A., De Bock, J., De Cooman, G.: Inference with choice functions made practical. In: Davis, J., Tabia, K. (eds.) SUM 2020. LNCS (LNAI), vol. 12322, pp. 113–127. Springer, Cham (2020). https://doi.org/10.1007/978-3-030-58449-8_8

Kikuti, D., Cozman, F.G., de Campos, C.P.: Partially ordered preferences in decision trees: Computing strategies with imprecision in probabilities. In: Proceedings of the IJCAI Workshop about Advances on Preference Handling, pp. 118–123 (2005)

Levi, I.: On indeterminate probabilities. In: Hooker, C.A., Leach, J.J., McClennen, E.F. (eds.) Foundations and Applications of Decision Theory. The University of Western Ontario Series in Philosophy of Science, vol. 13a, pp 233–261. Springer, Dordrecht (1978). https://doi.org/10.1007/978-94-009-9789-9_9

Matoušek, J., Gärtner, B.: Understanding and Using Linear Programming. Universitext (UTX). Springer, Heidelberg (2006). https://doi.org/10.1007/978-3-540-30717-4

Savage, L.J.: The Foundations of Statistics. Dover Publications Inc., New York (1972)

Seidenfeld, T., Schervish, M.J., Kadane, J.B.: Coherent choice functions under uncertainty. Synthese **172**, 157–176 (2010)

Troffaes, M.C.: Decision making under uncertainty using imprecise probabilities. Int. J. Approx. Reason. **45**(1), 17–29 (2007)

Utkin, L., Augustin, T.: Powerful algorithms for decision making under partial prior information and general ambiguity attitudes. In: Proceedings of the 3th International Symposium on Imprecise Probability: Theories and Applications, ISIPTA 2005, pp 349–358 ((2005))

Walley, P.: Statistical Reasoning with Imprecise Probabilities. Chapman and Hall, London (1991)

Copula-Based Divergence Measures for Dependence Between Random Vectors

Steven De Keyser and Irène Gijbels(✉)

Department of Mathematics, KU Leuven, Celestijnenlaan 200B, 3001 Leuven, Belgium
{steven.dekeyser,irene.gijbels}@kuleuven.be

Abstract. To measure the dependence between a finite number of random vectors, we propose to use the family of Φ-divergences. These measures quantify a statistical distance between arbitrary distributions and possess attractive properties that comply with typical axioms of overall dependence measures. We provide an extended axiomatic framework for the general setting of k random vectors. The univariate marginal distributions are assumed to be continuous, allowing copula theory to come into play and as such guarantee margin-freeness. Using basic measure theory, our definition is also applicable to copulas having a singular component.

1 Introduction

The Pearson correlation coefficient most likely comes to mind when trying to analyse the interconnectedness of two univariate random variables, along with its limitation of only detecting linear relationships. The preference might also go to rank-based association measures like Kendall's tau or Spearman's rho among others, detecting any kind of monotonic connection. In contrast to the Pearson correlation, the latter dependence measures are a functional of solely the copula, inspiring extensions to more than two univariate random variables, see e.g. Nelsen (1996), Schmid and Schmidt (2007) and Gijbels et al. (2020). When groups of variables are considered, one might wish to measure the strength of dependence between two multivariate random vectors, e.g. Grothe et al. (2014). Again, it seems natural to extend the purpose to more than two random vectors, which got little attention in the literature so far.

Definitely enjoying increasing attention in statistics, is the use of measures coming from information theory, including the works of Joe (1989) and Blumentritt and Schmid (2012), among others. Many of these measures are generalized by the Φ-divergences (also known as f-divergences) introduced by Rényi (1960), studied further by e.g. Ali and Silvey (1966) and utilized for dependence quantification in e.g. Micheas and Zografos (2006) and Geenens and De Micheaux (2020). The latter paper studied in detail the Hellinger correlation between two random variables.

L. A. García-Escudero et al. (Eds.): SMPS 2022, AISC 1433, pp. 104–111, 2023.
https://doi.org/10.1007/978-3-031-15509-3_14

With it comes the postulation of certain axioms required for a valid dependence measure, starting with Rényi (1959) and followed by e.g. Lancaster (1963), Schweizer and Wolff (1981) and Embrechts et al. (2002) for the case of two univariate random variables, and e.g. Wolff (1980), Nelsen (1996) and Gijbels et al. (2020) when more than two univariate variables are of interest. Possible axioms for association measures between two random vectors can be found in Grothe et al. (2014). The compliance with axioms is typically contingent on the specific aim of the dependence measure (e.g. measure concordance). Our aim is to detect any potential deviation from independence between k random vectors using Φ-divergence measures.

In addition, one cannot say dependence without thinking of copula theory, and successfully combining it with information theory. A recent paper adopting such a combination is e.g. Singh and Zhang (2018). In our paper we focus on divergences between copula distributions and properties.

The general setting of this paper is the following. We consider the measure space $(\mathbb{R}^q, \mathscr{B}(\mathbb{R}^q), \lambda^q)$ equipped with the q-dimensional Lebesgue measure λ^q and $\mathscr{B}(\mathbb{R}^q)$ the Borel sigma-algebra on $\mathbb{R}^q$. Let $\mathbb{P} : \mathscr{B}(\mathbb{R}^q) \to \mathbb{R}$ be a probability measure and $d_1, \dots, d_k \in \mathbb{Z}_{>0}$ such that $q = d_1 + \cdots + d_k$. Let $\mathscr{L}_q^{d_1,\dots,d_k}$ be the space of all (q-dimensional) random vectors $\mathbf{X} = (\mathbf{X}_1, \dots, \mathbf{X}_k)$ having k marginal vectors $\mathbf{X}_i = (X_{i1}, \dots, X_{id_i})$ for $i = 1, \dots, k$ having d_i continuous univariate marginal variables X_{ij} for $j = 1, \dots, d_i$. Our aim is then to measure the dependence between $\mathbf{X}_1, \dots, \mathbf{X}_k$. A dependence measure $\mathscr{D}^{d_1,\dots,d_k}$ is a map

$$\mathscr{D}^{d_1,\dots,d_k} : \mathscr{L}_q^{d_1,\dots,d_k} \to \mathbb{R} \cup \{-\infty, \infty\} : \mathbf{X} \mapsto \mathscr{D}^{d_1,\dots,d_k}(\mathbf{X}). \tag{1}$$

We also write $\mathscr{D}^{d_1,\dots,d_k}(\mathbf{X}) = \mathscr{D}(\mathbf{X}_1, \dots, \mathbf{X}_k)$, or just $\mathscr{D}$ when the context is clear. Given that we have continuous univariate marginals, Sklar's theorem guarantees the existence of a unique (q-dimensional) copula C of $\mathbf{X}$ and marginal (d_i-dimensional) copulas C_i of $\mathbf{X}_i$ for $i = 1, \dots, k$. They induce respective probability measures μ_C on $\mathscr{B}(\mathbb{I}^q)$ and μ_{C_i} on $\mathscr{B}(\mathbb{I}^{d_i})$ with $\mathbb{I} = [0, 1]$ the unit interval. The mutual independence of $\mathbf{X}_1, \dots, \mathbf{X}_k$ is equivalent with μ_C being equal to the product measure $\mu_{C_1} \times \cdots \times \mu_{C_k}$ and a natural way to measure dependence is to quantify a statistical distance between μ_C and $\mu_{C_1} \times \cdots \times \mu_{C_k}$. Before doing so, we formulate desirable properties for a dependence measure (1).

2 Axioms

The definition of (1) is comprehensive in the sense that dependence between two random vectors ($k = 2$) and dependence between univariate random variables ($q = k$ and $d_i = 1$ for all i) are both particular cases. Some overarching axioms for overall dependence measures are as follows.

(A1) For every permutation π of $\mathbf{X}_1, \dots, \mathbf{X}_k$: $\mathscr{D}^{d_1,\dots,d_k}(\mathbf{X}) = \mathscr{D}^{d_1,\dots,d_k}(\pi(\mathbf{X}))$; and for every permutation π_i of $X_{i1}, \dots, X_{id_i}$, for $i \in \{1, \dots, k\}$, it holds: $\mathscr{D}^{d_1,\dots,d_k}(\mathbf{X}) = \mathscr{D}\big(\mathbf{X}_1, \dots, \pi_i(\mathbf{X}_i), \dots, \mathbf{X}_k\big)$.

(A2) $0 \leq \mathscr{D}^{d_1,\dots,d_k}(\mathbf{X}) \leq 1$.

(A3) $\mathscr{D}^{d_1,\ldots,d_k}(\mathbf{X}) = 0$ if and only if $\mathbf{X}_1, \ldots, \mathbf{X}_k$ are mutually independent.
(A4) $\mathscr{D}(\mathbf{X}_1, \ldots, \mathbf{X}_k, \mathbf{X}_{k+1}) \geq \mathscr{D}(\mathbf{X}_1, \ldots, \mathbf{X}_k)$ with equality if and only if $\mathbf{X}_{k+1}$ is independent of $(\mathbf{X}_1, \ldots, \mathbf{X}_k)$.

Axiom (A1) is the typical permutation invariance, here with respect to the random vectors and to the components within one and the same vector. The normalization property (A2) is desirable for interpretation and can often be obtained by applying a certain transformation to the dependence measure. Zero measure in case of independence is obvious, and (A3) also imposes the reversed implication, which is often not valid (e.g. correlation measures), yet, as mentioned, one of our goals. Axiom (A4) states that dependence cannot decrease if an additional random vector is taken into account and remains unchanged when an independent random vector is added. When $\mathscr{D}$ is copula-based, additional relevant axioms might be the following ones.

(A5) $\mathscr{D}^{d_1,\ldots,d_k}(\mathbf{X})$ is well defined for any q-dimensional random vector $\mathbf{X}$ and is a functional of solely the copula C of $\mathbf{X}$.
(A6) Let T_{ij} for $i = 1, \ldots, k$ and $j = 1, \ldots, d_i$ be strictly increasing, continuous transformations. Then

$$\mathscr{D}\big(T_1(\mathbf{X}_1), \ldots, T_k(\mathbf{X}_k)\big) = \mathscr{D}(\mathbf{X}_1, \ldots, \mathbf{X}_k),$$

where $T_i(\mathbf{X}_i) = (T_{i1}(X_{i1}), \ldots, T_{id_i}(X_{id_i}))$ for $i = 1, \ldots, k$.
(A7) Let T_{ij} be a strictly decreasing, continuous transformation for a fixed $i \in \{1, \ldots, k\}$ and a fixed $j \in \{1, \ldots, d_i\}$. Then

$$\mathscr{D}\big(\mathbf{X}_1, \ldots, T_i(\mathbf{X}_i), \ldots, \mathbf{X}_k\big) = \mathscr{D}(\mathbf{X}_1, \ldots, \mathbf{X}_k),$$

where $T_i(\mathbf{X}_i) = (X_{i1}, \ldots, T_{ij}(X_{ij}), \ldots, X_{id_i})$.
(A8) Let $(\mathbf{X}_n)_{n\in\mathbb{N}}$ be a sequence of q-dimensional random vectors with corresponding copulas $(C_n)_{n\in\mathbb{N}}$, then

$$\lim_{n\to\infty} \mathscr{D}^{d_1,\ldots,d_k}(\mathbf{X}_n) = \mathscr{D}^{d_1,\ldots,d_k}(\mathbf{X})$$

if $C_n \to C$ uniformly, where C denotes the copula of $\mathbf{X}$.

Property (A6) tells us that we can replace one or more univariate components by a strictly increasing transformation applied to them. By the properties of copulas, we know that (A5) implies (A6). Property (A7) tells us that we can replace one single univariate component by a strictly decreasing transformation applied to it (and hence also more than one). This is definitely not a trivial property. The meaning of (A8) is somewhat ambiguous. It is for example possible to approach independence arbitrary close by perfect dependence in some sense, e.g. by looking at bivariate random vectors living on Peano curves, see discussion in Sect. 2.3 of Geenens and De Micheaux (2020). Instead of looking at the copula cumulative distribution functions in (A8), one can look at the convergence of the copula densities if they exist.

3 Φ-Dependence Measures

Using the family of Φ-divergences, measuring dependence boils down to computing

$$\mathscr{D}_\Phi(\mu_C, \mu_{C_1} \times \cdots \times \mu_{C_k}) = \int \Phi\left(\frac{d\mu_C}{d(\mu_{C_1} \times \cdots \times \mu_{C_k})}\right) d(\mu_{C_1} \times \cdots \times \mu_{C_k}) \quad (2)$$

for a continuous, convex function $\Phi : (0, \infty) \to \mathbb{R}$ with $\Phi(1) = 0$. The Radon-Nikodym derivative in (2) is allowed to be infinite in case of singularity. We will write (2) as $\mathscr{D}_\Phi(\mathbf{X}_1, \ldots, \mathbf{X}_k)$ or just $\mathscr{D}_\Phi$. If $\mu_{C_i} \ll \lambda^{d_i}$ (absolute continuity) for $i = 1, \ldots, k$ and $\mu_C \ll \mu_{C_1} \times \cdots \times \mu_{C_k}$, we have $\mu_C \ll \lambda^q$ and

$$\mathscr{D}_\Phi(\mathbf{X}_1, \ldots, \mathbf{X}_k) = \int_{\mathbb{I}^q} \prod_{i=1}^{k} c_i(\mathbf{u}_i) \Phi\left(\frac{c(\mathbf{u})}{\prod_{i=1}^{k} c_i(\mathbf{u}_i)}\right) d\mathbf{u}, \quad (3)$$

with c and c_i the copula densities (w.r.t. λ^q and λ^{d_i}) corresponding to C and C_i for $i = 1, \ldots, k$, and using the notation $\mathbf{u} = (\mathbf{u}_1, \ldots, \mathbf{u}_k)$. Note that if $d_i = 1$ for a certain i, μ_{C_i} can be seen as the univariate Lebesgue measure with as one dimensional copula a $\mathscr{U}[0, 1]$ distribution and density equal to 1. If μ_C is not absolutely continuous w.r.t. $\mu_{C_1} \times \cdots \times \mu_{C_k}$, we can use the Lebesgue decomposition theorem to write $\mu_C = \mu_C^{\mathrm{ac}} + \mu_C^{\mathrm{s}}$, where μ_C^{ac} is absolutely continuous with respect to $\mu_{C_1} \times \cdots \times \mu_{C_k}$ and μ_C^{s} is singular with respect to $\mu_{C_1} \times \cdots \times \mu_{C_k}$ (denoted as $\mu_C^{\mathrm{s}} \perp \mu_{C_1} \times \cdots \times \mu_{C_k}$). If $\mu_C^{\mathrm{s}} \neq 0$, we say that μ_C has a singular component w.r.t. $\mu_{C_1} \times \cdots \times \mu_{C_k}$ and if $\mu_C^{\mathrm{ac}} = 0$, we say that μ_C is singular w.r.t. $\mu_{C_1} \times \cdots \times \mu_{C_k}$. In line with Definition 4.1.2 of Agrawal and Horel (2021) (but restricted to positive measures), we get the following general definition.

Definition 1 (Φ-dependence measures). Consider a continuous, convex function $\Phi : (0, \infty) \to \mathbb{R}$ with $\Phi(1) = 0$. Extend Φ by defining

$$\Phi(0) = \lim_{\substack{t \to 0 \\ >}} \Phi(t), \qquad \text{and} \qquad \Phi^*(0) = \lim_{t \to \infty} \frac{\Phi(t)}{t}. \quad (4)$$

The *Φ-dependence* between $\mathbf{X}_1, \ldots, \mathbf{X}_k$ is the quantity $\mathscr{D}_\Phi = \mathscr{D}_\Phi(\mathbf{X}_1, \ldots, \mathbf{X}_k) \in [0, \infty]$ defined by

$$\mathscr{D}_\Phi = \int \Phi\left(\frac{d\mu_C^{\mathrm{ac}}}{d(\mu_{C_1} \times \cdots \times \mu_{C_k})}\right) d(\mu_{C_1} \times \cdots \times \mu_{C_k}) + \Phi^*(0)\mu_C^{\mathrm{s}}(B), \quad (5)$$

with B the set on which μ_C^s is concentrated. We use the convention $0 \cdot \infty = 0$.

We can now better see what the strength of dependence is in case a singularity is present. The extension (4) is straightforward with $\Phi(0)$ taking values in $[0, \infty]$. Defining Φ^* as in (4) is motivated from the convention that, based on continuous extensions, $0 \cdot \Phi(a/0) \approx a \cdot \Phi^*(0)$ and arises when $\mu_{C_1} \times \cdots \times \mu_{C_k} = 0$, but $\mu_C > 0$

(singularity). Suppose e.g. that μ_C has a singular component w.r.t. $\mu_{C_1}\times\cdots\times\mu_{C_k}$ concentrated on a set B. Then, the contribution to (2) is

$$\int_B \Phi^*(0)d\mu_C = \Phi^*(0)\mu_C(B) = \Phi^*(0)\mu_C^s(B),$$

motivating the second term in (5). What remains is the intensity of dependence propagated in the subspace where we have absolute continuity, i.e. the first term in (5). If $\mu_C \perp \mu_{C_1} \times \cdots \times \mu_{C_k}$, the first term in (5) becomes $\Phi(0)$, yielding $\mathscr{D}_\Phi(\mathbf{X}_1,\ldots,\mathbf{X}_k) = \Phi(0) + \Phi^*(0)$. The following theorem holds.

Theorem 1. *Let $\mathscr{D}_\Phi$ be defined by* (5). *Then, it holds that*

- $0 \leq \mathscr{D}_\Phi(\mathbf{X}_1,\ldots,\mathbf{X}_k) \leq \Phi(0) + \Phi^*(0)$.
- *If $\mathbf{X}_1,\ldots,\mathbf{X}_k$ are mutually independent, then $\mathscr{D}_\Phi(\mathbf{X}_1,\ldots,\mathbf{X}_k) = 0$.*
- *If $\mu_C \perp \mu_{C_1} \times \cdots \times \mu_{C_k}$, then $\mathscr{D}_\Phi(\mathbf{X}_1,\ldots,\mathbf{X}_k) = \Phi(0) + \Phi^*(0)$.*

Moreover, if Φ is strictly convex at 1*, it holds that*

- *If $\mathscr{D}_\Phi(\mathbf{X}_1,\ldots,\mathbf{X}_k) = 0$, then $\mathbf{X}_1,\ldots,\mathbf{X}_k$ are mutually independent.*
- *If $\mathscr{D}_\Phi(\mathbf{X}_1,\ldots,\mathbf{X}_k) = \Phi(0) + \Phi^*(0) < \infty$, then $\mu_C \perp \mu_{C_1} \times \cdots \times \mu_{C_k}$.*

Proof. This follows from Theorem 5 in Liese and Vajda (2006).

From Theorem 1, we see that if Φ is strictly convex at 1, property (A3) is fulfilled. Moreover, if $\Phi(0) + \Phi^*(0) < \infty$, then $\mathscr{D}_\Phi = \Phi(0) + \Phi^*(0)$ only if $\mu_C = \mu_C^s$, i.e. is completely singular w.r.t $\mu_{C_1} \times \cdots \times \mu_{C_k}$, and a natural normalized dependence measure satisfying (A2) is obtained through dividing by $\Phi(0) + \Phi^*(0)$. If $\Phi(0) + \Phi^*(0) = \infty$, then $\mathscr{D}_\Phi = \infty$ from the moment that μ_C has a singular component w.r.t. $\mu_{C_1} \times \cdots \times \mu_{C_k}$, and an artificial normalization, i.e. a continuous, strictly increasing map $N : [0, \infty] \to [0, 1]$ satisfying $N(0) = 0$ and $N(\infty) = 1$, can still be used to guarantee $N \circ \mathscr{D}_\Phi \in [0, 1]$.

In case $\mu_C \ll \lambda^q$, i.e. representation (3) holds, the ideas for a proof of the remaining axioms are straightforward. Property (A1) holds because of Fubini's theorem and knowing that permuting the components of a random vector, results in permuting the copula components accordingly. Property (A4) can be shown through Jensen's inequality. Obviously, by definition, (A5) and hence also (A6) are fulfilled. For proving (A7), it suffices to use the fact that applying a decreasing transformation to a certain X_{ij} results in a copula density with u_{ij} replaced by $1 - u_{ij}$, and afterwards perform this as integral substitution. Finally, from the continuity of Φ and the dominated convergence theorem, it follows that if copula densities $(c_n)_{n\in\mathbb{N}}$ converge uniformly to a certain copula density c and certain regularity conditions hold, then $\mathscr{D}_\Phi$ as in (3) of $(c_n)_{n\in\mathbb{N}}$ will converge to $\mathscr{D}_\Phi$ of c. In summary, the discussed Φ-dependences can be shown to satisfy the axioms of Sect. 2. A detailed study of Φ-dependence measures, including statistical inference, and the practical use of such measures, will be in an upcoming manuscript by the authors.

4 Example

The next example illustrates that certain Φ-dependence measures can still be descriptive in case the copula has a singular component. Consider a trivariate extreme value copula (see expression (15) in Durante and Salvadori (2010))

$$C(u_1, u_2, u_3) = \prod_{i=1}^{3} u_i^{1-\sum_{j=1,j\neq i}^{3}\theta_{ij}} \prod_{1\leq i<j\leq 3} (\min\{u_i, u_j\})^{\theta_{ij}}, \tag{6}$$

with dependence parameters $\theta_{ij} = \theta_{ji} \in \mathbb{I}$ satisfying $\sum_{j=1,j\neq i}^{3}\theta_{ij} \leq 1$, for every $i = 1, 2, 3$. Suppose that (X_1, X_2, X_3) has copula (6), and we are interested in the dependence between (X_1, X_2) and X_3. The marginal copula of (X_1, X_2) is given by $C_1(u_1, u_2) = (u_1u_2)^{1-\theta_{12}} \min\{u_1, u_2\}^{\theta_{12}}$, being a Cuadras-Augé copula (Cuadras and Augé (1981)), and similarly for (X_1, X_3) and (X_2, X_3). The copula C has a singular component with respect to the Lebesgue measure, with singularity concentrated on $\{(u_1, u_2, u_3) \in \mathbb{I}^3 : u_1 = u_2 \cup u_1 = u_3 \cup u_2 = u_3\}$. The mass of this singular component with respect to the Lebesgue measure λ^3 is computed to be

$$\begin{aligned}\mathbb{P}(U_1 = U_2 \cup U_1 = U_3 \cup U_2 = U_3) &= 1 - \frac{2(1-\theta_{12})(1-\theta_{13}-\theta_{23})}{(2-\theta_{12})(3-\theta_{12}-\theta_{13}-\theta_{23})}\\ &- \frac{2(1-\theta_{13})(1-\theta_{12}-\theta_{23})}{(2-\theta_{13})(3-\theta_{12}-\theta_{13}-\theta_{23})} - \frac{2(1-\theta_{23})(1-\theta_{12}-\theta_{13})}{(2-\theta_{23})(3-\theta_{12}-\theta_{13}-\theta_{23})},\end{aligned}$$

and the singular mass of the Cuadras-Augé copula of (X_1, X_2) is $\mathbb{P}(U_1 = U_2) = \theta_{12}/(2-\theta_{12})$, and similarly for (X_1, X_3) and (X_2, X_3). In this case, we apply the Lebesgue decomposition theorem to the measures μ_C and $\mu_{C_1} \times \lambda$. The singular component of μ_C with respect to $\mu_{C_1} \times \lambda$ has mass $\mathbb{P}(U_1 = U_3 \cup U_2 = U_3)$ which equals

$$\begin{aligned}&\frac{1}{2}\left[\mathbb{P}(U_1 = U_3) + \mathbb{P}(U_2 = U_3) - \mathbb{P}(U_1 = U_2) + \mathbb{P}(U_1 = U_2 \cup U_1 = U_3 \cup U_2 = U_3)\right]\\ &= \frac{\theta_{13}}{2(2-\theta_{13})} + \frac{\theta_{23}}{2(2-\theta_{23})} - \frac{\theta_{12}}{2(2-\theta_{12})} - \frac{(1-\theta_{12})(1-\theta_{13}-\theta_{23})}{(2-\theta_{12})(3-\theta_{12}-\theta_{13}-\theta_{23})}\\ &- \frac{(1-\theta_{13})(1-\theta_{12}-\theta_{23})}{(2-\theta_{13})(3-\theta_{12}-\theta_{13}-\theta_{23})} - \frac{(1-\theta_{23})(1-\theta_{12}-\theta_{13})}{(2-\theta_{23})(3-\theta_{12}-\theta_{13}-\theta_{23})} + \frac{1}{2}.\end{aligned} \tag{7}$$

We next want to compute the Hellinger distance ($\Phi(t) = (\sqrt{t}-1)^2$) between (X_1, X_2) and X_3. Here we have $\Phi^*(0) = 1$. Regarding the absolute continuity part (i.e. described by the measure μ_C^{ac}), one can find that

$$\begin{aligned}\int \Phi\left(\frac{d\mu_C^{\mathrm{ac}}}{d(\mu_{C_1}\times\lambda)}\right) d(\mu_{C_1}\times\lambda) &= \frac{2(1-\theta_{13}-\theta_{23})(1-\theta_{12})}{(2-\theta_{12})(3-\theta_{12}-\theta_{13}-\theta_{23})} - \frac{8(1-\theta_{13}-\theta_{23})^{1/2}(1-\theta_{12})}{(2-\theta_{12})(6-2\theta_{12}-\theta_{13}-\theta_{23})}\\ &+ \frac{2(1-\theta_{12})}{(2-\theta_{12})(3-\theta_{12})} + \frac{2(1-\theta_{13})(1-\theta_{12}-\theta_{23})}{(2-\theta_{13})(3-\theta_{12}-\theta_{13}-\theta_{23})} - \frac{16(1-\theta_{12})^{1/2}(1-\theta_{13})^{1/2}(1-\theta_{12}-\theta_{23})^{1/2}}{(4-\theta_{13})(6-2\theta_{12}-\theta_{13}-\theta_{23})}\\ &+ \frac{2(1-\theta_{23})(1-\theta_{12}-\theta_{13})}{(2-\theta_{23})(3-\theta_{12}-\theta_{13}-\theta_{23})} - \frac{16(1-\theta_{12})^{1/2}(1-\theta_{23})^{1/2}(1-\theta_{12}-\theta_{13})^{1/2}}{(4-\theta_{23})(6-2\theta_{12}-\theta_{13}-\theta_{23})} + \frac{2(1-\theta_{12})}{3-\theta_{12}}.\end{aligned} \tag{8}$$

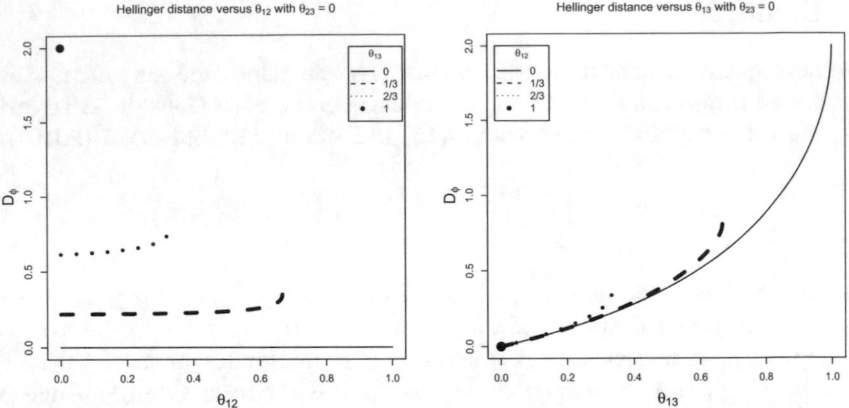

Fig. 1. The Hellinger distance $\mathscr{D}_\Phi\left((X_1, X_2); X_3\right)$, for $\theta_{23} = 0$, as a function of θ_{12} (left) and a function of θ_{13} (right), for various values of the other parameter

The Hellinger distance $\mathscr{D}_\Phi\left((X_1, X_2); X_3\right)$ between (X_1, X_2) and X_3 then equals the sum of the terms in (8) and (7). First note that the contribution of θ_{13} is the same as of θ_{23}, both parameters determining the inter-pairwise dependence between (X_1, X_2) and X_3. Figure 1 depicts the Hellinger distance, as a function of θ_{12} for different values of θ_{13}, and as a function of θ_{13} for different values of θ_{12}, assuming in both cases that $\theta_{23} = 0$. We observe a weak influence of the parameter θ_{12} on the strength of dependence, as we expect since θ_{12} mainly describes the dependence within (X_1, X_2). Furthermore, independence between (X_1, X_2) is equivalent with $\theta_{13} = \theta_{23} = 0$, even if we have singularity within (X_1, X_2) (i.e. $\theta_{12} = 1$). In general, the dependence becomes stronger as θ_{13} increases and reaches its maximum value of $\Phi(0) + \Phi^*(0) = 2$ if $\theta_{13} = 1$, meaning that $\mu_C = \mu_C^{\mathrm{s}}$.

Acknowledgements. The authors gratefully acknowledge support from the Research Fund KU Leuven [C16/20/002 project].

References

Agrawal, R., Horel, T.: Optimal bounds between f-divergences and integral probability metrics. J. Mach. Learn. Res. **22**, 1–59 (2021)

Ali, S.M., Silvey, S.D.: A general class of coefficients of divergence of one distribution from another. J. R. Stat. Soc. Ser. B (Methodol.) **28**, 131–142 (1966)

Blumentritt, T., Schmid, F.: Mutual information as a measure of multivariate association: analytical properties and statistical estimation. J. Stat. Comput. Simul. **82**, 1257–1274 (2012)

Cuadras, C.M., Augé, J.: A continuous general multivariate distribution and its properties. Commun. Stat. Theor. Meth. **10**, 339–353 (1981)

Durante, F., Salvadori, G.: On the construction of multivariate extreme value models via copulas. Environmetrics **21**, 143–161 (2010)

Embrechts, P., McNeil, A.J., Straumann, D.: Correlation and dependence in risk management: properties and pitfalls. In: Dempster, M.A.H. (ed.) Risk Management: Value at Risk and Beyond, pp. 176–223. Cambridge University Press (2002)

Geenens, G., De Micheaux, P.L.: The Hellinger correlation. J. Am. Stat. Assoc. **117**(538), 639–653 (2020). https://doi.org/10.1080/01621459.2020.1791132

Gijbels, I., Kika, V., Omelka, M.: On the specification of multivariate association measures and their behaviour with increasing dimension. J. Multivar. Anal. **182**, 104704 (2020)

Grothe, O., Schnieders, J., Segers, J.: Measuring association and dependence between random vectors. J. Multivar. Anal. **123**, 96–110 (2014)

Joe, H.: Relative entropy measures of multivariate dependence. J. Am. Stat. Assoc. **84**, 157–164 (1989)

Lancaster, H.O.: Correlation and complete dependence of random variables. Ann. Math. Stat. **34**, 1315–1321 (1963)

Liese, F., Vajda, I.: On divergences and informations in statistics and information theory. IEEE Trans. Inf. Theor. **52**, 4394–4412 (2006)

Micheas, A.C., Zografos, K.: Measuring stochastic dependence using φ-divergence. J. Multivar. Anal. **97**, 765–784 (2006)

Nelsen, R.B.: Nonparametric measures of multivariate association. In: Ruschendorf, L., Schweizer, B., Taylor, M.D. (eds.) Distributions with Fixed Marginals and Related Topics. IMS Lecture Notes - Monograph Series, vol. 28, pp. 223–232 (1996)

Rényi, A.: On measures of dependence. Acta Mathematica Academiae Scientiarum Hungaricae **10**, 441–451 (1959)

Rényi, A.: On measures of entropy and information. In: Proceedings of the 4th Berkeley Symposium on Mathematics. Statistics and Probability, vol. 1, pp 547–561. University of California Press, Berkeley (1960)

Schmid, F., Schmidt, R.: Multivariate extensions of Spearman's rho and related statistics. Stat. Probab. Lett. **77**, 407–416 (2007)

Schweizer, B., Wolff, E.F.: On nonparametric measures of dependence for random variables. Ann. Stat. **9**, 879–885 (1981)

Singh, V.P., Zhang, L.: Copula-entropy theory for multivariate stochastic modeling in water engineering. Geosci. Lett. **5**, 6 (2018)

Wolff, E.F.: n-dimensional measures of dependence. Stochastica **4**, 175–188 (1980)

The Winning Probability Relation of Parametrized Families of Random Vectors

Hans De Meyer[1(✉)] and Bernard De Baets[2]

[1] Department of Applied Mathematics, Computer Science and Statistics, Ghent University, Krijgslaan 281 S9, 9000 Gent, Belgium
hans.demeyer@ugent.be

[2] KERMIT, Department of Data Analysis and Mathematical Modelling, Ghent University, Coupure links 653, 9000 Gent, Belgium
bernard.debaets@ugent.be

Abstract. Pairwise winning probabilities are computed for the components of random vectors that are distributed according to certain parameterized families of multivariate location-scale distributions, such as the family of normal distributions and families of compounded normal distributions (Cauchy distribution, t-distribution, slash distribution). The reciprocal relation built from these pairwise winning probabilities is shown to be moderately stochastic transitive, allowing to associate a sequence of partial orders with a given random vector, providing alternatives to the omnipresent stochastic dominance order.

1 Introduction

In fields such as financial mathematics, risk management and decision theory, stochastic dominance is a commonly used tool for expressing a preference order between two random variables (RVs) (Levy 1998). Various types of stochastic dominance – e.g. first (FSD) and second (SSD) order stochastic dominance – are related to distinct recipes for comparing two cumulative distribution functions (CDFs), but share the property that they induce a (pre-)order on any set of RVs, turning this set into a (pre-)poset. This partial ordering is due to the fact that these recipes allow that of two RVs neither one stochastically dominates the other, whence these RVs become incomparable. A drawback is that the resulting (pre-)posets often show too many incomparabilities to be informative enough for decision making. To remedy this, sometimes weaker forms of stochastic dominance are considered, such as third or higher-order stochastic dominance, or concepts such as almost stochastic order and expectile order. They all tend towards the linearization of the posets. But yet another weakness of ordering methods based on stochastic dominance is that they solely rely on the knowledge of the marginal CDFs, and thus no account is ever taken of the influence of dependencies between the RVs on their pairwise ordering.

L. A. García-Escudero et al. (Eds.): SMPS 2022, AISC 1433, pp. 112–119, 2023.
https://doi.org/10.1007/978-3-031-15509-3_15

In the past, as an alternative to the concept of stochastic dominance, we have explored the use of so-called winning probabilities (De Meyer et al. 2007; De Schuymer et al. 2005) for the ordering of RVs. Roughly speaking, the winning probability $Q(X, Y)$ of a RV X w.r.t. a RV Y is the probability that X shows a higher outcome than Y in a random trial. Note that the winning probability not only depends on the marginal CDFs but also on the bivariate CDF, and thus implicitly on the way X and Y are coupled. The value of $Q(X, Y)$ can be regarded as a degree of preference of X over Y and therefore as a grading rather than as a weakening of the concept of stochastic dominance. Note that in reliability theory, the winning probability is known as the stress-strength reliability of X w.r.t. Y (Nadarajah and Kotz 2006).

For any given set A of RVs, the relation Q that is constructed from the winning probabilities $Q(X, Y)$, and which is called the winning probability relation on A, takes values in $[0, 1]$ and has the property of reciprocity, i.e. $Q(X, Y) + Q(Y, X) = 1$ for all X, Y. For a fixed threshold value $\lambda \geq 1/2$, we say that X is preferred to Y at level λ if $Q(X, Y) \geq \lambda$. Given a set of RVs and assuming that all bivariate CDFs are known, the preference relation at level $\lambda > 1/2$ is a partial order provided Q is moderately stochastic transitive (De Meyer et al. 2007; De Schuymer et al. 2005).

Some time ago, we have developed a general framework in which many types of transitivity of reciprocal relations can be characterized and their interrelationships clarified (De Baets et al. 2006). In this so-called cycle transitivity framework, moderate stochastic transitivity finds its place amongst transitivity types such as e.g. strong, weak and partial stochastic transitivity (Fishburn 1973).

The key question that concerns us here is what conditions, if any, still have to be imposed on a given set of RVs in order for the associated winning probability relation Q to be moderately stochastic transitive. Let us assume that the RVs $X_1, \ldots X_n$ are the components of a random vector X, i.e. they are characterized by a joint CDF. For the computation of the winning probabilities $Q(X_i, X_j)$, we need the two marginal CDFs and a copula C (Nelsen 2006). If X and Y are independent, then the copula is the product copula Π. Any copula C is situated between the two extreme copulas W and M, i.e. $W \leq C \leq M$. M is the minimum copula defined by $M(u, v) = \min(u, v)$, and W is the Łukasiewicz copula defined by $W(u, v) = \max(u + v - 1, 0)$. Two RVs coupled by M are called comonotone, while they are called countermonotone if they are coupled by W. We have previously shown that for a set of independent RVs with arbitrary marginal CDFs, the winning probability relation Q is dice-transitive. This is a type of transitivity that is weaker than moderate stochastic transitivity. Something analogous occurs when all RVs are pairwise comonotone, in which case Q is weakly stochastic transitive. In conclusion, the comparison in terms of winning probabilities of RVs with arbitrary marginal CDFs does not lead to a partial order, no matter how the RVs are coupled.

To guarantee a moderately stochastic transitive winning probability relation Q, we need to impose restrictions on the sets of RVs. To find out which restrictions might guarantee success is the principal aim of the present paper.

2 Preliminaries

Given two RVs X and Y with joint CDF $F_{X,Y}$, the winning probability $Q(X,Y)$ of X w.r.t. Y is defined as

$$Q(X,Y) = \text{Prob}(X > Y) + \frac{1}{2}\,\text{Prob}(X = Y). \tag{1}$$

with this definition, ties being resolved in an equally balanced way, Q satisfies the reciprocity property

$$Q(X,Y) = 1 - Q(Y,X). \tag{2}$$

Clearly, the computation of a winning probability takes into consideration the coupling of the random variables as expressed by the bivariate CDF $F_{X,Y}$. This coupling can be made explicit, thanks to Sklar's theorem, through the use of the copula decomposition of the CDF $F_{X,Y}$, namely $F_{X,Y}(x,y) = C(F_X(x), F_Y(y))$, where F_X and F_Y are the marginal CDFs of X and Y, and C is the copula describing the dependence structure. Stochastic dominance, on the other hand, only uses the information contained in the marginal CDFs and is ignorant of the coupling.

Starting from a random vector $(X_1, X_2, \ldots, X_d)$ with joint CDF $F_{X_1,\ldots,X_d}$, the pairwise comparison of the components X_i, $i = 1, \ldots, d$, yields a winning probability relation Q with elements $Q_{ij} = Q(X_i, X_j)$. The question now is how to build a partial order on the set $\{X_1, \ldots, X_d\}$ from such a Q. For any $\lambda \in [1/2, 1]$, we can define a crisp relation $\geq_\lambda$ as follows:

$$X_i \geq_\lambda X_j \Leftrightarrow Q_{ij} \geq \lambda, \tag{3}$$

which expresses that X_i is preferred to X_j with degree at least λ.

The transitivity of $\geq_\lambda$ is only guaranteed if it holds for all $i, j, k \in \{1, \ldots d\}$ that if $Q_{ij} \geq \lambda$ and $Q_{jk} \geq \lambda$ then $Q_{ik} \geq \lambda$. If we want to impose transitivity for all cutting levels $\lambda \in]1/2, 1]$, then it should hold that

$$(Q_{ij} > 1/2 \wedge Q_{jk} > 1/2) \Rightarrow Q_{ik} \geq \min(Q_{ij}, Q_{jk}).$$

Then, $\geq_\lambda$ induces a strict partial order for any $\lambda \in]1/2, 1]$. We finally include the cutting level $\lambda = 1/2$, with the sole difference that $\geq_{1/2}$ is reflexive and not necessarily anti-symmetric, so that if it is transitive then it is a pre-order. Hence, if it holds for all $i, j, k \in \{1, \ldots, d\}$ that

$$(Q_{ij} \geq 1/2 \wedge Q_{jk} \geq 1/2) \Rightarrow Q_{ik} \geq \min(Q_{ij}, Q_{jk}), \tag{4}$$

then $\geq_\lambda$ is a pre-order if $\lambda = 1/2$ and a strict partial order if $\lambda > 1/2$. Property (4) is known as moderate stochastic transitivity.

In case $\geq_\lambda$ it is not transitive, we can render it transitive by means of a standard operation called transitive closure (Freson et al. 2014). In some minimal sense arrows are added to the graph of $\geq_\lambda$ such that it becomes a strict partial order. It is, however, neither from a theoretical nor from a computational point of view, advantageous to deal with reciprocal relations that are merely weakly stochastic transitive, as for each λ-cut taken, it must first be decided whether $\geq_\lambda$ is a (strict) partial order (transitivity check) and, if this is not the case, a transitive closure operation must first be performed.

In the following sections, we investigate whether there exist families of multivariate RVs or random vectors that render this procedure redundant and for which the associated reciprocal relation Q is thus guaranteed to be moderately stochastic transitive.

3 Multivariate Normal Distribution

A d-dimensional random vector $\mathbf{X} = (X_1, \ldots, X_d)$ is a normal random vector, denoted as $\mathbf{X} \sim \mathcal{N}(d; \mu, \Sigma)$, where $\mu \in \mathbb{R}^d$ and $\Sigma \in \mathbb{R}^{d \times d}$ is a positive definite matrix, if its joint probability density function (PDF) is given by Fang et al. (1990), for any $\mathbf{x} \in \mathbb{R}^d$:

$$f_{\mathbf{X}}^N(\mathbf{x}; d; \mu, \Sigma) = \frac{1}{(2\,\pi)^{d/2}\sqrt{\det \Sigma}}\; e^{-\frac{1}{2}\,(\mathbf{x}-\mu)^\top \Sigma^{-1}(\mathbf{x}-\mu)}. \tag{5}$$

The vector μ is the vector of expectations, i.e. $\mu_i = \mathrm{E}[X_i]$, and Σ is the covariance matrix, i.e. $\Sigma_{ij} = \mathrm{E}[(X_i - \mu_i)(X_j - \mu_j)] = \mathrm{Cov}[X_i, X_j]$. Also, the matrix elements Σ_{ij} are factorizable as $\Sigma_{ij} = \rho_{ij}\,\sigma_i\,\sigma_j$, with $\sigma_i^2 = \mathrm{Var}[X_i]$, the variance of X_i, and $\rho_{ij} \in [-1, 1]$, the linear correlation coefficient of X_i and X_j.

All components of a normal random vector are normal RVs and the marginal PDF of X_i is given by, for any $\mathbf{x} \in \mathbb{R}^d$:

$$f_{X_i}^N(x; \mu_i, \sigma_i) = \frac{1}{\sqrt{2\,\pi\,\det \Sigma}}\; e^{-(x-\mu_i)^2/(2\,\sigma_i^2)}. \tag{6}$$

This is denoted as $X_i \sim \mathcal{N}(\mu_i, \sigma_i^2)$. There is no closed form for the CDF of X_i.

A random vector is called a standard normal random vector if all of its components X_i are independent and each is a zero-mean unit-variance normally distributed RV, i.e. if $X_i \sim \mathcal{N}(0, 1)$ for all $i = 1, 2, \ldots, d$. The CDF of a standard normal RV $X \sim \mathcal{N}(0, 1)$ is denoted as Φ. Note that $\Phi(0) = 1/2$.

The following proposition provides an answer to the question whether for any normal random vector $\mathbf{X} \sim \mathcal{N}(d; \mu, \Sigma)$, the winning probability relation Q is moderately stochastic transitive.

Proposition 1. *Let $\mathbf{X} \sim \mathcal{N}(d; \mu, \Sigma)$ and $Q_{ij} = Q(X_i, X_j)$ denote the winning probability of X_i w.r.t. X_j, then the winning probability relation Q is moderately stochastic transitive.*

4 Multivariate T-Distribution

A RV X is said to be t-distributed with $\nu \in \mathbb{N}_0$ degrees of freedom, location parameter $\mu \in \mathbb{R}$ and scale parameter $\sigma \in \mathbb{R}^+$, denoted as $X \sim t(\nu; \mu, \sigma)$, if its distribution has the following PDF (Fang et al. 1990):

$$f_X^t(x;\nu;\mu,\sigma) = \frac{1}{\sqrt{\nu}\,\sigma B\left(\frac{\nu}{2},\frac{1}{2}\right)}\left[1+\frac{1}{\nu}\frac{(x-\mu)^2}{\sigma^2}\right]^{-(1+\nu)/2}, \tag{7}$$

where B denotes Bernoulli's beta function.

A random vector $\mathbf{X} = (X_1,\ldots,X_d)^\top$ is said to be t-distributed with ν degrees of freedom, denoted $\mathbf{X} \sim t(d;\nu;\mu,\Sigma)$, where $\mu \in \mathbb{R}^d$ and $\Sigma \in \mathbb{R}^{d\times d}$ is a positive definite matrix, if its multivariate PDF is given by:

$$f_{\mathbf{X}}^t(\mathbf{x};d;\nu;\mu,\Sigma) = \frac{\Gamma((\nu+d)/2)}{\Gamma(\nu/2)\nu^{d/2}\pi^{d/2}\sqrt{\det\Sigma}}\left[\frac{1}{\nu}(\mathbf{x}-\mu)^\top\Sigma^{-1}(\mathbf{x}-\mu)\right]^{-(\nu+d)/2}. \tag{8}$$

For $\nu = 1$, the multivariate distribution is known as the Cauchy distribution, whereas in the limit of $\nu \to \infty$, the multivariate t-distribution tends towards the multivariate normal distribution. Also note that μ is still the vector of expectations (except for $\nu = 1$, in which case expectations are undefined), but Σ is no longer the covariance matrix (which is undefined for $\nu = 1$ and $\nu = 2$). In particular, if we set all $\rho_{ij} = 0$, then we do not retrieve the case of independent random vector components. One readily verifies that the multivariate PDF $f_{\mathbf{X}}^t(\mathbf{x};d;\nu;\mu,\Sigma)$ does not factorize into univariate PDFs. On the other hand, if we choose all $\rho_{ij} = 1$, then all underlying bivariate copulas are equal to M, and the components of the random vector are comonotone.

As far as the transitivity of the reciprocal relation Q is concerned, for a t-distributed random vector with ν degrees of freedom, the type of transitivity remains the same as for normal random vectors.

Proposition 2. *Let $\mathbf{X} \sim t(d;\nu;\mu,\Sigma)$ and $Q_{ij} = Q(X_i,X_j)$ denote the winning probability of X_i w.r.t. X_j, then the winning probability relation Q is moderately stochastic transitive.*

Note that the sequence of posets $\geq_\lambda$ one obtains by letting λ vary from $> 1/2$ to 1 is the same as the sequence of posets generated from a normal random vector with the same parameters μ and Σ.

The following proposition relates to the case of a random vector that consists of independent components, each t-distributed with the same number of degrees of freedom.

Proposition 3. *Let $X_1, X_2, \ldots, X_d$ be d independent t-distributed RVs with the same ν, i.e. $X_i \sim t(\nu;\mu_i,\sigma_i)$, then winning probability relation Q is moderately stochastic transitive.*

5 Compounding the Multivariate Normal Distribution

We investigate multivariate distributions that are obtained by mixing or scale compounding the multivariate normal distribution, namely the situation where the conditional distribution of $\mathbf{X}$ is the multivariate normal distribution in which a scaling parameter τ is treated as a random variable with PDF h. In general, the scale-compounded distribution has the form

$$f_{\mathbf{X}}(\mathbf{x};d) = \frac{1}{(2\pi)^{d/2}\sqrt{\det\Sigma}}\int_0^{+\infty} \tau^{-d/2}\, e^{-A/\tau}\, h(\tau)\, d\tau, \tag{9}$$

where

$$A = \frac{1}{2}(\mathbf{x}-\mu)^\top \Sigma^{-1}(\mathbf{x}-\mu). \tag{10}$$

In fact, the multivariate t-distribution with ν degrees of freedom can be regarded as an example of a scale-compounded multivariate normal distribution with as mixing distribution h the inverse gamma distribution which shape and scale parameter both equal to $\nu/2$, namely

$$h(t) = \frac{(\nu/2)^{(\nu/2)}}{\Gamma(\nu/2)}(1/t)^{\nu/2+1}\, e^{-\nu/(2t)}.$$

Proposition 4. *Let $\mathbf{X} = (X_1,\ldots,X_d)$ be a random vector whose distribution is a scale-compounded normal distribution, with PDF given by*

$$f_{\mathbf{X}}(\mathbf{x};d;\mu,\Sigma) = \frac{1}{(2\pi)^{d/2}\sqrt{\det\Sigma}}\int_0^{+\infty} \tau^{-d/2}\, e^{-(x-\mu)^\top\Sigma^{-1}(x-\mu)/(2\tau)}\, h(\tau)\, d\tau\,, \tag{11}$$

where h can be any univariate PDF with support $\mathbb{R}^+$, then the winning probability relation Q is moderately stochastic transitive.

Another example, different from the multivariate t-distribution, is obtained with the following choice of mixing distribution:

$$h(t) = \frac{q}{2}\, t^{-q/2-1}, \qquad t \geq 1, \tag{12}$$

where $q > 0$ is a real constant. The PDF of the scale-compounded multivariate normal distribution is given by:

$$f_{\mathbf{X}}(\mathbf{x};d;q;\mu,\Sigma) = \begin{cases} \dfrac{q}{2\,(2\pi)^{d/2}\sqrt{\det\Sigma}}\, A^{-(q+d)/2}\,\gamma((q+d)/2, A) & \text{, if } \mathbf{x}\neq\mu\,, \\[2ex] \dfrac{q}{(2\pi)^{d/2}\sqrt{\det\Sigma}\,(q+d)} & \text{, if } \mathbf{x}=\mu\,, \end{cases}$$

where $\gamma(a,z) = \int_0^z u^{a-1}\, e^{-u}\, du$ is the incomplete gamma function. This distribution is known as the multivariate slash distribution and denoted as $\mathcal{S}(d,q;\mu,\Sigma)$. Again, if $\mathbf{X} \sim S(d,q;\mu,\Sigma)$, then the winning probability relation Q is moderately stochastic transitive.

6 Discussion

We have explored whether it is feasible to construct a partial order on a given set of RVs using winning probabilities as an alternative to stochastic dominance. We have first shown that the winning probability relation, which is a reciprocal relation, should have the property of moderate stochastic transitivity. We then examined in which circumstances this property can be realized. We have shown that this is the case when the RVs are the components of a random vector with a multivariate normal distribution or that are derived from the multivariate normal distribution by scale-compounding.

Although there are quite a number of multivariate random vectors that are of interest in statistics and for which a sequence of partial orders can be constructed, it is not so that for any random vector the winning probability relation is moderately stochastic transitive. Note, however, that for any random vector we could try to find a threshold value Λ such that for all $\lambda \geq \Lambda$, the relation $\geq_\lambda$ is a partial order. In other words, for a given random vector we could find out which is the smallest value of Λ such that the winning probability relation Q has for all $\lambda \geq \Lambda$ the property that $Q_{ij} \geq \lambda$ and $Q_{jk} \geq \lambda$ imply that $Q_{ik} \geq \min(Q_{ij}, Q_{jk})$. This is a weaker form of moderate stochastic transitivity which to our knowledge has not been studied so far.

Another way of generating a sequence of partial orders for a given random vector is to build a relation R that instead of pairwise winning probabilities has elements $R_{ij} = (\mathrm{E}[X_i] - \mathrm{E}[X_j])/\sqrt{\mathrm{Var}[X_i - X_j]}$. Cutting this relation at values τ that vary from 0 to $+\infty$, we obtain a sequence of partial orders that is, in all the cases seen so far where the reciprocal relation Q is moderately stochastic transitive, identical to the sequence generated by Q. As for any $a, c \geq 0$ and $b, d > 0$, it holds that the inequality $a/b < c/d$ is equivalent to the inequality $a/(a+b) < c/(c+d)$, the same sequence of partial orders is also generated by the relation S where $S_{ij} = (\mathrm{E}[X_i] - \mathrm{E}[X_j])/\sqrt{\mathrm{Var}[X_i - X_j] + (\mathrm{E}[X_i] - \mathrm{E}[X_j])^2} = (\mathrm{E}[X_i] - \mathrm{E}[X_j])/\sqrt{\mathrm{E}[(X_i - X_j)^2]}$ when the cutting level runs from 0 to 1. A reciprocal relation that closely resembles Q and is for any random vector moderately stochastic transitive is the reciprocal relation $Q^{(2)} = (1 + S)/2$. We will report on this and similar reciprocal relations that generate a sequence of partial orders elsewhere.

References

De Baets, B., De Meyer, H., De Schuymer, B., Jenei, S.: Cyclic evaluation of transitivity of reciprocal relations. Soc. Choice Welfare **26**, 217–238 (2006)

De Meyer, H., De Baets, B., De Schuymer, B.: On the transitivity of the comonotonic and countermonotonic comparison of random variables. J. Multivar. Anal. **98**, 177–193 (2007)

De Schuymer, B., De Meyer, H., De Baets, B.: Cycle-transitive comparison of independent random variables. J. Multivar. Anal. **96**, 352–373 (2005)

Fang, K.-.T, Kotz, S., Ng, K.W.: Symmetric Multivariate and Related Distributions. Monographs on Statistics and Applied Probability, vol. 36. Chapman and Hall/CRC, New York (1990)

Fishburn, P.C.: Binary choice probabilities: on the varieties of stochastic transitivity. J. Math. Psychol. **10**, 321–352 (1973)

Freson, S., De Baets, B., De Meyer, H.: Closing reciprocal relations w.r.t. stochastic transitivity. Fuzzy Sets Syst. **241**, 2–26 (2014)

Levy, H.: Stochastic Dominance: Investment Decision Making under Uncertainty. Studies in Risk and Uncertainty, vol. 12. Springer, New York (1998). https://doi.org/10.1007/978-1-4757-2840-8

Nadarajah, S., Kotz, S.: Reliability for some exponential distributions. Math. Probl. Eng. **2006**, 1–14 (2006)

Nelsen, R.B.: An Introduction to Copulas, 2nd edn. Springer, New York (2006). https://doi.org/10.1007/0-387-28678-0

Tahir, M.H., Cordeiro, G.M.: Compounding of distributions: a survey and new generalized classes. J. Stat. Distrib. Appl. **3**(1), 1–35 (2016). https://doi.org/10.1186/s40488-016-0052-1

Convergence of Copulas Revisited: Different Notions of Convergence and Their Interrelations

Nicolas Dietrich[1(✉)], Juan Fernández-Sánchez[2], and Wolfgang Trutschnig[1]

[1] Universität Salzburg, Hellbrunner Straße 34, 5020 Salzburg, Austria
{nicolaspascal.dietrich,wolfgang.trutschnig}@plus.ac.at

[2] Grupo de investigación de Teoría de Cópulas y Aplicaciones, Universidad de Almería, Carretera de Sacramento s/n, 04120 Almería, Spain
juanfernandez@ual.es

Abstract. Building upon the one-to-one relation between the family $\mathscr{C}$ of bivariate copulas and Markov operators we consider the metric OP_p corresponding to the L_p, $p \in [1, \infty]$ operator norm and study its interrelation with other metrics on $\mathscr{C}$. In particular we prove the surprising result that OP_1 convergence implies weak conditional convergence of the transposed copulas and establish the fact that the topology induced by OP_∞ is strictly finer than the topology induced by weak conditional convergence.

1 Introduction

It is well known that the space of all bivariate copulas $\mathscr{C}$ equipped with the uniform metric d_∞ forms a compact metric space. Moreover, the family of shuffles of the minimum copula M is dense in $\mathscr{C}$ with respect to d_∞, see Durante et al. (2009), Li et al. (1998), Nelsen (2007). As an immediate consequence, the independence copula Π can be approximated by completely dependent copulas in the metric d_∞. In other words: The metric d_∞ is not suitable to distinguish between complete dependence and independence. Overcoming this problem, Trutschnig (2011) introduced the metrics D_1 and D_∞, defined via the Markov kernels corresponding to the respective copulas. It is straightforward to see that convergence of a sequence of copulas in D_1/D_∞ implies convergence in d_∞, but not vice versa. Furthermore, Trutschnig (2011) showed that the space $(\mathscr{C}, D_1)$ is indeed complete and separable, with the class of all checkerboard copulas $\mathscr{CB}$ being dense.

Again working with Markov kernels, Kasper et al. (2021) recently introduced an even stronger concept of convergence on the space $\mathscr{C}$, the so called weak conditional convergence (WCC), subsequently in Sect. 2. This concept of convergence induces a topology which is strictly finer than the topology induced by D_1.

In this article we introduce additional (seemingly novel and mostly Markov kernel based) metrics on the space of all bivariate copulas and study how they

L. A. García-Escudero et al. (Eds.): SMPS 2022, AISC 1433, pp. 120–127, 2023.
https://doi.org/10.1007/978-3-031-15509-3_16

are related, firstly, to each other, and, secondly, to the aforementioned metrics/topologies. To be more precise: We examine whether convergence in one of the metrics/topologies implies convergence in any of the other metrics/topologies. Applying Scheffé's Theorem, we are able to express the metric D_∞, in terms of the corresponding doubly stochastic measures (see Eq. (5)), which naturally leads to the definition of the metric D_R which turns out to be even stronger than the metrics D_1/D_∞. The so obtained distance D_R induces the same topology as the mixed operator metric ρ, see Eq. (12), first introduced by Kim (1965).

Using the one-to-one correspondence between $\mathscr{C}$ and the family of all Markov operators and considering the L_p operator norm induced the metric OP_p, the induced topology of OP_p will be shown to be even finer than the topology induced by D_R. It is to a certain amount surprising that convergence with respect to OP_1 implies weak conditional convergence of the transposed copulas, but does, however, not automatically imply weak conditional convergence of the original copulas. At last, applying a duality argument, convergence in the operator metric OP_∞ can be shown to imply convergence in WCC, showing that, in contrary to the metrics D_1 and D_∞ (see Trutschnig 2011), the metrics OP_1 and OP_∞ do not induce the same topology.

2 Notation and Preliminaries

Throughout this paper the space of all bivariate copulas will be denoted by $\mathscr{C}$. As we will show in this paper, the space of all copulas $\mathscr{C}$ can be equipped with various different metrics. One of the most obvious such metrics is the uniform metric d_∞ given by

$$d_\infty(A,B) := \max_{x,y\in[0,1]} |A(x,y) - B(x,y)|, \tag{1}$$

for every $A, B \in \mathscr{C}$. It can be shown that the space $(\mathscr{C}, d_\infty)$ is compact and therefore complete and separable, see Willard (2004).

Let $(\Omega_1, \mathscr{A}_1)$ and $(\Omega_2, \mathscr{A}_2)$ be measurable spaces. A Markov kernel is a map $K\colon \Omega_1 \times \mathscr{A}_2 \to [0,1]$, where $\omega_1 \mapsto K(\omega_1, E)$ is a measurable function for all $E \in \mathscr{A}_2$ and $E \mapsto K(\omega_1, E)$ is a probability measure for all $\omega_1 \in \Omega_1$. It is widely known, see e.g. Durante and Sempi (2016), that every copula A corresponds to a special Markov kernel (in fact, a Markov kernel having the Lebesgue measure on [0,1] as fixed point) $K_A\colon [0,1] \times \mathscr{B}([0,1]) \to [0,1]$, where $\mathscr{B}([0,1])$ is the Borel σ-algebra. Trutschnig (2011) defined new metrics on $\mathscr{C}$ using the Markov kernels linked to the copulas and, for $p \in [1,\infty)$ considered the metric

$$D_p(A,B) = \left(\int_{[0,1]}\int_{[0,1]} |K_A(x,[0,y]) - K_B(x,[0,y])|^p d\lambda(x) d\lambda(y)\right)^{\frac{1}{p}} \tag{2}$$

and for $p = \infty$ the metric

$$D_\infty(A,B) = \sup_{y\in[0,1]} \int_{[0,1]} |K_A(x,[0,y]) - K_B(x,[0,y])| d\lambda(x), \tag{3}$$

for all $A, B \in \mathscr{C}$. Following Kasper et al. (2021) we will say that a sequence of copulas $(A_n)_{n\in\mathbb{N}}$ converges weakly conditional (in WCC) to a copula A if there exists a set $\Lambda \in \mathscr{B}([0,1])$ with $\lambda(\Lambda) = 1$ such that for every $x \in \Lambda$ the conditional distributions $K_{A_n}(x,\cdot)$ converge weakly to $K_A(x,\cdot)$. Every copula A is associated with a unique doubly stochastic measure μ_A, see Durante and Sempi (2016). Obviously the uniform metric d_∞ can be expressed via the corresponding doubly stochastic measures as

$$d_\infty(A,B) = \sup_{x,y\in[0,1]} |\mu_A([0,x]\times[0,y]) - \mu_B([0,x]\times[0,y])|. \tag{4}$$

The subsequent first result (whose proof is a direct application of Scheffé's theorem) allows us to characterize D_∞ in a similar manner, which, in turn, will motivate the definition of a third metric D_R of this type.

Theorem 1. *For all $A, B \in \mathscr{C}$ the following identity holds:*

$$D_\infty(A,B) = 2 \sup_{E,[0,y]\in\mathscr{B}([0,1])} \big|\mu_A(E\times[0,y]) - \mu_B(E\times[0,y])\big|. \tag{5}$$

Taking the supremum over arbitrary rectangles $E \times F \in \mathscr{B}([0,1]^2)$ in (5), induces the new metric D_R, defined as

$$D_R(A,B) := \sup_{E,F\in\mathscr{B}([0,1])} \big|\mu_A(E\times F) - \mu_B(E\times F)\big|, \tag{6}$$

for all $A, B \in \mathscr{C}$. It is straightforward to verify that D_R really is a metric on $\mathscr{C}$. Bringing into play Markov kernels, the metric D_R can alternatively be expressed as stated in Lemma 1:

Lemma 1. *For every $A, B \in \mathscr{C}$ we have*

$$D_R(A,B) = \frac{1}{2} \sup_{F\in\mathscr{B}([0,1])} \int_{[0,1]} |K_A(x,F) - K_B(x,F)| d\lambda(x). \tag{7}$$

Taking the supremum over arbitrary sets $G \in \mathscr{B}([0,1]^2)$ in Eq. (6), we obtain the generally known total variation metric, given by

$$D_{TV}(A,B) := \sup_{G\in\mathscr{B}([0,1]^2)} |\mu_A(G) - \mu_B(G)| \tag{8}$$

for arbitrary $A, B \in \mathscr{C}$. By definition, the topology induced by D_{TV} is finer than the one induced by D_R. Concerning the other implication, a counter example can be found according to Kim (1965).

A Markov operator on $L^1 := L^1([0,1], \mathscr{B}([0,1]), \lambda)$ is a positive, linear operator $T\colon L^1 \to L^1$, fulfilling $T(\mathbb{1}_{[0,1]}) = \mathbb{1}_{[0,1]}$ as well as $\int_{[0,1]}(Tf)(x)d\lambda(x) = \int_{[0,1]} f(x)d\lambda(x)$ for every $f \in L^1$. We will write $\mathscr{M}$ for the class of all Markov operators on L^1. Using the one-to-one correspondence between the class of all bivariate copulas $\mathscr{C}$ and the space of all Markov operators $\mathscr{M}$ (see Darsow et al. 1992), we will denote the Markov operator associated with the copula A by

T_A. Trutschnig (2011) showed that the Markov operator T_A corresponding to the copula A can also be expressed in terms of the Markov kernel as

$$(T_A f)(x) = \int_{[0,1]} f(y) K_A(x, dy), \tag{9}$$

for λ-almost every $x \in [0,1]$. Moreover, we can interpret the Markov operator T_A linked to a bivariate copula A as a linear operator on $L^p := L^p([0,1], \mathscr{B}([0,1]), \lambda)$ for every $p \in [1, \infty]$ and extend the operator to L^1 in the standard way if necessary. For every $p \in [1, \infty]$, the p-operator norm is defined as

$$|||T_A|||_p := \sup \{ \|T_A f\|_p : f \in L^p, \|f\|_p \leq 1 \}. \tag{10}$$

In the following we will simply write $\sup_{\|f\|_p \leq 1} \|T_A f\|_p$ instead of $\sup \{ \|T_A f\|_p : f \in L^p, \|f\|_p \leq 1 \}$. Obviously for every $A \in \mathscr{C}$ we have

$$|||T_A|||_1 = |||T_A|||_\infty = 1,$$

Riesz Thorin's theorem (see Dunford and Schwartz 1988) therefore yields

$$|||T_A|||_p = 1,$$

for every $p \in [1, \infty]$. For all $A, B \in \mathscr{C}$ and arbitrary $p \in [1, \infty]$ we set

$$OP_p(A, B) := |||T_A - T_B|||_p = \sup_{\|f\|_p \leq 1} \|T_A f - T_B f\|_p \tag{11}$$

and will refer to OP_p as the p-operator distance of A and B. Following Kim (1965) and "mixing" the L_p metrics yields the metric $\rho \colon \mathscr{C}^2 \to [0, 2]$, defined by

$$\rho(A, B) := \sup_{\|f\|_\infty \leq 1} \|T_A f - T_B f\|_1, \tag{12}$$

for all $A, B \in \mathscr{C}$. Surprisingly, according to Kim (1965, Proposition 2.3.2) the metrics ρ and D_R induce the same topology.

In the following let A^t denote the transposed of a copula $A \in \mathscr{C}$, i.e., the copula fulfilling $A^t(x, y) := A(y, x)$ for all $x, y \in [0,1]$. Interpreting T_A as an operator on L^1, it is well known (see Darsow et al. 1996; Trutschnig 2013) that the adjoint operator $T_A^{adj} : L^\infty \to L^\infty$ is given by $T_A^{adj} = T_{A^t}$ which, in turn, yields the following identity we will use in the sequel

$$OP_1(A, B) = |||T_A - T_B|||_1 = |||T_{A^t} - T_{B^t}|||_\infty = OP_\infty(A^t, B^t). \tag{13}$$

3 Interrelation of Metrics and Topologies

In this section we will clarify some of the interrelations between the aforementioned metrics and their induced topologies, respectively.

3.1 Relations Between the Operator Metrics OP_p and D_R

It is well known that weak convergence in $\mathscr{M}$ coincides with convergence with respect to d_∞ in $\mathscr{C}$. Furthermore, for arbitrary copulas $A, A_1, A_2, \dots$ the condition $\lim_{n\to\infty} D_R(A_n, A) = 0$ obviously implies $\lim_{n\to\infty} D_\infty(A_n, A) = 0 = \lim_{n\to\infty} D_\infty(A_n^t, A^t)$.

Theorem 2. *Consider bivariate copulas $A, A_1, A_2, \dots$ and let $p \in [1,\infty]$ be arbitrary. The convergence with respect to OP_p implies convergence with respect to D_R*

Proof. Fix $p \in [1,\infty)$ and let $A, A_1, A_2, \dots$ be as in the theorem. Then by applying Lemma 1 and Hölder's inequality it follows that

$$\begin{aligned}
D_R(A_n, A) &= \sup_{E,F\in\mathscr{B}([0,1])} |\mu_{A_n}(E\times F) - \mu_A(E\times F)| \\
&= \frac{1}{2} \sup_{F\in\mathscr{B}([0,1])} \int_{[0,1]} |K_{A_n}(x,F) - K_A(x,F)| d\lambda(x) \\
&= \frac{1}{2} \sup_{F\in\mathscr{B}([0,1])} \int_0^1 \left| \int_0^1 \mathbb{1}_F(y) K_{A_n}(x,dy) - \int_0^1 \mathbb{1}_F(y) K_A(x,dy) \right| d\lambda(x) \\
&\leq \frac{1}{2} \sup_{F\in\mathscr{B}([0,1])} \left(\int_0^1 \left| \int_0^1 \mathbb{1}_F(y) K_{A_n}(x,dy) - \int_0^1 \mathbb{1}_F(y) K_A(x,dy) \right|^p d\lambda(x) \right)^{\frac{1}{p}} \\
&\leq \frac{1}{2} \sup_{\|f\|_p\leq 1} \left(\int_0^1 \left| \int_0^1 f(y) K_{A_n}(x,dy) - \int_0^1 f(y) K_A(x,dy) \right|^p d\lambda(x) \right)^{\frac{1}{p}} \\
&= \frac{1}{2} \sup_{\|f\|_p\leq 1} \|T_{A_n} f - T_A f\|_p \\
&= \frac{1}{2} OP_p(A_n, A) \overset{n\to\infty}{\longrightarrow} 0.
\end{aligned}$$

The case $p = \infty$ follows similarly.

Remark 1.

1. Alternatively, we can use the equivalence of D_R and ρ, going back to Kim (1965), to prove Theorem 2. The result is then a direct consequence of Hölder's inequality and the fact that the unit ball in L^∞ is contained in the unit ball of L^p, for every $p \in [1,\infty)$.
2. The converse of Theorem 2 does not hold in general, a counter example can be constructed.

3.2 Relations Between the Operator Metrics OP_p and Weak Conditional Convergence

According to (Trutschnig 2011, Lemma 7) weak conditional convergence implies convergence with respect to D_1 and with respect to D_∞. It is not difficult to construct an example showing that the reverse implication does not need to hold in general. Focusing on important subclasses of copulas, however, the two notions may coincide: According to Kasper et al. (2021) in the family of Archimedean copulas as well as in the family of all Extreme Value copulas standard pointwise/uniform convergence and weak conditional convergence coincide.

We are now going to show that convergence in the operator metric OP_1 implies weak conditional convergence of the transposed copulas and start with the following lemma.

Lemma 2. *Let $A, A_1, A_2, \ldots$ be bivariate copulas. Suppose that*

$$\lim_{n\to\infty} OP_1(A_n, A) = 0$$

and let $E \in \mathscr{B}([0,1])$ be arbitrary. Then there exists a set $\Lambda_E \in \mathscr{B}([0,1])$ with $\lambda(\Lambda_E) = 1$ such that for every $x \in \Lambda_E$ we have $\lim_{n\to\infty} K_{A_n^t}(x, E) = K_{A^t}(x, E)$.

Proof. Let Λ_E^0 denote the set of all Lebesgue points (see, Rudin 1987) of the function $x \mapsto K_{A^t}(x, E)$ and Λ_E^j the set of all Lebesgue points of the function $x \mapsto K_{A_j^t}(x, E)$ for $j \in \mathbb{N}$. Setting $\Lambda_E := \bigcap_{j=0}^{\infty} \Lambda_E^j \cap (0,1)$, then according to Rudin (1987) we have $\lambda(\Lambda_E) = 1$. Fix $x_0 \in \Lambda_E$ and for $m \in \mathbb{N}$ define the function $f_m : [0,1] \to \mathbb{R}$ by $f_m := m\mathbb{1}_{[x_0 - \frac{1}{2m}, x_0 + \frac{1}{2m}]}$. For sufficiently large $m \in \mathbb{N}$ we have $[x_0 - \frac{1}{2m}, x_0 + \frac{1}{2m}] \subseteq [0,1]$, hence considering $T_A f_m(x) = m\, K_A(x, [x_0 - \frac{1}{2m}, x_0 + \frac{1}{2m}])$ it follows that

$$\begin{aligned}\int_E T_A f_m \, d\lambda &= m \int_E K_A(x, [x_0 - \tfrac{1}{2m}, x_0 + \tfrac{1}{2m}]) \, d\lambda \\ &= m\mu_A\left(E \times [x_0 - \tfrac{1}{2m}, x_0 + \tfrac{1}{2m}]\right) \\ &= m\mu_{A^t}\left([x_0 - \tfrac{1}{2m}, x_0 + \tfrac{1}{2m}] \times E\right).\end{aligned}$$

Considering that x_0 is a Lebesgue point of $x \mapsto K_{A^t}(x, E)$ it follows that the last expression converges to $K_{A^t}(x_0, E)$ for $m \to \infty$. Since the same arguments apply to A replaced by A_n, setting $I_m = [x_0 - \frac{1}{2m}, x_0 + \frac{1}{2m}]$ we get

$$\begin{aligned}\left|K_{A_n^t}(x_0, E) - K_{A^t}(x_0, E)\right| &= \lim_{m\to\infty} \left(m \left|\mu_{A_n}(E \times I_m) - \mu_A(E \times I_m)\right|\right) \\ &\le \sup_{m\in\mathbb{N}} \left(m \left|\mu_{A_n}(E \times I_m) - \mu_A(E \times I_m)\right|\right) \\ &\le \sup_{m\in\mathbb{N}} \|T_{A_n} f_m - T_A f_m\|_1 \\ &\le |||T_{A_n} - T_A|||_1.\end{aligned}$$

We have therefore shown that $\lim_{n\to\infty} K_{A_n^t}(x, E) = K_{A^t}(x, E)$ holds for every $x \in \Lambda_E$.

The previous lemma enables us to prove weak conditional convergence of the transposed copulas as follows:

Theorem 3. *Let $A, A_1, A_2, \ldots \in \mathscr{C}$. If $\lim_{n\to\infty} OP_1(A_n, A) = 0$, then $(A_n^t)_{n\in\mathbb{N}}$ converges weakly conditional to A^t.*

Proof. Let $\mathbb{Q}_{0,1}$ denote the set of all rational numbers in $[0, 1]$. Then for every $y \in \mathbb{Q}_{0,1}$, according to Lemma 2, there exists a set $\Lambda_y := \Lambda_{[0,y]}$ with $\lambda(\Lambda_y) = 1$ such that we have $\lim_{n\to\infty} K_{A^t_n}(x, [0, y]) = K_{A^t}(x, [0, y])$ for every $x \in \Lambda_y$. Setting $\Lambda := \bigcap_{y\in\mathbb{Q}_{0,1}} \Lambda_y$, we get $\lambda(\Lambda) = 1$ and for every $x \in \Lambda$ we have $\lim_{n\to\infty} K_{A^t_n}(x, [0, y]) = K_{A^t}(x, [0, y])$ for every $y \in \mathbb{Q}_{0,1}$. In other words: For every $x \in \Lambda$ the conditional distribution functions $y \mapsto K_{A^t_n}(x, [0, y])$ converge on a dense set. As a direct consequence (see Billingsley 1968) the sequence $(K_{A^t_n}(x, \cdot))_{n\in\mathbb{N}}$ converges weakly to $K_{A^t}(x, \cdot)$ for every $x \in \Lambda$.

Using Eq. (13) yields the following complementing result:

Corollary 1. *Let $A, A_1, A_2, ...$ be bivariate copulas. If*

$$\lim_{n\to\infty} OP_\infty(A_n, A) = 0$$

then $(A_n)_{n\in\mathbb{N}}$ converges weakly conditional to A.

Proof. Direct consequence of Eq. (13) in combination with Theorem 3.

Remark 2.

1. Note that Corollary 1 also holds for the strong operator topology on L^∞. In other words: Taking bivariate copulas $A, A_1, A_2, ...$ with $\|T_{A_n} - T_A\|_\infty \overset{n\to\infty}{\longrightarrow} 0$ for all $f \in L^\infty$, implies $A_n \overset{n\to\infty}{\longrightarrow} A$ weakly conditional.
2. Corollary 1 does not hold for any other $p \in [1, \infty)$, i.e. we are able to find a sequence of copulas converging in the operator metric OP_p for any $p \in [1, \infty)$ but not weakly conditional.
3. As a consequence of the previous point and Corollary 1, we get that the metrics OP_1 and OP_∞ cannot induce the same topology.
4. It is also worth mentioning that neither does convergence in WCC imply convergence in D_R, nor vice versa. It is possible to find counter examples in both directions.

References

Billingsley, P.: Convergence of Probability Measures. Wiley Series in Probability and Mathematical Statistics. Wiley, New York (1968)

Darsow, W.F., Nguyen, B., Olsen, E.T.: Copulas and Markov processes. Ill. J. Math. **36**(4), 600–642 (1992)

Darsow, W.F., Nguyen, B., Olsen, E.T.: Copulas and Markov operators. In: Proceedings of the Conference on Distributions with Fixed Marginals and Related Topics. IMS Lecture Notes, Monograph Series, vol. 28, pp. 244–259 (1996)

Dunford, N., Schwartz, J.T.: Linear Operators. Parts I and II. Wiley, Hoboken (1988)

Durante, F., Sarkoci, P., Sempi, C.: Shuffles of copulas. J. Math. Anal. Appl. **352**(2), 914–921 (2009)

Durante, F., Sempi, C.: Principles of Copula Theory. CRC Press, Boca Raton (2016)

Kasper, T.M., Fuchs, S., Trutschnig, W.: On weak conditional convergence of bivariate Archimedean and Extreme Value copulas, and consequences to nonparametric estimation. Bernoulli **27**(4), 2217–2240 (2021)

Kim, C.: Uniform approximation and almost periodicity of doubly stochastic operators. Ph.D. thesis, University of Washington (1965)
Li, X., Mikusinski, P., Taylor, M.D.: Strong approximation of Copulas. J. Math. Anal. Appl. **225**, 608–623 (1998)
Nelsen, R.: An Introduction to Copulas. Springer Series in Statistics. Springer, New York (2007). https://doi.org/10.1007/0-387-28678-0
Rudin, W.: Real and Complex Analysis. McGraw-Hill Education, London (1987)
Trutschnig, W.: On a strong metric on the space of copulas and its induced dependence measure. J. Math. Anal. Appl. **384**, 690–705 (2011)
Trutschnig, W.: On Cesáro convergence of iterates of the Star Product of copulas. Stat. Probab. Lett. **83**, 357–365 (2013)
Willard, S.: General Topology. Addison-Wesley Series in Mathematics. Dover Publications Inc., New York (2004)

An INDSCAL-Type Approach for Three-Way Spectral Clustering

Cinzia Di Nuzzo[1(✉)], Salvatore Ingrassia[2], and Donatella Vicari[1]

[1] Department of Statistics, Sapienza University of Rome, Piazzale Aldo Moro 5, 00185 Rome, Italy
{cinzia.dinuzzo,donatella.vicari}@uniroma1.it

[2] Department of Economics and Business, University of Catania, Corso Italia, 55, 95129 Catania, Italy
salvatore.ingrassia@unict.it

Abstract. A spectral clustering method for three-way data using an INDSCAL-type approach is introduced. Our proposal takes into account the single geometrical structures for each occasion and returns a common low-dimensional space which is representative for all occasions. Finally, a real application to three-way data is presented to illustrate the potential of the method.

1 Introduction

The spectral clustering algorithm is a technique based on the properties of the Laplacian matrix of a pairwise similarity matrix. This is also a useful approach for high-dimensional data since the units are clustered not in the original space but in a suitable feature space with a reduced number of dimensions, see e.g. von Luxburg (2007) for an introductive review. Relationships between spectral clustering and multidimensional scaling has been investigated in Bavaud (2006) that formalizes the intuition that the embedding subspace generated from the spectral clustering is equivalent to the subspace generated from classical multidimensional scaling; specifically, the link between these two subspaces can be expressed through a linear relation that connects the Laplacian matrix used in spectral clustering to the double-centered matrix used in multidimensional scaling.

In this paper, we focus on the extension of spectral clustering to three-way data. Spectral clustering for three-way data has been recently introduced in Di Nuzzo and Ingrassia (2022b). From a different perspective, we provide here a tensorial structure for three-way spectral clustering through an INDSCAL-type approach since the INDSCAL model can be considered as a natural extension of the classical multidimensional scaling to three-way data (Carroll and Chang 1970). We propose a spectral clustering method for three-way data that, starting from all Laplacian matrices of the single occasions, returns only one common embedding space. Considering all the Laplacian matrices allows to take into

L. A. García-Escudero et al. (Eds.): SMPS 2022, AISC 1433, pp. 128–135, 2023.
https://doi.org/10.1007/978-3-031-15509-3_17

account the differences between occasions and the INDSCAL model allows to synthesize such differences in a single common configuration.

The rest of the paper is structured as follows. In Sect. 2 we summarize some background results; Sect. 3 introduces our proposal. Finally, in Sect. 4 a real application to three-way data is presented.

2 Background Results

In this section we present background theory concerning four main topics: spectral clustering, multidimensional scaling, the equivalence between spectral clustering and multidimensional scaling, the INDSCAL approach.

Spectral Clustering. Let $V = \{\boldsymbol{x}_1, \dots \boldsymbol{x}_n\}$ be a set of points in $\mathbb{R}^p$. Spectral clustering is a flexible method based on the graph theory, where the units $\boldsymbol{x}_i$ are the nodes of the graph and the edges represent the similarities w_{ij} between pairs of units $\boldsymbol{x}_i$ and $\boldsymbol{x}_j$; finally, let $W = (w_{ij})$ be a *similarity matrix*. The main steps of the spectral clustering can be summarized as follows, (see, e.g., von Luxburg (2007) and Ng et al. (2002)).

1. Create the $(n \times n)$ *normalized symmetric Laplacian* matrix
$$L_{\text{sym}} = D^{-1/2} W D^{-1/2}, \tag{1}$$
where D =diag$(d_1, \dots, d_n)$ is the degree matrix and d_i is the degree of a vertex $\boldsymbol{x}_i$ defined as $d_i = \sum_{j=1}^{n} w_{ij}$, for $i = 1, \dots, n$.
2. Introduce the *normalized Laplacian embedding* as the map
$$\phi_\Gamma : \{\boldsymbol{x}_1, \dots, \boldsymbol{x}_n\} \to \mathbb{R}^K, \qquad \phi_\Gamma(\boldsymbol{x}_i) = (\gamma_{1i}, \dots, \gamma_{Ki}), \text{ for } i = 1, \dots, n,$$
where $\gamma_{1i}, \dots, \gamma_{Ki}$ are the *i-th* components of the first largest K eigenvectors of L_{sym}. Afterwards, create the embedding space as the $(n \times K)$ matrix $\tilde{B} = (\phi_\Gamma(\boldsymbol{x}_1), \dots, \phi_\Gamma(\boldsymbol{x}_n))'$, and normalize the rows of Y to have unit lengths.
3. Run a clustering algorithm on the normalized matrix Y.

The last step is usually carried out using the k-means algorithm; another approach is based on taking into account mixture models (Di Nuzzo and Ingrassia 2022a).

Multidimensional Scaling. Given a pairwise $(n \times n)$ dissimilarity matrix Δ, the goal of MultiDimensional scaling (MDS) is to reconstruct a low-dimensional map that preserves distances, (Borg and Gröenen 2005). In practice, MDS finds the centered configuration such that the pairwise Euclidean distances between units best approximate the corresponding proximities in Δ. The classical MDS solution provides the coordinate matrix L which is derived by eigendecomposition. The main steps to perform MDS are summarized as follows.

1. Set up the matrix of squared dissimilarity Δ.
2. Consider the double centering $S = -\frac{1}{2}(I - \frac{1}{n}J_n)\Delta(I - \frac{1}{n}J_n)$, where I is the identity matrix of order n, and J_n is an $n \times n$ matrix of all ones.

3. Compute the first K largest eigenvalues of S, $\lambda_1, \ldots, \lambda_K$ and the corresponding eigenvectors $\boldsymbol{u}_1, \ldots, \boldsymbol{u}_K$.
4. Create the low-dimensional configuration $L = U_K \Lambda_K^{1/2}$, where U_K is the matrix of the K eigenvectors, and Λ_K is the diagonal matrix of the K eigenvalues of S, respectively.

Equivalence Between Spectral Clustering and Multidimensional Scaling. From a geometrical point of view, spectral clustering is a method for partitioning an undirect graph using the eigen-decomposition of the symmetric Laplacian matrix derived from the similarity matrix between pairs of units; in this framework, spectral clustering performs dimensionality reduction. Similarly, multidimensional scaling uses the spectral decomposition of a matrix to perform dimensionality reduction. The relationship providing the link between these two methods has been introduced by Bavaud (2006), where a linear relation that connects the Laplacian matrix used in spectral clustering to the centered matrix used in multidimensional scaling is described. In particular, the following theorem holds.

Theorem 1. *Let W be an $(n \times n)$ weighted matrix with associated symmetric Laplacian matrix $L_{\text{sym}} = U\Gamma U'$, and the degree vector $d = (d_1, \ldots, d_n)$. Then, any $(n \times n)$ matrix S of the form*

$$S := (a-b)L_{\text{sym}} + (a+b)I - 2a\sqrt{d}\sqrt{d}' \tag{2}$$

is, for $a, b \geq 0$, a centered matrix with spectral decomposition $U\Lambda U'$.

In practice, the subspace spanned by the eigenvectors of S is equal to the subspace spanned by the eigenvectors of L_{sym}.

INDSCAL Model. MDS is a methodology where dissimilarity data are modeled to fit distances in a low-dimensional space. When H subjects evaluate dissimilarities between pairs of n units, a two-mode three-way array derives. A multidimensional scaling model for three-way data called INDSCAL (INDividual SCALing) has been introduced by Carroll and Chang (1970).

The INDSCAL model simultaneously analyzes H similarity matrices between n units. For example, H subjects may be asked to evaluate the similarities between pairs of products. The INDSCAL model relies on the assumption that a single representation of the n units (called *common space*) exists and each occasion h ($h = 1, \ldots, H$) differs in the weights that are attached to these dimensions. For example, albeit in a common perception of the similarities between products, each subject has his own configuration space of the n products. In the ideal case, in the absence of heterogeneity the common space is fully representative of the same H individual configurations.

Let S_h $(h = 1, \ldots, H)$ be an $(n \times n)$ symmetric similarity matrix, the INDSCAL model decomposes each matrix as

$$S_h = BC_hB' + E_h, \tag{3}$$

where B is the $(n \times K)$ matrix of the coordinates of the units in the common space with K dimensions, C_h is the $(K \times K)$ diagonal matrix of the weights of occasion h, and E_h is the error term. Note that the common space B is unique for all occasions, while C_h provides individual weights for each occasion.

The INDSCAL model is fitted in a least-squares framework by an ALS algorithm which minimizes the following function (see Carroll and Chang 1970, for details):

$$\sum_{h=1}^{H} \|S_h - BC_hB'\|^2. \tag{4}$$

3 A Three-Way Extension of Spectral Clustering Method Through an INDSCAL-Type Approach

Exploiting the results from Theorem 1, which establishes the link between SC and MDS, and considering that INDSCAL can be considered as a three-way extension of classical MDS, we propose an extension of spectral clustering for three-way data by adopting an INDSCAL-type approach.

Let $\boldsymbol{X} = \{x_{ijh} : i = 1, \ldots, n, j = 1, \ldots, p, h = 1, \ldots, H\}$ be a three-way data set, where x_{ijh} describes the value of the j-th variable observed on the i-th unit, at the h-th occasion. We denote with X_h the h-th $(n \times p)$ unit-variable matrix or the h-th frontal slice of the three-way array.

For each X_h $(h = 1, \ldots, H)$, we consider a $(n \times n)$ symmetric similarity matrix W_h. Moreover, we consider the degree matrices D_h $(h = 1, \ldots, H)$. Likewise, let $\mathbf{L}_{\text{sym}}$ be the normalized symmetric Laplacian tensor, whose frontal slices are defined by $L_{\text{sym}_h} = D_h^{-1/2} W_h D_h^{-1/2}$ $(h = 1, \ldots, H)$. Finally, we consider the tensor $\boldsymbol{S}$, whose frontal slices are defined as in (2), i.e., $S_h = (a-b)L_{\text{sym}_h} + (a+b)I - 2a\sqrt{d_h}\sqrt{d_h}'$, where d_h represents the degree vector for each $h = 1, \ldots, H$.

Therefore, we have a set of $(n \times n)$ matrices S_h $(h = 1, \ldots, H)$ and we can fit the INDSCAL model (3). Moreover, since in the spectral clustering method the resulting low-embedding is orthogonal, we impose the orthogonality constraint for B in (3). Finally, we apply the k-means algorithm on the $(n \times K)$ common space B. The method is summarized in Fig. 1.

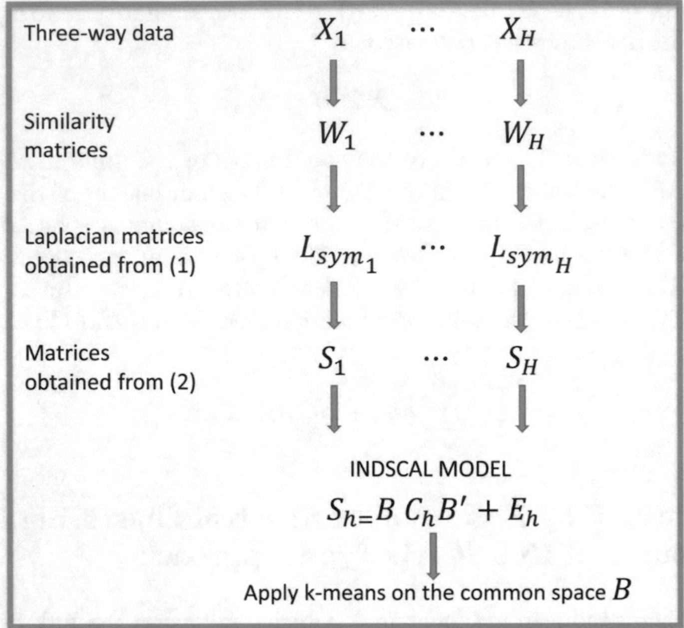

Fig. 1. Graphical sketch of the proposed method for three-way data

4 A Real Illustrative Application: Blue Crabs Data

Blue crabs data are available in https://three-mode.leidenuniv.nl/ (see Gemperline et al. (1992) and Kroonenberg et al. (2004)). The data refer to gill, hepatopancreas, and muscle tissue samples taken from blue crabs which, according to their origin and health status, belong to three categories: Albermarle Sound (Healthy), Pamlico River (Healthy), Pamlico River (Diseased). The data consist of 48 units (tissue samples) by 3 variables (tissue types) by 25 occasions (trace elements).

The three-way spectral clustering method described in Sect. 3 has been applied to this data set.

The similarity tensor $\boldsymbol{W}$ has been computed according to the symmetric kernel function (see Zelnik-Manor and Perona 2004)

$$w_{ij} = \exp\left(-\frac{\|\boldsymbol{x}_i - \boldsymbol{x}_j\|^2}{\epsilon_i \epsilon_j}\right), \tag{5}$$

with $\epsilon_i = \|\boldsymbol{x}_i - \boldsymbol{x}_m\|$, where $\boldsymbol{x}_m$ is the m-neighbor of the point $\boldsymbol{x}_i$, for $i, j = 1, \ldots, n$, and $m = 2$, while matrices S_h have been computed as in (2) by setting $a = 2$ and $b = 1.5$. Our algorithm was run in R; the INDSCAL model was fitted by using the `multiway` R package (see Helwig 2019) and setting the orthogonality constraint on the common space B and 50 starting points. By setting $K = 3$ as the number of dimensions for the low-dimensional space, we obtain exactly the real classes, as displayed in Fig. 2 where the common space has been represented and the colors identify the membership to the clusters. Moreover, the INDSCAL model returns the following weights: C_1 =diag$(0.0786, 0.0792, 0.0773)$, C_2 =diag$(0.0799, 0.0807, 0.0804)$ and C_3 =diag$(0.0787, 0.0812, 0.0813)$ for some individual spaces corresponding to $h = 1$, 6 and 24, respectively.

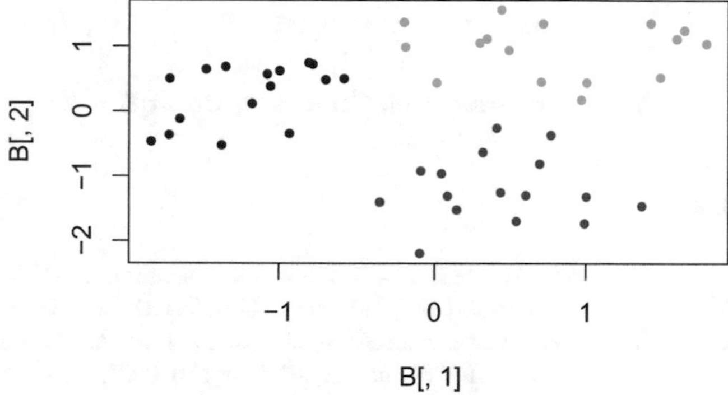

Fig. 2. *Blue Crabs data.* Common space for $K = 3$

Table 1. *Blue Crabs data.* PseudoF indices for different numbers of dimensions

K = 2	K = 3	K = 4	K = 5	K = 6
30.799	**31.882**	18.520	16.285	15.89

As far as the selection of the number of groups, the PseudoF index (Calinski and Harabasz 1974) has been considered. From Table 1, where the PseudoF values for $K = 2, 3, 4, 5, 6$ are listed, it emerges that the maximum value is attained at $K = 3$, even if the value for $K = 2$ is very close to the maximum. Therefore, we carried out our method again by setting $K = 2$. In Fig. 3 the common space has been represented and the colors highlight the k-means clustering result. It is worth noting that the clustering result identifies exactly two classes according to the origin of the crabs: 16 crabs belong to Albemarle Sound and 32 crabs from Pamlico River regardless of the health status.

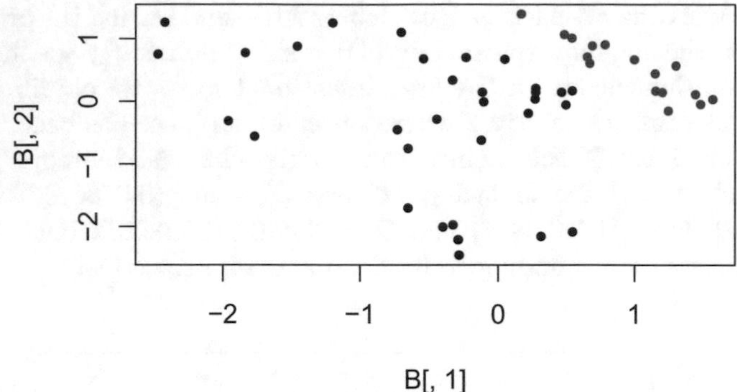

Fig. 3. *Blue Crabs data.* Common space for $K = 2$

References

Bavaud, F.: Spectral clustering and multidimensional scaling: a unified view. In: Batagelj, V., Bock, H.H., Ferligoj, A., Žiberna, A. (eds.) Data Science and Classification. Studies in Classification, Data Analysis, and Knowledge Organization, pp. 131–139. Springer, Heidelberg (2006). https://doi.org/10.1007/3-540-34416-0_15

Borg, I., Gröenen, P.J.F.: Modern Multidimensional Scaling: Theory and Applications, 2nd edn. Springer, New York (2005). https://doi.org/10.1007/0-387-28981-X

Calinski, T., Harabasz, J.: A dendrite method for cluster analysis. Commun. Stat. **3**, 1–27 (1974)

Carroll, J.D., Chang, J.J.: Analysis of individual differences in multidimensional scaling via an N-generalization of the Eckart-Young decomposition. Psychometrika **35**, 283–319 (1970)

Di Nuzzo, C., Ingrassia, S.: A mixture model approach to spectral clustering and application to textual data. Stat. Methods Appl. (2022). https://doi.org/10.1007/s10260-022-00635-4

Di Nuzzo, C., Ingrassia, S.: Three-way spectral clustering. In: Brito, P., Dias, J.G., Lausen, B., Montanari, A., Nugent, R. (eds.) Classification and Data Science in the Digital Age. Springer, Cham (2022, forthcoming)

Gemperline, P.J., Miller, K.H., West, T.L., Weinstein, J.E., Hamilton, J.C., Bray, J.T.: Principal component analysis, trace elements, and blue crab shell disease. Anal. Chem. **64**, 523–531 (1992)

Helwig, N.E.: multiway: Component Models for Multi-Way Data. R package version 1.0-6 (2019). https://cran.r-project.org/package=multiway

Kroonenberg, P.M., Basford, K.E., Gemperline, P.J.: Grouping three-mode data with mixture methods: the case of the diseased blue crabs. J. Chemom. **18**, 508–518 (2004)

Ng, A.Y., Jordan, M., Weiss, Y.: On spectral clustering: analysis and an algorithm. In: Dietterich, T., Becker, S., Ghahramani, Z. (eds.) Advances in Neural Information Processing Systems (NIPS 2001), vol. 14. MIT Press (2002). https://proceedings.neurips.cc/paper/2001/file/801272ee79cfde7fa5960571fee36b9b-Paper.pdf

von Luxburg, U.: A tutorial on spectral clustering. Stat. Comput. **17**(4), 395–416 (2007)
Zelnik-Manor, L., Perona, P.: Self-tuning spectral clustering. In: Saul, L., Weiss, Y., Bottou, L. (eds.) Advances in Neural Information Processing Systems (NIPS 2004), vol. 17. MIT Press (2004). https://proceedings.neurips.cc/paper/2004/file/40173ea48d9567f1f393b20c855bb40b-Paper.pdf

On Clustering of Star-Shaped Sets with a Fuzzy Approach: An Application to the Clasts in the Cantabrian Coast

Maria Brigida Ferraro[1(✉)], Elena Fernández Iglesias[2], Ana Belén Ramos-Guajardo[3], and Gil González-Rodríguez[3]

[1] Department of Statistical Sciences, Sapienza University of Rome, Rome, Italy
mariabrigida.ferraro@uniroma1.it

[2] INDUROT, University of Oviedo, Mieres, Spain
elena.indurot@uniovi.es

[3] INDUROT/Department of Statistics, OR and MD, University of Oviedo, Mieres/Oviedo, Spain
ramosana@uniovi.es, gil@uniovi.es

Abstract. Star-shaped sets represent a large class of sets that contains compact and convex ones as a particular case. The center-radial characterization of these sets is very useful to identify their original shape. The complexity of these data lies in the fact that several aspects of the data are considered in a single representation. Since in some situations it may be very useful to identify groups of these sets, a fuzzy clustering method has been proposed. It is a generalization of the well-know fuzzy k-means that takes into account the nature of the data through an appropriate distance measure. The adequacy of the proposal has been checked by means of a real application on the clasts in the Cantabrian Coast.

1 Preliminaries on Star-Shaped Sets

Before introducing the fuzzy clustering proposal, some preliminaries on the type of data that will be used are necessary.

Let the space $\mathbb{R}^p$ be endowed with the Euclidean norm $\|\cdot\|$ and the corresponding inner product $\langle\cdot,\cdot\rangle$ and let $\mathbb{S}^{p-1} = \{u \in \mathbb{R}^p : \|u\| = 1\}$ be the hypersphere with radius 1. The theory of the well-known space of all non-empty compact and convex subsets of $\mathbb{R}^p$, denoted by $\mathscr{K}_c(\mathbb{R}^p)$, has been deeply studied in the literature in works as Matheron (2018), Molchanov (2014) and Simó (2004). The tool that allows to embed such a space into a linear functional space is the support function, which is also useful for preserving some metric and arithmetic properties. However, as it has been shown in Ramos-Guajardo et al. (2018), the support function identifies the boundary of the set, but the obtained result is not easy to relate with the original shape of the set. To overcome this drawback, a general alternative based on star-shaped sets can be considered (see, for instance, González-Rodríguez et al. 2018) whose employment, according to Klain (1997), relaxes the condition of convexity to directional convexity from a given point.

L. A. García-Escudero et al. (Eds.): SMPS 2022, AISC 1433, pp. 136–143, 2023.
https://doi.org/10.1007/978-3-031-15509-3_18

Thus, star-shaped sets are parameterized by means of a crisp centre (related to the location of the set) and a radial function (related to the imprecision of the set) and, in addition, uncertainty spreading is given directionally, in contrast to the general dilation produced by the Minkowski sum.

Formally, the space of general star-shaped sets of $\mathbb{R}^p$ is defined as $\mathbb{X}(\mathbb{R}^p) = (\mathbb{R}^p \times \mathbb{X}_0(\mathbb{R}^p))$, where $\mathbb{X}_0(\mathbb{R}^p)$ is the space of star-shaped sets of $\mathbb{R}^p$ w.r.t. 0, defined so that

$$\mathbb{X}_0(\mathbb{R}^p) = \left\{A \subset \mathbb{R}^p \mid \gamma a \in A \text{ for all } a \in A \text{ and all } \gamma \in [0,1]\right\}.$$

A star-shaped set $A \in \mathbb{X}(\mathbb{R}^p)$ is characterized by means of two elements:

- its center, $c_A \in \mathbb{R}^p$, which is its location point such that for all $a \in A$ $\lambda c_A + (1-\lambda)a \in A$ for all $\lambda \in [0,1]$;
- the radial function (see Schneider 1993) of the associated set centred at 0, i.e., the radial function of $A^c = A - c_A$, which is defined as $\rho_{A^c} : \mathbb{S}^{p-1} \to \mathbb{R}^+$ so that $\rho_{A^c}(u) = \sup\{\lambda \geq 0 : \lambda u \in A^c\}$ and which is related to the imprecision of the set.

Thus, by considered the so-called *center-radial characterization* of a star-shaped set $A \in \mathbb{X}(\mathbb{R}^p)$, denoted by (c_A, ρ_{A^c}), A can be expressed as follows:

$$A = \{c_A + \gamma \rho_{A_c}(u) | \gamma \in [0,1], u \in \mathbb{S}^{p-1}\},$$

where $\rho_A = \sup\{\lambda \geq 0 : \lambda u \in A\}$.

The arithmetic between star-shaped sets is associated with a directional propagation of the imprecision, and extends the interval Minkowski arithmetic as follows: $A^c + B^c = \{\gamma(\rho_{A^c}(u) + \rho_{B^c}(u)) | \gamma \in [0,1], u \in \mathbb{S}^{p-1}\}$ and $\gamma \cdot A^c = \{\gamma(|\lambda| \rho_{A^c}(sign(\lambda)u)) | \gamma \in [0,1], u \in \mathbb{S}^{p-1}\}$. Therefore, it is clear that $A + B = (c_A + c_B, \rho_{A^c} + \rho_{B^c})$, and $\lambda(c_A, \rho_{A^c}(\cdot)) = (\lambda c_A, |\lambda| \rho_{A^c}(sign(\lambda)))$.

On the other hand, the space $\mathbb{X}(\mathbb{R}^p)$ can be embedded into a cone on the Hilbert space $\mathscr{H}_r = \mathbb{R}^p \times \mathscr{L}^2(\mathbb{S}^{p-1})$ through the center-radial characterization. Thus, for theoretical purposes, star-shaped sets belonging to the cone $\mathbb{X}^*(\mathbb{R}^p)$ will be considered, where $\mathbb{X}^*(\mathbb{R}^p) = \left\{A \in \mathbb{X}(\mathbb{R}^p) | \rho_A \in \mathscr{L}^2(\mathbb{S}^{p-1})\right\}$.

Concerning the metric structure in $\mathbb{X}^*(\mathbb{R}^p)$, the center-radial characterization induces a natural family of distances from the corresponding one in the associated Hilbert space. Thus, the distance between two star-shaped sets $A, B \in \mathbb{X}^*(\mathbb{R}^p)$ was introduced in González-Rodríguez et al. (2018) and it is given by

$$d_\tau(A, B) = \sqrt{\tau \|c_A - c_B\|^2 + (1-\tau) \|\rho_A - \rho_B\|^2_{\mathscr{L}^2}}, \tag{1}$$

where $\tau \in (0,1)$ determines the importance given to the location in contrast to the imprecision, $\|\cdot\|$ denotes the usual norm in $\mathbb{R}^p$ and $\|\cdot\|_{\mathscr{L}^2}$ is the usual L_2-type norm in $\mathscr{L}^2(\mathbb{S}^{p-1})$.

2 Fuzzy Clustering Proposal

In this section a fuzzy k-means method for star-shaped sets is provided. Given n star-shaped sets $\mathbf{x}_i = (c_{\mathbf{x}_i}, \rho_{\mathbf{x}_i})$ in $\mathbb{R}^p$, the aim is to partition them in k clusters such that sets characterized by similar features are assigned to the same clusters (inner cohesion) and the dissimilar ones to different clusters (separation between clusters). In order to do that, a proper distance/dissimilarity measure for the considered object space has to be adopted. In this case, we use the distance $d_\tau(\cdot,\cdot)$, defined in the previous section. The fuzzy k-means method (Bezdek 1981) can be generalized to the case of star-shaped sets (FkM-SSS) and the resulting minimization problem can be formalized as:

$$\begin{aligned} \min_{\mathbf{U},\mathbf{h}} J_{\mathrm{F}k\mathrm{M\text{-}SSS}} &= \sum_{i=1}^{n}\sum_{g=1}^{k} u_{ig}^m d_\tau^2\left((c_{\mathbf{x}_i}, \rho_{\mathbf{x}_i}), (c_{\mathbf{h}_g}, \rho_{\mathbf{h}_g})\right) \\ &= \sum_{i=1}^{n}\sum_{g=1}^{k} u_{ig}^m \left(\tau \left\|c_{\mathbf{x}_i} - c_{\mathbf{h}_g}\right\|^2 + (1-\tau)\|\rho_{\mathbf{x}_i} - \rho_{\mathbf{h}_g}\|_{\mathscr{L}^2}\right) \\ \text{s.t.} \quad & u_{ig} \in [0,1], \sum_{g=1}^{k} u_{ig} = 1. \end{aligned} \tag{2}$$

where $\mathbf{U} = [u_{ig}]$ is the membership degree matrix of order $(n \times k)$, $\mathbf{h} = [h_1, \cdots h_g, \cdots, h_k]$, with $h_g = (c_{\mathbf{h}_g}, \rho_{\mathbf{h}_g})$, is the vector of k prototypes, and $m(> 1)$ is the parameter of fuzziness (usually $m = 2$). If m tends to 1, the membership degrees are close to 0 and 1, hence FkM reduces to the usual k-means algorithm. On the other hand, as the values of m increase, the membership degrees move away from 0 and 1 and the partition is more fuzzy.

The optimal solution is obtained by means of an iterative algorithm, where two steps alternate until convergence. In particular, the update equation for the membership degrees is

$$u_{ig} = \frac{1}{\sum_{g'=1}^{k} \left(\frac{d_\tau^2\left((c_{\mathbf{x}_i}, \rho_{\mathbf{x}_i}), (c_{\mathbf{h}_g}, \rho_{\mathbf{h}_g})\right)}{d_\tau^2\left((c_{\mathbf{x}_i}, \rho_{\mathbf{x}_i}), (c_{\mathbf{h}_{g'}}, \rho_{\mathbf{h}_{g'}})\right)} \right)^{\frac{1}{m-1}}} \tag{3}$$

whilst the centroids are parametrized by means of the following crisp center

$$c_{\mathbf{h}_g} = \frac{\sum_{i=1}^{n} u_{ig}^m c_{\mathbf{x}_i}}{\sum_{i=1}^{n} u_{ig}^m}, \qquad g = 1, \cdots, k, \tag{4}$$

and this radial function

$$\rho_{\mathbf{h}_g}(u) = \frac{\sum_{i=1}^{n} u_{ig}^m \rho_{\mathbf{x}_i}(u)}{\sum_{i=1}^{n} u_{ig}^m}, \qquad g = 1, \cdots, k. \tag{5}$$

3 Empirical Analysis

The cliffy rocky coast in the Cantabrian Coast located in NW of Spain contains several sandy, gravelly and mixed pocket beaches bounded by rocky outcrop. Rivers provide important sediment inputs to these beaches. Most of the material deposited in these coarse sediment beaches is supplied from running water of the river that flow into the beach. Besides, another part of sediment comes from cliffs nearby.

Processes such as ocean climate and rising sea levels are causing an erosional tendency in many beaches and dunes around the world, so climatic change is expected to reduce the sediment availability, in Cantabrian coast too. Besides, anthropogenic interventions in the fluvial basins such as dams, channelization or increased forested areas due to land use changes are reducing the sediment supply to the littoral area.

The rule of bedload sediment is of fundamental importance in biodiversity or socioeconomic development but the knowledge about their movement and availability is scarce due to the complex methods to quantify and follow them. In order to analyse the amount of sediment that might come from the cliffs or from the rivers, in this study some geometric properties of each individual clasts are analyzed. The basic premise is that the transport distance and frequent water processes on sedimentary particles erode the shape, so the erosive action of the waves and tides two times per day round their shape in contrast to the young fluvial sediment supplied from the short rivers in the Cantabrian coast.

Several clasts from the two mentioned sources, continental or littoral, have been collected. Each individual clast has been digitalized by a photogrametry approach and has been encoded as a star-shaped set after a convenient wrapping. Different shape-selection indexes have been computed on the basis of the star-shaped sets such as the short (S) and long (L) axes among others. The proposed fuzzy clustering algorithm was then applied by searching for two clusters by considering the shape and different combinations of shape-selection indexes. The aim is to search for similarities between the obtained groups and the source of the particles, short fluvial transport versus coastal ones.

The dataset that was analysed refers to 41 stones, of which 19 are river stones and 22 are sea stones, on which the shape (X_1) and roundness (X_2) were measured. In particular, the first feature was parameterised through star-shaped random sets. The second one is a crisp measurement of roundness of each stone computed as a comparison of the cosine of the normal at each point with respect to the corresponding one for the sphere. The stones are reported in Fig. 1.

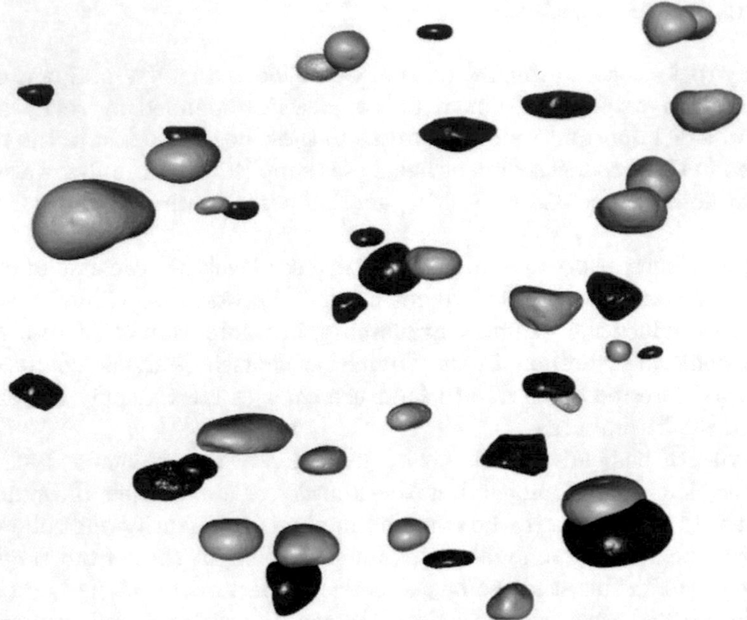

Fig. 1. 41 stones, of which 19 are river stones (dark grey) and 22 are sea stones (light grey)

As we can note, most of the sea stones are more rounded than the river ones.

Before applying the proper fuzzy k-means algorithm, the data was standardized. In particular, firstly X_1 has been centred, since the location is not relevant for the analysis, and divided by the means of the squared spread (in this way the stones have the same volume), and then the obtained data has been divided by the corresponding standard deviations. X_2 has been standardized has a usual crisp variable. It is important to note that, for the same volume, lower values of X_2 correspond to higher roundness.

In this application, the value of τ used for the distance is the default one, 0.5. The obtained centroids are reported in Table 1.

Table 1. Centroids

Cluster	h_{g1}	h_{g2}
1	$(0, 0, 0) \pm 4.4$	0.9480
2	$(0, 0, 0) \pm 3.0$	−0.6177

The notation $(x, y, z) \pm r$ denotes the center (location), (x, y, z), of each stone and the mean r of the directions (spreads). In this case, since the stones have been centred, (x, y, z) is always equal to $(0, 0, 0)$. Cluster 1 is characterized by the highest values of both variables, that is, it contains less rounded stones. On the other hand, stones in Cluster 2 have the lowest values and this means that they are generally more rounded stones.

In Table 2 some details on the obtained partition are reported. In particular, the type, the belonging cluster, the membership degrees and the standardized values of X_1 and X_2 can be found for each of the 41 stones in the Cantabrian Coast.

The results are consistent with what was expected. In particular, most of the sea stones are rounder and therefore belong to Cluster 2. In the case of the river stones, the largest group characterised by higher X_2 values (and therefore less round) are in Cluster 1 and the rest in the other cluster. In Fig. 2 two examples of stones are gathered. The dark grey one (on the left) is a river stone not very rounded belonging to Cluster 1. On the contrary, the light grey one (on the right) is a sea stone assigned to Cluster 2, characterized by high roundness.

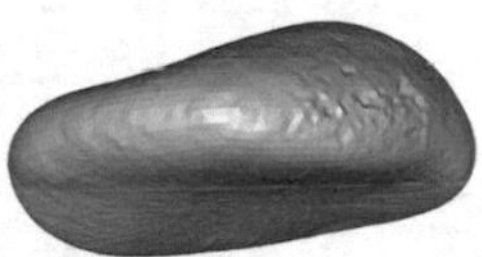

Fig. 2. A river stone (dark grey) belonging to Cluster 1 on the left and a sea stone (light grey) contained in Cluster 2 on the right

Table 2. Type, cluster, membership degrees and standardized values of X_1 and X_2 of 41 stones in the Cantabrian Coast

Type	Cluster	u_{i1}	u_{i2}	X_1	X_2
River	2	0.014	0.986	(0, 0, 0) ± 1.732351	−0.802
River	2	0.179	0.821	(0, 0, 0) ± 2.185047	0.412
River	1	0.924	0.076	(0, 0, 0) ± 3.106206	1.301
River	1	0.921	0.079	(0, 0, 0) ± 4.343241	0.410
River	1	0.873	0.127	(0, 0, 0) ± 2.686426	2.185
River	2	0.078	0.922	(0, 0, 0) ± 3.399997	−0.128
River	1	0.516	0.484	(0, 0, 0) ± 5.055569	−0.391
River	1	0.876	0.124	(0, 0, 0) ± 6.567304	0.367
River	1	0.960	0.040	(0, 0, 0) ± 4.185450	0.628
River	1	0.966	0.034	(0, 0, 0) ± 3.376670	1.658
River	1	0.979	0.021	(0, 0, 0) ± 4.144758	0.588
River	2	0.106	0.894	(0, 0, 0) ± 4.399970	−0.770
River	2	0.046	0.954	(0, 0, 0) ± 2.475345	0.055
River	2	0.000	1.000	(0, 0, 0) ± 2.935276	−0.557

(*continued*)

Table 2. (*continued*)

Type	Cluster	u_{i1}	u_{i2}	X_1	X_2
River	1	0.737	0.263	(0, 0, 0) ± 3.947412	0.284
River	1	0.957	0.043	(0, 0, 0) ± 5.765528	0.576
River	1	0.964	0.036	(0, 0, 0) ± 5.352079	1.359
River	1	0.967	0.033	(0, 0, 0) ± 5.158370	0.934
River	1	0.963	0.037	(0, 0, 0) ± 4.785521	2.183
Sea	2	0.016	0.984	(0, 0, 0) ± 2.448187	−0.147
Sea	2	0.004	0.996	(0, 0, 0) ± 2.267412	−0.991
Sea	2	0.099	0.901	(0, 0, 0) ± 2.423822	0.210
Sea	1	0.639	0.361	(0, 0, 0) ± 2.403939	1.010
Sea	2	0.007	0.993	(0, 0, 0) ± 3.050143	−1.297
Sea	2	0.019	0.981	(0, 0, 0) ± 2.546451	−1.662
Sea	2	0.017	0.983	(0, 0, 0) ± 1.645430	−0.749
Sea	1	0.770	0.230	(0, 0, 0) ± 2.203187	1.616
Sea	2	0.001	0.999	(0, 0, 0) ± 2.517363	−0.674
Sea	2	0.443	0.557	(0, 0, 0) ± 4.780993	−0.414
Sea	2	0.046	0.954	(0, 0, 0) ± 4.044563	−1.540
Sea	2	0.004	0.996	(0, 0, 0) ± 2.528714	−1.156
Sea	2	0.001	0.999	(0, 0, 0) ± 2.665708	−0.414
Sea	2	0.003	0.997	(0, 0, 0) ± 3.460816	−0.878
Sea	2	0.001	0.999	(0, 0, 0) ± 3.262299	−0.878
Sea	2	0.001	0.999	(0, 0, 0) ± 3.146093	−0.631
Sea	2	0.264	0.736	(0, 0, 0) ± 3.463897	0.124
Sea	2	0.014	0.986	(0, 0, 0) ± 3.063523	−1.542
Sea	1	0.977	0.023	(0, 0, 0) ± 5.447317	0.948
Sea	2	0.485	0.515	(0, 0, 0) ± 4.854299	−0.348
Sea	2	0.008	0.992	(0, 0, 0) ± 3.227399	−0.315
Sea	2	0.000	1.000	(0, 0, 0) ± 3.053607	−0.562

4 Concluding Remarks

A fuzzy clustering method for star-shaped sets has been introduced. Starting from the well-known fuzzy k-means, through an appropriate distance measure for the above sets, the optimization problem has been formalized and the updating equations to be used in an iterative algorithm have been provided. Furthermore, the proposal has been applied to a real-case study. In detail, a set of river and sea stones of the Cantabrian Coast have been partitioned by taking into account their shape and roundness. The obtained results are consistent with what was expected. In the future, it will be interesting to generalize this approach to the case of mixed complex data, besides the star-shaped ones.

Acknowledgements. The research in this paper has been supported by the Spanish Government through MINECO-18-MTM2017-89632-P Grant and by the Sapienza Grant "New developments on clustering of text, functional and other complex data" (Ateneo 2021). Their financial support is gratefully acknowledged.

References

Bezdek, J.C.: Pattern Recognition with Fuzzy Objective Function Algorithms. Plenum Press, New York (1981)

González-Rodríguez, G., Ramos-Guajardo, A.B., Colubi, A., Blanco-Fernández, Á.: A new framework for the statistical analysis of set-valued random elements. Int. J. Approx. Reason. **92**, 279–294 (2018)

Klain, D.: Invariant valuations on star-shaped sets. Adv. Math. **125**(1), 95–113 (1997)

Matheron, G.: Random Sets and Integral Geometry. Wiley, New York (2018)

Molchanov, I., Molinari, F.: Applications of random set theory in econometrics. Ann. Rev. Econ. **6**, 229–251 (2014)

Ramos-Guajardo, A.B., González-Rodríguez, G., Colubi, A., Ferraro, M.B., Blanco-Fernández, Á.: On some concepts related to star-shaped sets. In: Gil, E., Gil, E., Gil, J., Gil, M.Á. (eds.) The Mathematics of the Uncertain. SSDC, vol. 142, pp. 699–708. Springer, Cham (2018). https://doi.org/10.1007/978-3-319-73848-2_64

Schneider, R.: Convex Bodies: The Brunn-Minkowski Theory. Cambridge University Press, Cambridge (1993)

Simó, A., De Ves, E., Ayala, G.: Resuming shapes with applications. J. Math. Imaging Vis. **20**(3), 209–222 (2004)

Cluster Validity Measures for Fuzzy Two-Mode Clustering

Maria Brigida Ferraro(✉), Paolo Giordani, and Maurizio Vichi

Department of Statistical Sciences, Sapienza University of Rome, p.le A. Moro 5, 00185 Rome, Italy
{mariabrigida.ferraro,paolo.giordani,maurizio.vichi}@uniroma1.it

Abstract. Two-mode clustering consists in simultaneously partitioning rows (mode 1, e.g. objects) and columns (mode 2, e.g., variables) of a data matrix. Recently, several soft two-mode clustering techniques have been developed according to the fuzzy approach, but how to determine the optimal numbers of clusters for objects and variables is an open problem not yet investigated. In this paper some new cluster validity measures for fuzzy two-mode clustering are introduced. Such measures, defined in terms of the compactness within each cluster and separation between clusters, can be seen as generalizations of well-known indices widely used in the standard fuzzy clustering framework. The adequacy of these proposals is assessed by means of a simulation study.

1 Background on Fuzzy Two-Mode Clustering

Clustering usually refers to the problem of grouping the rows (mode 1, e.g., objects) of a data matrix. Sometimes, however, the research interest relies on synthesizing not only the rows, but also the columns (mode 2, e.g., variables) of a matrix. Simultaneous partitioning of the rows and columns of a matrix can be performed by two-mode clustering methods. Two-mode clustering is also known as bi-clustering, co-clustering or double clustering (Hartigan 1972).

Most of the existing two-mode clustering methods adopt the hard approach to clustering where the rows and columns either belong or do not to the clusters. In the last years, soft extensions have been proposed. In Ferraro et al. (2021) a class of two-mode clustering methods has been introduced in a fuzzy setting: the Fuzzy Double k-Means with polynomial fuzzifiers (FDkMpf). Let $\mathbf{X} = [x_{ij}]$ be a data matrix of order $(n \times p)$, the FDkMpf algorithm consists in simultaneously partitioning the objects into k clusters and the variables into c clusters by solving the following constrained problem:

$$\begin{aligned} &\min_{\mathbf{U},\mathbf{V},\mathbf{Y}} J_{\mathrm{FD}k\mathrm{Mpf}} = \sum_{i=1}^{n}\sum_{j=1}^{p}\sum_{g=1}^{k}\sum_{h=1}^{c} (x_{ij} - y_{gh})^2 f(u_{ig}) f(v_{jh}), \\ &\text{s.t.} \quad u_{ig}, v_{jh} \in [0,1], \sum_{g=1}^{k} u_{ig} = 1, \sum_{h=1}^{c} v_{jh} = 1, \end{aligned} \tag{1}$$

L. A. García-Escudero et al. (Eds.): SMPS 2022, AISC 1433, pp. 144–150, 2023.
https://doi.org/10.1007/978-3-031-15509-3_19

where $f(u_{ig}) = \left(\frac{1-\beta_1}{1+\beta_1}u_{ig}^2 + \frac{2\beta_1}{1+\beta_1}u_{ig}\right)$ and $f(v_{jh}) = \left(\frac{1-\beta_2}{1+\beta_2}v_{jh}^2 + \frac{2\beta_2}{1+\beta_2}v_{jh}\right)$ with β_1 and $\beta_2 \in [0, 1]$ (a common choice is $\beta_1 = \beta_2 = 0.5$). The resulting output is a block decomposition of the observed data matrix into kc blocks, characterized by the centroids $y_{gh}, g = 1, \ldots, k, h = 1, \ldots, c$. To assess the cluster memberships of objects and variables, the fuzzy membership degrees u_{ig} and v_{jh} are introduced, respectively. They range in the interval $[0, 1]$ to exploit the soft (fuzzy) approach to clustering.

FDkMpf contains the Fuzzy Double k-Means algorithm (Ferraro and Vichi 2015) with fuzziness parameters equal to 2, a generalization of the Fuzzy k-Means algorithm (Bezdek 1981), as a special case when $\beta_1 = \beta_2 = 0$. Furthermore, if $\beta_1 = \beta_2 = 1$, FDkMpf coincides with the Double k-Means algorithm (Vichi 2001). Finally, when the number of clusters for the variables c is equal to that of variables p, FDkMpf reduces to a standard one-mode clustering method for partitioning the objects. In particular, by suitable choices of β_1, we can obtain the (one-mode) fuzzy polynomial k-means (Klawonn and Hoppner 2003), the fuzzy k-means and the standard k-Means (Mac Queen 1967) algorithms.

2 Cluster Validity Measures

This section is devoted to new cluster validity indices which measure the overall compactness and separation of a fuzzy two-mode partition. Our proposals are generalizations of measures used in standard fuzzy clustering (fuzzy one-mode clustering). In particular, we consider the Xie and Beni (XB) index (Xie and Beni 1991), the Silhouette (S) index (Rousseeuw 1987; Kaufman and Rousseeuw 1990) and the Fuzzy Silhouette (FS) index (Campello and Hruschka 2006), that will be briefly recalled.

The XB index for one-mode clustering is defined as the ratio between compactness and separation. Compactness is measured by the variation of the fuzzy partition, whilst the squared distance between pairs of centroids acts as the measure of separation between clusters. In the fuzzy two-mode setting, the following measure of compactness can be introduced:

$$C = \frac{\sum_{i=1}^{n}\sum_{j=1}^{p}\sum_{g=1}^{k}\sum_{h=1}^{c}\left(x_{ij} - y_{gh}\right)^2 u_{ig}^2 v_{jh}^2}{np}, \tag{2}$$

where the numerator indicates the sum of the variations of each cluster. To measure the separation, three alternatives are suggested. To this purpose, first we define the minimum squared distance between the row vectors of the centroid matrix ($\mathbf{Y}$), referring to the row (object) clusters, d_{row}^2, and the minimum squared distance between the column vectors of $\mathbf{Y}$, referring to the column (variable) clusters, d_{col}^2. In detail, we have, respectively,

$$d_{row}^2 = \min_{g,g'(g \neq g')} \frac{d^2(\mathbf{y}_g, \mathbf{y}_{g'})}{c} \tag{3}$$

and

$$d^2_{col} = \min_{h,h'(h \neq h')} \frac{d^2(\mathbf{y}_h, \mathbf{y}_{h'})}{k}, \tag{4}$$

with $\mathbf{y}_g = (y_{g1}, \cdots y_{gc})^T$ and $\mathbf{y}_h = (y_{1h}, \cdots y_{kh})^T$. The separation can then be evaluated by using the minimum between d^2_{row} and d^2_{col}, the sample mean of d^2_{row} and d^2_{col}, or the geometric one. The corresponding three cluster validity measures are, respectively:

$$\text{XB-2a} = \frac{\sum_{i=1}^{n} \sum_{j=1}^{p} \sum_{g=1}^{k} \sum_{h=1}^{c} (x_{ij} - y_{gh})^2 u_{ig}^2 v_{jh}^2}{np \min(d^2_{row}, d^2_{col})}, \tag{5}$$

$$\text{XB-2b} = \frac{\sum_{i=1}^{n} \sum_{j=1}^{p} \sum_{g=1}^{k} \sum_{h=1}^{c} (x_{ij} - y_{gh})^2 u_{ig}^2 v_{jh}^2}{np \frac{(d^2_{row} + d^2_{col})}{2}} \tag{6}$$

and

$$\text{XB-2c} = \frac{\sum_{i=1}^{n} \sum_{j=1}^{p} \sum_{g=1}^{k} \sum_{h=1}^{c} (x_{ij} - y_{gh})^2 u_{ig}^2 v_{jh}^2}{np \sqrt{(d^2_{row} \cdot d^2_{col})}}. \tag{7}$$

The above indices depend on both the number of row clusters and the number of column clusters, but their values are not affected by the specific algorithm used to obtain the membership degrees and the centroids. As in one-mode clustering, a smaller value of the numerator indicates more compact clusters, whilst a larger value of the denominator denotes more separate clusters. Hence, the optimal values of k and c are obtained by minimizing the indices, that is, by maximizing the compactness (minimizing the numerator) and maximizing the separation between clusters (maximizing the denominator).

An alternative proposal is based on the Silhouette (S) index for one-mode clustering. The generalized silhouette to the two-mode case, S-2, is defined as the sample mean of the silhouette index for the row partition, SR, and the silhouette index for the column partition, SC. The former is formalized as

$$\text{SR} = \frac{\sum_{i=1}^{n} s_{ri}}{n}, \tag{8}$$

where s_{ri} denotes the silhouette of row i:

$$s_{ri} = \frac{b_{ri} - a_{ri}}{\max(b_{ri}, a_{ri})}, i = 1, \ldots, n, \tag{9}$$

where a_{ri} and b_{ri} are, respectively, the average distance between the involved row and all the rows belonging to the same cluster and the lowest average distance of i to any other cluster of which i is not a member. Note that the memberships of the rows to the clusters are determined by assuming that every row belongs to the cluster with the highest membership degree. The cluster used for calculating b_{ri} is the second best fit cluster for row i and is usually referred to as the neighbouring cluster. Therefore, s_{ri} is essentially the standard silhouette index S applied to the partition of the rows. It follows that, similarly to S, s_{ri} takes values in the interval $[-1, 1]$. When it is close to 1 it follows that the row is well assigned to the cluster. On the other hand, when it is close to -1 it implies a wrong assignment of the row to the cluster.

With regard to the partition of the columns, SC is expressed as

$$\text{SC} = \frac{\sum_{j=1}^{p} s_{cj}}{p}, \tag{10}$$

with

$$s_{cj} = \frac{b_{cj} - a_{cj}}{\max(b_{cj}, a_{cj})}, j = 1, \ldots, p, \tag{11}$$

In this case, s_{cj} denotes the silhouette value corresponding to column j, a_{cj} and b_{cj} represent, respectively, the average distance between the involved column and all the columns belonging to the same cluster and the lowest average distance of j to any other cluster of which j is not a member. Again, the memberships of the columns to the clusters are found by considering the highest membership degree. As for the rows, s_{cj} takes values in the interval $[-1, 1]$.

The generalized silhouette index is formalized as

$$\text{S-2} = \frac{SR + SC}{2}. \tag{12}$$

It takes values in the interval $[-1, 1]$. The best values of k and c are achieved when S-2 is maximized. It implies the maximization of the compactness and the maximization of the separation. If S-2 is close to 1, the obtained row and column partitions are appropriate. On the contrary, when S-2 takes values close to -1, the rows and columns are not well clustered.

The fuzzy counterpart of S-2, is the generalized fuzzy silhouette index for fuzzy two-mode clustering, FS-2, defined as the sample mean of the fuzzy silhouette index for the row partition, FSR, and the fuzzy silhouette index for the column partition, FSC. FS-2 is built upon the fuzzy silhouette (FS) index for one-mode clustering that, differently from S, explicitly takes into account the fuzzy membership degree information. Specifically, FS-2 is formalized as

$$\text{FS-2} = \frac{FSR + FSC}{2}, \tag{13}$$

where, bearing in mind the standard FS index, FSR and FSC are expressed as

$$\text{FSR} = \frac{\sum_{i=1}^{n} (u_{ig} - u_{ig'})^{\alpha} s_{ri}}{\sum_{i=1}^{n} (u_{ig} - u_{ig'})^{\alpha}}, \tag{14}$$

and

$$\text{FSC} = \frac{\sum_{j=1}^{p} (v_{jh} - v_{jh'})^{\alpha} s_{cj}}{\sum_{j=1}^{p} (v_{jh} - v_{jh'})^{\alpha}}, \tag{15}$$

respectively. Notice that s_{ri} is defined in (9) and u_{ig} and $u_{ig'}$ are the first and the second largest elements of the i-th row of $\mathbf{U}$, while s_{cj} is defined in (11) and v_{jh} and $v_{jh'}$ are the first and the second largest elements of the j-th row of $\mathbf{V}$. The parameter α is a weighting coefficient (usually $\alpha = 1$). As for S-2, the best values of k and c are obtained by maximizing FS-2.

3 Empirical Results

To evaluate the performance of the proposed cluster validity measures a simulation study was carried out. We considered several scenarios by varying the number of objects, the number of variables, the number of clusters for both the object and variable partitions and the level of separation of the clusters. For each scenario, a number of datasets were randomly generated.

For illustrative purposes, we report the result of a specific scenario. In particular, we consider a randomly generated dataset characterized by $n = 120$ objects, $p = 30$ variables, $k = 3$ row clusters of equal size ($\frac{n}{k} = 40$) and $c = 3$ column clusters of equal size ($\frac{c}{p} = 10$), and low level of separation between clusters. The object and variable clusters are found by applying the FDkMpf algorithm setting $\beta_1 = \beta_2 = 0.5$ and ranging c and k from 2 to 7. For each pair (c, k), the values of the XB-2a, XB-2b, XB-2c, S-2 and FS-2 indices are recorded.

For every index, a 3D scatterplot is displayed, where the x- and y-axes refer to the k and c values and the z-axis reports the index value. Figure 1 gives the 3D scatterplots concerning S-2 (left side) and FS-2 (right side), whilst Fig. 2 gives the 3D scatterplots concerning XB-2a (left side), XB-2b (center) and XB-2c (right side).

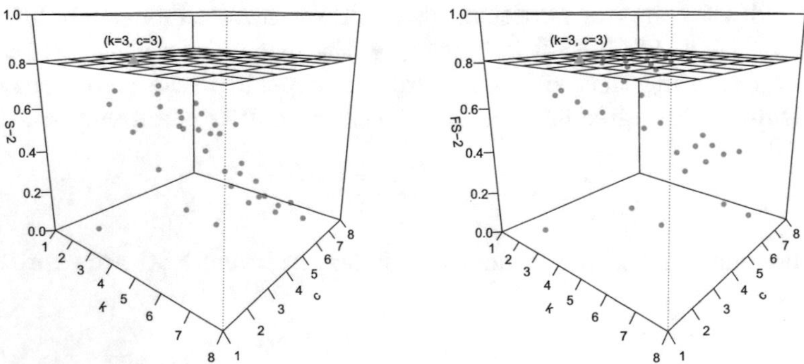

Fig. 1. 3D scatterplots for S-2 and FS-2. The optimal/estimated values of k and c are reported as solid triangle points. The horizontal planes correspond to the maximum values of each index

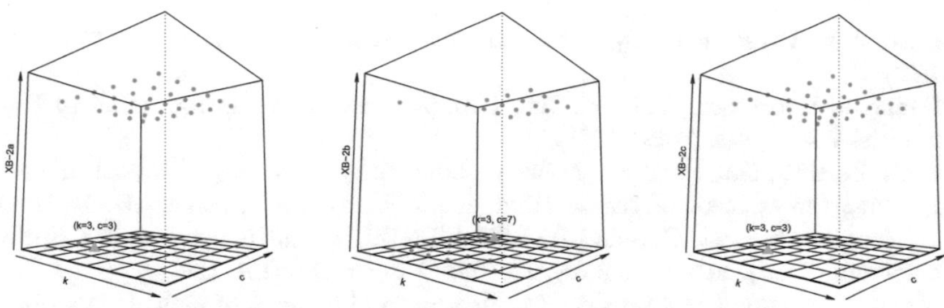

Fig. 2. 3D scatterplots for XB-2a, XB-2b and XB-2c. The optimal/estimated values of k and c are reported as solid triangle points. The horizontal planes correspond to the minimum values of each index

We can see that the optimal values of k and c obtained by S-2, FS-2, XB-2a and XB-2c correspond to the true values. Only the XB-2b is minimized for $k = 3$ and $c = 7$ (solid triangle point on the horizontal plane). However, we note that the index value for $k = 3$ and $c = 3$, 0.1254, is very close to the lowest one for $k = 3$ and $c = 7$, 0.1137.

4 Concluding Remarks

In this paper we have proposed some fuzzy cluster validity measures in order to select the optimal numbers of clusters in fuzzy two-mode clustering. A simulation study was conducted in order to evaluate the effectiveness of our proposals by considering randomly generated datasets with given object and variable cluster structures. We have found that the measures have recovered the true numbers of clusters in a satisfactory way. Here, we have presented the results on an illustrative dataset.

Acknowledgements. The research in this paper has been supported by the Sapienza Grant "New developments on clustering of text, functional and other complex data" (Ateneo 2021).

References

Bezdek, J.C.: Pattern Recognition with Fuzzy Objective Function Algorithm. Plenum Press, New York (1981)

Campello, R.J.G.B., Hruschka, E.R.: A fuzzy extension of the silhouette width criterion for cluster analysis. Fuzzy Sets Syst. **157**, 2858–2875 (2006)

Ferraro, M.B., Giordani, P., Vichi, M.: A class of two-mode clustering algorithms in a fuzzy setting. Econ. Stat. **18**, 63–78 (2021)

Ferraro, M.B., Vichi, M.: Fuzzy double clustering: a robust proposal. In: Grzegorzewski, P., Gagolewski, M., Hryniewicz, O., Gil, M.Á. (eds.) Strengthening Links Between Data Analysis and Soft Computing. AISC, vol. 315, pp. 225–232. Springer, Cham (2015). https://doi.org/10.1007/978-3-319-10765-3_27

Hartigan, J.A.: Direct clustering of a data matrix. J. Am. Stat. Assoc. **67**, 123–129 (1972)

Kaufman, L., Rousseeuw, P.J.: Finding Groups in Data: An Introduction to Cluster Analysis. Wiley, New York (1990)

Klawonn, F., Höppner, F.: What is fuzzy about fuzzy clustering? Understanding and improving the concept of the fuzzifier. In: R. Berthold, M., Lenz, H.-J., Bradley, E., Kruse, R., Borgelt, C. (eds.) IDA 2003. LNCS, vol. 2810, pp. 254–264. Springer, Heidelberg (2003). https://doi.org/10.1007/978-3-540-45231-7_24

Mac Queen, J.B.: Some methods for classification and analysis of multivariate observations. In: Proceedings of the Fifth Berkeley Symposium on Mathematical Statistics and Probability, vol. 2, pp. 281–297 (1967)

Rousseeuw, P.J.: Silhouettes: a graphical aid to the interpretation and validation of cluster analysis. J. Comput. Appl. Math. **20**, 53–65 (1987)

Vichi, M.: Double *k*-means clustering for simultaneous classification of objects and variables. In: Borra, S., Rocci, R., Vichi, M., Schader, M. (eds.) Advances in Classification and Data Analysis. Studies in Classification, Data Analysis, and Knowledge Organization, pp. 43–52. Springer, Heidelberg (2001). https://doi.org/10.1007/978-3-642-59471-7_6

Xie, X.L., Beni, G.: A validity measure for fuzzy clustering. IEEE Trans. Pattern Anal. Mach. Intell. **13**, 841–847 (1991)

The Simplifying Assumption in Pair-Copula Constructions from an Analytic Perspective

Sebastian Fuchs(✉)

University of Salzburg, Hellbrunner Straße 34, 5020 Salzburg, Austria
sebastian.fuchs@plus.ac.at

Abstract. Pair-copula constructions are a very popular bottom-up approach for constructing high-dimensional copulas out of several bivariate ones; they have a handy graphical representation and can be considered as an ordered sequence of trees. In numerous applications, it is assumed that the so-called "simplifying assumption" holds, a condition that allows for a significant reduction of complexity. Motivated by the broad applicability of pair-copula constructions, we here present a short and concise summary of the article *How simplifying and flexible is the simplifying assumption in pair-copula constructions - analytic answers in dimension three and a glimpse beyond* by Mroz et al. (2021) which explores the limitations of the simplifying assumption in pair-copula constructions from an analytic perspective.

1 Introduction

More that 700 scientific contributions working with and applying simplified pair-copulas have been published within the last decade. In this respect, it is quite surprising that, apart from a few critical voices (see, e.g., Acar et al. 2012; Derumigny and Fermanian 2017; Gijbels et al. 2017; Spanhel and Kurz 2019), no analytic and systematic study on the approximation quality and flexibility of these concepts seems to have been published so far.

In the present review paper we first show that, on the one hand, simplified copulas are dense in the family of all three-dimensional copulas with respect to the uniform metric d_∞ indicating a high flexibility. However, considering stronger notions of convergence the family turns out to be nowhere dense and hence insufficient for any kind of flexible approximation which is why, for the remainder of the analysis, we can restrict ourselves to the metric d_∞.

We then show that the partial vine copula (special simplified pair-copulas whose conditional distribution functions follow a certain intuitive construction principle) is never the optimal simplified copula approximation of a given, non-simplified copula C, and illustrate that the corresponding approximation error can be strikingly large. We further focus on continuity properties of the mapping ψ assigning each three-dimensional copula its unique partial vine copula and

L. A. García-Escudero et al. (Eds.): SMPS 2022, AISC 1433, pp. 151–158, 2023.
https://doi.org/10.1007/978-3-031-15509-3_20

show that this mapping is *not* continuous with respect to d_∞. In other words: if $d_\infty(A,B)$ is small then in general we can not infer that the corresponding vines have small d_∞ distance too. As a direct consequence, although simplified pair-copulas are "highly flexible" (Killiches et al. 2017) and partial vine copulas "can yield an approximation that is superior to competing approaches" (Spanhel and Kurz 2019), approximations in terms of partial vine copulas can be of very poor quality and lead to wrong conclusions - implying a surprising sensitivity of partial vine copula approximations.

Aiming at a simplest possible setup we restrict ourselves on the three-dimensional setting. However, the main results concerning d_∞ can be extended to the general multivariate setting. For the proofs and further findings and insights we refer to Mroz et al. (2021).

Throughout this paper we will write $\mathbb{I} := [0,1]$ and let $\mathscr{C}^d$ denote the family of all d-dimensional copulas. According to (Durante and Sempi 2016, Theorem 3.4.3) and due to disintegration, every copula C fulfills

$$C(\mathbf{u},v) = \int_{[0,v]} K_C(t,[\mathbf{0},\mathbf{u}])\,\mathrm{d}\lambda(t)$$

where λ denotes the Lebesgue measure on the Borel σ-field $\mathscr{B}(\mathbb{I})$ and K_C is (a version of) the Markov kernel of C: A *Markov kernel* from $\mathbb{I}$ to $\mathscr{B}(\mathbb{I}^{d-1})$ is a mapping $K : \mathbb{I}\times\mathscr{B}(\mathbb{I}^{d-1}) \to \mathbb{I}$ such that for every fixed $F\in\mathscr{B}(\mathbb{I}^{d-1})$ the mapping $v\mapsto K(v,F)$ is measurable and for every fixed $v\in\mathbb{I}$ the mapping $F\mapsto K(v,F)$ is a probability measure. Given a uniformly distributed random variable V and a $(d-1)$-dimensional random vector $\mathbf{U}$ with uniform marginals on a probability space $(\Omega,\mathscr{A},P)$ we say that a Markov kernel K is a *regular conditional distribution* of $\mathbf{U}$ given V if $K(V(\omega),F) = P(\mathbf{U}\in F\,|\,V)(\omega)$ holds P-almost surely for every $F\in\mathscr{B}(\mathbb{I}^{d-1})$. It is well-known that for each such random vector $(\mathbf{U},V)$ a regular conditional distribution $K(.,.)$ of $\mathbf{U}$ given V always exists and is unique for P^V-a.e. $v\in\mathbb{I}$, where P^V denotes the push-forward of P under V. For more background on conditional expectation and general disintegration we refer to Kallenberg (1997), Klenke (2007).

Markov kernels can be used to define metrics stronger than the standard *uniform metric* d_∞. Following Fernández-Sánchez and Trutschnig (2015) and defining

$$D_1(C_1,C_2) := \int_{\mathbb{I}^{d-1}}\int_{\mathbb{I}} \left|K_{C_1}(v,[\mathbf{0},\mathbf{u}]) - K_{C_2}(v,[\mathbf{0},\mathbf{u}])\right|\,\mathrm{d}\lambda(v)\mathrm{d}\lambda^{d-1}(\mathbf{u})$$

it can be shown that D_1 is a metric. Finally, we consider two additional notions of convergence, the total variation metric TV and the Kullback-Leibler divergence (distance) KL on $\mathscr{C}^d$ given by

$$TV(C_1,C_2) := \sup_{G\in\mathscr{B}(\mathbb{I}^d)} |\mu_{C_1}(G) - \mu_{C_2}(G)|$$

$$KL(C_1,C_2) := \int_{\mathbb{I}^d} c_1(\mathbf{u},v)\,\log\left(\frac{c_1(\mathbf{u},v)}{c_2(\mathbf{u},v)}\right)\,\mathrm{d}\lambda^d(\mathbf{u},v)$$

where c_1 and c_2 denote the Lebesgue densities of the absolutely continuous copulas C_1 and C_2. It is well-known that KL divergence (which is not a metric and only well-defined for absolutely continuous copulas whose density is positive λ^d-almost everywhere) is stronger than TV (see the generalized Pinsker inequality in, e.g., Reid and Williamson 2009). Altogether we have the following interrelation, where $a \Longrightarrow b$ indicates the convergence with respect to a implies convergence with respect to b (and the first implication is restricted to those copulas for which KL divergence is well-defined): $KL \Longrightarrow TV \Longrightarrow D_1 \Longrightarrow d_\infty$.

For any subset $J = \{j_1, ..., j_{|J|}\} \subseteq \{1, \ldots, d\}$ with $2 \le |J| \le d$ such that $j_k < j_l$ for all $k, l \in \{1, ..., |J|\}$ with $k < l$ we let C_J denote the *marginal* copula of C with respect to the coordinates in J. If J only contains two indices i, j then we will sometimes also write C_{ij} instead of $C_{\{i,j\}}$ (no confusion will arise).

2 Simplified Copulas

We start with the introduction of 3-dimensional simplified copulas and show that simplified pair-copula constructions may fail to approximate a given dependence structure w.r.t. d_∞ reasonably well.

With very few exceptions, in literature pair-copula constructions are introduced by assuming absolute continuity. Ensuring that no key idea of the underlying concept is left out and aiming at a setting as general as possible we deviate from this approach and work with (the more general concept of) Markov kernels instead.

In this and the subsequent sections all conditioning will be done with respect to the last coordinate which does not impose any restrictions.

According to disintegration for every copula $C \in \mathscr{C}^3$ there exists some Markov kernel K_C such that C can be expressed as

$$C(\mathbf{u}, v) = \int_{[0,v]} K_C(t, [\mathbf{0}, \mathbf{u}]) \, \mathrm{d}\lambda(t)$$

for all $(\mathbf{u}, v) \in \mathbb{I}^2 \times \mathbb{I}$. Since K_C is a Markov kernel, for every $\mathbf{u} \in \mathbb{I}^2$ the mapping $t \mapsto K_C(t, [\mathbf{0}, \mathbf{u}])$ is measurable and for every $t \in \mathbb{I}$ the mapping $\mathbf{u} \mapsto K_C(t, [\mathbf{0}, \mathbf{u}])$ is a bivariate distribution function with (*conditional*) univariate marginal distribution functions $F_{1|3}(\cdot|t)$ and $F_{2|3}(\cdot|t)$ (conditional on t). Sklar's Theorem implies that for every $t \in \mathbb{I}$ there exists some (*conditional*) bivariate copula $C_{12;3}^t$ (conditional on t) satisfying

$$K_C(t, [\mathbf{0}, \mathbf{u}]) = C_{12;3}^t\big(F_{1|3}(u_1|t), F_{2|3}(u_2|t)\big)$$

for all $\mathbf{u} \in \mathbb{I}^2$ such that the identity

$$C(\mathbf{u}, v) = \int_{[0,v]} C_{12;3}^t\big(F_{1|3}(u_1|t), F_{2|3}(u_2|t)\big) \, \mathrm{d}\lambda(t) \tag{1}$$

holds for all $(\mathbf{u}, v) \in \mathbb{I}^2 \times \mathbb{I}$. The following two observations are key:

- the (conditional) bivariate copulas $C^t_{12;3}$ may depend on t;
- since the (conditional) univariate marginal distribution functions $F_{1|3}(.|t)$ and $F_{2|3}(.|t)$ may fail to be continuous the (conditional) bivariate copulas $C^t_{12;3}$ are not necessarily unique.

We will refer to a copula $C \in \mathscr{C}^3$ as *simplified* (with respect to the third coordinate) if it belongs to $\mathscr{C}^3_{\mathrm{c}}$ (denoting the class of copulas having continuous (conditional) univariate marginal distribution functions $F_{1|3}(.|t)$ and $F_{2|3}(.|t)$) and there exists some (hence unique) bivariate copula $A \in \mathscr{C}^2$ such that the identity

$$C(\mathbf{u}, v) = \int_{[0,v]} A\left(F_{1|3}(u_1|t), F_{2|3}(u_2|t)\right) \mathrm{d}\lambda(t) \tag{2}$$

holds for all $(\mathbf{u}, v) \in \mathbb{I}^2 \times \mathbb{I}$. In the sequel $\mathscr{C}^3_{\mathrm{S}}$ will denote the family of all three-dimensional simplified copulas.

Example 1 (Class of three-dimensional copulas C satisfying $C_{13} = \Pi = C_{23}$)

- The independence copula Π is simplified.
- The copula C^{Cube} which distributes mass uniformly within the four cubes

  $$\begin{array}{ll} (0, \tfrac{1}{2}) \times (0, \tfrac{1}{2}) \times (0, \tfrac{1}{2}) & (\tfrac{1}{2}, 1) \times (\tfrac{1}{2}, 1) \times (0, \tfrac{1}{2}) \\ (0, \tfrac{1}{2}) \times (\tfrac{1}{2}, 1) \times (\tfrac{1}{2}, 1) & (\tfrac{1}{2}, 1) \times (0, \tfrac{1}{2}) \times (\tfrac{1}{2}, 1) \end{array}$$

  and has no mass outside these cubes satisfies

  $$C^{\mathrm{Cube}}(\mathbf{u}, v) = \int_{[0,v]} (C^{\mathrm{Cube}})^t_{12;3} \left(F_{1|3}(u_1|t), F_{2|3}(u_2|t)\right) \mathrm{d}\lambda(t)$$

  for all $(\mathbf{u}, v) \in \mathbb{I}^2 \times \mathbb{I}$, where $(C^{\mathrm{Cube}})^t_{12;3} = A^1$ for almost all $t \in (0, \tfrac{1}{2})$ and $(C^{\mathrm{Cube}})^t_{12;3} = A^2$ for almost all $t \in (\tfrac{1}{2}, 1)$, and the copulas A^1 and A^2 are checkerboard copulas whose density is depicted in Fig. 1. As a direct consequence C^{Cube} is non-simplified.

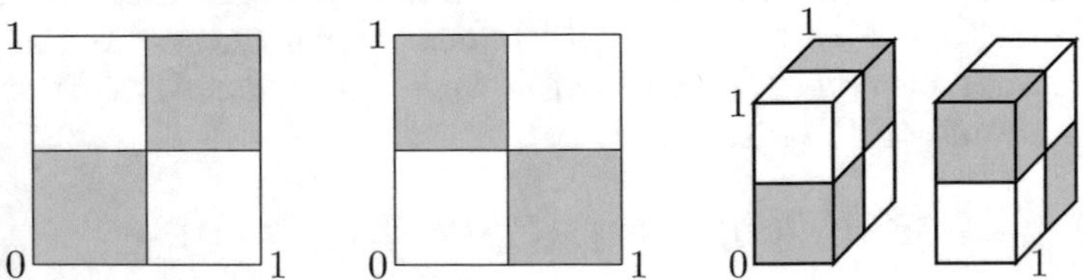

Fig. 1. Mass distribution of the copulas A^1, A^2 and C^{Cube} from Example 1

Since empirical copulas are dense in $(\mathscr{C}^3, d_\infty)$ and simplified, we conclude:

Theorem 1. *The collection of all simplified copulas* $\mathscr{C}^3_S$ *is dense in* $(\mathscr{C}^3, d_\infty)$.

Although every copula can be approximated arbitrarily well by simplified copulas a reasonable approximation from the same Fréchet class might not be possible as the following example illustrates:

Example 2 (Class of three-dimensional copulas C satisfying $C_{13} = \Pi = C_{23}$) For the non–simplified copula C^{Cube} there exists some $\varepsilon > 0$ such that for every simplified copula D with $D_{13} = \Pi = D_{23}$ we have $d_\infty\big(C^{\text{Cube}}, D\big) > \varepsilon$.

We now focus on stronger metrics or finer topologies on $\mathscr{C}$. To simplify notation we will write $\mathscr{C}^3_{\text{ac},>0}$ for the collection of all absolutely continuous copulas with positive density.

Theorem 2.

1. *The collection of all simplified copulas is nowhere dense in* $(\mathscr{C}^3, D_1)$.
2. *The collection of all simplified copulas is nowhere dense in* $(\mathscr{C}^3, TV)$.
3. *The collection of all simplified copulas with positive density is nowhere dense in* $(\mathscr{C}^3_{\text{ac},>0}, KL)$.

The above Theorem 2 answers the question "How dense does the set of simplified densities lie in the set of all densities?" posed by Nagler and Czado (2016) in a complete and definitive manner. In the same article the authors also pose the question on "how far off can we be by assuming a simplified model?" - one of the main objectives of the subsequent sections is to answer this very question. Notice that, for this purpose, we can restrict ourselves to the metric d_∞ since simplified copulas are nowhere dense w.r.t. D_1, TV and KL.

3 Optimality and Continuity Results of Partial Vine Copulas (PVCs)

Equation (1) suggests the construction of a three-dimensional copula in terms of two families of (conditional) univariate marginal distribution functions characterizing the dependence structure between coordinates 1&3 and coordinates 2&3, respectively, and (conditional) bivariate copulas representing the dependence structure between coordinates 1&2 conditional on the third variable. This just-mentioned construction principle is called *vine decomposition* or *pair-copula construction* (see Aas et al. 2009; Bedford and Cooke 2002). In case the conditioning variable only enters indirectly through the conditional marginals (as it is the case in Eq. (2); see, e.g., Joe (1996) for an early reference), the pair-copula construction is said to be *simplified* (see Hobæk Haff et al. 2010).

Aiming at obtaining analytic results concerning the optimality of simplified pair-copula constructions, in what follows we discuss the concept of partial vine copulas. The basic idea behind a partial vine copula is that the conditional bivariate copulas of the original three-dimensional copula are averaged (see Spanhel and Kurz 2019): Considering that for every $C \in \mathscr{C}_c^3$ the copula $C_{12;3}^t$ is unique for almost every $t \in \mathbb{I}$ it follows that the function $C_p : \mathbb{I}^2 \to \mathbb{I}$, given by

$$C_p(\mathbf{s}) := \int_{\mathbb{I}} C_{12;3}^t(\mathbf{s}) \, \mathrm{d}\lambda(t)$$

is well–defined. In the sequel we will refer to C_p as the *partial copula* of C (see also Bergsma 2011). Coinciding with the expected conditional copula, the partial copula is often used as an approximation of the conditional copula (see Spanhel and Kurz 2019, for more information). Given C_p in the above setting the mapping $\psi : \mathscr{C}_c^3 \to \mathscr{C}_c^3$, given by

$$(\psi(C))(\mathbf{u}, v) := \int_{[0,v]} C_p\big(F_{1|3}(u_1|t), F_{2|3}(u_2|t)\big) \, \mathrm{d}\lambda(t)$$

is well-defined and assigns to every copula C a simplified copula $\psi(C)$. The copula $\psi(C)$ is referred to as the *partial vine copula* of C (with respect to the third coordinate) in the sequel. It is obvious that every partial vine copula is simplified, and it is straightforward to verify that ψ preserves the dependence structure between coordinates 1&3 as well as between coordinates 2&3, i.e. $(\psi(C))_{13} = C_{13}$ as well as $(\psi(C))_{23} = C_{23}$.

Example 3. (Class of three-dimensional copulas C satisfying $C_{13} = \Pi = C_{23}$) The copula C^{Cube} is non-simplified, satisfies $C_{12}^{\mathrm{Cube}} = \Pi$ and $C_p^{\mathrm{Cube}} = \Pi$, and hence $\psi(C^{\mathrm{Cube}}) = \Pi \neq C^{\mathrm{Cube}}$.

In Spanhel and Kurz (2019) the authors showed that "under regularity conditions, stepwise estimators of pair-copula constructions converge to the PVC irrespective of whether the simplifying assumption holds or not" (see Spanhel and Kurz 2019, Corollary 6.1). Nevertheless, this does not need to be true if the estimation is done jointly in a non-simplified setting (see Spanhel and Kurz 2019, Corollary 6.1). The authors further proved that "if one sequentially minimizes the Kullback-Leibler divergence related to each tree then the optimal SVC is the PVC" (see Spanhel and Kurz 2019, Theorem 5.1). Since, again, this is not necessarily true if the estimation is done jointly in a non-simplified setting (see Spanhel and Kurz 2019, Theorem 5.2) the authors conclude that PVCs "may not be the best approximation in the space of SVCs" but are "often the best feasible SVC approximation in practice."

Motivated by these results in what follows we discuss analytic properties and optimality of simplified pair-copula constructions and focus mainly on partial vine copulas. Main objective of this section is to provide an answer to the question "how far off can we be by assuming a simplified model?" posed by Nagler and Czado (2016).

We first show that partial vine copulas are never the best simplified copula approximation (with respect to d_∞) if the true copula is non-simplified:

Theorem 3. *Suppose that $C \in \mathscr{C}_c^3$ is non-simplified. Then there exists some simplified copula $D \in \mathscr{C}_S^3$ satisfying $d_\infty(C, D) < d_\infty(C, \psi(C))$.*

Theorem 3 implies that if C does not fulfill the simplifying assumption then the partial vine copula fails to be optimal with respect to d_∞.

We now compare non-simplified copulas C with their unique partial vine copulas $\psi(C)$ and calculate their d_∞-distance. It turns out that the maximal d_∞-distance of a copula $C \in \mathscr{C}_c^3$ and its assigned partial vine copula $\psi(C)$ is at least $3/16$ which corresponds to 28.125% of the diameter of the metric space $(\mathscr{C}_c^3, d_\infty)$:

Theorem 4. *There exists a copula $C \in \mathscr{C}_c^3$ fulfilling $d_\infty(C, \psi(C)) \geq \frac{3}{16}$ and we have*

$$\sup_{C \in \mathscr{C}_c^3} d_\infty(C, \psi(C)) \geq \frac{3}{16}.$$

In other words, $\psi(C)$ can be far away from C, so working with PVCs must be done with care.

Intuitively, one might interpret ψ as projection and therefore think that ψ has to be continuous with respect to d_∞. It turns out, however, that this interpretation is wrong: ψ is *not* continuous with respect to d_∞:

Theorem 5. *Suppose that $C \in \mathscr{C}_c^3$ satisfies $d_\infty(C, \psi(C)) \neq 0$. Then C is a discontinuity point of thc mapping $\psi : \mathscr{C}_c^3 \to \mathscr{C}_c^3$. In other words: Every non-simplified $C \in \mathscr{C}_c^3$ is a discontinuity point of ψ.*

It is straightforward to verify that the set of all $C \in \mathscr{C}_c^3$ that are non-simplified is dense in $(\mathscr{C}_c^3, d_\infty)$ - Theorem 5 therefore has the following corollary:

Corollary 1. *The mapping $\psi : \mathscr{C}_c^3 \to \mathscr{C}_c^3$ is discontinuous on a dense subset of $(\mathscr{C}_c^3, d_\infty)$.*

Acknowledgements. SF gratefully acknowledges the support of the WISS 2025 project 'IDA-lab Salzburg' (20204-WISS/225/197-2019 and 20102-F1901166-KZP).

References

Aas, K., Czado, C., Frigessi, A., Bakken, H.: Pair-copula constructions of multiple dependence. Insurance Math. Econom. **44**(2), 182–198 (2009)

Acar, E., Genest, C., Nešlehová, J.: Beyond simplified pair-copula constructions. J. Multivar. Anal. **110**, 74–90 (2012)

Bedford, T., Cooke, R.: Vines: a new graphical model for dependent random variables. Ann. Stat. **30**(4), 1031–1068 (2002)

Bergsma, I.: Nonparametric testing of conditional independence by means of the partial copula (2011). arXiv preprint https://doi.org/10.48550/arXiv.1101.4607

Derumigny, A., Fermanian, J.-D.: About tests of the "simplifying" assumption for conditional copulas. Depend. Model. **5**(1), 154–197 (2017)

Durante, F., Sempi, C.: Principles of Copula Theory. CRC Press, Boca Raton, FL (2016)

Fernández-Sánchez, J., Trutschnig, W.: Conditioning based metrics on the space of multivariate copulas and their interrelation with uniform and levelwise convergence and iterated function systems. J. Theor. Probab. **28**, 1311–1336 (2015)

Gijbels, I., Omelka, M., Veraverbeke, N.: Nonparametric testing for no covariate effects in conditional copulas. Statistics **51**, 475–509 (2017)

Hobæk Haff, I., Aas, K., Frigessi, A.: On the simplified pair-copula construction - simply useful or too simplistic? J. Multivar. Anal. **101**, 1296–1310 (2010)

Joe, H.: Families of m-variate distributions with given margins and $m(m-1)/2$ bivariate dependence parameters. In: Lecture Notes-Monograph Series, vol. 28, pp. 120–141 (1996)

Kallenberg, O.: Foundations of Modern Probability. Springer, New York (1997). https://doi.org/10.1007/978-3-030-61871-1

Killiches, M., Kraus, D., Czado, C.: Examination and visualisation of the simplifying assumption for vine copulas in three dimensions. Aust. NZ J. Stat. **59**(1), 95–117 (2017). https://doi.org/10.1111/anzs.12182

Klenke, A.: Probability Theory - A Comprehensive Course. Springer, London (2008). https://doi.org/10.1007/978-1-84800-048-3

Mroz, T., Fuchs, S., Trutschnig, W.: How simplifying and flexible is the simplifying assumption in pair-copula constructions - analytic answers in dimension three and a glimpse beyond. Electron. J. Stat. **15**(1), 1951–1992 (2021)

Nagler, T., Czado, C.: Evading the curse of dimensionality in nonparametric density estimation with simplified vine copulas. J. Multivar. Anal. **151**, 69–89 (2016)

Reid, M.D., Williamson, R.C.: Generalised Pinsker Inequalities (2009). arXiv preprint https://doi.org/10.48550/arxiv.0906.1244

Spanhel, F., Kurz, M.: Simplified vine copula models: approximations based on the simplifying assumption. Electron. J. Stat. **13**, 1254–1291 (2019)

On Positive Dependence Properties for Archimedean Copulas

Sebastian Fuchs and Marco Tschimpke(✉)

University of Salzburg, Hellbrunner Straße 34, 5020 Salzburg, Austria
{sebastian.fuchs,marco.tschimpke}@plus.ac.at

Abstract. Due to their appealing characteristics, Archimedean copulas are frequently used in applications and have been thoroughly investigated in the literature. Their very restrictive, exchangeable structure leads to certain positive dependence properties such as lower tail decreasingness (LTD) and total positivity of order 2 of an Archimedean copula being equivalent. In the present paper we show that stochastical increasingness (SI) is equivalent to total positivity of order 2 where the latter now refers to the Archimedean copula's conditional distribution. Thus, each of the coordinate-wise defined properties LTD and SI has an equivalent counterpart formulated in terms of total positivity of order 2.

1 Introduction and Preliminaries

For the quantification of dependence, numerous indices and measures are available that examine dependence from a wide range of perspectives. Apart from the option of assigning a single value to a dependence structure, an alternative way consists in checking whether a copula fulfills certain (positive) dependence properties such as positive quadrant dependence (PQD), left tail decreasingness (LTD), stochastically increasingness (SI), or total positivity of order 2, the latter usually considered for a copula (TP2) or (if existent) its density (d-TP2 for short). According to Joe (2014) and Nelsen (2006) the different notions of positive dependence are linked as illustrated in Fig. 1.

The property *total positivity of order* 2 has been extensively studied for copulas (TP2) and their densities (d-TP2); see, e.g., Hürlimann (2003), Joe (1997, 2014) and Lehmann (1966). In this paper we investigate total positivity of order 2 for a copula's Markov kernel (MK-TP2 for short), a positive dependence property that is stronger than TP2 and SI, weaker than d-TP2 but, unlike d-TP2,

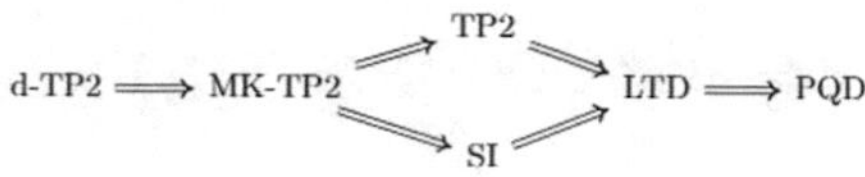

Fig. 1. Relations between the different notions of positive dependence

L. A. García-Escudero et al. (Eds.): SMPS 2022, AISC 1433, pp. 159–165, 2023.
https://doi.org/10.1007/978-3-031-15509-3_21

is not restricted to absolutely continuous copulas, making it, to the best of our knowledge, the strongest dependence property defined for any copula. From a conditional distribution point of view the MK-TP2 property has been studied by Capéraà and Genest (1990) and Guillem (2000) showing its interrelation with the other above-mentioned dependence properties.

In the present paper we show that, for Archimedean copulas, the properties MK-TP2 and SI are equivalent, and can be characterized by the Archimedean generator. The equivalence of MK-TP2 and SI is consistent with the very restrictive, exchangeable structure of Archimedean copulas and the well-known equivalence of TP2 and LTD property (see, e.g., Nelsen 2006).

Throughout this paper we will write $\mathbb{I} := [0, 1]$ and let $\mathscr{C}$ denote the family of all bivariate copulas; M will denote the comonotonicity copula, Π the independence copula and W will denote the countermonotonicity copula. According to Durante and Sempi (2016, Theorem 3.4.3) and due to disintegration, every copula C fulfills

$$C(u, v) = \int_{[0,u]} K_C(s, [0, v]) \,\mathrm{d}\lambda(u)$$

where λ denotes the Lebesgue measure on the Borel σ-field $\mathscr{B}(\mathbb{I})$ and K_C is (a version of) the Markov kernel of C: A *Markov kernel* from $\mathbb{I}$ to $\mathscr{B}(\mathbb{I})$ is a mapping $K : \mathbb{I} \times \mathscr{B}(\mathbb{I}) \to \mathbb{I}$ such that for every fixed $F \in \mathscr{B}(\mathbb{I})$ the mapping $u \mapsto K(u, F)$ is measurable and for every fixed $u \in \mathbb{I}$ the mapping $F \mapsto K(u, F)$ is a probability measure. Given two uniformly distributed random variables U and V on a probability space $(\Omega, \mathscr{A}, P)$ we say that a Markov kernel K is a *regular conditional distribution* of V given U if $K(V(\omega), F) = P(V \in F \,|\, X)(\omega)$ holds P-almost surely for every $F \in \mathscr{B}(\mathbb{I})$. It is well-known that for each such random vector (U, V) a regular conditional distribution $K(., .)$ of V given U always exists and is unique for P^U-a.e. $u \in \mathbb{I}$, where P^U denotes the push-forward of P under U. For more background on conditional expectation and general disintegration we refer to Kallenberg (2002) and Klenke (2008) for more information on Markov kernels. In the context of copulas we refer to Durante and Sempi (2016), Kasper et al. (2021) and Mroz et al. (2021).

2 Dependence Properties

In this section we resume some well-known concepts of positive dependence viewed from a copula perspective; their probabilistic interpretation is presented in Fig. 2 below.

A copula $C \in \mathscr{C}$ is said to be

- *positively quadrant dependent* (PQD) if $C(u, v) \geq \Pi(u, v)$ holds for all $(u, v) \in (0, 1)^2$.

- *left tail decreasing* (LTD) if, for any $v \in (0,1)$, the mapping $(0,1) \to \mathbb{R}$ given by $u \mapsto \frac{C(u,v)}{u}$ is non-increasing.
- *stochastically increasing* (SI) if, for (a version of) the Markov kernel K_C and any $v \in (0,1)$, the mapping $u \mapsto K_C(u,[0,v])$ is non-increasing.

According to Nelsen (2006) these three dependence properties are related as presented in Fig. 1.

Another notion of positive dependence that differs from the above mentioned dependence properties is *total positivity of order 2* - a property applicable to various copula-related objects: the copula itself, its Markov kernel and its density, leading to three different but related positive dependence properties. In general, a function $f : [0,1]^2 \to \mathbb{R}$ is said to be *totally positive of order 2* (TP2) if the inequality

$$f(u_1,v_1)f(u_2,v_2) - f(u_1,v_2)f(u_2,v_1) \geq 0 \tag{1}$$

holds for all $0 < u_1 \leq u_2 < 1$ and all $0 < v_1 \leq v_2 < 1$. The TP2 property has been extensively discussed for copulas and their densities (see, e.g., Capéraà and Genest 1993; Joe 1997, 2014; Nelsen 2006) and, to a certain extent, also for their partial derivatives respectively their conditional distribution functions (see, e.g., Capéraà and Genest, 1990; Guillem 2000). To avoid problems concerning well-definedness, which inevitably occur when using partial derivatives, we here use Markov kernels: A copula $C \in \mathscr{C}$ is said to be

- *TP2* if the copula itself is TP2.
- *MK-TP2* (short for Markov kernel TP2) if (a version of) its Markov kernel is TP2.

If the copula has a density another dependence property can be formulated: C is said to be

- *d-TP2* (short for density TP2) if (a version of) its density c is TP2.

The property d-TP2 is also referred to as *positively likelihood ratio dependence* (see, e.g., Lehmann (1966)).

Altogether the different notions of positive dependence are linked as depicted in Fig. 1 (compare Guillem 2000; Nelsen 2006). Notice that the MK-TP2 property, unlike d-TP2, can be determined for any copula, making it presumably the strongest dependence property defined for any copula.

We now provide a probabilistic interpretation of the above-mentioned positive dependence properties in terms of continuous random variables X and Y with connecting copula C:

Dependence property	Probabilistic interpretation
PQD	$\mathbb{P}(Y \le y \mid X \le x) \ge \mathbb{P}(Y \le y)$
LTD	$x \mapsto \mathbb{P}(Y \le y \mid X \le x)$ is non-increasing for any y
SI	$x \mapsto \mathbb{P}(Y \le y \mid X = x)$ is non-increasing (a.s.) for any y
TP2	$x \mapsto \dfrac{\mathbb{P}(Y \le y \mid X \le x)}{\mathbb{P}(Y \le y' \mid X \le x)}$ is non-increasing for any $y \le y'$
MK-TP2	$x \mapsto \dfrac{\mathbb{P}(Y \le y \mid X = x)}{\mathbb{P}(Y \le y' \mid X = x)}$ is non-increasing (a.s.) for any $y \le y'$
d-TP2	$x \mapsto \dfrac{f_{Y\mid X}(y\mid x)}{f_{Y\mid X}(y'\mid x)}$ is non-increasing (a.s.) for any $y \le y'$

Fig. 2. Probabilistic interpretation of the positive dependence properties

We conclude this section by examining the MK-TP2 property for simple examples:

1. The independence copula Π is d-TP2 and hence MK-TP2.
2. The comonotonicity copulas M is MK-TP2; since M is singular, it cannot be d-TP2.
3. The countermonotonicity copula W fulfills $W(0.5, 0.5) < \Pi(0.5, 0.5)$. Thus it fails to be PQD and hence fails to satisfy any of the above-mentioned positive dependence properties.

3 Archimedean Copulas

In this section we study under which conditions on the Archimedean generator (or its pseudo-inverse) the corresponding Archimedean copula is MK-TP2. It is well-known that an Archimedean copula C is TP2 if and only if it is LTD (see, e.g., Nelsen 2006). In what follows, we add a second characterisation to Fig. 1 by showing that an Archimedean copula C is MK-TP2 if and only if it is SI.

Recall that a generator of a bivariate Archimedean copula (see Durante and Sempi 2016; Nelsen 2006) is a convex, strictly decreasing function $\varphi : \mathbb{I} \to [0, \infty]$ with $\varphi(1) = 0$. According to Kasper et al. (2021) we may, w.l.o.g., assume that all generators are right-continuous at 0. Every generator φ induces a symmetric copula C via

$$C(u, v) = \psi(\varphi(u) + \varphi(v))$$

for all $(u, v) \in \mathbb{I}^2$ where $\psi : [0, \infty] \to \mathbb{I}$ denotes the pseudo-inverse of φ defined by

$$\psi(x) := \begin{cases} \varphi^{-1}(x) & \text{if } x \in [0, \varphi(0)) \\ 0 & \text{if } x \geq \varphi(0). \end{cases}$$

The pseudo-inverse ψ is convex, non-increasing, strictly decreasing on $[0, \varphi(0))$ and fulfills $\psi(0) = 1$. Since most of the subsequent results are formulated in terms of ψ, we call ψ *co-generator*. If $\varphi(0) = \infty$ the induced copula C is called strict, otherwise it is referred to as non-strict.

In the sequel we restrict ourselves to strict copulas whose co-generator ψ is continuously differentiable. Convexity then allows to view ψ' as non-decreasing and continuous function on $(0, \infty)$.

To investigate under which conditions an Archimedean copula C with generator φ and co-generator ψ is MK-TP2 we need to establish (a version of) the Markov kernel of C. According to Kasper et al. (2021) K_C defined by

$$K_C(u, [0, v]) = \begin{cases} 1 & \text{if } u \in \{0, 1\} \\ \frac{\varphi'(u)}{\varphi'(C(u,v))} & \text{if } u \in (0, 1) \end{cases}$$

is (a version of) the Markov kernel of C. Differentiability of φ and the rules for differentiation of inverse functions yield that K_C fulfills

$$K_C(u, [0, v]) = \begin{cases} 1 & \text{if } u \in \{0, 1\} \\ \frac{\psi'(\varphi(u)+\varphi(v))}{\psi'(\varphi(u))} & \text{if } u \in (0, 1) \end{cases}$$

For completeness, we repeat the following necessary and sufficient condition for an Archimedean copula to be TP2 which is well known, see, e.g., Nelsen (2006): For every Archimedean copula C with generator φ and co-generator ψ, the following statements are equivalent:

1. C is LTD.
2. C is TP2.
3. ψ is log-convex.

Recall that a function $f : [0, \infty] \to (0, \infty)$ is log-convex if $\log f|_{(0,\infty)}$ is convex.

In what follows we prove that the dependence properties SI and MK-TP2 are equivalent as well. It has been shown in Capéraà and Genest (1993) that C is SI if and only if $-\psi'$ is log-convex. Therefore, it remains to verify that log-convexity of $-\psi'$ implies MK-TP2:

Lemma 1. *Let C be a strict Archimedean copula with continuously differentiable co-generator ψ. If $-\psi'$ is log-convex, then C is MK-TP2.*

Proof. Defining the functions $f : (-\infty, 0) \to \mathbb{R}$ and $G : (0,1)^2 \to (-\infty, 0)$ by $f(x) := \log(-\psi'(-x))$ and $G(u, v) := -(\varphi(u) + \varphi(v))$, the copula C is MK-TP2 if

$$
\begin{aligned}
0 &\le \log\left(\frac{K_C(u_1,[0,v_1])\,K_C(u_2,[0,v_2])}{K_C(u_1,[0,v_2])\,K_C(u_2,[0,v_1])}\right)\\
&= \log\left(\frac{\psi'(-G(u_1,v_1))\,\psi'(-G(u_2,v_2))}{\psi'(-G(u_1,v_2))\,\psi'(-G(u_2,v_1))}\right)\\
&= (f\circ G)(u_1,v_1) + (f\circ G)(u_2,v_2) - (f\circ G)(u_1,v_2) - (f\circ G)(u_2,v_1)
\end{aligned}
$$

for all $0 < u_1 \le u_2 < 1$ and all $0 < v_1 \le v_2 < 1$. Since G is a 2-increasing and coordinatewise non-decreasing function and f is non-decreasing and, by assumption, convex (recall that convexity of $\log(-\psi')$ implies convexity of f), it follows from Marshall et al. (2011, p.219) that the composition $f \circ G$ is 2-increasing. This proves the assertion. □

Combining Fig. 1 and, Lemma 1, we are now in the position to state the main result of this section:

Theorem 1. *Let C be a strict Archimedean copula with continuously differentiable co-generator ψ. Then the following statements are equivalent:*

1. *C is MK-TP2.*
2. *C is SI.*
3. *$-\psi'$ is log-convex.*

According to Capéraà and Genest (1993), for an Archimedean copula with twice differentiable co-generator ψ, the following statements are equivalent:

1. C is d-TP2.
2. ψ'' is log-convex.

In sum, for Archimedean copulas, Fig. 1 reduces to

$$
\text{d-TP2} \overset{(1)}{\Longrightarrow} \text{MK-TP2} \Longleftrightarrow \text{SI} \overset{(2)}{\Longrightarrow} \text{TP2} \Longleftrightarrow \text{LTD} \Longrightarrow \text{PQD}
$$

Remark 2.13 in Müller and Scarsini (2005) provides an Archimedean copula that is MK-TP2 and SI, but not d-TP2, i.e., the reverse of (1) does not hold, in general. An Archimedean copula that is LTD and TP2 but not MK-TP2 is given in Spreeuw (2013, Example 19): The mapping $\psi : [0,\infty] \to \mathbb{I}$ given by $\psi(x) := \left(x + \sqrt{1+x^2}\right)^{-1/10}$ is an Archimedean co-generator which is log-convex. However, since $-\psi'$ is not log-convex, the reverse of (2) does not hold, in general.

An Archimedean co-generator ψ is said to be *completely monotone* if its restiction $\psi|_{(0,\infty)}$ has derivatives of all orders and satisfies $(-1)^n\psi^{(n)}(x) \ge 0$ for all $x \in (0,\infty)$ and all $n \in \mathbb{N}$, and it follows from Niculescu and Persson (2006) that completely monotone functions are log-convex. If ψ is completely monotone then its negative derivative is completely monotone as well implying that $(-1)^n\psi^{(n)}$ is log-convex for all $n \in \mathbb{N}$. Therefore, every Archimedean copula with completely monotone co-generator (this includes Frank copulas, Gumbel copulas, and Clayton copulas for certain parameters) is d-TP2 and hence MK-TP2.

Acknowledgements. The first author gratefully acknowledges the support of the WISS 2025 project 'IDA-lab Salzburg' (20204-WISS/225/197-2019 and 20102-F1901166-KZP). The second author gratefully acknowledges the financial support from AMAG Austria Metall AG within the project ProSa.

References

Capéraà, P., Genest, C.: Concepts de dépendance et ordres stochastiques pour des lois bidimensionnelles. Can. J. Stat. **18**(4), 315–326 (1990)

Capéraà, P., Genest, C.: Spearman's ρ is larger than Kendall's τ for positively dependent random variables. J. Nonparametric Stat. **2**(2), 183–194 (1993)

Durante, F., Sempi, C.: Principles of copula theory. CRC Press, Boca Raton (2016)

Guillem AIG Structure de dépendance des lois de valeurs extrêmes bivariées. Comptes Rendus Mathématiques des l'Académie des Sciences. La Société Royale du Canada 330(7), 593–596 (2000)

Hürlimann, W.: Hutchinson-Lai's conjecture for bivariate extreme value copulas. Stat. Probab. Lett. **61**, 191–198 (2003)

Joe, H.: Multivariate Models and Multivariate Dependence Concepts. CRC Press, Boca Raton (1997)

Joe, H.: Dependence Modeling with Copulas. CRC Press, Boca Raton (2014)

Kallenberg, O.: Foundations of Modern Probability. Springer-Verlag, New York (2002), https://doi.org/10.1007/978-3-030-61871-1

Kasper, T., Fuchs, S., Trutschnig, W.: On weak conditional convergence of bivariate Archimedean and Extreme Value copulas, and consequences to nonparametric estimation. Bernoulli **27**, 2217–2240 (2021)

Klenke, A.: Wahrscheinlichkeitstheorie. Springer, Heidelberg (2008). https://doi.org/10.1007/978-3-642-36018-3

Lehmann, E.: Some concepts of dependence. Ann. Math. Stat. **37**, 1137–1153 (1966)

Marshall, A., Olkin, I., Arnold, B.: Inequalities: Theory of Majorization and Its Applications, 2nd edn. Springer, New York (2011). https://doi.org/10.1007/978-0-387-68276-1

Mroz, T., Fuchs, S., Trutschnig, W.: How simplifying and flexible is the simplifying assumption in pair-copula constructions - analytic answers in dimension three and a glimpse beyond. Electron. J. Stat. **15**(1), 1951–1992 (2021)

Müller, A., Scarsini, M.: Archimedean copulae and positive dependence. J. Multivar. Anal. **93**, 434–445 (2005)

Nelsen, R.: An Introduction to Copulas. Springer, Heidelberg (2006). https://doi.org/10.1007/0-387-28678-0

Niculescu, C., Persson, L.E.: Convex Functions and Their Applications. A Contemporary Approach. Springer, New York (2006). https://doi.org/10.1007/0-387-31077-0

Spreeuw, J.: Archimedean copulas derived from Morgenstern utility functions (2013). https://ssrn.com/abstract=2215701

Advances in Robust Constrained Model Based Clustering

Luis A. García-Escudero[1], Agustín Mayo-Iscar[1], Gianluca Morelli[2(✉)], and Marco Riani[2]

[1] Department of Statistics and Operational Research and IMUVA, University of Valladolid, Valladolid, Spain
{lagarcia,agustin.mayo.iscar}@uva.es

[2] Department of Economics and Management and Interdepartmental Centre of Robust Statistics, University of Parma, Parma, Italy
{gianluca.morelli,mriani}@unipr.it

Abstract. Model-based approaches to cluster analysis and mixture modelling often involve maximizing classification and mixture likelihoods. Robust clustering and mixture modelling procedures, that can resist certain amount of contaminating data, can be introduced by considering trimmed versions of those classification and mixture likelihoods. Without appropriate constrains on the scatter matrices of the components, these trimmed likelihood maximizations result in ill-posed problems. Moreover, non-interesting or "spurious" clusters are often detected by unconstrained algorithms aimed at maximizing these trimmed likelihood criteria.

A useful approach to avoid spurious solutions is to restrict relative components scatter by prespecified tuning constants. Recently new methodologies for constrained parsimonious model-based clustering have been introduced which include, in the untrimmed case, the 14 parsimonious models that are often applied in model-based clustering when assuming normal components as limit cases. In this paper we extend this approach to cope with the presence of atypical observations and discuss two viable strategies for automatically estimating the restriction parameters.

1 Trimmed Mixture Likelihood with Constraints

The problem of estimating the parameters in finite mixture models has received a lot of attention. This interest is mainly due to the flexibility of this kind of models to adapt to different settings, as well as to the existence of feasible EM-type algorithms to provide approximate solutions. However, it is well known that a few outlying observations can produce undesirable effects in the determination of the fitted mixture. Moreover, the classical maximum-likelihood estimation approach for these models often leads to ill-posed problems because of the unboundedness of the objective function to maximize, which favors the appearance of non-interesting local maximizers and degenerate or spurious solutions. This can be seen as another kind of lack of robustness.

L. A. García-Escudero et al. (Eds.): SMPS 2022, AISC 1433, pp. 166–173, 2023.
https://doi.org/10.1007/978-3-031-15509-3_22

The troubles of lack of robustness in mixture fitting appear when the sample contains a certain proportion of observations that have been generated by some strange mechanism and do not follow the underlying population model. Moreover, practitioners and users of these models may not be aware of the presence of contaminated data.

Robust proposals for mixture modeling are based on trimming, which has been shown to be a simple, flexible, powerful and computationally feasible way to robustify statistical methods in many different problems and different frameworks. In this work, we will focus on this trimming approach and we concentrate on the problem of fitting a mixture of k normal components to a given data set $\{x_1, \ldots, x_n\}$ in $\mathbb{R}^p$ by maximizing a "trimmed mixture likelihood".

$$L_k(\alpha) = \sum_{i=1}^{n} z(x_i) \log \left[\sum_{j=1}^{k} \pi_j \varphi(x_i; \mu_j, \Sigma_j) \right], \tag{1}$$

where $\varphi(\cdot; \mu, \Sigma)$ stands for the probability density function of the p-variate normal distribution with mean μ and covariance matrix Σ and where z is a trimming indicator function that tells us whether observation x_i is trimmed off ($z(x_i) = 0$) or not ($z(x_i) = 1$). A fraction α of the observations are allowed to be unassigned or trimmed off by imposing that $\sum_{i=1}^{n} z(x_i) = [n(1-\alpha)]$. It is also common to enforce constraints on the Σ_j scatter matrices when maximizing (1). Among them, the use of "parsimonious" models (Celeux and Govaert 1995; Banfield and Raftery 1993) is one of the most popular and widely applied approaches in practice. These parsimonious models follow from a decomposition of the Σ_j scatter matrices as

$$\Sigma_j = \lambda_j \Omega_j \Gamma_j \Omega_j', \tag{2}$$

with $\lambda_j = |\Sigma_j|^{1/p}$ (volume parameters),

$$\Gamma_j = \mathsf{diag}(\gamma_{j1}, ..., \gamma_{jl}, ..., \gamma_{jp}) \text{ with } \mathsf{det}(\Gamma_j) = \prod_{l=1}^{p} \gamma_{jl} = 1$$

(shape matrices), and Ω_j (rotation matrices) with $\Omega_j \Omega_j' = \mathsf{I}_p$. 14 parsimonious models are traditionally defined depending on the constraints imposed on the λ_j, Ω_j and Γ_j elements across components. These constraints reduce notably the number of parameters to be estimated and serve to improve their efficiency and interpretability. Constraints also serve to get well-defined likelihood maximizations and to avoid the detection of spurious solutions. However, many of these 14 models are still associated to unbounded likelihood maximizations (those with unconstrained λ_j parameters). Although these problems could be partially solved by resorting to good initializations or the early stopping of non-converging, we prefer to depend on well-posed mathematical problem definitions.

A procedure to obtain well-defined problems is to constrain the Σ_j scatter matrices by specifying some tuning constants that control the strength of the constraints (Hathaway 1985). For instance, we can force the ratio between the largest and the smallest of the $k \times p$ eigenvalues of the Σ_j matrices to be smaller than a given fixed constant $c^* \geq 1$ (Ingrassia and Rocci 2007; García-Escudero et al. 2008, 2011, 2014, 2015). In this direction, we obtain the TCLUST procedure when the maximization of (1) is done under the (more simple) constraint:

$$\max_{jl} \lambda_l(\Sigma_j) / \min_{jl} \lambda_l(\Sigma_j) \leq c^*, \tag{3}$$

where $\{\lambda_l(\Sigma_j)\}_{l=1}^p$ are the set of eigenvalues of the Σ_j matrix, $j = 1, ..., k$. Notice that, through the use of the trimmed likelihood in (1), we are also considering the possibility of trimming a fixed fraction α of the (hopefully) most outlying observations. With this eigenvalue-ratio approach, we need a very high c^* value to be close to affine equivariance. Unfortunately, such a high c^* value does not always successfully prevent us from incurring into spurious solutions.

2 The New Constraints and Algorithm

García-Escudero et al. (2022) have recently introduced three different types of constraints on the Σ_j matrices which depend on three constants c_{det}, c_{shw} and c_{shb} all of them being greater than or equal to 1. The first type of constraint controls the maximal ratio among determinants or, in other words, the maximum allowed difference between component volumes:

$$\text{“deter”:} \qquad \frac{\max_{j=1,\ldots,k} |\Sigma_j|}{\min_{j=1,\ldots,k} |\Sigma_j|} = \frac{\max_{j=1,\ldots,k} \lambda_j^p}{\min_{j=1,\ldots,k} \lambda_j^p} \leq c_{\mathsf{det}}. \tag{4}$$

The second type of constraint controls departures from sphericity “within” each component:

$$\text{shape-“within”:} \qquad \frac{\max_{l=1,\ldots,p} \gamma_{jl}}{\min_{l=1,\ldots,p} \gamma_{jl}} \leq c_{\mathsf{shw}} \text{ for } j = 1, ..., k. \tag{5}$$

We thus get a set of k constraints that, in the most constrained $c_{\mathsf{shw}} = 1$ case imposes $\Gamma_1 = ... = \Gamma_p = \mathsf{I}_p$, where I_p is the identity matrix of size p.

This third type of constraint controls the maximum allowed difference between shape elements “between” components:

$$\text{shape-“between”:} \qquad \frac{\max_{j=1,\ldots,k} \gamma_{jl}}{\min_{j=1,\ldots,k} \gamma_{jl}} \leq c_{\mathsf{shb}} \text{ for } l = 1, ..., p. \tag{6}$$

Notice that the same shape matrices are imposed, as $\Gamma_1 = ... = \Gamma_k$, in the most constrained $c_{\mathsf{shb}} = 1$ case.

The algorithm for the trimmed $\alpha > 0$ case follows the same lines as the algorithm in the untrimmed $\alpha = 0$ case presented in García-Escudero et al. (2022). It is only needed to be included a “trimming” step. If $\theta^{(t)} = (\pi_1^{(t)}, \cdots, \pi_k^{(t)}, \mu_1^{(t)}, \cdots, \mu_k^{(t)}, \Sigma_1^{(t)}, \cdots, \Sigma_k^{(t)})$ denotes the values of the parameters in t-th step of the iterative process then we just need to update the observation weights $\tau_j(x_i; \theta^{(t)})$ exactly as in García-Escudero et al. (2022) but setting them equal to 0 for all $i \in \mathscr{I}$ where $\mathscr{I} = \{i : W(x_i; \theta^{(t)}) < W(x_{([n\alpha])}; \theta^{(t)})\}$ for $W_j(x; \theta^{(t)}) = \pi_j^{(t)} \phi(x; \mu_j^{(t)}, \Sigma_j^{(t)})$, $W(x; \theta^{(t)}) = \sum_{j=1}^k W_j(x; \theta^{(t)})$ and $W_j(x_{(1)}; \theta^{(t)}) \leq ... \leq W_j(x_{(n)}; \theta^{(t)})$.

3 Automatic Choice of the Constraints

In this section we initially recall the new BIC introduced in García-Escudero et al. (2022) and then we discuss two viable procedures to automatically estimate the values of the constraints.

In order to automatically choose the number of mixture components k and $\mathsf{pars} = [c_{\mathsf{det}}, c_{\mathsf{shw}}, c_{\mathsf{swb}}, \mathsf{rot}]$, the proposal is to choose $[\widehat{k}, \widehat{\mathsf{pars}}] = \arg\min_{k,\mathsf{pars}}$ $\mathsf{BIC}[k, \mathsf{pars}]$, for

$$\mathsf{BIC}[k, \mathsf{pars}] = -2L_k^{\mathsf{pars}}(\alpha) + v_k^{\mathsf{pars}} \log n[1-\alpha], \tag{7}$$

where $L_k^{\mathsf{pars}}(\alpha)$ is the maximum value achieved in the constrained maximization of (1), the trimmed likelihood, under constraints defined by pars, and where v_k^{pars} is a penalty term defined as

$$v_k^{\mathsf{pars}} = kp + k - 1 + (k-1)\left(1 - \frac{1}{c_{\mathsf{det}}^{1/p}}\right) + 1 + \\ + (p-1)\left(1 - \frac{1}{c_{\mathsf{shw}}}\right)\left[(k-1)\left(1 - \frac{1}{c_{\mathsf{shb}}}\right) + 1\right] + k(\mathsf{rot})\frac{p(p-1)}{2},$$

and $k(\mathsf{rot})$ takes the values 0, 1 and k depending on the considered rotation (in the order `rot=I`, `rot=E` and `rot=V`). More complex models are allowed to be fitted with larger values of c_{det}, c_{shw} and c_{swb}, given that less restricted Σ_j scatter matrices can be chosen. "Model complexity" here does not imply an increased number of parameters, but we are actually considering higher flexibility for the Σ_j matrices. In the limit cases (restriction constants equal to 1 or tending to ∞), v_k^{pars} exactly coincides with the number of free parameters in the traditional 14 parsimonious models.

It is not computationally easy to perform the minimization of the criterion (7) over all the possible combinations of k and pars parameters. Consequently, we just focus on powers of 2 for the restriction constants, $2^0, 2^1, \ldots, 2^7$. We will assume that k and α are fixed an we discuss two strategies to automatically estimate the other parameters. In García-Escudero et al. (2022) it was proposed an initial preliminary step where all the 14 GPCM are applied but imposing a large 2^7 values for the free parameters. For example when `EVE` is investigated the constraints for c_{shw} and c_{swb} are set to 2^7. This gives a restricted parametric space which can be further refined to obtain the best solution. For example, suppose that `VEE` is found as the best specification, a further set of maximizations using $c_{\mathsf{det}} = 2^0, 2^1, ..., 2^7$ is employed in order to understand which is the best value of this parameter.

The new alternative strategy, which is explored in this paper, is based on the maximization over the $2^0, 2^1, ..., 2^7$ for c_{shw} and c_{det} keeping $c_{\mathsf{swb}} = 128$ set to 128 and the type of rotation set to `V`. Once the optimal combination of c_{shw} and c_{det} has been found, it is possible to refine for c_{swb} and, as final step, to investigate the best among the 3 types of rotation. As an illustration of the suggested approach we consider the data in Fig. 1.

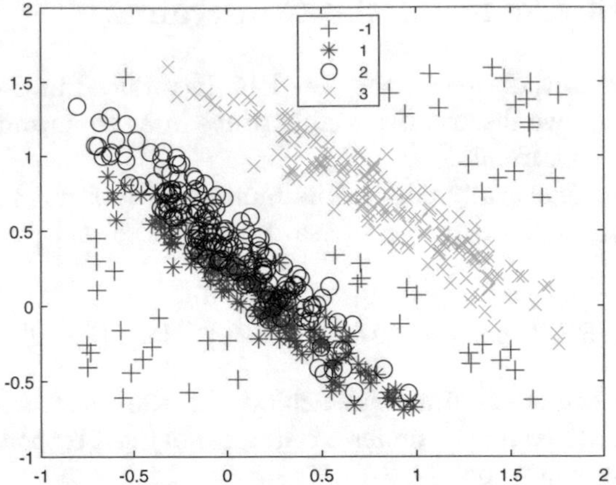

Fig. 1. Data generating process. The units shown with symbol '+' are 50 outlying units

The data have been generated by drawing 500 observations from a mixture with $k = 3$, $p = 2$, $\pi_1 = \pi_2 = \pi_3 = 1/3$, $c_{\mathsf{det}} = 2$ (a sensible departure from sphericity "within" each component), $c_{\mathsf{shw}} = 30$ and $c_{\mathsf{shb}} = 1.1$ (a very moderate difference "between" shape elements components). An average overlap of 0.15 has been imposed through the `MixSim` method of Maitra and Melnykov (2010), as extended by Riani et al. (2015) and incorporated into the `FSDA` Matlab toolbox (Riani et al. 2012, 2015). Additional 50 units from the uniform distribution are generated in such a way that their squared Mahalanobis distance from the centroids of each existing group is larger then the quantile 0.999 of the χ^2_2 distribution. The final sample size is therefore equal to $n = 550$.

The application of traditional TCLUST approach (reviewed in Sect. 1) with maximum ratio between eigenvalues (c^*) respectively equal to 16 and 10^{10} produces the classifications shown in Fig. 2. When c^* is too small it is impossible to

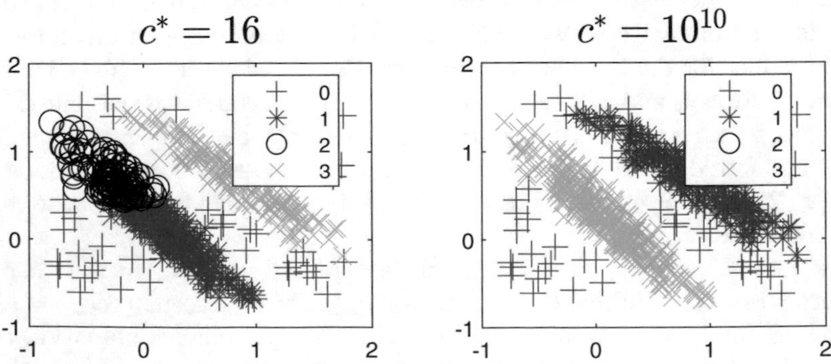

Fig. 2. Results derived by the application of the traditional tclust procedure using $c^* = 16$ (left panel) and $c^* = 10^{10}$ (right panel) and $\alpha = 0.10$

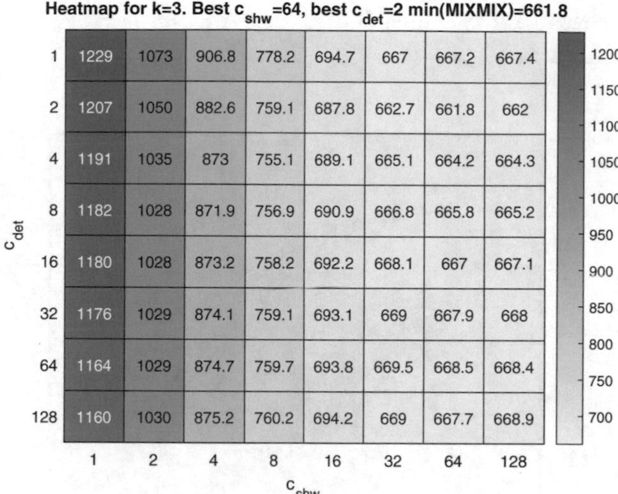

Fig. 3. Heatmap of BIC in Eq. 7. The best combination of values is associated with $c_{\mathsf{shw}} = 64$ and $c_{\mathsf{det}} = 2$

find those elongated elliptical clusters. On the other hand, the choice $c^* = 10^{10}$ results in an unbounded likelihood and the detection of a spurious group.

Figure 3 shows the heatmap of the BIC (Eq. 7) in the grid of $2^0, 2^1, \ldots, 2^7$ for c_{shw} and c_{swb}. The optimal combination of these two parameters gives $c_{\mathsf{shw}} = 64$ and $c_{\mathsf{det}} = 2$. The subsequent refinement "between" shape elements components and type of rotation gives $c_{\mathsf{shb}} = 1$ and unconstrainted rotation (see Fig. 4).

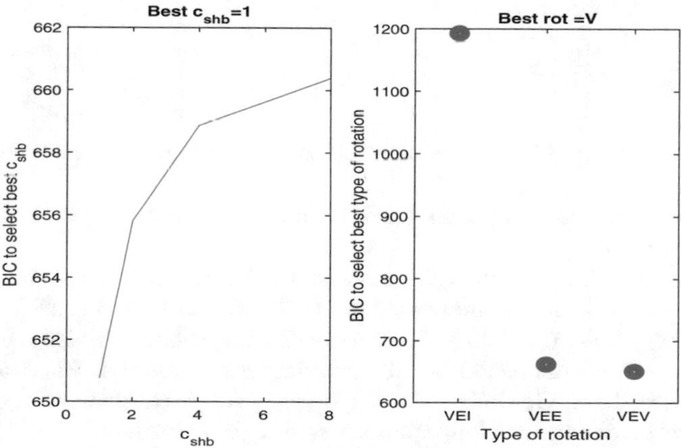

Fig. 4. Refinement step in order to decide the optimal value of c_{shb} and optimal type of rotation. Note that the first two letters in the x-axis of the right panel are 'VE' because $c_{\mathsf{det}} > 1$ and $c_{\mathsf{shb}} = 1$

The final classification which is obtained by the application of the above procedure is shown in Fig. 5.

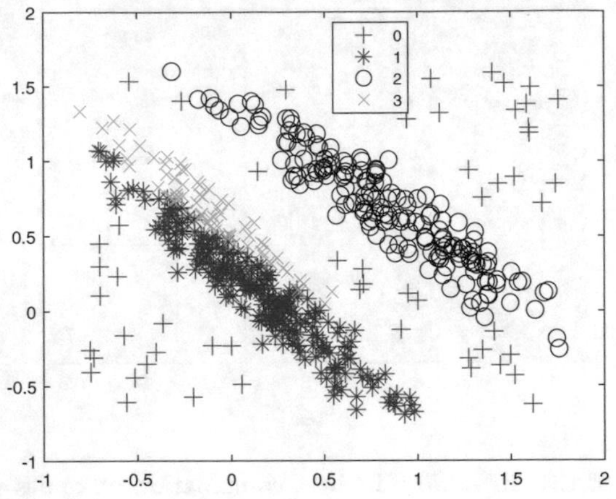

Fig. 5. Final classification obtained using $c_{\mathsf{shw}} = 64$, $c_{\mathsf{det}} = 2$, $c_{\mathsf{shb}} = 1$, and $\alpha = 0.10$

Acknowledgements. The work benefits from the High Performance Computing (HPC) facility of the University of Parma. We also acknowledge financial support from the "Statistics for fraud detection, with applications to trade data and financial statements" project of the University of Parma. All the calculations in this paper have used the Flexible Statistics and Data Analysis (FSDA) MATLAB toolbox, which is freely downloadable from GitHub at the web address https://github.com/UniprJRC/FSDA.

References

Banfield, J.D., Raftery, A.E.: Model-based Gaussian and non-Gaussian clustering. Biometrics **49**, 803–821 (1993)

Celeux, G., Govaert, G.: Gaussian parsimonious clustering models. Pattern Recogn. **28**, 781–793 (1995)

García-Escudero, L.A., Gordaliza, A., Matrán, C., Mayo-Iscar, A.: A general trimming approach to robust cluster analysis. Ann. Stat. **36**, 1324–1345 (2008)

García-Escudero, L.A., Gordaliza, A., Matrán, C., Mayo-Iscar, A.: Exploring the number of groups in robust model-based clustering. Stat. Comput. **21**, 585–599 (2011)

García-Escudero, L.A., Gordaliza, A., Mayo-Iscar, A.: A constrained robust proposal for mixture modeling avoiding spurious solutions. Adv. Data Anal. Classif. **8**(1), 27–43 (2013). https://doi.org/10.1007/s11634-013-0153-3

García-Escudero, L.A., Gordaliza, A., Matrán, C., Mayo-Iscar, A.: Avoiding spurious local maximizers in mixture modeling. Stat. Comput. **25**(3), 619–633 (2014). https://doi.org/10.1007/s11222-014-9455-3

García-Escudero, L.A., Mayo-Iscar, A., Riani, M.: Constrained parsimonious model-based clustering. Stat. Comput. **32**(1), 1–15 (2021). https://doi.org/10.1007/s11222-021-10061-3

Hathaway, R.: A constrained formulation of maximum likelihood estimation for normal mixture distributions. Ann. Stat. **13**, 795–800 (1985)

Ingrassia, S., Rocci, R.: Constrained monotone EM algorithms for finite mixture of multivariate Gaussians. Comput. Stat. Data Anal. **51**, 5339–5351 (2007)

Maitra, R., Melnykov, V.: Simulating data to study performance of finite mixture modeling and clustering algorithms. J. Comput. Graph. Stat. **19**, 354–376 (2010)

Riani, M., Cerioli, A., Perrotta, D., Torti, F.: Simulating mixtures of multivariate data with fixed cluster overlap in FSDA library. Adv. Data Anal. Classif. **9**(4), 461–481 (2015). https://doi.org/10.1007/s11634-015-0223-9

Riani, M., Perrotta, D., Torti, F.: FSDA: a MATLAB toolbox for robust analysis and interactive data exploration. Chemom. Intell. Lab. Syst. **116**, 17–32 (2012)

Trimmed Spatio-Temporal Variogram Estimator

Alfonso García-Pérez(✉)

Departamento de Estadística, I.O. y C.N., Universidad Nacional de Educación a Distancia (UNED), Paseo Senda del Rey 9, 28040 Madrid, Spain
agar-per@ccia.uned.es

Abstract. The spatio-temporal variogram is the key element in spatio-temporal prediction based on kriging, but the classical estimator of this parameter is very sensitive to outliers. In this contributed paper we propose a trimmed estimator of the spatio-temporal variogram as a robust estimator. We obtain an accurate approximation of its distribution with small samples sizes and a scale contaminated normal model. We conclude with an example with real data.

1 Introduction

Let us suppose that we have a spatio-temporal random field $Z(\mathbf{s}, t)$, $(\mathbf{s}, t) \in D \times T$, where $D \subset \mathbb{R}^d$ and $T \subset \mathbb{R}$, which is intrinsically stationary in space and time, i.e., with zero mean in their increments in space and time, and with variance that depends only on displacements in space and differences in time.

The parameter in which we are interested in this paper is the spatio-temporal variogram of Z, defined as

$$2\,\gamma_z(\mathbf{h};\tau) = var(Z(\mathbf{s}+\mathbf{h};t+\tau) - Z(\mathbf{s};t))$$

where var is the variance of Z, $\mathbf{h}$ a spatial lag and τ a temporal lag. Furthermore, we shall assume that Z is spatially isotropic, i.e., that the variogram depends on the spatial lag $\mathbf{h}$ only through the Euclidean norm $\|\mathbf{h}\|$.

To estimate $2\,\gamma_z(\mathbf{h};\tau)$, we shall consider observations Z_u, $u = 1, ..., n$, of $Z(\mathbf{s}, t)$ at spatial locations $\{\mathbf{s}_i : i = 1, ..., m\}$ and time moments $\{t_j : j = 1, ..., T\}$, where $n = m \cdot T$ is the sample size.

The spatio-temporal variogram is usually estimated with the classical method-of-moments estimator, also called empirical spatio-temporal variogram, (Wikle et al. 2019; Varouchakis and Hristopulos 2019; Cressie 1993).

$$2\,\widehat{\gamma}_z(\mathbf{h};\tau) = \frac{1}{|N_\mathbf{s}(\mathbf{h})|}\frac{1}{|N_t(\tau)|}\sum_{\mathbf{s}_i,\mathbf{s}_k \in N_\mathbf{s}(\mathbf{h})}\;\sum_{t_j,t_l \in N_t(\tau)} (Z(\mathbf{s}_i;t_j) - Z(\mathbf{s}_k;t_l))^2$$

where $N_\mathbf{s}(\mathbf{h})$ refers to the set containing all pairs of spatial locations with spatial lag $\mathbf{h}$ and $N_t(\tau)$ refers to the set containing all pairs of time points with time lag τ. Also, $|N(\cdot)|$ will refer to the number of elements in the set $N(\cdot)$.

L. A. García-Escudero et al. (Eds.): SMPS 2022, AISC 1433, pp. 174–179, 2023.
https://doi.org/10.1007/978-3-031-15509-3_23

Let us observe that this estimator is a sample mean of $n(\mathbf{h},\tau) = |N_{\mathbf{s}}(\mathbf{h})| \cdot |N_t(\tau)|$ terms and therefore, very sensitive to outliers.

In García-Pérez (2020) robust estimators of the spatial variogram and accurate approximations for their distributions were obtained. In García-Pérez (2021) these results were extended to the multivariate case with robust estimators for the cross-variogram. In García-Pérez (2022b) the temporal component was included, obtaining robust M-estimators of the spatio-temporal variogram. In this paper we propose, in Sect. 2, α-trimmed estimators of the spatio-temporal variogram. In Sect. 3 we obtain accurate approximations for the distribution of these new estimators. We conclude the paper, in Sect. 4, with a real-world application.

2 α-Trimmed Spatio-Temporal Variogram Estimator

All over the paper we shall assume that the observations come from a scale contaminated normal model (Huber and Ronchetti 2009, p. 2).

$$(1-\epsilon)N(\mu,\sigma^2) + \epsilon N(\mu, g^2\sigma^2)$$

$\epsilon \in (0,1)$ and $g > 1$. This class of distributions is considered the usual model class in robustness studies because it establishes a neighborhood of the standard model distribution, the *contamination neighborhood*, within which the underlying model lies (Huber and Ronchetti 2009, p. 12).

Let us consider the transformation $X_{ij} = (Z(\mathbf{s}_i + \mathbf{h}; t_j + \tau) - Z(\mathbf{s}_i; t_j))^2$, $\forall \mathbf{s}_i, t_j$. These new variables will be shortened by X_u, $u = 1, ...n$, and will be considered as a sample of a new variable $X = (Z(\mathbf{s} + \mathbf{h}; t + \tau) - Z(\mathbf{s}; t))^2$, defined from the lags of Z in space and time. Now, the parameter of interest is $2\,\gamma_z(\mathbf{h};\tau) = E[X]$, and the problem proposed in the paper is now the problem of estimating the expectation of the random variable X, obtained from the original Z through this transformation.

If can accept a linear semivariogram for the n original observations Z_u and linear cross-variograms for each pair (Z_i, Z_k), then we can admit independence in the X_u (García-Pérez 2022b, Sect. 4).

Considering a scale contaminated normal model for the original Z_u observations, the distribution of the transformed variables X_u is, (García-Pérez 2022b, Sect. 2.3)

$$F = (1-\epsilon)\,2\,\gamma_z(\mathbf{h};\tau)\,\chi_1^2 + \epsilon\, g^2\, 2\,\gamma_z(\mathbf{h};\tau)\,\chi_1^2$$

where χ_1^2 is a Chi-Square distribution with 1 degree of freedom.

When we think in trimming the data we have, mainly, two possibilities: First, to consider all the X_u, $u = 1, ..., n(\mathbf{h},\tau)$, observations as homogenous to be trimmed, or second, to trim by time moments.

From a robustness point of view, trimming by time moments can hide outliers if there is a time moment with many of them; so, it is worse than considering all the observations at once. Hence, we shall choose the first option.

For this reason, we define the *α-trimmed spatio-temporal variogram estimator*, $2\,\widehat{\gamma}_\alpha(\mathbf{h};\tau)$, defined again for the transformed $n(\mathbf{h},\tau)$ variables X_u as follows:

If we trim the $100 \cdot \alpha\%$ of the smallest and the $100 \cdot \alpha\%$ of the largest ordered data $X_{(u)}$, the (symmetrically) sample *α-trimmed spatio-temporal variogram estimator* is defined as

$$2\,\widehat{\gamma}_\alpha(\mathbf{h};\tau) = \frac{1}{n(\mathbf{h},\tau) - 2r}\left(X_{(r+1)} + ... + X_{(n(\mathbf{h},\tau)-r)}\right) = \overline{X}_\alpha$$

where $r = [n(\mathbf{h},\tau)\alpha]$ if $[\,.\,]$ stands for the integer part.

An asymmetric trimmed spatio-temporal variogram estimator could also be a good option because the observations X_u are positive, since they come from squared differences of the original Z_u.

3 VOM+SAD Approximation of the Distribution of the α-Trimmed Spatio-Temporal Variogram Estimator

Obtaining the distribution of the estimator is necessary to be able to assess its statistical properties and, specifically, its robustness properties. In addition, we can make robust inferences with it such as intervals and robust tests. Moreover, knowing its distribution is useful to reduce the number of temporal lags, as we do in García-Pérez (2022b, Sect. 8) for M-estimators. Even, it would be possible to choose a variogram model in a robust way, as we do in García-Pérez (2022a).

With a von Mises expansion (Von Mises 1947) we can obtain an approximation (VOM approximation) for the distribution of the estimator under a contaminated normal model, computing the approximation under just a normal model, doing in this way the problem easier. This approximation depends on the Hampel's influence function of the tail probability functional. If we have a small sample size, we can use a saddlepoint approximation (SAD approximation) to approximate this functional. Combining both approximations we obtain a VOM+SAD approximation for the distribution of the estimator.

An accurate VOM+SAD approximation of the distribution of the sample α-trimmed mean is obtained in García-Pérez (2016). We see there we can base our approximation for the trimmed mean distribution on an approximation for the classical sample mean. Hence, we can approximate the small sample distribution of the α-trimmed spatio-temporal variogram estimator, using the small sample distribution of the empirical spatio-temporal estimator, $P_F\{2\widehat{\gamma}_z(\mathbf{h};\tau) > a\}$. Namely, if the number of iterations k is large and also the approximation stabilizes when we increase k, the approximation is

$$P_F\{2\widehat{\gamma}_\alpha(\mathbf{h};\tau) > a\} \simeq (1 + n(\mathbf{h},\tau)\, c_1)^{k+1}\,(1 + n(\mathbf{h},\tau)\, c_2)^{k+1}\, P_F\{2\widehat{\gamma}_z(\mathbf{h};\tau) > a\}$$

where $c_1 = \left[(1-2\alpha)^{1/(k+1)} - 1\right]$ and $c_2 = \left[1/(1-2\alpha)^{1/(k+1)} - 1\right]$.

In García-Pérez (2022b) an accurate approximation for the distribution of the empirical spatio-temporal estimator is obtained; hence, an accurate approximation for the tail probability of the sample α-trimmed spatio-temporal variogram estimator $2\widehat{\gamma}_\alpha(\mathbf{h};\tau)$, under a scale contaminated normal model, is

$$\begin{aligned} P_F\{2\widehat{\gamma}_\alpha(\mathbf{h};\tau) > a\} \simeq & (1 + n(\mathbf{h},\tau)\, c_1)^{k+1}\,(1 + n(\mathbf{h},\tau)\, c_2)^{k+1} \left[P\left\{ \chi^2_{n(\mathbf{h},\tau)} > \frac{a\, n(\mathbf{h},\tau)}{2\gamma_z(\mathbf{h};\tau)} \right\} \right. \\ & + \epsilon\, \sqrt{n(\mathbf{h},\tau)}\, \frac{2\gamma_z(\mathbf{h};\tau)}{\sqrt{\pi}(a - 2\gamma_z(\mathbf{h};\tau))} \\ & \cdot \exp\left\{ -\frac{n(\mathbf{h},\tau)}{2} \left(\frac{a}{2\gamma_z(\mathbf{h};\tau)} - 1 - \log \frac{a}{2\gamma_z(\mathbf{h};\tau)} \right) \right\} \\ & \left. \cdot \left(\frac{\sqrt{2\gamma_z(\mathbf{h};\tau)}}{\sqrt{a - ag^2 + 2\, g^2 \gamma_z(\mathbf{h};\tau)}} - 1 \right) \right]. \end{aligned}$$

4 Example

Let us consider daily weather data, obtained in the US National Oceanic and Atmospheric Administration (NOAA) National Climatic Data Center (Wikle et al. 2019). In this data set we shall consider the variable Tmax, the daily maximum temperature in Fahrenheit degrees.

The values of the Classical Spatio-Temporal Variogram Estimator, the 0.05-trimmed spatio-temporal variogram estimator, the 0.1-trimmed spatio-temporal variogram estimator and the 0.2-trimmed spatio-temporal variogram estimator are given in the Supplementary Material, at https://www2.uned.es/pea-metodos-estadisticos-aplicados/trimmed-spa-temp-variogram.htm.

In Fig. 1 here we plot the values of the 0.1-trimmed spatio-temporal variogram estimator obtained in this example.

It is possible to see in the Supplementary Material, p. 15, that there is some differences between the 0.1-trimmed spatio-temporal variogram estimator and the classical one, at some spatial and temporal lags, differences that could be attributed to outliers.

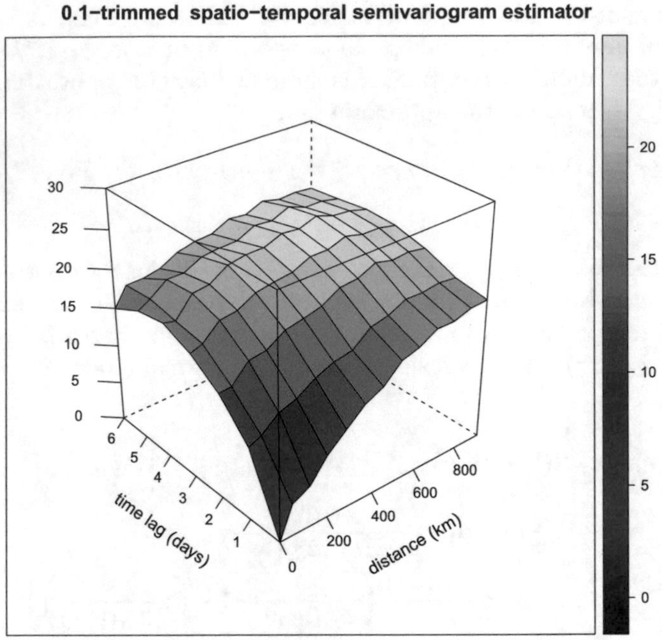

Fig. 1. Three-dimensional picture of the 0.1-trimmed spatio-temporal semivariogram estimator of daily Tmax from the NOAA data during July 2003

5 Conclusions and Further Research

In this paper we define a new robust Trimmed Spatio-Temporal Variogram Estimator and we give an accurate approximation of its distribution. As further research we think that, with this approximation, we could test if it is possible to reduce the number of temporal lags. We also think that it would be possible to obtain an approximation to the distribution of the difference of two α-trimmed spatio-temporal variogram estimators with which we could detect spatio-temporal outliers as it is done in García-Pérez (2022b) for M-estimators.

Acknowledgements. The author is very grateful to the referee and to the *Ministerio de Ciencia e Innovación*.

References

Cressie, N.A.C.: Statistics for Spatial Data. Wiley, New York (1993)

García-Pérez, A.: A von Mises approximation to the small sample distribution of the trimmed mean. Metrika **79**(4), 369–388 (2016)

García-Pérez, A.: Saddlepoint approximations for the distribution of some robust estimators of the variogram. Metrika **83**, 69–91 (2020)

García-Pérez, A.: New robust cross-variogram estimators and approximations for their distributions based on saddlepoint techniques. Mathematics **9**, 762 (2021)

García-Pérez, A.: Variogram model selection. In: Balakrishnan, N., Gil, M.A., Martin, N., Morales, D., Pardo, M.C. (eds.) Trends in Mathematical, Information and Data Sciences, Studies in Systems, Decision and Control, vol. 445. Springer, Heidelberg (2022a). https://doi.org/10.1007/978-3-031-04137-2_3

García-Pérez, A.: On robustness for spatio-temporal data. Mathematics **10**, 1785 (2022)

Huber, P.J., Ronchetti, E.M.: Robust Statistics, 2nd edn. Wiley, New York (2009)

Varouchakis, E.A., Hristopulos, D.T.: Comparison of spatiotemporal variogram functions based on a sparse dataset of groundwater level variations. Spat. Stat. **34**, 1–18 (2019)

von Mises, R.: On the asymptotic distribution of differentiable statistical functions. Ann. Math. Stat. **18**, 309–348 (1947)

Wikle, C.K., Zammit-Mangion, A., Cressie, N.: Spatio-Temporal Statistics with R. Chapman & Hall/CRC, New York (2019)

Two Notions of Depth in the Fuzzy Setting

Luis González-De La Fuente[1], Alicia Nieto-Reyes[1(✉)], and Pedro Terán[2]

[1] Departamento de Matemáticas, Estadística y Computación, Universidad de Cantabria, Avd/Los Castros s/n., 39005 Santander, Spain
{gdelafuentel,alicia.nieto}@unican.es

[2] Universidad de Oviedo, 33071 Oviedo, Spain
teranpedro@uniovi.es

Abstract. Statistical depth functions order the elements in a space with respect to their centrality in a probability distribution. Their study has substantially grown since the notion of depth for multivariate data was introduced in 2000 and nowadays it is an important tool in non-parametric statistics. González-de la Fuente et al. (2022) propose two general notions of depth in the fuzzy setting. We comment here on them, studying their similarities and differences.

1 Introduction

The concept of a fuzzy set was introduced by Zadeh (1965) to formalize some classes of objects which do not have a clear-cut membership criteria, e.g., the set of all numbers *much greater* than 0. In ordinary (crisp) subsets of $\mathbb{R}^p$, each element of the space belongs or not to the set. In contrast, a fuzzy subset on $\mathbb{R}^p$ is a function $U : \mathbb{R}^p \to [0,1]$, a *membership function* that maps each element $x \in \mathbb{R}^p$ to its *degree* of membership $U(x)$ in the fuzzy set.

Thus, taking into account the above example of the real numbers much greater than 0, its membership function must be 0 at every negative real number. It is clear that numbers such as 1/2 or 1 must have positive degrees but not far from 0, while other numbers such as 50 or 100 must have membership values closer to 1. Considering that, a crisp set $V \subseteq \mathbb{R}^p$ can be viewed as a fuzzy set, by identifying it with the indicator function, $\mathrm{I}_V : \mathbb{R}^p \to [0,1]$, where $\mathrm{I}_V(x) = 1$ if $x \in V$ and $\mathrm{I}_V(x) = 0$ otherwise.

This work is focused on elements of $\mathscr{F}_c(\mathbb{R}^p)$, the set of all fuzzy sets U on $\mathbb{R}^p$ such that every α-level is a non-empty, compact and convex subsets of $\mathbb{R}^p$. The α-level of U is defined to be $U_\alpha := \{x \in \mathbb{R}^p : U(x) \geq \alpha\}$, for every $\alpha \in (0,1]$ and $U_0 := \mathrm{clo}\{x \in \mathbb{R}^p : U(x) > 0\}$, where clo denotes the closure of a set. From now on, when speaking of fuzzy sets we will refer to elements of $\mathscr{F}_c(\mathbb{R}^p)$.

Tukey (1975) conceived the notion of a depth function for the multivariate case when introducing the halfspace, or Tukey, depth. For $p > 1$ there is not a *natural* ordering of $\mathbb{R}^p$, as there exists in $\mathbb{R}$, where the order is total. The essence of statistical depth functions is to order a space in the sense that if a point moves

L. A. García-Escudero et al. (Eds.): SMPS 2022, AISC 1433, pp. 180–185, 2023.
https://doi.org/10.1007/978-3-031-15509-3_24

towards the center of the data cloud (or distribution) then its depth increases. Correspondingly, if a point moves toward the external part of the distribution, then its depth decreases.

An axiomatic definition of a statistical depth function was proposed by Zuo and Serfling (2000), which is considered to be the conventional notion of depth in the multivariate setting. The use of depth functions has considerably increased and many other instances of multivariate depth functions have been proposed; the other most well known is simplicial depth (Liu 1990). However, it does not satisfy all the defining properties proposed by Zuo and Serfling (2000). Nieto-Reyes and Battey (2016) proposed a formal notion of statistical depth function in the functional (metric) setting and Nieto-Reyes and Battey (2021) proposed a depth function satisfying all the requirements. It was later used in Nieto-Reyes et al. (2021) for a real data application.

Here we will comment on the two existing general notions of depth function for the fuzzy setting (González-de la Fuente et al. 2022). The first one, *semilinear depth*, is based on the operations of sum and product by a scalar of fuzzy sets, while the second one, *geometric depth*, regards the space of fuzzy sets as a metric space. The semilinear notion is based on the multivariate notion of depth (Zuo and Serfling 2000) and the geometric notion is closer to the notion of depth for the functional case (Nieto-Reyes and Battey 2016) in the sense that both of them involve a metric.

The following notation will be used throughout. Let $\mathbb{S}^{p-1} := \{x \in \mathbb{R}^{p-1} : \|x\| = 1\}$ be the unit sphere of $\mathbb{R}^p$, with $\|\cdot\|$ denotes the Euclidean norm. The *support function* of a fuzzy set U is $s_U : \mathbb{S}^{p-1} \times [0,1] \to \mathbb{R}$ defined by $s_U(u,\alpha) = \sup_{v \in U_\alpha} \langle u, v\rangle$, for every $(u,\alpha) \in \mathbb{S}^{p-1} \times [0,1]$. An important tool to work with fuzzy sets is the (*mid*, *spr*)-decomposition: the support function of a fuzzy set U can be expressed as $s_U(u,\alpha) = mid(s_U)(u,\alpha) + spr(s_U)(u,\alpha)$, where

$$\begin{aligned} mid(s_U)(u,\alpha) &= (s_U(u,\alpha) - s_U(-u,\alpha))/2 \\ spr(s_U)(u,\alpha) &= (s_U(u,\alpha) + s_U(-u,\alpha))/2 \end{aligned}$$

for every $(u,\alpha) \in \mathbb{S}^{p-1} \times [0,1]$.

The paper is organized as follows. Section 2 contains both notions of depth for fuzzy sets. A discussion between the similarities and differences of both axiomatic definitions is in Sect. 3. Some final remarks are in Sect. 4.

2 Definition

In this section we provide the two axiomatic definitions of statistical depth functions for fuzzy sets. These are not specific instances of depth functions, but general proposals of what plausible properties a function should satisfy in order to be called a depth function.

The first notion, is based on the operations defined in $\mathscr{F}_c(\mathbb{R}^p)$. It is called *semilinear* since $\mathscr{F}_c(\mathbb{R}^p)$ satisfies the axiomatic properties of a linear space only partially and is often called a semilinear space. The second notion takes into account that the space $\mathscr{F}_c(\mathbb{R}^p)$ can be viewed as a metric space. It is called *geometric* because $\mathscr{F}_c(\mathbb{R}^p)$ can be endowed with different metrics and some defining properties depend on the geometry of the resulting space.

Definition 1 (Semilinear notion, González-de la Fuente et al. 2022**).** *Let $\mathscr{H} \subseteq L^0[\mathscr{F}_c(\mathbb{R}^p)]$ and $\mathscr{J} \subseteq \mathscr{F}_c(\mathbb{R}^p)$. A function $D(\cdot;\cdot) : \mathscr{J} \times \mathscr{H} \to [0,\infty)$ is a semilinear depth function if it satisfies the following properties*

P1. *$D(M \cdot U + V; M \cdot \mathscr{Y} + V) = D(U;\mathscr{Y})$ for any non-singular matrix $M \in \mathscr{M}_{p\times p}(\mathbb{R})$, any $U, V \in \mathscr{J}$ and any $\mathscr{Y} \in \mathscr{H}$.*

P2. *For (some notion of symmetry and) any symmetric fuzzy random variable $\mathscr{Y} \in \mathscr{H}$,*

$$D(U;\mathscr{Y}) = \sup_{B\in\mathscr{F}_c(\mathbb{R}^p)} D(B;\mathscr{Y}),$$

where $U \in \mathscr{J}$ is the center of symmetry of $\mathscr{Y}$.

P3a. *$D(U;\mathscr{Y}) \geq D((1-\gamma)\cdot U + \gamma \cdot V;\mathscr{Y})$ for all $\gamma \in [0,1]$, all $V \in \mathscr{J}$ and any $U \in \mathscr{J}$ such that $D(U;\mathscr{Y}) = \sup\{D(B;\mathscr{Y}) : B \in \mathscr{F}_c(\mathbb{R}^p)\}$ for all $\mathscr{Y} \in \mathscr{H}$.*

P4a. *$\lim_{\lambda\to\infty} D(U+\gamma\cdot V;\mathscr{Y}) = 0$ for all $V \in \mathscr{F}_c(\mathbb{R}^p)$ and any $U \in \mathscr{F}_c(\mathbb{R}^p)$ such that $D(U;\mathscr{Y}) = \sup\{D(B;\mathscr{Y}) : B \in \mathscr{F}_c(\mathbb{R}^p)\}$ for all $\mathscr{Y} \in \mathscr{H}$.*

Definition 2 (Geometric notion, González-de la Fuente et al. 2022**).** *Let $\mathscr{H} \subseteq L^0[\mathscr{F}_c(\mathbb{R}^p)]$, $\mathscr{J} \subseteq \mathscr{F}_c(\mathbb{R}^p)$ and m a metric defined in $\mathscr{F}_c(\mathbb{R}^p)$. A function $D(\cdot;\cdot) : \mathscr{J} \times \mathscr{H} \to [0,\infty)$ is a geometric depth function with respect to m if it satisfies properties P1. and P2. above and also*

P3b. *$D(U;\mathscr{Y}) \geq D(V;\mathscr{Y}) \geq D(W;\mathscr{Y})$ for all $V, W \in \mathscr{J}$ satisfying $m(U,T) = m(U,V) + m(V,T)$ and any $U \in \mathscr{J}$ such that $D(U;\mathscr{Y}) = \sup\{D(B;\mathscr{Y}) : B \in \mathscr{F}_c(\mathbb{R}^p)\}$ for all $\mathscr{Y} \in \mathscr{H}$.*

P4b. *$\lim_{n\to\infty} D(U_n;\mathscr{Y}) = 0$ for every sequence of fuzzy sets $\{U_n\}_n$ such that $\lim_{n\to\infty} m(U_n,U) = \infty$ and any $U \in \mathscr{F}_c(\mathbb{R}^p)$ such that $D(U;\mathscr{Y}) = \sup\{D(B;\mathscr{Y}) : B \in \mathscr{F}_c(\mathbb{R}^p)\}$ for all $\mathscr{Y} \in \mathscr{H}$.*

3 Relationship Between the Two Notions of Depth

In this section, we will discuss the choice of properties in the definitions above. We also present some results connecting both notions and show some differences between them.

P1. Affine Invariance

This property is common in both definitions and it is an adaptation to the fuzzy setting of the affine invariance property in the multivariate and functional cases. Using *Zadeh's extension principle* (Zadeh 1975) we can apply a non-fuzzy function to fuzzy sets. In our case, let $M \in \mathscr{M}_{p\times p}(\mathbb{R})$ be a non-singular matrix, we consider the function $f : \mathbb{R}^p \to \mathbb{R}^p$ defined by $f(x) = M \cdot x^T$ for every $x \in \mathbb{R}^p$. This property concerns the fact that any fuzzy depth must remain the same under the transformation $M \cdot U + V$ with $U, V \in \mathscr{F}_c(\mathbb{R}^p)$. The key idea is that fuzzy sets are objects constructed upon $\mathbb{R}^p$ and so depth should invariant with respect to affine transformations of the underlying space (which represent coordinate changes) rather than transformations of all $\mathscr{F}_c(\mathbb{R}^p)$.

P2. Maximality at the Center of Symmetry

This property is again common to both definitions and the same as in the multivariate and functional cases. To make sense of this property, it is necessary to define a symmetry notion in the fuzzy space. That means a notion of symmetric fuzzy random variables, not a notion of fuzzy random variables taking on symmetric values. In González-de la Fuente et al. (2022) we propose two notions of symmetry, called F-symmetry and (mid, spr)-symmetry. The first one is based on the support function: we say that a fuzzy random variable $\mathscr{Y} \in L^0[\mathscr{F}_c(\mathbb{R}^p)]$ is *F-symmetric* with respect to $U \in \mathscr{F}_c(\mathbb{R}^p)$ if

$$s_U(u,\alpha) - s_{\mathscr{Y}}(u,\alpha) =^{\mathscr{L}} s_{\mathscr{Y}}(u,\alpha) - s_U(u,\alpha),$$

for every $(u,\alpha) \in \mathbb{S}^{p-1} \times [0,1]$, where $=^{\mathscr{L}}$ denotes equality in distribution. Equivalently, each $s_{\mathscr{Y}}(u,\alpha)$ is a symmetric random variable with respect to the point $s_U(u,\alpha)$.

The (mid, spr)-symmetry notion is based on the (mid, spr)-decomposition: we say that a fuzzy random variable $\mathscr{Y} \in L^0[\mathscr{F}_c(\mathbb{R}^p)]$ is (mid, spr)*-symmetric* with respect to $U \in \mathscr{F}_c(\mathbb{R}^p)$ if

$$mid(s_U)(u,\alpha) - mid(s_{\mathscr{Y}})(u,\alpha) =^{\mathscr{L}} mid(s_{\mathscr{Y}})(u,\alpha) - mid(s_U)(u,\alpha),$$
$$spr(s_U)(u,\alpha) - spr(s_{\mathscr{Y}})(u,\alpha) =^{\mathscr{L}} spr(s_{\mathscr{Y}})(u,\alpha) - spr(s_U)(u,\alpha),$$

for every $(u,\alpha) \in \mathbb{S}^{p-1} \times [0,1]$. Hence both the mid and spread functions are separately required to be symmetric.

In González-de la Fuente et al. (2022) we prove that they are logically independent notions of symmetry (none of them implies the other), thus they generate in some cases different centers of symmetry.

P3. Strictly Decreasing with Respect to the Deepest Point

The property P3a. of the semilinear notion is analogous to the multivariate one, using the operations sum and product by a scalar defined in the space of fuzzy sets. In its turn, property P3b. is an adaptation of the third property of the axiomatic definition for functional (metric) spaces, while our fuzzy property it is not as restrictive as the functional one.

P4. Vanishing at Infinity

Property P4a. of the semilinear notion is an adaptation to the fuzzy setting of the multivariate property, but in our case we consider the sequence $U + n \cdot V$, considering only operations defined in the fuzzy space. Property P4b. takes into account that the space of fuzzy sets can be viewed as a metric space and it is a property analogous to the one of the multivariate case. This property is not based on any property of the functional case.

3.1 Comparison

Next, we present some results to compare the third and fourth properties of both notions. Let $m : \mathscr{F}_c(\mathbb{R}^p) \times \mathscr{F}_c(\mathbb{R}^p) \to [0, \infty)$ be a metric and consider the conditions

(A1.) $m(\beta \cdot U, \beta \cdot V) = \beta \cdot m(U, V)$ for every $U, V \in \mathscr{F}_c(\mathbb{R}^p)$ and $\beta \in [0, \infty)$.
(A2.) $m(U + W, V + W) = m(U, V)$ for all $U, V, W \in \mathscr{F}_c(\mathbb{R}^p)$.

Let $\mathscr{Y}$ be a fuzzy random variable, consider a metric m defined over $\mathscr{F}_c(\mathbb{R}^p)$ fulfilling (A1.) and (A2.), and $D(\cdot; \mathscr{Y}) : \mathscr{F}_c(\mathbb{R}^p) \to [0, \infty)$ a function. The following statements are proved in González-de la Fuente et al. (2022).

1. If $D(\cdot; \mathscr{Y})$ satisfies P3b. with respect to m, then it also satisfies P3a.
2. If there exists a strictly convex Banach space $(\mathbb{E}, \|\cdot\|)$ such that $(\mathscr{F}_c(\mathbb{R}^p), m)$ embeds isometrically into $(\mathbb{E}, \|\cdot\|)$ then $D(\cdot; \mathscr{Y})$ satisfies P3a. if and only if it satisfies P3b.
3. If $D(\cdot; \mathscr{Y})$ satisfies P4b. with respect to m, then it also satisfies P4a.

In this regard, notice that a number of metrics in the literature are known to embed into Banach spaces or are explicitly defined as the distance between the support functions in some Banach space of functions. Recall that all Hilbert spaces and all L^p-spaces, for $p \in (1, \infty)$, are strictly convex.

Let us denote by ν the Lebesgue measure, and by $\mathscr{V}_p$ the normalized Haar measure over $\mathbb{S}^{p-1}$. The ρ_r metric between fuzzy sets U and V was defined by Diamond and Kloeden (1990) in the following way

$$\rho_r(U, V) = \left(\int_{\mathbb{S}^{p-1}} \int_{[0,1]} |s_U(u, \alpha) - s_V(u, \alpha)|^r \mathrm{d}\nu(\alpha) \mathrm{d}\mathscr{V}_p(u) \right)^{1/r},$$

with $r \in [1, \infty)$. The conditions in the second statement are satisfied for the ρ_r metric for $r \in (1, \infty)$, thus we have that properties P3a. and P3b. are equivalent for ρ_r metric when $r \in (1, \infty)$.

A direct implication of the above results is that a geometric depth function with respect to a metric m which fulfils (A1.) and (A2.) is also a semilinear depth function.

4 Concluding Remarks

We present two different notions of depth for the fuzzy setting, trying to adapt the preexistent concept from the multivariate and functional cases to the specificities of fuzzy sets. The semilinear notion is based on the operations defined over the space of fuzzy sets. For this reason, it is easy to prove whether a given function satisfies or not the semilinear axioms. On the other hand, the geometric notion is based on the particular fact that the fuzzy space can be interpreted as a metric space. In some cases, it could be a difficult task to prove whether or not a function is a geometric depth. However, for some subfamilies of fuzzy sets or specific practical situations, one metric could be considered more appropriate than another and for this reason it is interesting to introduce the concept of metric in the definition of a fuzzy depth function.

The viability of the two notions is shown also in González-de la Fuente et al. (2022) by defining the first instance of a semilinear and geometric depth function, the Tukey depth for fuzzy sets. Therefore there exist generalizations of important depth functions, like Tukey depth, which are both semilinear and geometric depth functions when generalized to fuzzy data.

Acknowledgements. A. Nieto–Reyes and L. González–De La Fuente are supported by grant MTM2017-86061-C2-2-P funded by MCIN/AEI/ 10.13039/501100011033 and "ERDF A way of making Europe". P. Terán is supported by the Ministerio de Ciencia, Innovación y Universidades grant PID2019–104486GB–100 and the Consejería de Empleo, Industria y Turismo del Principado de Asturias grant GRUPIN AYUD/2021/50897.

References

Diamond, P., Kloeden, P.: Metric spaces of fuzzy sets. Fuzzy Sets Syst. **35**(2), 241–249 (1990)

Gónzalez-de la Fuente, L., Nieto-Reyes, A., Terán, P.: Statistical depth for fuzzy sets. Fuzzy Sets and Syst. **443**(Part A), 58–86 (2022). https://doi.org/10.1016/j.fss.2021.09.015

Liu, R.Y.: On a notion of data depth based on random simplices. Ann. Stat. **18**(1), 405–414 (1990)

Nieto-Reyes, A., Battey, H.: A topologically valid definition of depth for functional data. Stat. Sci. **31**(1), 61–79 (2016)

Nieto-Reyes, A., Battey, H.: A topologically valid construction of depth for functional data. J. Multivar. Anal. **184**, 104738 (2021)

Nieto-Reyes, A., Battey, H., Francisci, G.: Functional symmetry and statistical depth for the analysis of movement patterns in Alzheimer's Patients. Mathematics **9**(8), 820 (2021)

Tukey, J.W.: Mathematics and Picturing Data. In: Proceedings of the International Congress of Mathematicians (ICM 1974), vol. 2, pp 523–531 (1975)

Zadeh, L.A.: Fuzzy sets. Inf. Control **8**(3), 338–353 (1965)

Zadeh, L.A.: The concept of a linguistic variable and its application to approximate reasoning. Part I. Inf. Sci. **8**(3), 199–249 (1975)

Zuo, Y., Serfling, R.: General notions of statistical depth function. Ann. Stat. **28**(2), 461–482 (2000)

Tukey Depth for Fuzzy Sets

Luis González-De La Fuente[1(✉)], Alicia Nieto-Reyes[1], and Pedro Terán[2]

[1] Departamento de Matemáticas, Estadística y Computación, Universidad de Cantabria, Avd/Los Castros s/n, 39005 Santander, Spain
{gdelafuentel,alicia.nieto}@unican.es

[2] Universidad de Oviedo, 33071 Oviedo, Spain
teranpedro@uniovi.es

Abstract. Tukey depth is the first and one of the most notorious statistical depth functions. It has very good properties and measures in a proper way the centrality of points with respect to a distribution. In a forthcoming paper by the authors, the Tukey depth function is extended to the fuzzy setting. Here, we comment on this extension and some interesting properties satisfied by this function. Additionally, we include a novel illustration of this extension on a real dataset on the luminous intensities of light emitting diodes (LED) sources.

1 Introduction and Preliminaries

The term *depth function* was coined by Tukey (1975). He aimed at measuring how *deep* a point in $\mathbb{R}^p$ is with respect to a data cloud of points in the space. In general, this would result in ordering $\mathbb{R}^p$ with respect to a distribution on it. Let $(\Omega, \mathscr{A}, \mathbb{P})$ be a probabilistic space defined over $\mathbb{R}^p$. In the univariate case $p = 1$, a natural way of ordering the points in $\mathbb{R}$ with respect to their depth in a distribution $\mathbb{P}$ is by considering the function

$$x \longmapsto \min\{\mathbb{P}((-\infty, x]), \mathbb{P}([x, \infty))\}, \tag{1}$$

for every $x \in \mathbb{R}$. That ranks points according to the difference between their quantile rank and $1/2$, with the maximum being attained at the medians. When $p \geq 2$, the definition of *Tukey depth* (Tukey 1975) generalizes that formula using halfspaces instead of halflines. The *Tukey depth*, also called *halfspace depth*, is defined as

$$HD(x; \mathbb{P}) := \inf\{\mathbb{P}(H) : \text{ H is a closed halfspace and } x \in H\}. \tag{2}$$

In fact, for the real case the function defined in (2) is exactly the function in (1). Tukey depth is relevant because of its theoretical properties (Cuesta-Albertos and Nieto-Reyes 2008a) and because it can be easily approximated in different settings (Cuesta-Albertos and Nieto-Reyes 2008b, 2010). Moreover, depth functions are widely used in non-parametric statistics, and, in particular, for real data applications (Nieto-Reyes et al. 2021a,b).

L. A. García-Escudero et al. (Eds.): SMPS 2022, AISC 1433, pp. 186–193, 2023.
https://doi.org/10.1007/978-3-031-15509-3_25

Here, we focus on a part of our paper González-de la Fuente et al. (2022) concerning an adaptation of the definition of Tukey depth to the fuzzy setting. We will comment on the relationship between this notion and the multivariate Tukey depth, as the multivariate case could be seen as a particular case of the Tukey depth for fuzzy sets. We also remark on some interesting properties satisfied by the fuzzy Tukey depth, showing that the Tukey depth for fuzzy sets enjoys the properties of being both a semilinear depth function and a geometric depth function (with respect to L^r-type metrics). The semilinear and geometric notions are axiomatic definition of statistical depth functions for fuzzy sets González-de la Fuente et al. (2022). The literature also contains axiomatic definitions for multivariate (Zuo and Serfling 2000) and functional (metric) spaces (Nieto-Reyes and Battey 2016). Moreover, this chapter is dedicated to introduce a new illustration of the Tukey depth on fuzzy data. For this, we make use of the dataset provided in Pak et al. (2013). There, a LED lighting manufacturing process is studied under the assumption of following a Weibull distribution.

The notation used is that of Chapter "Two notions of depth in the fuzzy setting" of this book.

2 On the Extension of the Tukey Depth for Fuzzy Spaces

In this section we comment on the formal definition of the Tukey depth for fuzzy sets. We also study a result that connects it with the halfspace depth for the multivariate case, HD.

Let us consider a probability space $(\Omega, \mathscr{A}, \mathbb{P})$. We denote by $L^0[\mathscr{F}_c(\mathbb{R}^p)]$ the class of all fuzzy random variables with values in $\mathscr{F}_c(\mathbb{R}^p)$ defined over the measurable space $(\Omega, \mathscr{A})$.

Let $\mathscr{H} \subseteq L^0[\mathscr{F}_c(\mathbb{R}^p)]$ and $\mathscr{J} \subseteq \mathscr{F}_c(\mathbb{R}^p)$. According to González-de la Fuente et al. (2022), the *Tukey depth function* is

$$D_{FT}(\cdot;\cdot) : \mathscr{H} \times \mathscr{J} \to [0,1]$$

with

$$\begin{aligned} D_{FT}(U;\mathscr{X}) = \inf_{u \in \mathbb{S}^{p-1}, \alpha \in [0,1]} \min(&\mathbb{P}[\omega \in \Omega : \mathscr{X}(\omega) \in S^{-}_{u,\alpha}], \\ &\mathbb{P}[\omega \in \Omega : \mathscr{X}(\omega) \in S^{+}_{u,\alpha}]), \end{aligned}$$

for every $U \in \mathscr{J}$ and $\mathscr{X} \in \mathscr{H}$, where

$$\begin{aligned} S^{-}_{u,\alpha} &:= \{V \in \mathscr{F}_c(\mathbb{R}^p) : s_V(u,\alpha) - s_U(u,\alpha) \leq 0\} \\ S^{+}_{u,\alpha} &:= \{V \in \mathscr{F}_c(\mathbb{R}^p) : s_V(u,\alpha) - s_U(u,\alpha) \geq 0\}. \end{aligned}$$

Thus $D_{FT}(U;\mathscr{X})$ represents the depth of a fuzzy set U in the probability distribution of a fuzzy random variable $\mathscr{X}$.

Note that this results in

$$D_{FT}(U;\mathscr{X}) = \inf_{u\in\mathbb{S}^{p-1},\alpha\in[0,1]} \min(\mathbb{P}[\omega\in\Omega : s_{\mathscr{X}(\omega)}(u,\alpha) \le s_U(u,\alpha)], \\ \mathbb{P}[\omega\in\Omega : s_{\mathscr{X}(\omega)}(u,\alpha) \ge s_U(u,\alpha)]).$$

Taking into account that the support function of a fuzzy set takes on values in the real line, and using the expression of the halfspace depth for the real case given by Eq. (1), we have then the formula

$$D_{FT}(U;\mathscr{X}) = \inf_{u\in\mathbb{S}^{p-1},\alpha\in[0,1]} HD(s_U(u,\alpha); s_{\mathscr{X}}(u,\alpha)) \tag{3}$$

which explicitly writes D_{FT} in terms of HD.

Any ordinary (crisp) set $U \subseteq \mathbb{R}^p$ can be identified with a fuzzy set, by considering its membership function $\mathrm{I}_U : \mathbb{R}^p \to [0,1]$, where $\mathrm{I}_U(\cdot)$ denotes the indicator function of a set U defined as $\mathrm{I}_U(x) = 1$ if $x \in U$ and $\mathrm{I}_U(x) = 0$ otherwise. On this setting, we can view the space $\mathbb{R}^p$ of multivariate points as a particular subset of $\mathscr{F}_c(\mathbb{R}^p)$, considering the set

$$\mathscr{R}^p := \{\mathrm{I}_{\{z\}} \in \mathscr{F}_c(\mathbb{R}^p) : z \in \mathbb{R}^p\}.$$

The next result writes the halfspace depth HD in terms of D_{FT}.

Proposition 1 (González-de la Fuente et al. 2022**).** *Let Z be a random vector in $\mathbb{R}^p$, and let $\mathscr{X} \in L^0[\mathscr{F}_c(\mathbb{R}^p)]$ take on values in $\mathscr{R}^p$ given by $\mathscr{X} = I_{\{Z\}}$. Then*

$$HD(z;Z) = D_{FT}(I_{\{z\}};\mathscr{X}) \text{ for all } z \in \mathbb{R}^p.$$

This shows that D_{FT} is a genuine generalization of HD while applying to general fuzzy random variables. Note that we have made an abuse of notation writing $HD(\cdot;Z)$ instead of $HD(\cdot;\mathbb{P})$, as in Eq. (2), where $\mathbb{P}$ would be the distribution of the random variable Z.

Next, we provide the results which show that the Tukey depth for fuzzy sets is the first instance in the literature to be a semilinear and a geometric depth function, for a certain family of metrics. The notions of semilinear and a geometric depth function have been included on Chapter "Two notions of depth in the fuzzy setting" of this book. There, the reader can also find the definitions of F-symmetry and metric ρ_r.

Theorem 1 (González-de la Fuente et al. 2022**).** *When symmetry is understood in the sense of F-symmetry, D_{FT} is a semilinear depth function.*

Theorem 2 (González-de la Fuente et al. 2022**).** *When symmetry is understood in the sense of F-symmetry, D_{FT} is a geometric depth function with respect to the metric ρ_r for any $r \in (1,\infty)$.*

3 Real Data Illustration

This section is dedicated to observe the functioning of the Tukey depth on a real dataset reported in Pak et al. (2013), which consists of the luminous intensities of LED sources obtained from a LED manufacturing process. As it occurs that the luminous intensity of each LED light has a degree of imprecision, each element of the dataset is provided in terms of three values: the lower and upper bound of the intensities as well as a point estimate. Thus, each element of the dataset is easily modeled as a triangular fuzzy number.

Given $l, e, u \in \mathbb{R}$ with $l < e < u$, the triangular fuzzy number determined by them is

$$\mathrm{Tri}(l, e, u)(x) := \begin{cases} \dfrac{x-l}{e-l}, & \text{if } \; l < x \leq e, \\ \dfrac{u-x}{u-e}, & \text{if } \; e < v < u, \\ 0, & \qquad \text{otherwise.} \end{cases} \tag{4}$$

The dataset under study consists of 30 triangular fuzzy sets, that we denote by

$$T_i := \mathrm{Tri}(l_i, e_i, u_i), \text{ for } i = 1, \ldots, 30.$$

Together with the resulting Tukey depth values, the triplets $\{(l_i, e_i, u_i)\}_{i=1}^{30}$ are provided in Table 1, for the sake of completeness and a better interpretability of the Tukey depth results we illustrate in this section. In each panel of the table, the first row represents the lower value of the triangular fuzzy sets, namely l_i, the second row the point estimation, that is e_i, and the third row is the upper bound, u_i. The dataset in Table 1 is organized so that

$$l_1 < l_2 < \ldots < l_{30}. \tag{5}$$

Additionally, the dataset is displayed in the left panel of Fig. 1.

In González-de la Fuente et al. (2022), the Tukey depth is illustrated in the more general case of a trapezoidal dataset. We consider that the triangular dataset used here is, however, of the most interest due to we can assume the elements of the dataset are drawn from an asymmetric continuous distribution that results in triangular data of different wides. We know this because the dataset was introduced in Pak et al. (2013) to illustrate the estimation of the parameters of a Weibull distribution due to, as stated there, the dataset seems to have been drawn form a distribution fairly close to a Weibull one.

Table 1. Triangular fuzzy elements of the LED dataset, given by the lower bound, estimation point and upper bound, and corresponding depth value. The elements are ordered increasingly by their lower bound

Data ID	1	2	3	4	5	6
Lower	0.618	0.628	0.830	0.974	1.027	1.032
Estimation	0.839	0.964	1.288	1.839	1.218	1.971
Upper	2.217	1.735	2.541	2.045	3.116	2.642
Depth	$0.0\widehat{3}$	$0.0\widehat{3}$	0.1	$0.0\widehat{6}$	0.1	0.2
Data ID	7	8	9	10	11	12
Lower	1.336	1.375	1.766	2.163	2.247	2.821
Estimation	2.750	2.195	2.190	2.738	2.990	3.409
Upper	3.284	3.086	2.638	3.068	4.128	5.272
Depth	$0.2\widehat{3}$	$0.2\widehat{6}$	$0.1\widehat{6}$	$0.2\widehat{3}$	$0.3\widehat{6}$	0.4
Data ID	13	14	15	16	17	18
Lower	2.995	3.177	3.590	3.766	3.920	4.065
Estimation	3.442	3.600	4.169	5.814	4.762	5.312
Upper	5.066	4.213	4.899	6.212	6.121	7.480
Depth	$0.4\widehat{3}$	0.4	$0.4\widehat{3}$	$0.3\widehat{6}$	$0.4\widehat{6}$	0.3
Data ID	19	20	21	22	23	24
Lower	4.352	4.575	4.634	5.434	5.443	5.965
Estimation	5.206	6.050	5.780	7.093	6.231	7.344
Upper	5.988	6.734	7.058	7.655	7.395	8.019
Depth	0.4	$0.\widehat{3}$	$0.\widehat{3}$	$0.2\widehat{3}$	$0.2\widehat{6}$	0.2
Data ID	25	26	27	28	29	30
Lower	5.972	6.032	6.279	7.125	7.155	7.261
Estimation	6.353	7.746	8.156	8.470	8.013	8.325
Upper	8.150	8.529	9.435	9.044	8.352	8.871
Depth	0.2	$0.1\widehat{3}$	$0.0\widehat{3}$	$0.0\widehat{3}$	$0.0\widehat{6}$	$0.0\widehat{3}$

Let us denote by $\mathscr{T}$ the fuzzy random variable corresponding to the empirical distribution associated to $\{T_i\}_{i=1}^{30}$; that is, each fuzzy value has the probability given by its relative frequency in the dataset, which is, in this case, 1/30. Analogously, let us denote by $\mathscr{L}$, $\mathscr{E}$ and $\mathscr{U}$ the real random variables corresponding to the empirical distribution associated to $\{l_i\}_{i=1}^{30}$, $\{e_i\}_{i=1}^{30}$ and $\{u_i\}_{i=1}^{30}$, respectively. Then, taking into account Eqs. (3) and (4), we have that the Tukey depth of T_i with respect to $\mathscr{T}$ is

$$D_{FT}(T_i; \mathscr{T}) = \min\{HD(l_i; \mathscr{L}), HD(e_i; \mathscr{E}), HD(u_i; \mathscr{U})\},$$

for each $i = 1, \ldots, 30$.

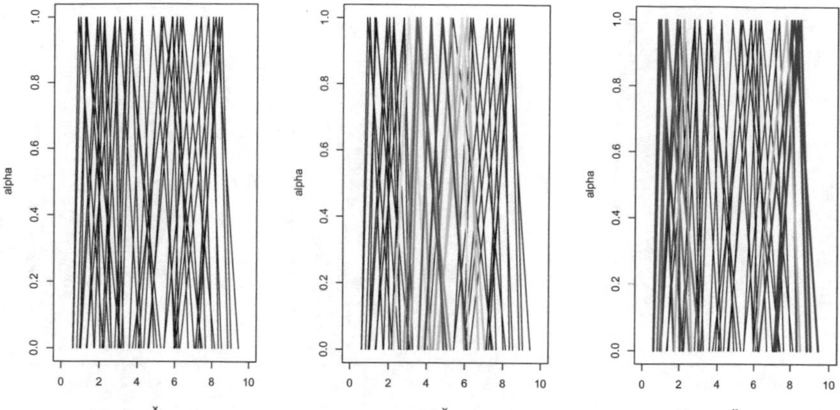

Fig. 1. Illustration of the studied LED dataset. The left panel represents the dataset. The deepest elements are depicted on the central panel by the color, which ranges from red (highest depth) to lime green (high depth) through orange and yellow (medium high depth). The least deep elements are depicted on the right panel, also by the color, which ranges from navy blue (least depth) to cyan (small depth) through baby blue (medium small depth)

Table 1 provides the obtained depth values for each of the elements in the dataset, $\{D_{FT}(T_i; \mathscr{T})\}_{i=1}^{30}$. They are displayed in the fourth row of each panel in the table. From there, we observe that T_{17} has deepest Tukey depth, that the depth values generally increase from T_1 to T_{16} and that generally decrease from T_{18} to T_{30}. Thus, the ordering of the Tukey depth is consistent with (5) and the fact that $\{e_i\}_{i=1}^{30}$ and $\{u_i\}_{i=1}^{30}$ generally follow the same order. This is also elucidated from Fig. 2 and, in particular, from the central and right panels where the lines from the lower bounds to the estimation point and those from the estimation point to the upper bounds are respectively highlighted with color according to their depth values. From the central panel of Fig. 1, we observe that the deepest elements are the T_i's whose point estimation is around 5, which corresponds with the center of the x-axis. The T_i's with minimal depths are those located furthest to the right and to the left in the right plot of Fig. 1. Thus, the consistency of the ordering produced by the Tukey depth is also derived from Fig. 1.

The ordering disagreements indicate some interesting features of the Tukey depth ordering. T_8 and T_9 have almost the same core but rather different depths; T_9 is less vague than the smaller data and that difference in shape is penalized. T_{16} (resp. T_{18}) is very left-skewed (resp. right-skewed), which gets penalized in the context of the surrounding data which are not. Another interesting case is T_{14} (light orange), which has both the lower bound and estimate point between those of the very deep T_{13} (dark orange), and T_{15}, but the behavior of the right estimate differs from the surrounding data. In fact, as $u_{14} < u_{13}$ despite $e_{13} < e_{14}$, $D_{FT}(T_{14}; \mathscr{T}) = HD(u_{14}; \mathscr{U}) = 0.4$ while

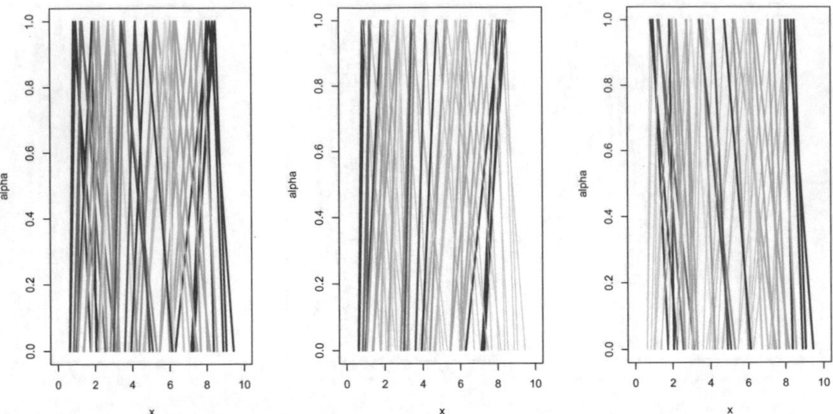

Fig. 2. Illustration of the Tukey depth values of the elements in the studied LED dataset. The left panel represents those depth values in the color while the central (resp. right) panel only highlights the left (resp. right) side of the T_i's. The color ranges from red (highest depth) to navy blue (least depth), through orange, yellow and green colors

$$D_{FT}(T_{13}; \mathscr{T}) = HD(l_{13}; \mathscr{L}) = HD(e_{13}; \mathscr{E}) = 0.4\widehat{3}.$$

Another example is that of T_3, T_5 (both light blue) and T_4 (dark blue):

$$D_{FT}(T_3; \mathscr{T}) = HD(u_3; \mathscr{U}) = 0.1 = D_{FT}(T_5; \mathscr{T}) = HD(e_5; \mathscr{E})$$

while $D_{FT}(T_4; \mathscr{T}) = HD(u_4; \mathscr{U}) = 0.0\widehat{6}$, due to $e_5 < e_3 < e_4$ and $u_4 < u_3 < u_5$. All that suggests that the Tukey depth accommodates a number of different features (does not collapse to a naive ordering of the core points, for example; see T_{17}) and gives sensible results in a variety of situations, reducing the depth of the data that show significant differences of behavior from their neighbors.

Acknowledgements. A. Nieto–Reyes and L. González–De La Fuente are supported by grant MTM2017-86061-C2-2-P funded by MCIN/AEI/ 10.13039/501100011033 and "ERDF A way of making Europe". P. Terán is supported by the Ministerio de Ciencia, Innovación y Universidades grant PID2019–104486GB–100 and the Consejería de Empleo, Industria y Turismo del Principado de Asturias grant GRUPIN AYUD/2021/50897.

References

Cuesta-Albertos, J.A., Nieto-Reyes, A.: The Tukey and the random Tukey depths characterize discrete distributions. J. Multivar. Anal. **99**(10), 2304–2311 (2008)

Cuesta-Albertos, J.A., Nieto-Reyes, A.: The random Tukey depth. Comput. Stat. Data Anal. **52**(11), 4979–4988 (2008)

Cuesta-Albertos, J.A., Nieto-Reyes, A.: Functional classification and the random Tukey depth. Practical issues. In: Borgelt, C., et al. (eds.) Combining Soft Computing and Statistical Methods in Data Analysis. AISC, vol. 77, pp. 123–130. Springer, Heidelberg (2010). https://doi.org/10.1007/978-3-642-14746-3_16

Gónzalez-de la Fuente, L., Nieto-Reyes, A., Terán, P.: Statistical depth for fuzzy sets. Fuzzy Sets Syst. **443**(Part A), 58–86 (2022). https://doi.org/10.1016/j.fss.2021.09.015

Nieto-Reyes, A., Battey, H.: A topologically valid definition of depth for functional data. Stat. Sci. **31**(1), 61–79 (2016)

Nieto-Reyes, A., Battey, H., Francisci, G.: Functional symmetry and statistical depth for the analysis of movement patterns in Alzheimer's patients. Mathematics **9**(8), 820 (2021)

Nieto-Reyes, A., Duque, R., Francisci, G.: A method to automate the prediction of student academic performance from early stages of the course. Mathematics **9**(21), 2677 (2021)

Pak, A., Parham, G.A., Saraj, M.: Inference for the Weibull distribution based on fuzzy data. Rev. Colomb. Estad. **36**(2), 337–356 (2013)

Tukey, J.W.: Mathematics and picturing data. In: Proceedings of the International Congress of Mathematicians (ICM 1974), vol. 2, pp. 523–531 (1975)

Zuo, Y., Serfling, R.: General notions of statistical depth function. Ann. Stat. **28**(2), 461–482 (2000)

Making Data Fair Through Optimal Trimmed Matching

Paula Gordaliza(✉) and Hristo Inouzhe

Basque Center for Applied Mathematics, Bilbao, Spain
{pgordaliza,hinouzhe}@bcamath.org

Abstract. Algorithmic fairness is one of the main concerns of today's scientific society due to the generalization of predictive algorithms in all aspects of human life. The aim of this work is to check if there is group bias in the response variable Y with respect to a sensitive information S present in the data. However, not all individuals in S are comparable, and some differences in the target Y may arise from genuine differences in the data. We propose to eliminate such cases by trimming an α proportion of the input data as a pre-processing step to any further learning mechanism in order to obtain the two closest possible marginal distributions (with respect to S). On this population that is 'similar enough' we can check for discrimination, in the sense of Demographic Parity. We solve a trimmed matching problem subject to fairness constraints that is a linear program that can be addressed with well-known techniques. We present some successful results of application to synthetic and real data.

1 Introduction and Notations

Today, all facets of human life are heavily dependent on a technology that is rapidly being improved by the implementation of ever better machine learning (ML) models. The generalization of Artificial Intelligence (AI) based-systems in the decision-making process in a wide variety of fields, particularly in the everyday and professional life, have raised serious ethical concerns about the implications of the adoption of such technologies. The scope of this problem has made fair learning one of the most important topics among the ML scientific community lately. Therefore, the number of works proposed to ensure algorithmic fairness from many different and varied points of view has been spectacular in recent years.

However, there is no consensus on the definition of algorithmic fairness and a wide array of different metrics have been proposed. In this paper, bias is understood as any dependence between the prediction and a certain variable, usually called protected variable in the literature, that has sensitive information about individuals. The presence of such a bias in algorithmic outcomes would lead to unfairness and the crucial aim in fair learning is designing procedures to remove it.

L. A. García-Escudero et al. (Eds.): SMPS 2022, AISC 1433, pp. 194–199, 2023.
https://doi.org/10.1007/978-3-031-15509-3_26

In a supervised fair learning setting, the aim of an algorithm is to learn the relationships between characteristic variables X and a target variable Y with the additional challenge of dealing with the presence of a protected attribute S, that conveys sensitive information about the observations X that should not be used for the prediction of $\hat{Y}$. In this sense, this variable S models the bias and we assume in the following that it is observed. A similar situation can be described in an unsupervised learning problem (fair clustering), where the goal is to hide sensitive attributes during data partition by balancing the distribution of protected subgroups in each cluster.

Following the independence-based approach, an algorithm is called fair or unbiased when its outcome $\hat{Y}$ does not depend on S. The well-known criterion Demographic Parity (DP) requires the statistical independence between the outcome and the protected attribute $\hat{Y} \perp\!\!\!\perp S$. In certain scenarios, the ground truth is available and the above definition could be weaken into a conditional independence that is given by the Equalized Odds (EO) criterion $\hat{Y} \perp\!\!\!\perp S \mid Y$. We are interested here in a general framework, so DP is the metric chosen to quantify fairness in algorithmic outcomes. For further details on fairness metrics, we refer to the comprehensive study Barocas et al. (2019).

Yet, building fair algorithms may lead to poor generalization errors with respect to the unfair case. While in some fields of application it is desirable to ensure the highest level of fairness, in others, including Health Care or Criminal Justice, performance should not be decreased since the decisions would have serious implications for individuals and society. Therefore, it is of great interest to set a trade-off between fairness and accuracy. From a procedural viewpoint, fairness-enhancing methods are roughly divided into pre, in or post-processing procedures, depending on the time when constraints are imposed Oneto et al. (2020).

Realistically, in most situations, algorithms are inaccessible and the most practical approach is trying to obtain fairness by constraining the training sample. More precisely, in this work we propose to remove the influence of the sensitive variable by trimming (partially or fully removing) a proportion of the input data as a pre-processing step to any further learning mechanism. Let us consider a binary protected attribute $S \in \{0, 1\}$, meaning that the population is supposed to be divided into two categories, taking the value $S = 0$ for the minority (assumed to be the unfavored class), and $S = 1$ for the majority (and usually favored class). The idea is to check if there is group bias in Y with respect to S. However, not all individuals in S are comparable, therefore some differences in the response variable Y may arise from genuine differences in the data. In order to eliminate these cases we want to trim an α proportion of the data in order to obtain the two closest possible marginal distributions (with respect to S). On this population that is 'similar enough' we can check for discrimination, in the sense of DP. A similar strategy can be used to obtain clusters that preserve spatial information and produce more fair partitions. Therefore, the tool we propose below can be used in both unsupervised and supervised problems as a pre-processing step that transforms the training sample so that the two conditional subsets with respect to S are more similar, without modifying the data that are not trimmed. Further analysis could be done on the also unmodified trimmed data.

The model behind the procedure we propose is the popular contamination model:

$$\begin{aligned} \mu_0 &= \mathcal{L}(X|S=0) = (1-\alpha_0)P_S + \alpha_0 Q_0 \\ \mu_1 &= \mathcal{L}(X|S=1) = (1-\alpha_1)P_S + \alpha_1 Q_1 \end{aligned} \tag{1}$$

where P_S represents the common structure between the marginals, Q_0 and Q_1 the structure that produces the differences between them, and $\alpha_0, \alpha_1 \geq 0$ are the levels of mixing for each marginal, respectively. Hence, there are, not necessarily unique, μ_0^* and μ_1^* which are α-trimmed versions of the marginals μ_0 and μ_1 and which have the same distribution, i.e., $d_{TV}(\mu_0^*, \mu_1^*) = 0$. Indeed, this characterizes the complete absence of bias in the training sample in the DP sense as proven in Gordaliza et al. (2019). Thus, any sensible (supervised or unsupervised) learning algorithm that learns on these α-trimmed versions of the marginals should produce results that are independent from S.

Notice that many fair learning procedures (Feldman et al. 2015; Gordaliza et al. 2019) implicitly assume something similar to (1). In other words, applying any fairness correction when the marginals are 'quite different' is very likely to produce a significant worsening in the objective (loss) function of interest. Hence, sensible fairness corrections and performance depend on a degree of similarity between the conditional distributions. We cite here the work Barrio et al. (2019) that proposes an application of similarity tests based on Wasserstein distances to fair learning. Therefore (1) is a rather strong but reasonable assumption.

Our goal is to find an α-trimmed version of the marginal distributions corresponding to each protected class, with the corresponding trimmed marginal data, and perform learning on these trimmed data. This problem can be stated as a matching problem for which trimming is allowed. We refer to Álvarez-Esteban et al. (2008) for a complete study on the comparison of trimmed distributions. We select a trimming level α and a cost function c. We denote by $\{x_i\}_{i=1}^{N_0}$ the conditional sample $X_{N_0+N_1}|S=0$ and by $\{y_j\}_{j=1}^{N_1}$ the conditional sample $X_{N_0+N_1}|S=1$. We have $\boldsymbol{\alpha} = (\alpha_1, \ldots, \alpha_{N_0})$, $\boldsymbol{\beta} = (\beta_1, \ldots, \beta_{N_1})$, $\mathbf{W} = [w_{ij}]$. We want to solve the following trimmed matching problem

$$\begin{array}{lll} \text{minimize} & \sum_{i=1}^{N_0} \sum_{j=1}^{N_1} w_{ij} c(x_i, y_j) & \\ \text{subject to} & w_{ij}, \alpha_i, \beta_j \geq 0, & 1 \leq i \leq N_0, 1 \leq j \leq N_1 \\ & \alpha_i, \beta_j \leq 1, & 1 \leq i \leq N_0, 1 \leq j \leq N_1 \\ & \sum_{j=1}^{N_1} w_{ij} = \alpha_i p_i, & 1 \leq i \leq N_0 \\ & \sum_{i=1}^{N_0} w_{ij} = \beta_j q_j, & 1 \leq j \leq N_1 \\ & \sum_{i=1}^{N_0} \alpha_i p_i = 1 - \alpha & \\ & \sum_{j=1}^{N_1} \beta_j q_j = 1 - \alpha & \end{array} \tag{2}$$

which is a linear program that can be solved with well known techniques. The results of solving the linear problem are $\boldsymbol{\alpha}^*$ an α-trimming of $\{x_i\}_{i=1}^{N_0}$ and $\boldsymbol{\beta}^*$

an α-trimming of $\{y_i\}_{i=1}^{N_0}$ that minimize the cost of matching the two samples, alongside the optimal matching $\mathbf{W}^*$.

Finally, it should be noted that a crucial task is how to select an appropriate trimming level α. A simple strategy is to explore different levels, starting with α close to 0 and increasing it until α reaches 0.5, since usually trimming more than 50% of the sample is inadmissible. Next, we keep the smallest trimming level that significantly increases DP for the learning algorithm at hand. This heuristic is the typical approach since extracting the theoretical value of α is usually quite challenging.

2 Experiments and Discussion

In this section we present some simple numerical examples. We will fix the cost function to be the squared Euclidean distance. We start with a synthetic example that fulfills (1) with

$$\begin{aligned} X|S=0 &\sim \frac{2}{5}N\left((-1,0),\operatorname{diag}(0.1,0.1)\right) \\ &\quad + \frac{3}{5}N((1,0),\operatorname{diag}(0.1,0.1)) = \left(1-\frac{1}{5}\right)X_S + \frac{1}{5}X_0, \\ X|S=1 &\sim \frac{3}{5}N\left((-1,0),\operatorname{diag}(0.1,0.1)\right) \\ &\quad + \frac{2}{5}N((1,0),\operatorname{diag}(0.1,0.1)) = \left(1-\frac{1}{5}\right)X_S + \frac{1}{5}X_1, \end{aligned} \tag{3}$$

where, $X_S \sim \frac{1}{2}N\left((-1,0),\operatorname{diag}(0.1,0.1)\right) + \frac{1}{2}N\left((1,0),\operatorname{diag}(0.1,0.1)\right)$, $X_0 \sim N\left((1,0),\operatorname{diag}(0.1,0.1)\right)$ and $X_1 \sim N\left((-1,0),\operatorname{diag}(0.1,0.1)\right)$.

A sample of 500 points from this model, assuming $P(S=0)=P(S=1)=0.5$, is shown in a) of Fig. 1. What we see in top row of Fig. 1 is that by performing a trimming of level $\alpha = 1/5$ we make the conditional distributions almost indistinguishable. An example of how things work in the Ricci data is given in the bottom row of Fig. 1. We clearly see that after solving a matching with a trimming of 0.25 we get a trimmed sample where dependence on 'race' is much lower.

Any learning algorithm, supervised or unsupervised, will produce fairer results, in the sense of DP, in the trimmed data than in the original data shown in Fig. 1. Furthermore, removed elements can be treated separately to asses discrimination or to improve accuracy. Finally, the trimmed samples are actual subsamples of the data in contrast to other pre-processing fairness methods.

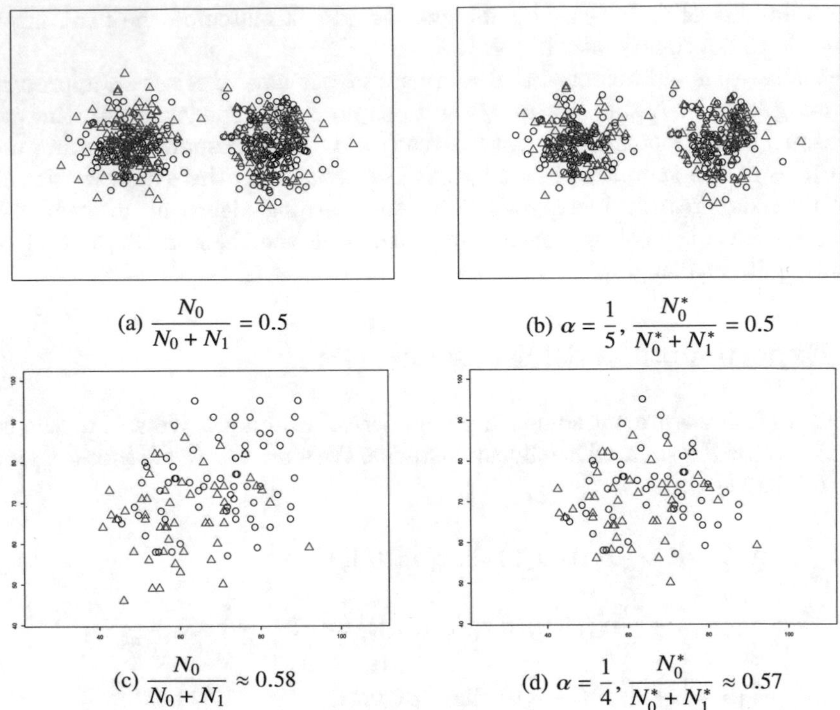

(a) $\frac{N_0}{N_0 + N_1} = 0.5$ (b) $\alpha = \frac{1}{5}, \frac{N_0^*}{N_0^* + N_1^*} = 0.5$

(c) $\frac{N_0}{N_0 + N_1} \approx 0.58$ (d) $\alpha = \frac{1}{4}, \frac{N_0^*}{N_0^* + N_1^*} \approx 0.57$

Fig. 1. Top: Example of an 1/5-trimmed optimal matching from solving (2) for a sample of 500 points from (3). Left: the original data with class $S = 0$ as circles and class $S = 1$ as triangles. Right: the trimmed sample obtained by solving (2) for $\alpha = 1/5$. Bottom: Trimmed matching in the Ricci data. Circles are white individuals and triangles are black and Hispanic individuals. Left: original data. Right: 1/4 trimmed sample. N_S^* are the proportions of the protected classes $S = 0, 1$ in the α-trimmed samples

Acknowledgements. This research was funded by the Basque Government through the BERC 2022-2025 program and Elkartek project 3KIA (KK-2020/00049), and by the Spanish Ministry of Science, Innovation, and Universities (BCAM Severo Ochoa accreditation SEV-2017-0718).

References

Álvarez-Esteban, P.C., Del Barrio, E., Cuesta-Albertos, J.A., Matran, C.: Trimmed comparison of distributions. J. Am. Stat. Assoc. **103**(482), 697–704 (2008)

Barocas, S., Hardt, M., Narayanan, A.: Fairness and Machine Learning (2019). https://www.fairmlbook.org

Del Barrio, E., Gordaliza, P., Loubes, J.-M.: A central limit theorem for L_p transportation cost on the real line with application to fairness assessment in machine learning. Inf. Inference J. IMA **8**(4), 817–849 (2019)

Feldman, M., Friedler, S.A., Moeller, J., Scheidegger, C., Venkatasubramanian, S.: Certifying and removing disparate impact. In: Proceedings of the 21st ACM SIGKDD International Conference on Knowledge Discovery and Data Mining, pp. 259–268. ACM (2015)

Gordaliza, P., Del Barrio, E., Gamboa, F., Loubes, J.-M.: Obtaining fairness using optimal transport theory. In: International Conference on Machine Learning, pp. 2357–2365. ACM (2019)

Oneto, L., Navarin, N., Sperduti, A., Anguita, D.: Recent Trends in Learning from Data. Tutorials from the INNS Big Data and Deep Learning Conference (INNSBDDL2019), vol. 896. Springer, New York (2020). https://doi.org/10.1007/978-3-030-43883-8

Paired Sample Test for Fuzzy Data

Przemyslaw Grzegorzewski[1,2](✉)

[1] Systems Research Institute, Polish Academy of Sciences, Newelska 6, 01-447 Warsaw, Poland
pgrzeg@ibspan.waw.pl

[2] Faculty of Mathematics and Information Science, Warsaw University of Technology, Koszykowa 75, 00-662 Warsaw, Poland

Abstract. The paired sample problem denotes situations in which two measurements are taken from the same object or individual to determine whether there is a statistical evidence that the mean difference between paired observations is significantly different from zero. In this contribution a new statistical test for the paired sample problem with imprecise measurements is proposed. Its idea goes back to permutation tests. It is worth noting that the suggested test is completely distribution-free which seems to be of extreme importance in statistical reasoning with fuzzy data.

1 Introduction

Two-sample tests designed to conclude whether two populations or two features differ significantly are commonly used in statistics. Here we may consider a hypothesis that two samples are drawn from the same population against the general alternative that these populations somehow differ. Such tests, although quite general, are not very efficient in detecting any specific type of the difference between distributions. In most cases we are interested in the so-called two-sample location problem which comes down to deciding whether the means or other location parameters of these samples differ significantly. A typical example is a comparison of two different treatments with one possibly being a control or placebo. If samples are independent and normally distributed we can apply the t-test for the equality of means. Sometimes we are faced with the so-called **paired sample** problem, where each object (person) is measured twice and a statistical test is used to determine whether the difference between two means is zero. Such case appears when measurement are taken at two different times (pre-test and post-test) or under two different conditions ('experimental' and 'control'). Assuming normality the paired sample t-test can be used to solve the problem.

If the assumptions on distributions are not satisfied we may consider distribution-free methods (Gibbons and Chakraborti 2003), like the Mann-Whitney-Wilcoxon test for the difference in location of independent samples and the sign test or the Wilcoxon signed-rank test for paired data. However,

L. A. García-Escudero et al. (Eds.): SMPS 2022, AISC 1433, pp. 200–207, 2023.
https://doi.org/10.1007/978-3-031-15509-3_27

classical nonparametric tests usually turn out to be insufficient when samples contain imprecise data. Fuzzy set theory appears very useful and effective in modeling such data but, unfortunately, fuzzy structures are hardly susceptible to direct application of classical statistical methods. This, in turn, prompts the search for new methods and analytical tools construction that can support decisions in fuzzy environment (some approaches can be found in, e.g., Blanco-Fernández et al. 2014; González-Rodríguez et al. 2006; Grzegorzewski 2020a,b; Grzegorzewski and Gadomska 2021; Lubiano et al. 2016; Montenegro et al. 2004; Ramos-Guajardo et al. 2010).

In this paper we suggest a new statistical test designed for the **paired fuzzy sample** problem. Its design is based on the idea used in permutation tests. Both its simple construction (free from assumptions that are difficult to check) and the results of preliminary studies suggest that the proposed test may prove useful for researchers and practitioners who encounter imprecise data in their work.

The paper is organized as follows: in Sect. 2 we recall basic concepts related to fuzzy numbers. Section 3 is devoted to fuzzy random variables. In Sect. 4 we propose a new two-sample permutation test for paired fuzzy data while some of its properties are shown in Sect. 5.

2 Fuzzy Data

Given imprecise results of experiments we need a tool for describing such observations. A common tool for modeling imprecise values is a **fuzzy number** identified by a mapping $\widetilde{A} : \mathbb{R} \to [0,1]$, called a membership function, such that its α-cuts

$$\widetilde{A}_\alpha = \begin{cases} \{x \in \mathbb{R} : \widetilde{A}(x) \geqslant \alpha\} & \text{if } \alpha \in (0,1], \\ cl\{x \in \mathbb{R} : \widetilde{A}(x) > 0\} & \text{if } \alpha = 0, \end{cases} \tag{1}$$

are nonempty compact intervals for each $\alpha \in [0,1]$, where cl is the closure operator. Each fuzzy number is completely characterized by its membership function $\widetilde{A}(x)$ or by a family of all its α-cuts $\{\widetilde{A}_\alpha\}_{\alpha\in[0,1]}$. A family of all fuzzy numbers will be denoted by $\mathbb{F}(\mathbb{R})$.

The most often used fuzzy numbers are the **trapezoidal fuzzy numbers** with membership functions of the form

$$\widetilde{A}(x) = \begin{cases} \frac{x-a_1}{a_2-a_1} & \text{if } a_1 \leqslant x < a_2, \\ 1 & \text{if } a_2 \leqslant x \leqslant a_3, \\ \frac{a_4-x}{a_4-a_3} & \text{if } a_3 < x \leqslant a_4, \\ 0 & \text{otherwise,} \end{cases} \tag{2}$$

where $a_1, a_2, a_3, a_4 \in \mathbb{R}$ such that $a_1 \leqslant a_2 \leqslant a_3 \leqslant a_4$. Thus a trapezoidal fuzzy number is completely described by its support and core, so a trapezoidal fuzzy number given by (2) is often denoted as $\widetilde{A} = \text{Tra}(a_1, a_2, a_3, a_4)$.

Basic operations in $\mathbb{F}(\mathbb{R})$ are defined with α-cut-wise operations on intervals. For instance, the sum of $\widetilde{A} \in \mathbb{F}(\mathbb{R})$ and $\widetilde{B} \in \mathbb{F}(\mathbb{R})$ is given by the Minkowski addition of α-cuts, i.e. for all $\alpha \in [0,1]$ we have

$$(\widetilde{A} + \widetilde{B})_\alpha = \left[\inf \widetilde{A}_\alpha + \inf \widetilde{B}_\alpha, \sup \widetilde{A}_\alpha + \sup \widetilde{B}_\alpha\right],$$

while the product of $\widetilde{A} \in \mathbb{F}(\mathbb{R})$ by a scalar $\gamma \in \mathbb{R}$ is defined by the Minkowski scalar product for intervals, i.e. for all $\alpha \in [0, 1]$

$$(\gamma \cdot \widetilde{A})_\alpha = \left[\min\{\gamma \inf \widetilde{A}_\alpha, \gamma \sup \widetilde{A}_\alpha\}, \max\{\gamma \inf \widetilde{A}_\alpha, \gamma \sup \widetilde{A}_\alpha\}\right].$$

It should be stressed that $\left(\mathbb{F}(\mathbb{R}), +, \cdot\right)$ has not linear but a semilinear structure since in general $\widetilde{A} + (-1 \cdot \widetilde{A}) \neq \mathbb{1}_{\{0\}}$. Consequently, the Minkowski-based difference does not satisfy the addition/subtraction property, i.e. $(\widetilde{A} + (-1 \cdot \widetilde{B})) + \widetilde{B} \neq \widetilde{A}$. To overcome this problem the so-called Hukuhara difference can be used but it does not always exist.

One can define many metrics in $\mathbb{F}(\mathbb{R})$ but the most often used in statistical context is the distance defined for any $A, B \in \mathbb{F}(\mathbb{R})$ as follows (see Trutschnig et al. 2009).

$$D_\theta^\lambda(\widetilde{A}, \widetilde{B}) = \left(\int_0^1 \left[(\operatorname{mid} \widetilde{A}_\alpha - \operatorname{mid} \widetilde{B}_\alpha)^2 + \theta \cdot (\operatorname{spr} \widetilde{A}_\alpha - \operatorname{spr} \widetilde{B}_\alpha)^2\right] d\lambda(\alpha)\right)^{1/2}, \quad (3)$$

where λ is a normalized measure associated with a continuous distribution on $[0, 1]$, θ is a positive constant, $\operatorname{mid} \widetilde{A}_\alpha = \frac{1}{2}(\inf \widetilde{A}_\alpha + \sup \widetilde{A}_\alpha)$ and $\operatorname{spr} \widetilde{A}_\alpha = \frac{1}{2}(\sup \widetilde{A}_\alpha - \inf \widetilde{A}_\alpha)$ denote the mid-point and the radius of the α-cut $\widetilde{A}_\alpha$, respectively. Here λ allows to weight the α-cut's influence, while θ may be used to weight the impact of the distance between spreads of the α-cuts and the distance between their mid-points. Usually λ is the Lebesgue measure on $[0, 1]$, but whatever (λ, θ) is chosen D_θ^λ is an L^2-type metric in $\mathbb{F}(\mathbb{R})$, is invariant to translations and rotations. Moreover, $(\mathbb{F}(\mathbb{R}), D_\theta^\lambda)$ is a separable metric space and for each fixed λ all metrics D_θ^λ are topologically equivalent.

3 Fuzzy Random Variables

For statistical inference we need a model which allows to grasp both aspects of uncertainty that appear in fuzzy data: fuzziness connected with data imprecision and randomness. To handle fuzzy data Puri and Ralescu (1986) introduced **fuzzy random variables**, sometimes called **random fuzzy numbers**.

Let $(\Omega, \mathcal{A}, P)$ be a probability space. A mapping $\widetilde{X} : \Omega \to \mathbb{F}(\mathbb{R})$ is a **random fuzzy number** if for all $\alpha \in [0, 1]$ the α-level function is a compact random interval. It can be shown that $\widetilde{X}$ is a random fuzzy number if and only if $\widetilde{X}$ is a Borel measurable function w.r.t. the Borel σ-field generated by the topology induced by D_θ^λ.

The Aumann-type mean of a fuzzy random variable $\widetilde{X}$ is the fuzzy number $E(\widetilde{X}) \in \mathbb{F}(\mathbb{R})$ such that $\forall \alpha \in [0, 1]$ the α-cut $\left(E(\widetilde{X})\right)_\alpha$ is equal to the Aumann integral of $\widetilde{X}_\alpha$, i.e. $\left(E(\widetilde{X})\right)_\alpha = \left[\mathbb{E}(\operatorname{mid} \widetilde{X}_\alpha) - \mathbb{E}(\operatorname{spr} \widetilde{X}_\alpha), \mathbb{E}(\operatorname{mid} \widetilde{X}_\alpha) + \mathbb{E}(\operatorname{spr} \widetilde{X}_\alpha)\right]$.

A fuzzy random sample $\widetilde{\mathbb{X}} = (\widetilde{X}_1, \ldots, \widetilde{X}_n)$ is a finite sequence of independent random fuzzy numbers from the same distribution. Given $\widetilde{\mathbb{X}}$ we can determine the sample mean $\overline{\widetilde{X}} \in \mathbb{F}(\mathbb{R})$ defined by the following α-cuts

$$\overline{\widetilde{X}}_\alpha = \left[\frac{1}{n}\sum_{i=1}^{n} \text{mid}\,(\widetilde{X}_i)_\alpha - \frac{1}{n}\sum_{i=1}^{n} \text{spr}\,(\widetilde{X}_i)_\alpha, \frac{1}{n}\sum_{i=1}^{n} \text{mid}\,(\widetilde{X}_i)_\alpha + \frac{1}{n}\sum_{i=1}^{n} \text{spr}\,(\widetilde{X}_i)_\alpha\right].$$

Unfortunately, it is not easy to transfer facts from the classical statistics to fuzzy domain. Firstly, problems with subtraction and division of fuzzy numbers, mentioned in Sect. 2, lead to the conclusion that avoiding these operations whenever it is possible is strongly recommended. Moreover, since fuzzy numbers are not linearly ordered, the lack of a commonly accepted total ranking precludes the application of the rank methods popular in nonparametric statistics. Next, the absence of suitable models for the distribution of fuzzy random variables makes the statistical reasoning really hard. Finally, there are not yet satisfying Central Limit Theorems for fuzzy random variables that can be applied directly in decision making. Hence, inference with fuzzy data requires innovative solutions like the bootstrap (González-Rodríguez et al. 2006; Lubiano et al. 2016; Montenegro et al. 2004) or permutation tests (Grzegorzewski 2020a,b; Grzegorzewski and Gadomska 2021). The last methodology is applied also in this contribution.

4 Two-Sample Permutation Test for Paired Fuzzy Data

Consider two random samples $\mathbb{X} = (X_1, \ldots, X_n)$ and $\mathbb{Y} = (Y_1, \ldots, Y_m)$ drawn from distributions with unknown means μ_1 and μ_2, respectively. Our goal is to test the null hypothesis $H_0 : \mu_1 = \mu_2$ against $H_0 : \mu_1 \neq \mu_2$. Usually we assume that the $\mathbb{X}$ and $\mathbb{Y}$ are independent of one another. However, there is a number of experimental situations in which only one set of n individuals or experimental objects is available but all objects are examined twice, so actually we have $\mathbb{X} = (X_1, \ldots, X_n)$ and $\mathbb{Y} = (Y_1, \ldots, Y_n)$, with $\mathbb{X}$ corresponding to the 'pre-treatment' and $\mathbb{Y}$ to the 'post-treatment'. In this case, obviously, $\mathbb{X}$ and $\mathbb{Y}$ are no longer independent. Making two observations on each object provides a natural pairing $(X_1, Y_1), \ldots, (X_n, Y_n)$ with independent data pairs of dependent random variables in each pair. The null hypothesis H_0 indicates no 'treatment' while the general two-sided alternative H_1 means that some 'treatment' effect exists. If our data are normally distributed then the testing problem can be solved using the paired t-test based on the following statistic

$$t = \frac{\frac{1}{n}\sum_{i=1}^{n}(X_i - Y_i)}{s_D}\sqrt{n}, \tag{4}$$

where s_D is the sample standard deviation of the differences $X_i - Y_i$, where $i = 1, \ldots, n$. If H_0 holds then (4) has the t-Student distribution with $n - 1$ degrees of freedom and we reject H_0 if $|t|$ exceeds an adequate critical level.

Unfortunately, if our data consists of the pairs $(\widetilde{X}_1, \widetilde{Y}_1), \ldots, (\widetilde{X}_n, \widetilde{Y}_n)$, where both $\widetilde{X}_i$ and $\widetilde{Y}_i$ are random fuzzy numbers, we cannot generalize directly the paired t-test and it becomes obvious that a desired test for paired fuzzy data must be constructed on completely different principles than the classical paired t-test. To cope with the two-sample location problem with fuzzy data we propose a new permutation test. The specificity of paired data means that we have to use a different method of randomization than in other permutation test discussed in Grzegorzewski (2020a,b) and Grzegorzewski and Gadomska (2021).

Suppose that $(\widetilde{x}_1, \widetilde{y}_1), \ldots, (\widetilde{x}_n, \widetilde{y}_n)$, where $\widetilde{x}_i, \widetilde{y}_i \in \mathbb{F}(\mathbb{R})$, for $i = 1, \ldots, n$, is a particular outcome of the random experiment, i.e. a realization of $(\widetilde{X}_1, \widetilde{Y}_1), \ldots, (\widetilde{X}_n, \widetilde{Y}_n)$.

Firstly, let us define a test statistic based on $(\widetilde{X}_1, \widetilde{Y}_1), \ldots, (\widetilde{X}_n, \widetilde{Y}_n)$. For a fixed distance D_θ^λ described in (3) our test statistic $T = T(\widetilde{\mathbb{X}}, \widetilde{\mathbb{Y}})$ will be defined by

$$T(\widetilde{\mathbb{X}}, \widetilde{\mathbb{Y}}) = D_\theta^\lambda(\overline{\widetilde{X}}, \overline{\widetilde{Y}}), \tag{5}$$

where $\overline{\widetilde{X}}$ and $\overline{\widetilde{Y}}$ stand for the average of $\widetilde{X}_1, \ldots, \widetilde{X}_n$ and $\widetilde{Y}_1, \ldots, \widetilde{Y}_n$, respectively. One can see that in contrast to (4) we replace the difference between averages by the distance between them. Now, for given results we compute a value of our test statistic for the observed data

$$T_0 = T(\widetilde{\mathbb{x}}, \widetilde{\mathbb{y}}) = D_\theta^\lambda(\overline{\widetilde{x}}, \overline{\widetilde{y}}), \tag{6}$$

where $\overline{\widetilde{x}}$ and $\overline{\widetilde{y}}$ stand for the average of $\widetilde{\mathbb{x}} = (\widetilde{x}_1, \ldots, \widetilde{x}_n)$ and $\widetilde{\mathbb{y}} = (\widetilde{y}_1, \ldots, \widetilde{y}_n)$, respectively.

Secondly, we have to propose a rejection rule. Keeping in mind the problems with determining an adequate probability distribution of the considered test statistic we will use a permutation test appropriate for the considered paired fuzzy data. Thus we need a randomization method appropriate for the nature of paired data. Starting from the initial set of pairs $(\widetilde{x}_1, \widetilde{y}_1), \ldots, (\widetilde{x}_n, \widetilde{y}_n)$ we will generate a sequence $(\widetilde{x}_1^*, \widetilde{y}_1^*), \ldots, (\widetilde{x}_n^*, \widetilde{y}_n^*)$ such that for each $i = 1, \ldots, n$

$$\widetilde{x}_i^* := \begin{cases} \widetilde{x}_i & \text{if} \quad \eta_i = 1, \\ \widetilde{y}_i & \text{if} \quad \eta_i = 0, \end{cases} \qquad \widetilde{y}_i^* := \begin{cases} \widetilde{y}_i & \text{if} \quad \eta_i = 1, \\ \widetilde{x}_i & \text{if} \quad \eta_i = 0, \end{cases} \tag{7}$$

where $\eta_1, \ldots, \eta_n$ are i.i.d. random variable from the Bernoulli distribution with probability 0.5.

For given $(\widetilde{x}_1, \widetilde{y}_1), \ldots, (\widetilde{x}_n, \widetilde{y}_n)$ we determine the corresponding value of the test statistic (5), i.e.

$$T(\widetilde{\mathbb{x}}^*, \widetilde{\mathbb{y}}^*) = D_\theta^\lambda(\overline{\widetilde{x}^*}, \overline{\widetilde{y}^*}), \tag{8}$$

where $\overline{\widetilde{x}^*}$ and $\overline{\widetilde{y}^*}$ denote the average of $\widetilde{\mathbb{x}^*} = (\widetilde{x}_1^*, \ldots, \widetilde{x}_n^*)$ and $\widetilde{\mathbb{y}^*} = (\widetilde{y}_1^*, \ldots, \widetilde{y}_n^*)$, respectively.

We repeat the whole procedure by considering various permutations of observations in each pair and compute a value $T(\widetilde{\mathbb{x}}_b^*, \widetilde{\mathbb{y}}_b^*)$ of the test statistic (8), B times, i.e. $b = 1, \ldots, B$. A typical number of repetitions is $B = 999$. Then we compute the approximate p-value of our test as follows

$$\text{p-value} \approx \frac{1 + \sum_{b=1}^{B} \mathbb{1}\big(T(\widetilde{\mathbb{x}}_b^*, \widetilde{\mathbb{y}}_b^*) \geqslant T_0\big)}{B+1}, \tag{9}$$

where T_0 stands for the test statistic value (6) obtained for the original fuzzy sample. Finally, we reject the null hypothesis H_0 on independence at significance level δ if p-value $\leqslant \delta$.

It should be stressed that the proposed permutation test for paired data, as other permutation tests, requires very limited assumptions. Indeed, the only requirement is the so-called *exchangeability*, i.e. if the null hypothesis holds then we can exchange the labels of some observations within pairs without affecting the results. Obviously, this assumption is satisfied in the considered case.

5 Simulation Study

To examine the properties of the proposed test we conducted a simulation study. Due to the limited space in the contribution we only show a fragment of the research carried out.

Let us consider trapezoidal observations generated using the notation convention indicating the center of the core c, the half of the core's length s and the spread of the left and right arm l and r, respectively, i.e. for $\widetilde{A} = \text{Tra}(a_1, a_2, a_3, a_4)$ we have $\widetilde{A} = \langle c_A, s_A, l_A, r_A \rangle$, where

$$c_A = \tfrac{1}{2}(a_2 + a_3), \quad s_A = \tfrac{1}{2}(a_3 - a_2), \quad l_A = a_2 - a_1, \quad r_A = a_4 - a_3. \tag{10}$$

Therefore, the considered pairs $(\widetilde{x}_1, \widetilde{y}_1), \ldots, (\widetilde{x}_n, \widetilde{y}_n)$ are obtained by simulating four independent real-valued random variables for each $\widetilde{x}_i = \langle c_{Xi}, s_{Xi}, l_{Xi}, r_{Xi} \rangle$ and four random variables for each $\widetilde{y}_j = \langle c_{Yj}, s_{Yj}, l_{Yj}, r_{Yj} \rangle$, respectively, with the last three random variables in each quartet being nonnegative. In particular, we generated random fuzzy numbers using the following real-valued random variables: c_{Xi}, c_{Yj} from the standard normal distribution and $s_{Xi}, s_{Yj}, l_{Xi}, l_{Yj}, r_{Xi}, r_{Yj}$ from the uniform distribution.

Figure 1 shows a histogram illustrating the null distribution of the test statistic (5) and obtained for a fuzzy sample of size $n = 10$ generated by independent random variables c_X and c_Y from the standard normal distribution $\text{N}(0, 1)$ and s_X, s_Y, l_X, l_Y and r_X, r_Y from the uniform distribution $\text{U}(0.1)$. In this case we have obtained $t_0 = 0.2119$, marked on the graph by the black dot. The corresponding p-value $= 0.3397$ leads to the following decision: do not reject H_0.

We also examined the proposed test with respect to its size. were performed under H_0. In each of 1000 repetitions of the test at 5% significance level $K = 1000$ permutations were drawn and the empirical percentages of rejections were determined. The results obtained for a few sample sizes n gathered in Table 1 show that the size of our test is stable at the desired level.

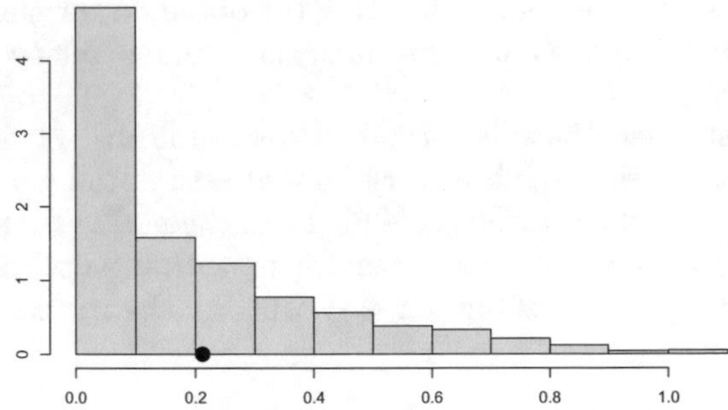

Fig. 1. Empirical null distribution of the permutation test with a black dot indicating the value T_0 of the test statistic.

Table 1. Empirical size of the test for various sample sizes

n	10	25	50	100
Empirical size	0.041	0.052	0.051	0.047

6 Conclusions

Hypothesis testing with imprecise data usually cannot be generalize straightforwardly from the classical approach oriented on crisp data. This is due to certain difficulties connected with fuzzy modeling of such data. It appears that some of those difficulties in test constructions might be solved by applying some specific nonparametric methodology based of permutations. The aforementioned idea was also applied in this paper to propose a statistical test for the two-sample problem with paired fuzzy data. As a result we have obtained a test which is completely distribution-free. The preliminary results seem to be promising. Obviously, further research including an extensive power study is strongly desired and is intended in the nearest future.

References

Blanco-Fernández, A., et al.: A distance-based statistic analysis of fuzzy number-valued data (with Rejoinder). Int. J. Approx. Reason. **55**, 1487–1501 (1601–1605) (2014)

Gibbons, J.D., Chakraborti, S.: Nonparametric Statistical Inference. Marcel Dekker Inc., New York (2003)

González-Rodríguez, G., Montenegro, M., Colubi, A., Gil, M.A.: Bootstrap techniques and fuzzy random variables: synergy in hypothesis testing with fuzzy data. Fuzzy Sets Syst. **157**, 2608–2613 (2006)

Grzegorzewski, P.: Two-sample dispersion problem for fuzzy data. In: Lesot, M.J., et al. (eds.) IPMU 2020. CCIS, vol. 1239, pp. 82–96. Springer, Cham (2020). https://doi.org/10.1007/978-3-030-50153-2_7

Grzegorzewski, P.: Permutation k-sample goodness-of-fit test for fuzzy data. In: Proceedings of the 2020 IEEE International Conference on Fuzzy Systems (FUZZ-IEEE), pp. 1–8 (2020)

Grzegorzewski, P., Gadomska, O.: Nearest neighbor tests for fuzzy data. In: Proceedings of the 2021 IEEE International Conference on Fuzzy Systems (FUZZ-IEEE), pp. 1–6 (2021)

Lubiano, M.A., Montenegro, M., Sinova, B., de la Rosa de Sáa, S., Gil M.A.: Hypothesis testing for means in connection with fuzzy rating scale-based data: algorithms and applications. Eur. J. Oper. Res. **251**, 918–929 (2016)

Montenegro, M., Colubi, A., Casals, M.R., Gil, M.A.: Asymptotic and Bootstrap techniques for testing the expected value of a fuzzy random variable. Metrika **59**, 31–49 (2004)

Puri, M.L., Ralescu, D.A.: Fuzzy random variables. J. Math. Anal. Appl. **114**, 409–422 (1986)

Ramos-Guajardo, A.B., Colubi, A., González-Rodríguez, G., Gil, M.A.: One-sample tests for a generalized Fréchet variance of a fuzzy random variable. Metrika **71**, 185–202 (2010)

Trutschnig, W., González-Rodríguez, G., Colubi, A., Gil, M.A.: A new family of metrics for compact, convex (fuzzy) sets based on a generalized concept of mid and spread. Inf. Sci. **179**, 3964–3972 (2009)

Monitoring of Possibilisticaly Aggregated Complex Time Series

Olgierd Hryniewicz(✉) and Katarzyna Kaczmarek-Majer

Systems Research Institute, Newelska 6, 01-447 Warsaw, Poland
{hryniewi,kkaczmarek}@ibspan.waw.pl

Abstract. Monitoring of inhomogeneous processes characterized by complex structures has been considered. The application of well-known methods of Statistical Process Control (SPC) for such processes may be questionable, as the interpretation of the results of monitoring of such complex processes using the SPC methods is difficult. In the paper, we consider processes consisted of segments and subsegments. The data from subsegments belonging to respective segments are aggregated using a possibilistic methodology. Using computer simulations we show that the monitoring of aggregated data is efficient and interpretable. Moreover, it is possible now to evaluate statistical properties of monitoring processes that use different methods of data aggregation.

1 Introduction

Activities that are represented by the series of measurements may be considered as processes. The behavior of a process may change in time. When such change happens, we say that the process has changed its state. States of a process may be considered as desirable or undesirable. The main aim of monitoring of processes using Statistical Process Control (SPC) methods is to identify transitions from a desirable to an undesirable state. In 1924 W. Shewhart introduced a simple tool for monitoring processes - a control chart. According to Shewhart's proposal, in an initial stage of a monitored process, which is assumed to be in a desirable state, process characteristics are measured and recorded. The values of these statistics are used for the design of a control chart which consists of one or many control lines. The process is considered as remaining in a desirable state when its observations, plotted on a control chart, are located inside control lines (limits). When an observation falls outside the control lines, an alarm signal is generated, and the process is considered as being possibly in an undesirable state.

Since the time when Shewhart proposed a control chart, many control charts have been proposed and applied in many areas, such as industry, insurance, banking, health services, and many others. In his original proposal, Shewhart assumed that recorded values of process characteristics (variables) are mutually independent and described by the normal distribution. This assumption is still widely used in many univariate and multivariate control charts. However, in many practical cases, especially when individual process observations

L. A. García-Escudero et al. (Eds.): SMPS 2022, AISC 1433, pp. 208–215, 2023.
https://doi.org/10.1007/978-3-031-15509-3_28

are monitored, these assumptions are not fulfilled. Thus, since the years 1970s, many inspection procedures that do not rely on these assumptions have been proposed (for more information see, Montgomery 2011). However, in nearly all new proposals it is assumed that data coming from a process being in a certain state are homogeneous, i.e., described by the same, usually known, probability distribution.

The assumption of data homogeneity is usually not valid for data acquired by data collecting devices such as sensors, smartphones, network bots, etc. For example, continuously measured medical data whose characteristics may be different at different times of a day or on different days (depending on the state of health of a monitored person). In this example, data collected at different times of a day can be considered as subsegments of larger segments, consisted of data collected on consecutive days. Hryniewicz and Kaczmarek-Majer (2021) considered such strongly inhomogeneous data coming from the measurements of voice characteristics of patients suffering from Bipolar Disorder - a severe mental illness. An example of such process data is presented in Fig. 1 (Hryniewicz and Kaczmarek-Majer 2021).

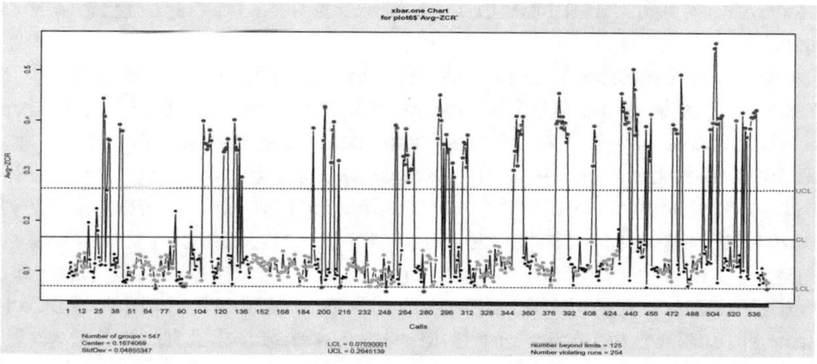

Fig. 1. BD calls data - average values of the voice *zcr* characteristic

The process displayed in Fig. 1 looks quite inhomogeneous, even though physicians considered the mental state of the patient as stable. This problem motivated Hryniewicz and Kaczmarek-Majer (2021) to aggregate subsegment data (i.e., call data) for each segment (i.e., each day), and further consider the process consisted of aggregated data. After the performed aggregation the process from Fig. 1 looks now as (Hryniewicz and Kaczmarek-Majer 2021).

From the analysis of data presented in Figs. 1 and 2, we can see that despite apparent instability of the data stream representing consecutive calls the aggregated data stream that represents consecutive days exhibits apparent stability that confirms psychiatrists' assessments. Therefore, the method of data stream aggregation proposed in Hryniewicz and Kaczmarek-Majer (2021) confirms its applicability for the monitoring of BD patients.

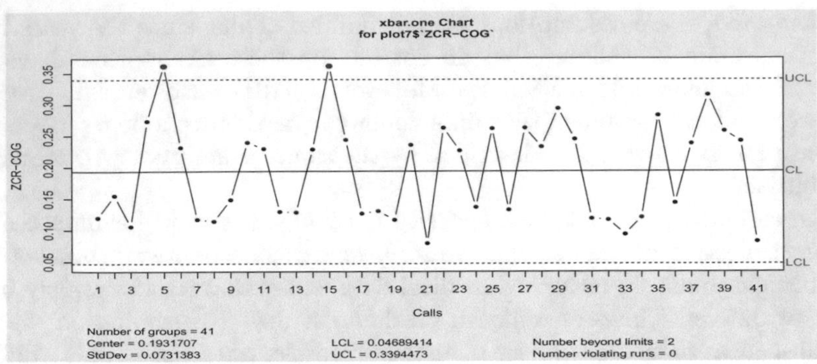

Fig. 2. BD calls aggregated for days

This finding is rather of a qualitative character and does not provide information, expected in the community of SPC practitioners, how the monitoring of aggregated data is effective in monitoring of a complex process that may change its state. Therefore, the aim and the novelty of this paper is the demonstration this effectiveness using statistical characteristics used by the community of SPC practitioners.

The paper is organized as follows. In the second section we briefly present the method that is often used in monitoring processes - the Cumulative Sum (CUSUM) control chart. We will use this chart for monitoring processes considered in this paper. In the third section we briefly present the possibilistic methods of aggregation proposed in Hryniewicz and Kaczmarek-Majer (2021). The main and original results of this research are presented in the fourth section of the paper where we demonstrate the properties of the CUSUM control charts designed to monitor aggregated time series. These properties are evaluated using computer simulation experiments. The paper is concluded in its last section.

2 Monitoring of Processes with a CUSUM Control Chart

Control charts are popular tools for monitoring processes. Their usage can be considered as a special case of a more general problem, namely the detection of a *change point*. The problem of the change point can be formally introduced as follows. Let $y_1, \ldots, y_n$ be a sequence of of observed random variables with conditional density $p_\theta(p_k|p_{k-1}, \ldots, p_1)$. Before the unknown change time t_0 the parameter θ is constant and equal to θ_0. After the change, θ is equal to θ_1. The problem is to detect this change as quickly as possible, with a fixed rate of false alarms before t_0. There are many methods for solving this problem, and some of them are given in a form of other control charts. A good introduction to this area of statistics is a monograph of Basseville and Nikiforov (1993).

A very important control chart is the Cumulative Sum (CUSUM) control chart introduced by Page (1954). The general theory of control charts of this type is presented in Basseville and Nikiforov (1993). In this paper, we use its special case, dedicated for the normal distribution with the expected value μ and standard deviation σ. We assume that these values are either known or precisely estimated. Thus, we can transform the original observations $y_1, \ldots, y_n$ into their standardized version $x_i = (y_i - \mu)/\sigma, i = 1, \ldots, n$. Then, we calculate recursively the cumulative sums $S_k^+ = \max(0, x_k - K + S_{k-1}^+)$, when the upward shift in the expected value has to be detected, or $S_k^- = \max(0, -x_k - K + S_{k-1}^-)$, when the downward shift in the expected value has to be detected. When any shift is undesirable, the cumulative sums are calculated as $S_k = \min(S_k^-, S_k^+)$, The constant K is equal to one half of the magnitude of the detectable shift for the standardized variable, i.e., expressed as the multiplicity of σ, and $S_0^- = S_0^+ = 0$. The existence of the change point is detected, when the value of the cumulative sum exceeds the decision value h. The values of K and h may be calculated for given values of the required values of $ARL's$, but practitioners usually take $K = 0.5$ and $h = 5$.

Statistical properties of control charts are evaluated using the characteristic known as the Average Run Length (ARL). The ARL is defined as the expected value (average) between consecutive alarms. When the monitored process remains in a desirable stable state these alarms are called false alarms, and the respective value of ARL in this paper is denoted as $ARL(0)$. However, the main purpose of a chart is to trigger alarms when the process goes out-of-control, e.g., when its average increases (decreases) by a certain value s, usually expressed as a multiplicity of the process standard deviation. The average waiting time to alarm is then denoted as $ARL(s)$. CUSUM control charts, considered in this paper, are known to be very efficient in the detection of relatively small shifts in the process level (process average).

3 Possibilistic Aggregation of Segmented Processes

Let us assume that we observe a stream of numerical data $x_1, \ldots, x_N$ consisted of a large number of elements. We assume that this stream is divided into n segments, and each of these segments is further divided into $n_i, i = 1, \ldots, n$ subsegments. The structure of this segmentation may be either known or determined unequivocally using some independent statistical or machine learning methods. We also assume that the number of observations $n_{i,j}$ in each of the subsegments is sufficiently large for purposes of probability distribution estimation. In each of these subsegments, the observed data are described by a joint probability distribution $F_{i,j}(x_{i,j,1}, \ldots, x_{i,j,n_{i,j}}), i = 1, \ldots, n, j = 1, \ldots, n_i$. These distributions are unknown, but may be consistently estimated from data. Now, for purposes of the proposed data aggregation, we assume that within a given segment all subsegment data are classified into one of $m_i, i = 1, \ldots, n$ classes, defined by the set of class limits $(c_{i,0}, c_{i,1}, \ldots, c_{i,m_i})$.

Having this partition of data values, we can represent the data from subsegments by the respective histograms $k_{i,j,l}, i = 1, \ldots, n, j = 1, \ldots, n_i, l = 1, \ldots, m_i$, where $k_{i,j,l}$ is the number of observations in the i-th segment and its j-th subsegment that belong to the l-th class. Let $k_{i,j}$ be the the number of observations in the (i, j)-th subsegment. The observed histograms may be transformed into empirical probability density functions (epdf's) $p_{i,j,l}, i = 1, \ldots, n, j = 1, \ldots, n_i, l = 1, \ldots, m$, where $p_{i,j,l} = k_{i,j,l}/n_{i,j}$, i.e., by piece-wise constant density functions. These distributions can be used as the approximations of the subsegment distributions $F_{i,j}(x), i = 1, \ldots, n, j = 1, \ldots, n_i$.

In many practical applications the segmented stream of data is interpretable on the level of segments. Therefore, there is a need to aggregate all data coming from subsegments belonging to the same segment into one probability distribution describing the whole segment. Hryniewicz and Kaczmarek-Majer (2021) considered two approaches to this aggregation problem. First, they proposed a simple probabilistic method leading to the construction of a probability distribution that describes a given segment. They proposed to combine subsegment epdf's by the calculation of their weighted sum with the weights proportional to the relative sizes of their respective data sets. Simple arithmetics shows that this approach is equivalent to building for each segment a combined histogram $(n^0_{i,1}, \ldots, n^0_{i,m}), i = 1, \ldots, n$, such that $n^0_{i,l} = k_{i,1,l} + \cdots + k_{i,n_i,l}, i = 1, \ldots, n, l = 1, \ldots, m$. This combined histogram can be easily transformed into a combined epdf. Second, they proposed a possibilistic method that describes all segment data using a possibility distribution. The concept of the possibility has many different interpretations. According to one of them, see Dubois and Prade (1988), the possibility distribution can be understood as an upper envelope for all ordinary discrete probability distributions compatible with a value imprecisely described by a fuzzy number. In our case, the epdf's describing data are, in fact, discrete probability distributions, so the attempt to represent them by a respective possibility distribution seems to be fully justified.

The first step in the attempt to build a possibility distribution is the construction of an upper envelope. For this purpose, we propose to use the concept of t-conorms (or s-norms). Let $x, y \in [0, 1]$, then for aggregation purposes we can use different popular t-conorms, such as: maximum t-conorm $AG1(x, y) = \max(x, y)$, probabilistic sum $AG2(x, y) = x + y - x * y$, and maximum bounded sum $AG3(x, y) = min(x + y, 1)$. It is easy to show that $AG1(x, y) \leq AG2(x, y) \leq AG3(x, y)$. Therefore, all these t-conorms can be used for building an upper envelope of the considered epdf's. To do so we sequentially apply the chosen aggregation function (t-conorm) to probabilities assigned to each interval that describes the empirical probability density function arriving at the upper envelope $(\tilde{p}_{i,1}, \ldots, \tilde{p}_{i,m}), i = 1, \ldots, n$ of the aggregated epdf's.

The upper possibilistic envelope described above is not a proper possibility distribution, as it is not represented as a fuzzy number whose membership function must be convex and normalized. Let us find an interval (class) for whom the upper envelope $(\tilde{p}_{i,1}, \ldots, \tilde{p}_{i,m}), i = 1, \ldots, n$ attains its maximum. Denote the index of this interval by

$$L_i = \arg \max_{l=1,\ldots,m} \tilde{p}_{i,l}, \ i = 1, \ldots, n. \tag{1}$$

Note, that this maximum may be attained in more than one interval, and in this case the choice of L_i may be arbitrary. Let $\tilde{p}^M_i = \tilde{p}_{i,L_i}, i = 1, \ldots, n$. The upper possibilistic envelope may be transformed to the proper possibility distribution $(\pi_{i,1}, \ldots, \pi_{i,m}), i = 1, \ldots, n$ defined on the set of intervals defined by class limits $(c_0, c_1, \ldots, c_m)$ using the following transformation

$$\pi_{i,l} = \begin{cases} \max(\tilde{p}_{i,1}, \ldots, \tilde{p}_{i,l})/\tilde{p}_i^M & \text{for } l = 1, \ldots, L_i \\ \max(\tilde{p}_{i,l}, \ldots, \tilde{p}_{i,m})/\tilde{p}_i^M & \text{for } l = L_i, \ldots, m \end{cases} \quad (2)$$

for $i = 1, \ldots, n$.

4 Statistical Properties of the CUSUM Chart for Aggregated Process Data

The values of $ARL0$ and $ARL(s)$ characteristics can be precisely computed for only very few special cases of control charts and monitoring processes. For more complicated procedures, such as CUSUM charts, these values can be computed numerically by solving integral equations or using approximations by Markov chains. However, for such cases, as that considered in this paper, the statistical characteristics of control charts are usually evaluated in computer simulation experiments. In these simulations it is necessary to assume the probability distribution that governs the monitored process. For processes described in Sect. 4 it is hardly possible. Therefore, we have built a simulation model that produces series of observations that look like series of observations encountered in the monitoring of BD patients.

In our system, subsegment data are generated from a possibly multivariate distribution having the same marginal distribution for observed data, and the possible dependence between consecutive observations described by a given copula. The marginal distribution of observations is randomly chosen for one subsegment from the set of (normal, exponential, Weibull, gamma, and beta) distributions whose parameters are either set or randomly chosen from given intervals. Dependence models are randomly chosen from the set of copulae (independent, Gaussian, Clayton, Frank, FGM) whose parameters are related to the strength of dependence described by the Pearson autocorrelation (Gaussian copula) or Kendall's coefficient of dependence. The values describing the strength of dependence are either set in advance or randomly chosen from given intervals. The types of probability distributions and the types of dependence are chosen in order to different statistical characteristics of distributions (e.g., skewness and kurtosis) and different types of dependence (symmetric or asymmetric). Additionally, in our system we have introduced mechanisms for the generation of dependence between subsegments within a one segment, and between whole segments. Formally, each datum simulated by our system is represented by an additive model

$$X_{i,j,k} = Y_{i,j,k} + Z_{i,j} + W_i, i = 1, \ldots, n, j = 1, \ldots, n_i, k = 1, \ldots, n_{i,j}, \quad (3)$$

where $Y_{i,j,k}$ is the observation generated for a subsegment, $Z_{i,j}$ is a random component generated from a given $AR(1)$ model for the ith segment, and W_i is a random component generated from a given *AR*(1) model for all simulated segments. The numbers of subsegments within one segment (n_i), and the numbers of observations within one subsegment (n_{ij}) were also chosen randomly from a given distribution.

The streams of data generated according to (3) were aggregated according to one of the aggregation methods described in Sect. 3, and then the CUSUM control chart was used for the monitoring of the stream of aggregated process data. The control chart was designed, according to the methodology described in Sect. 2, using the data from the first 25 segments, and control chart characteristics (times to alarms) were evaluated using the next 275 segments. When the impact of the shift of the process level was evaluated, the shift s (see Sect. 2) was introduced after additional (warm up) segments. The whole process was repeated several times using different parameters of the simulation model, and the average values *ARL*(0) and *ARL*(s) were calculated and presented in Table 1 for the cases of $s = 0$ and $s = 0.2$.

Table 1. Values of ARL ($K = 0.5$, $h = 5$)

ARL	AG1	AG2	AG3	AG0
ARL(0)	185.2	209.3	212.9	214.5
ARL(0.2)	36.2	31.5	30.8	38.6

The results of simulations presented in the first row of Table 1 show that the considered aggregation methods have very good properties for non-shifted processes, i.e. we observe similar large values of *ARL*(0). Note, that the presented values have been obtained from the analysis of curtailed (at 275) samples, and the actual values of *ARL*(0) are much better. When the waiting times for real alarms are considered (*ARL*(0.2)), the results are also similar but less satisfactory. The methods AG2 and AG3 give better results, especially in cases for which subsegment data distributions within one segment are more diversified. Thus, when we are interested in the early detection of a process shift, the aggregation using the probabilistic conorm (AG2) or Lukasiewicz conorm (AG3) is preferred. When we are interested in even earlier detection of a process shift, we can decrease the value of the design parameter K (see Sect. 2). However, we have to bear in mind that in this case the value of *ARL*(0) will decrease. In the considered problem of the monitoring BD patients this is, according to psychiatrists, not a serious problem.

One has to note, however, that the numbers presented in Table 1 represent averages. In particular experiments, the observed preference order was different. The question arises then, if it is possible to choose the appropriate aggregation method using the statistical analysis of all data collected during the period used for the construction of a control chart. The solution to this problem needs further research.

5 Conclusions

In this paper we have demonstrated that the data aggregation for complex segmented data series not only gives better interpretable representation of the considered time series (as it was demonstrated in Hryniewicz and Kaczmarek-Majer (2021)), but allows to use efficiently monitoring procedures of SPC, such as the CUSUM control charts. The possibilistic aggregation methods, originally proposed in Hryniewicz and Kaczmarek-Majer (2021), which are based on the probabilistic and Lukasiewicz co-norms are better, on average, when the monitoring goal is to detect the change of a process level as quickly as possible. When the main goal is to avoid frequent false alarms all considered methods of aggregation (a simple probabilistic method included) have similar good properties.

When the possiblistic methods of aggregation are used, the process segments are described by fuzzy numbers. For the purpose of monitoring using a classical control chart we represented these fuzzy numbers by the respective crisp numbers. In future research, it seems to be worth to investigate the possibility to use fuzzy control charts, such as the CUSUM chart proposed in Wang and Hryniewicz (2013) for the purpose of monitoring.

References

Basseville, M., Nikiforov, I.V.: Detection of Abrupt Changes: Theory and Applications. Prentice-Hall Inc., Englewood Cliffs, N.J. (1993)

Hryniewicz, O., Kaczmarek-Majer, K.: Possibilistic aggregation of inhomogeneous streams of data. In: Proceedings of the 2021 IEEE International Conference on Fuzzy Systems (FUZZ-IEEE), pp. 1–6 (2021)

Dubois, D., Prade, H.: Possibility Theory: An Approach to Computerized Processing of Uncertainty. Plenum Press, New York (1988)

Montgomery, D.C.: Introduction to Statistical Quality Control, 6th edn. Wiley, New York (2011)

Page, E.S.: Continuous inspection schemes. Biometrika **41**, 100–115 (1954)

Wang, D., Hryniewicz, O.: The design of a CUSUM control chart for LR-fuzzy data. In: Proceedings of the 2013 Joint IFSA World Congress and NAFIPS Annual Meeting (IFSA/NAFIPS), pp. 175–180 (2013)

Tk-Merge: Computationally Efficient Robust Clustering Under General Assumptions

Luca Insolia[1] and Domenico Perrotta[2(✉)]

[1] Sant'Anna School of Advanced Studies, 56127 Pisa, Italy
luca.insolia@santannapisa.it

[2] Joint Research Centre (JRC), European Commission, 21027 Ispra, Italy
domenico.perrotta@ec.europa.eu

Abstract. We address general-shaped clustering problems under very weak parametric assumptions with a two-step hybrid robust clustering algorithm based on trimmed k-means and hierarchical agglomeration. The algorithm has low computational complexity and effectively identifies the clusters also in the presence of data contamination. Its generalizations and an adaptive procedure to estimate the amount of contamination are also presented.

1 Introduction

Cluster analysis aggregates "similar" objects under same groups according to a similarity measure. It is widely used as an exploratory tool across different domains. The relevant literature has proposed algorithms of various degrees of sophistication. The simplest ones rely on assumptions, sometimes unexpressed albeit strong, which restrict considerably their applicability to complex data. On the other hand, also the more flexible algorithms are often not practicable, because of their computational and technical complexities. Moreover, real-world data are often contaminated and detecting outliers becomes essential to avoid distorted outcomes.

We aim at reaching flexibility without sacrificing algorithmic simplicity, computational efficiency and practicality of use in real case studies. Specifically, we combine and extend existing methodologies in a framework that relies on very weak and general assumptions. Our proposal identifies general-shaped clusters and tolerates data contamination in a computationally efficient manner. It builds upon existing two-step hybrid clustering where a preliminary model-based algorithm is followed by a hierarchical agglomeration phase (Melnykov and Michael 2019; Peterson et al. 2018). It thus inherits their proven properties but, unlike existing hybrid methods, it can detect and discard arbitrary forms of contamination. In the first step, we exploit the robustness and computational efficiency of trimmed k-means (Cuesta-Albertos et al. 1997), which is used to identify an inflated number of clusters and detect outlying units and/or noise. The second

L. A. García-Escudero et al. (Eds.): SMPS 2022, AISC 1433, pp. 216–223, 2023.
https://doi.org/10.1007/978-3-031-15509-3_29

step performs hierarchical clustering based on the robust centroids computed in the first step. We also generalise the approach to tackle more complex data structures, as well as to estimate the amount of contamination following Torti et al. (2021).

2 Methodology

In model-based clustering, there are two main strategies in dealing with data contamination: mixture models (Fraley and Raftery 1998) and trimming approaches (Cuesta-Albertos et al. 1997). While the former is used to tolerate and fit contaminants, the latter aims at excluding them from the model, which is the focus of this work. In this setting, existing methods are very effective but also computationally intensive and require mild assumptions on the shape and/or volume of the clusters (García-Escudero et al. 2008). The *trimmed k-means* algorithm (tk-means, by Cuesta-Albertos et al. 1997) relies on "hard-trimming" to discard contaminants in k-means. Its objective function minimizes the trimmed sum, over all clusters, of the within-cluster sums of point-to-cluster-centroid distances, where an additional parameter α controls the trimming proportion (García-Escudero and Gordaliza 1999; García-Escudero et al. 2003). More generally, *TCLUST* is considered the state-of-the-art robust algorithm to identify heterogeneous and non-spherical clusters (García-Escudero et al. 2008; 2010). Its objective function assumes a mixture of elliptical distributions: $Q = \min_{C_j,\pi_j,\mu_j,\Sigma_j} \prod_{j=1}^{K} \prod_{i\in C_j:|i\in C_j|\leq n(1-\alpha)} \pi_j \phi(\boldsymbol{x}_i; \boldsymbol{\mu}_j, \boldsymbol{\Sigma}_j)$, where $\phi(\cdot; \boldsymbol{\mu}_j, \boldsymbol{\Sigma}_j)$ are densities for multivariate normal distributions with mean vector $\boldsymbol{\mu}_j \in \mathbb{R}^p$ and covariance matrix $\boldsymbol{\Sigma}_j \in \mathbb{R}^{p\times p}$, and it provides non-crisp assignments for the "most typical" $\lfloor n(1-\alpha)\rfloor$ points based on mixing proportions π_j. Similarity between clusters' shape and variability is controlled through a restriction factor r constraining the determinant or eigenvalues of $\boldsymbol{\Sigma}_j$'s, including tk-means as a special (spherical) case. Recent developments now include more general constraints (García-Escudero et al. 2021).

Trimming methods for general-shaped components have not received much attention in the literature, as a general elegant theory providing such flexibility is not easy to derive. However, (non-robust) probabilistic and hierarchical procedures have been combined to identify general-shaped clusters in a computationally efficient manner (Melnykov 2016; Melnykov and Michael 2019; Peterson et al. 2018). The former finds an inflated number of components with k-means and the latter merges them according to a hierarchical agglomeration strategy. This exploits the advantages of the two methods: efficiency and flexibility. These hybrid approaches are very effective in the presence of well-separated clusters with complex shapes, but they are not robust since even a single atypical observation can severely affect the resulting partitions.

3 Tk-Merge

We focus on the identification of heterogeneous general-shaped clusters in the presence of data contamination. Unlike existing robust clustering methods based on trimming, our proposal does not use restrictive parametric assumptions and is computationally more efficient, which is a key requirement in modern applications involving ever increasing sample sizes. Our assumptions concern three aspects: (i) the overlap between the K clusters is not "very large"; (ii) noise and/or outliers are "distinguishable" from good data; (iii) the contamination percentage m/n is smaller than the fraction associated with the cluster with the smallest size and, when it is not known even approximately, it is estimated from the data (the *actual* outlying cases remaining unknown). We have no knowledge on the data generating process or clusters' shapes, but we rely on assumption (i) to control their degree of separability. Assumption (ii) is used to avoid that contaminants form clusters themselves or get too close to the true ones. Assumption (iii) is also quite standard in robust statistics: assuming that the amount of contamination is known simplifies the exposition, but in practice its level can be unknown and should be estimated.

Algorithm 1 tk/TC-merge(X, K, k, α, r, d, L)

Input: data X, number of clusters K, inflated number of clusters $k > K$, trimming rate α, restriction factor r, distance $d(\cdot,\cdot)$, linkage $L(\cdot,\cdot)$
Output: a partition $\mathscr{C}_1, \ldots, \mathscr{C}_K$ of $\approx n(1-\alpha)$ points into K clusters

if $r = 1$ **then**
  apply tk-means with α% trimming to detect $C_1^*, \ldots, C_k^*$ groups;
else
  apply TCLUST with restriction factor r and with α% trimming to detect $C_1^*, \ldots, C_k^*$ groups;
end if
initialize a $(k \times k)$ dissimilarity matrix D_k
if $d(\cdot,\cdot) = L_2$ **then**
  compute D_k using Euclidean distance between the k centroids of $C_1^*, \ldots, C_k^*$;
else
  use any other metric to compute D_k (e.g., directly estimated misclassification probabilities; Melnykov and Michael (2019));
end if
initialize $k^* = k$
while $k^* \geq K$ **do**
  do agglomerative hierarchical clustering using D_K and linkage $L(\cdot,\cdot)$;
end while

Algorithm 1 details our proposal. Key ingredients are:

1. Use TCLUST, or tk-means as its special case, to identify an inflated number of components $k > K$ with a trimming level $\alpha \approx m/n$.
2. Compute a $(k \times k)$ dissimilarity matrix D_k between the k components identified in Step 1 based on a distance $d(\cdot,\cdot)$.
3. Do hierarchical aggregation based on D_k from Step 2 using a linkage $L(\cdot,\cdot)$.
4. Cut the hierarchical tree from Step 3 to identify K groups.

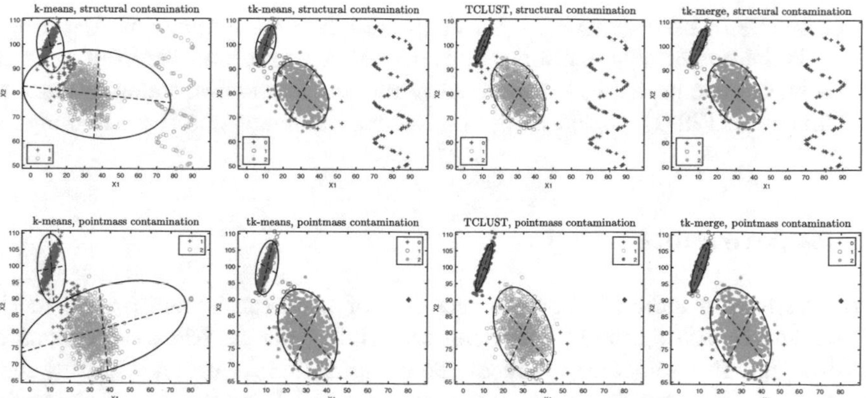

Fig. 1. Clustering partitions for k-means, tk-means, TCLUST, and tk-merge (from left to right) for data with structural (top) and point-mass (bottom) contamination. Trimmed points labeled as '0' are in red

Figure 1 shows $K = 2$ elliptical heterogeneous clusters affected by different forms of adversarial contamination, namely: structural (top panels) and point-mass (bottom panels). We consider a setting with $n_j = 800$ points belonging to C_j (for $j = 1, 2$) and $m = 100$ points are in fact contaminated. Here k-means provides very poor solutions as it breaks down. On the other hand, robust methods perform very well in detecting the true clusters, as well as contaminated points. TCLUST provides optimal partitions since the underlying assumptions are perfectly met, tk-means is negatively affected by the presence of non-spherical components, and our proposal – denoted as *tk-merge* when it relies on tk-means – performs between these two. All robust methods use a trimming proportion $\alpha \approx 6\%$. Despite our proposal suffers particularly in the presence of point-mass contamination, where assumption (ii) can be violated, it still performs comparably to state-of-the-art algorithms in this setting that does not comprise general-shaped clusters.

We now specialize our proposal to a few typical settings. In Step 1 we tend to prefer tk-means to achieve computational efficiency, but in specific scenarios we rely on the greater flexibility offered by TCLUST – denoted as *TC-merge* – since it can identify elliptical components of different size (see Example 3 below). In presence of general-shaped components, we empirically observed that $k = 2\log(n)$ and $k = \log(n)$ are often reasonable choices for tk-means and TC-merge, respectively; the latter is reduced due to higher flexibility of TCLUST. However, specific applications might require a more careful choice. The distance metric and linkage function for Steps 3 and 4 can be chosen based on clusters characteristics (e.g., shapes and overlaps). We focus on Euclidean distances (between clusters' centroids) and single linkage to exploit their efficiency. Our implementation also offers estimated misclassification probabilities (Melnykov and Michael 2019) to build an initial dissimilarity matrix, which provides more stable results in settings with higher degrees of overlap. If the trimming proportion α is unknown, diagnostic methods are extremely useful for Step 1, as this choice affects the next steps: we currently rely on the semi-automated approach by Torti et al. (2021), which monitors how results change as the trimming rate α varies.

4 Experiments

Our first results indicate that the accuracy of tk-merge is comparable to that of state-of-the-art methods when the underlying parametric assumptions are satisfied, but it significantly reduces the computational burden. Figure 2 shows the median with confidence bands for the *adjusted Rand index* and computing time over 100 replications. Here we set $K = 3$, $p = 2$, $n_1, \ldots, n_K = n/K$, and we draw heterogeneous, non-spherical Gaussian components with overlap values $\omega = 0.005$, for n increasing from 1000 to 45000 with 10 equispaced values. Contamination is injected through additional $m/n \approx 0.2$ uniform noisy points. Notably, the percentage gain in computing time for tk-merge and tk-means with respect to TCLUST varies between 50–70% and 70–85%, respectively, but tk-merge has a larger more stable ARI.

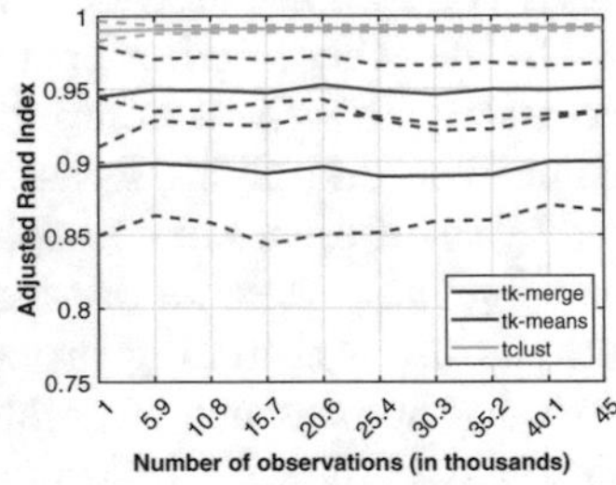

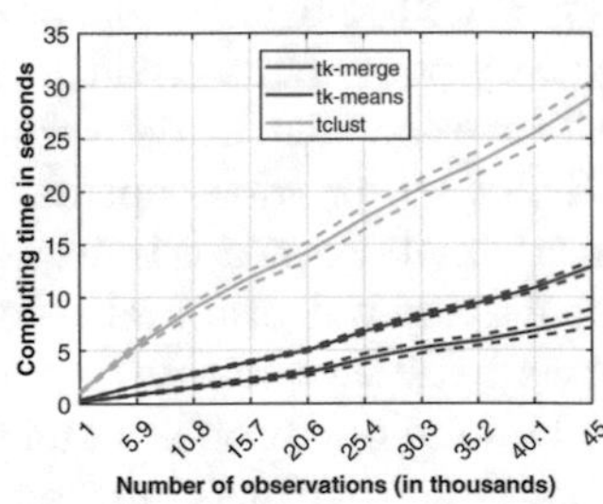

Fig. 2. Gaussian setting. Median ARI (left) and computing time in seconds (right) for tk-merge, tk-means and TCLUST as a function of the sample size across 100 replications. Dashed lines represent median $\pm S_n$ confidence bands

Figure 3 shows the clustering partitions for TCLUST (top panels) and tk-merge (bottom panels) for three synthetic datasets with general-shaped clusters; results for tk-means were worse and they are not reported. Each j-th cluster contains an equal number of points, and we set: $n_j = 1000$ with $K = 2$ (left panels), $n_j = 3000$ with $K = 3$ (central panels), $n_j = 5000$ with $K = 4$ (right panels). We further included a 5% uniform contamination $m \approx 0.05 \sum_j^K n_j$, and each method trims $\alpha \approx m/(n+m) - m/\{10(n+m)\}$; the last term is used since noise overlaps with true clusters in this setting. For tk-merge we used $k \approx 2K \log n$, and $r = 1000$ for TCLUST. In each example our proposal effectively detects the true clusters and discards the uniform noise therein. On the other hand,

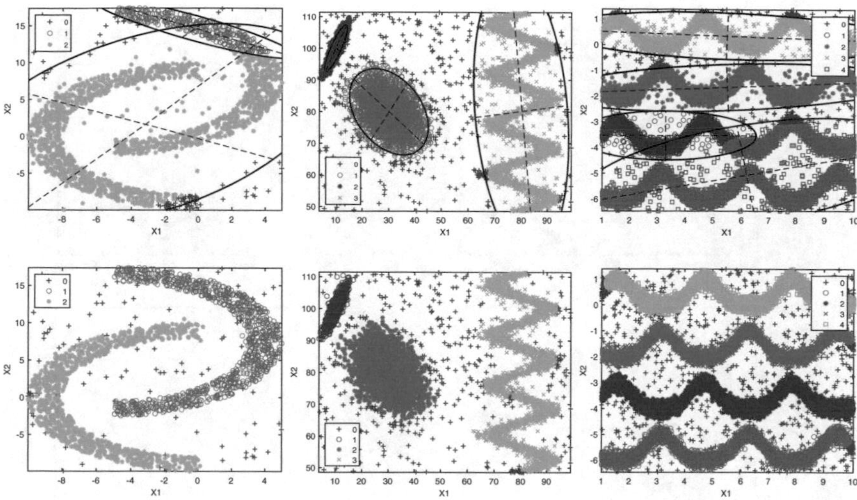

Fig. 3. Non-Gaussian setting. TCLUST (top panels) and tk-merge (bottom panels) solutions for three simulated datasets with general-shaped clusters

TCLUST performs poorly, since its assumptions are violated (i.e., the presence of non-elliptical components).

As a real-world application, we consider the modeling strategy described in García-Escudero et al. (2017) to analyze binarized retinographic images. Diabetic retinopathy is one of the major causes of blindness and vision defects in developed countries, and its early detection can play a significant role is mitigating these risks. Major blood vessels arch above and below the posterior pole of the retina, which can be approximated by parabola shapes, and it is important to detect lesions that appear in this area of the retina. The dataset $\boldsymbol{X}$ results in a matrix of dimension 1210×2. The top panels of Fig. 4 compare tk-merge (left) and TC-merge (right) solutions. We set $k \approx 2 \log n$ with $\alpha \approx 15\%$ for the former, and $k \approx \log n$ with $\alpha \approx 25\%$ for the latter. The choice of these trimming levels was dictated by our "semi-automated" procedure. Namely, for a decreasing range of trimming proportions (from 0.35 to 0), we computed the ARI and chose the solution with maximum ARI; see bottom panels of Fig. 4 for tk-means (left) and TCLUST (right). In this challenging setting, due to the presence of a "very thin" single cluster, TC-merge outperforms tk-merge due to its flexibility in accommodating a larger range of data at the expenses of a higher computing time, which is however quite negligible for the considered sample size. Nevertheless, tk-merge provides a fairly accurate and comparable partition.

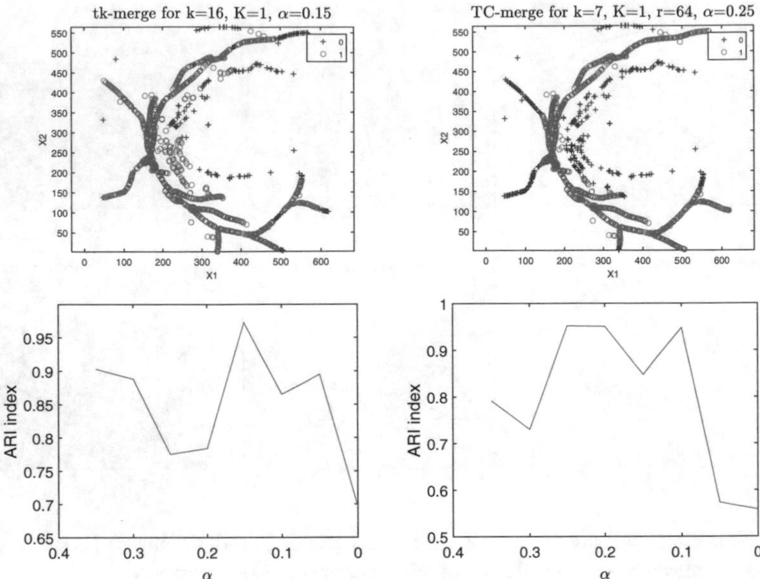

Fig. 4. Example 3. ARI across different trimming levels (top) and merged components with maximum ARI (bottom, where red points denote the estimated noise/outlying cases) for tk-means (left) and TCLUST (right)

5 Next Steps

There are theoretical and practical developments that need to be addressed to understand the limitations induced by our assumptions (stronger overlap, contamination structure, agglomeration phase, etc.), the sensitivity of our algorithm to these choices and the parameters tuning effect. For this, we are now studying extensions of the "semi-automatic" procedure used in this work, in comparison with classification trimmed likelihood curves (García-Escudero et al. 2011) and the recent approaches of Cappozzo et al. (2021).

References

Cappozzo, A., García Escudero, L.A., Greselin, F., Mayo-Iscar, A.: Parameter choice, stability and validity for robust cluster weighted modeling. Stats **4**(3), 602–615 (2021)

Cuesta-Albertos, J.A., Gordaliza, A., Matrán, C.: Trimmed k-means: an attempt to robustify quantizers. Ann. Stat. **25**(2), 553–576 (1997)

Fraley, C., Raftery, A.E.: How many clusters? Which clustering method? Answers via model-based cluster analysis. Comput. J. **41**(8), 578–588 (1998)

García-Escudero, L.A., Gordaliza, A.: Robustness properties of k means and trimmed k means. J. Am. Stat. Assoc. **94**(447), 956–969 (1999)

García-Escudero, L.A., Gordaliza, A., Matrán, C.: Trimming tools in exploratory data analysis. J. Comput. Graph. Stat. **12**(2), 434–449 (2003)

García-Escudero, L.A., Gordaliza, A., Matrán, C., Mayo-Iscar, A.: A general trimming approach to robust cluster analysis. Ann. Stat. **36**(3), 1324–1345 (2008)

García-Escudero, L.A., Gordaliza, A., Matrán, C., Mayo-Iscar, A.: A review of robust clustering methods. Adv. Data Anal. Classif. **4**(2–3), 89–109 (2010)

García-Escudero, L.A., Gordaliza, A., Matrán, C., Mayo-Iscar, A.: Exploring the number of groups in robust model-based clustering. Stat. Comput. **21**(4), 585–599 (2011)

García-Escudero, L.A., Mayo-Iscar, A., Riani, M.: Constrained parsimonious model-based clustering. Stat. Comput. **32**(1), 1–15 (2021). https://doi.org/10.1007/s11222-021-10061-3

García-Escudero, L.A., Mayo-Iscar, A., Sánchez-Gutiérrez, C.I.: Fitting parabolas in noisy images. Comput. Stat. Data Anal. **112**, 80–87 (2017)

Melnykov, V.: Merging mixture components for clustering through pairwise overlap. J. Comput. Graph. Stat. **25**(1), 66–90 (2016)

Melnykov, V., Michael, S.: Clustering large datasets by merging k-means solutions. J. Classif. **37**(1), 1–27 (2019)

Peterson, A.D., Ghosh, A.P., Maitra, R.: Merging k-means with hierarchical clustering for identifying general-shaped groups. Stat **7**(1), e172 (2018)

Torti, F., Riani, M., Morelli, G.: Semiautomatic robust regression clustering of international trade data. Stat. Meth. App. **30**(3), 863–894 (2021)

A Markov Kernel Approach to Multivariate Archimedean Copulas

Thimo M. Kasper and Wolfgang Trutschnig(✉)

Department for Artificial Intelligence and Human Interfaces, University of Salzburg, Hellbrunnerstraße 34, 5020 Salzburg, Austria
{thimo.kasper,wolfgang.trutschnig}@plus.ac.at

Abstract. This contribution deals with describing the mass distribution of multivariate Archimedean copulas via Markov kernels, i.e. regular conditional distributions. After establishing an explicit expression for the Markov kernel of an Archimedean copula we use it to provide alternative derivations of the formulas for the Kendall distribution function as well as for the mass of the level sets. The described approach is purely based on Markov kernels and does not build upon ℓ_1-norm symmetric distributions which seem to be the current standard.

1 Introduction

Archimedean copulas are a well-known family of copulas whose popularity is mainly due to their simple algebraic form. Indeed, Archimedean copulas are generated by a single univariate function $\varphi : [0,1] \to [0,\infty]$ which is strictly decreasing, convex and sufficiently smooth, and it is well-known that up to a multiplicative constant there is a one-to-one correspondence between Archimedean copulas and the family of generators. As consequence, analytic as well as dependence properties of Archimedean copulas can be fully characterized in terms of properties of the generator φ. Furthermore, formulas for important quantities can be derived explicitly and represented solely through the generator. Formulas for the Kendall distribution function and the way Archimedean copulas distribute mass on their level sets in terms of φ are well-known, see Nelsen (2006) for the two-dimensional and McNeil and Nešlehová (2009) for the multivariate scenario. To the best of the authors knowledge working with so-called ℓ_1-norm symmetric distributions seem to be the standard approach to deriving these formulas in the multivariate setting (see McNeil and Nešlehová (2009)). In the present paper we show how to derive the known formulas easily by focusing solely on the Markov kernels of Archimedean copulas

The rest of this note is organized as follows: Sect. 2 contains notation and preliminaries that are used throughout the text. Section 3 starts with proving an explicit expression for the Markov kernels of multivariate Archimedean copulas and then derives the known formulas for the mass distribution and the Kendall distribution via Markov kernels.

L. A. García-Escudero et al. (Eds.): SMPS 2022, AISC 1433, pp. 224–230, 2023.
https://doi.org/10.1007/978-3-031-15509-3_30

2 Notation and Preliminaries

In the sequel we will let $\mathscr{C}^d$ denote the family of all d-variate copulas for some fixed integer $d \geq 2$. For each copula $C \in \mathscr{C}^d$ the corresponding d-stochastic measure will be denoted by μ_C, i.e. $\mu_C([\mathbf{0}, \mathbf{x}]) = C(\mathbf{x})$ for all $\mathbf{x} \in \mathbb{I}^d$, where we set $[\mathbf{0}, \mathbf{x}] := [0, x_1] \times [0, x_2] \times \ldots \times [0, x_d]$ and $\mathbb{I} := [0, 1]$. Considering $1 \leq i < j \leq d$, the i-j-marginal of C is written as C^{ij}, i.e., $C^{ij}(x_i, x_j) = C(1, \ldots, 1, x_i, 1, \ldots, 1, x_j, 1, \ldots, 1)$. Similarly, for $k < d$ the marginal copula of the first k coordinates is denoted by $C^{1:k}$ and is given by $C^{1:k} = C(x_1, x_2, \ldots, x_k, 1, \ldots, 1)$. For more background on copulas and d-stochastic probability measures we refer to Nelsen (2006) and Durante and Sempi (2016).

For every metric space (S, d) the Borel σ-field on S will be denoted by $\mathscr{B}(S)$. The multivariate Lebesgue measure on $\mathscr{B}(\mathbb{I}^d)$ is simply written as λ as dimensionality will be clear from the context. The key objects of the present paper are so-called Markov kernels. For $k < d$ a *k-Markov kernel* from $\mathbb{R}^k$ to $\mathbb{R}^{d-k}$ is a mapping $K : \mathbb{R}^k \times \mathscr{B}(\mathbb{R}^{d-k}) \to \mathbb{I}$ such that for every fixed $E \in \mathscr{B}(\mathbb{R}^{d-k})$ the mapping $\mathbf{x} \mapsto K(\mathbf{x}, E)$ is $\mathscr{B}(\mathbb{R}^k)$-$\mathscr{B}(\mathbb{R}^{d-k})$-measurable and for every fixed $\mathbf{x} \in \mathbb{R}^k$ the mapping $E \mapsto K(\mathbf{x}, E)$ is a probability measure on $\mathscr{B}(\mathbb{R}^{d-k})$.

Given a $(d-k)$-dimensional random vector $\mathbf{Y}$ and a k-dimensional random vector $\mathbf{X}$ on a probability space $(\Omega, \mathscr{A}, \mathbb{P})$ we say that a Markov kernel K is a regular conditional distribution of $\mathbf{Y}$ given $\mathbf{X}$ if $K(\mathbf{X}(\omega), E) = \mathbb{E}(\mathbf{1}_E \circ \mathbf{Y} | \mathbf{X})(\omega)$ holds $\mathbb{P}$-almost surely for every $E \in \mathscr{B}(\mathbb{R}^{d-k})$. It is well-known that for each pair of random vectors $(\mathbf{X}, \mathbf{Y})$ as above, a regular conditional distribution $K(\cdot, \cdot)$ of $\mathbf{Y}$ given $\mathbf{X}$ always exists and is unique for $\mathbb{P}^{\mathbf{X}}$-a.e. $\mathbf{x} \in \mathbb{R}^k$ whereby $\mathbb{P}^{\mathbf{X}}$ denotes the push-forward of $\mathbb{P}$ via $\mathbf{X}$. In case $(\mathbf{X}, \mathbf{Y})$ has distribution function $C \in \mathscr{C}^d$ we let $K_C : \mathbb{I}^k \times \mathscr{B}(\mathbb{I}^{d-k}) \to \mathbb{I}$ denote (a version of) the regular conditional distribution of $\mathbf{Y}$ given $\mathbf{X}$ and refer to it as *k-Markov kernel* (or simply Markov kernel) of C. Defining the $\mathbf{x}$-section of a set $G \in \mathscr{B}(\mathbb{I}^d)$ w.r.t. the first k coordinates by $G_{\mathbf{x}} := \{\mathbf{y} \in \mathbb{I}^{d-k} : (\mathbf{x}, \mathbf{y}) \in G\} \in \mathscr{B}(\mathbb{I}^{d-k})$ the so-called *disintegration theorem* implies

$$\mu_C(G) = \int_{\mathbb{I}^k} K_C(\mathbf{x}, G_{\mathbf{x}}) \, \mathrm{d}\mu_{C^k}(\mathbf{x}).$$

For more background on conditional expectation and general disintegration we refer to Kallenberg (2002) and Rüschendorf (2008).

An Archimedean generator is a continuous, strictly decreasing function $\varphi : [0, 1] \to [0, \infty]$ that is right-continuous at 0 and satisfies $\varphi(1) = 0$. We call a copula $C \in \mathscr{C}^d$ *Archimedean* (and write $C \in \mathscr{C}^d_{\mathrm{ar}}$) if there exists some generator φ with

$$C(\mathbf{x}) = \varphi^{\leftarrow}(\varphi(x_1) + \cdots + \varphi(x_d)), \quad \mathbf{x} \in \mathbb{I}^d,$$

where $\varphi^{\leftarrow} : [0, \infty] \to [0, 1]$ denotes the pseudoinverse of φ defined by $\varphi^{\leftarrow}(x) := \varphi^{-1}(x)$ for $x \in [0, \varphi(0))$ and $\varphi^{\leftarrow}(x) := 0$ for $x \geq \varphi(0)$. On the other hand, for $\mathbf{x} \in \mathbb{I}^d$, the function $\varphi^{\leftarrow}(\varphi(x_1) + \cdots + \varphi(x_d))$ is a d-variate copula if and only if $\varphi^{\leftarrow}$ is

d-monotone on $[0, \infty)$ (see McNeil and Nešlehová 2009), i.e., $\varphi^{\leftarrow}$ is continuous in $[0, \infty)$, differentiable up to order $(d-2)$ on $(0, \infty)$, satisfies $(-1)^m(\varphi^{\leftarrow})^{(m)}(x) \geq 0$ for $x \in (0, \infty)$ and $m = 0, 1, \ldots, d-2$, and $(-1)^{d-2}(\varphi^{\leftarrow})^{(d-2)}$ is non-increasing and convex on $(0, \infty)$, whereby as usual we let $g^{(m)}$ denote the m-th derivative of a function g. Additionally, we write $g_{\pm}^{(m)}$ to indicate right (left) derivatives. For $d = 2$ the above preliminaries reduce to the well-known definitions in the 2-dimensional setting (see, e.g., Genest and Rivest 1993; Nelsen 2006; Charpentier and Segers 2008). Notice further that fixing $\varphi(0.5) = 1$ yields a unique generator (see Kasper et al. 2021).

Furthermore, again following McNeil and Nešlehová (2009), if $C \in \mathscr{C}^d_{\text{ar}}$ is absolutely continuous then

$$c(\mathbf{x}) = \prod_{i=1}^{d} \varphi'(x_i) \cdot (\varphi^{\leftarrow})_{-}^{(d)}(\varphi(x_1) + \cdots + \varphi(x_d)) \tag{1}$$

is the density of C. In the sequel we will also use the nice property that lower dimensional marginals of Archimedean copulas are absolutely continuous (see Proposition 4.1 in McNeil and Nešlehová 2009).

3 Results

We derive a formula for (a version of) the $(d-1)$-Markov kernel of an Archimedean copula and then utilize it to give alternative proofs of the well-known formulas for the mass of the level sets and the Kendall distribution function of multivariate Archimedean copulas. To this end, let $C \in \mathscr{C}^d_{\text{ar}}$, $k < d$ and $t \in (0, 1]$. Then the set

$$L_t := \{(\mathbf{x}, \mathbf{y}) \in \mathbb{I}^k \times \mathbb{I}^{d-k} : \sum_{i=1}^{k} \varphi(x_i) + \sum_{m=1}^{d-k} \varphi(y_m) = \varphi(t)\}$$

is called the t-level set of C whereas the zero set is given by $L_0 = \{(\mathbf{x}, \mathbf{y}) \in \mathbb{I}^k \times \mathbb{I}^{d-k} : \sum_{i=1}^{k} \varphi(x_i) + \sum_{m=1}^{d-k} \varphi(y_m) \geq \varphi(0)\}$. We will make use of the zero set of the $(d-1)$-margin of C defined by $L_0^{d-1} := \{\mathbf{x} \in \mathbb{I} : \sum_{i=1}^{d-1} \varphi(x_i) \geq \varphi(0)\}$. Additionally, $\text{int}(S)$ denotes the interior of a set S and $M(\mathbf{x}) := \min(\mathbf{x})$.

Theorem 1. *Let $d \geq 2$ and $C \in \mathscr{C}^d_{ar}$ with generator φ. Setting*

$$K_C(\mathbf{x}, [0, y]) := \begin{cases} 1, & M(\mathbf{x}) \in \{0, 1\}, y \in [0, 1] \\ 1, & M(\mathbf{x}) \in (0, 1), \mathbf{x} \in \text{int}L_0^{d-1}, y \in [0, 1] \\ 0, & M(\mathbf{x}) \in (0, 1), \mathbf{x} \notin \text{int}L_0^{d-1}, (\mathbf{x}, y) \in \text{int}L_0^C \\ \dfrac{(\varphi^{\leftarrow})_{-}^{(d-1)}\left(\sum\limits_{i=1}^{d-1} \varphi(x_i) + \varphi(y)\right)}{(\varphi^{\leftarrow})_{-}^{(d-1)}\left(\sum\limits_{i=1}^{d-1} \varphi(x_i)\right)}, & M(\mathbf{x}) \in (0, 1), \mathbf{x} \notin \text{int}L_0^{d-1}, (\mathbf{x}, y) \notin \text{int}L_0^C \end{cases}$$

yields (a version of) the $(d-1)$-Markov kernel of C.

Notice that for $d = 2$ the fraction in the last case of the above equation simplifies to the same expression already established in Fernández Sánchez and Trutschnig (2015). In fact, for $(x, y) \in (0,1)^2 \setminus L_0$ we have

$$K_C(x,[0,y]) = \frac{(\varphi^{\leftarrow})^{(2-1)}_{-}(\varphi(x)+\varphi(y))}{(\varphi^{\leftarrow})^{(2-1)}_{-}(\varphi(x))} = \frac{\frac{1}{\varphi^{(1)}_{+}(\varphi^{\leftarrow}(\varphi(x)+\varphi(y)))}}{\frac{1}{\varphi^{(1)}_{+}(\varphi^{\leftarrow}(\varphi(x)))}} = \frac{\varphi^{(1)}_{+}(x)}{\varphi^{(1)}_{+}(C(x,y))}$$

where we used the identity $(\varphi^{\leftarrow})^{(1)}_{-}(z_0) = 1/\varphi^{(1)}_{+}(\varphi^{\leftarrow}(z_0))$ which holds for every $z_0 \in (0, \infty)$.

Proof. We first show that K_C defined as above really is a $(d-1)$-Markov kernel. As $\varphi^{\leftarrow}$ is d-monotone, $(-1)^{(d-2)}(\varphi^{\leftarrow})^{(d-2)}$ is decreasing and convex. Hence, the left-derivative of $(\varphi^{\leftarrow})^{(d-2)}$ exists on $(0, \infty)$ and measurability of the kernel as a function of $\mathbf{x}$ for given y follows (see, e.g., Fernández Sánchez and Trutschnig (2015) for the underlying Dynkin argument). Since $(-1)^{d-2}(\varphi^{\leftarrow})^{(d-2)}$ is convex and decreasing, $(-1)^{d-2}(\varphi^{\leftarrow})^{(d-1)}_{-}$ is increasing and monotonicity of $y \mapsto K_C(\cdot, [0, y])$ follows. Right-continuity is a direct consequence of left-continuity of $(\varphi^{\leftarrow}_{-})^{(d-1)}$. Altogether it follows that K_C is a $(d-1)$-Markov kernel and we can focus on showing that it is the Markov kernel of C.
As mentioned in Sect. 2 all lower dimensional marginals of C are absolutely continuous and the density of the $(d-1)$-marginal is given by

$$c^{1:d-1}(\mathbf{x}) = \prod_{i=1}^{d-1} \varphi'(x_i) \cdot (\varphi^{\leftarrow})^{(d-1)}_{-}\Big(\sum_{i=1}^{d-1} \varphi(x_i)\Big). \tag{2}$$

Notice first that for any $y \in \mathbb{I}$ we have

$$\int_{\operatorname{int} L_0^{d-1}} K_C(\mathbf{s}, [0, y]) \, \mathrm{d}\mu_{C^{1:d-1}}(\mathbf{s}) = 0.$$

Hence, it remains to show the following identity for the last case of the definition:

$$C(\mathbf{x}, y) = \int_{[\mathbf{0},\mathbf{x}]} K_C(\mathbf{s}, [0, y]) \, \mathrm{d}\mu_{C^{1:d-1}}(\mathbf{s}) = \int_{[\mathbf{0},\mathbf{x}]} K_C(\mathbf{s}, [0, y]) \cdot c^{1:d-1}(\mathbf{s}) \, \mathrm{d}\lambda(\mathbf{s}) =: (I)$$

Canceling out and using Fubini's theorem yields

$$\begin{aligned}(I) &= \int_{[\mathbf{0},\mathbf{x}]} \prod_{j=1}^{d-1} \varphi'(s_j)(\varphi^{\leftarrow})^{(d-1)}_{-}\Big(\sum_{i=1}^{d-1}\varphi(s_i) + \varphi(y)\Big) \, \mathrm{d}\lambda(\mathbf{s}) \\ &= \int_{[\mathbf{0},\mathbf{x}]} \prod_{j=1}^{d-2} \varphi'(s_j) \int_{[0,x_{d-1}]} \varphi'(s_{d-1})(\varphi^{\leftarrow})^{(d-1)}_{-}\Big(\sum_{i=1}^{d-1}\varphi(s_i) + \varphi(y)\Big) \, \mathrm{d}\lambda(s_{d-1}) \, \mathrm{d}\lambda(\mathbf{s}_{d-2})\end{aligned}$$

where $\mathbf{s}_l := (s_1, s_2, \ldots, s_l)$ for $l < d-1$. Applying substitution with the bounds $u_0 := \varphi(s_1) + \cdots + \varphi(0) + \varphi(y)$ and $u_{d-1} := \varphi(s_1) + \cdots + \varphi(x_{d-1}) + \varphi(y)$ we get

$$(I) = \int_{[\mathbf{0},\mathbf{x}_{d-2}]} \prod_{j=1}^{d-2} \varphi'(s_j) \cdot (\varphi^{\leftarrow})^{(d-2)}(u_{d-1}) \, \mathrm{d}\lambda(\mathbf{s}_{d-2}).$$

Notice that $\varphi^{\leftarrow}$ is 0 in an open interval around u_0, so every derivative of $\varphi^{\leftarrow}$ vanishes in u_0. Therefore, proceeding analogously for each of the remaining $d-2$ variables finally leads to

$$(I) = \int_{[0,x_1]} \varphi'(s_1) \cdot (\varphi^{\leftarrow})^{((d-1)-(d-2))} \big(\varphi(s_1) + \varphi(x_2) \cdots + \varphi(x_{d-1}) + \varphi(y)\big) \, \mathrm{d}\lambda(s_1)$$

which obviously coincides with $C(\mathbf{x}, y)$. □

As first application of Theorem 1 we prove a formula describing how d-variate Archimedean copulas distribute mass on their level sets. This result was already established in McNeil and Nešlehová (2009) via ℓ_1-norm symmetric distributions, as we will show subsequently Markov kernels allow for an alternative quick proof. We will make use of the t-level hypersurfaces of Archimedean copulas defined as follows: Let $t \in [0,1]$ and $\mathbf{x} \in \mathbb{I}^{d-1}$ such that $x_1 \geq t$, $x_2 \geq \varphi^{-1}(\varphi(t) - \varphi(x_1)) =: f^t(x_1)$ and iteratively, $x_{d-1} \geq \varphi^{-1}(\varphi(t) - \sum_{i=1}^{d-2} \varphi(x_i)) =: f^t(\mathbf{x}_{d-2})$. Then

$$f^t(\mathbf{x}) := \varphi^{-1}\big(\varphi(t) - \sum_{i=1}^{d-1} \varphi(x_i)\big)$$

is called the t-level $(d-1)$-hypersurface of C.

Theorem 2. *Let $C \in \mathscr{C}_{ar}^d$ have generator φ and let μ_C denote the corresponding d-stochastic measure. Then for every $t > 0$ we have*

$$\mu_C(L_t) = \frac{(-\varphi(t))^{d-1}}{(d-1)!} \cdot \big((\varphi^{\leftarrow})_-^{(d-1)}(\varphi(t)) - (\varphi^{\leftarrow})_-^{(d-1)}(\varphi(t-))\big). \tag{3}$$

If C is strict then $\mu_C(L_0) = 0$ and for non-strict C,

$$\mu_C(L_0) = \tfrac{(-\varphi(0))^{d-1}}{(d-1)!} \cdot (\varphi^{\leftarrow})_-^{(d-1)}(\varphi(0)). \tag{4}$$

Proof. We start with $t > 0$. Using the definition of the t-level hypersurfaces, disintegration and the fact that $C^{1:d-1}$ is absolutely continuous with density $c^{1:d-1}$ we get

$$\begin{aligned}
\mu_C(L_t) &= \int_{[t,1]\times[f^t(s_1),1]\times\ldots\times[f^t(\mathbf{s}_{d-2}),1]} K_C(\mathbf{s}, \{f^t(\mathbf{s})\}) \, \mathrm{d}\mu_{C^{1:d-1}}(\mathbf{s}) \\
&= \int_{[t,1]} \cdots \int_{[f^t(\mathbf{s}_{d-2}),1]} \prod_{i=1}^{d-1} \varphi'(s_i) \cdot \Big[(\varphi^{\leftarrow})_-^{(d-1)}(\varphi(t)) - (\varphi^{\leftarrow})_-^{(d-1)}(\varphi(t-))\Big] \, \mathrm{d}\lambda(\mathbf{s}) \\
&= \Big[(\varphi^{\leftarrow})_-^{(d-1)}(\varphi(t)) - (\varphi^{\leftarrow})_-^{(d-1)}(\varphi(t-))\Big] \cdot \int_{[t,1]} \cdots \int_{[f^t(\mathbf{s}_{d-2}),1]} \prod_{i=1}^{d-1} \varphi'(s_i) \, \mathrm{d}\lambda(\mathbf{s}).
\end{aligned}$$

Letting (II) denote the iterated integrals in the previous line we have

$$(II) = \int\limits_{[t,1]} \int\limits_{[f^t(s_1),1]} \cdots \prod_{i=1}^{d-2} \varphi'(s_i) \int\limits_{[f^t(\mathbf{s}_{d-2}),1]} \varphi'(s_{d-1})(-1)^0 \big[\varphi(t) - \sum_{i=1}^{d-1-0} \varphi(s_i)\big]^0 \, \mathrm{d}\lambda(\mathbf{s})$$

and the chain rule directly yields

$$(II) = \int_{[t,1]} \cdots \int_{[f^t(\mathbf{s}_{d-4}),1]} \prod_{i=1}^{d-3} \varphi'(s_i)$$

$$\int_{[f^t(\mathbf{s}_{d-3}),1]} \varphi'(s_{d-2}) \cdot \frac{(-1)^1}{1} \cdot \left[\varphi(t) - \sum_{i=1}^{d-1-1} \varphi(s_i)\right]^1 \mathrm{d}\lambda(s_{d-2}) \mathrm{d}\lambda(\mathbf{s}_{d-3}).$$

Proceeding analogously for s_{d-2} gives

$$(II) = \int_{[t,1]} \cdots \int_{[f^t(\mathbf{s}_{d-4}),1]} \prod_{i=1}^{d-3} \varphi'(s_i) \cdot \frac{(-1)^2}{1 \cdot 2} \cdot \left[\varphi(t) - \sum_{i=1}^{d-1-2} \varphi(s_i)\right]^2 \mathrm{d}\lambda(\mathbf{s}_{d-3})$$

and after finitely many steps we obtain

$$(II) = \int_{[t,1]} \varphi'(s_1) \cdot \frac{(-1)^{d-2}}{1 \cdot 2 \cdots (d-2)} \left[\varphi(t) - \varphi(s_1)\right]^{d-2} \mathrm{d}\lambda(s_1) = \frac{(-1)^{d-1}}{(d-1)!} \cdot \varphi(t)^{d-1}$$

as desired. For $t = 0$ and strict C we obviously have $\mu_C(L_0) = 0$. For non-strict C we have $K_C(\mathbf{s}, \{f^0(\mathbf{s})\}) = K_C(\mathbf{s}, [0, f^0(\mathbf{s})])$ and calculations as those above yield the result. □

As final step we prove the formula for the Kendall distribution function again only utilizing the Markov kernel. Doing so, instead of the level set L_t, we now consider the set $E_t := \{(\mathbf{x}, y) \in \mathbb{I}^{d-1} \times \mathbb{I} : C(\mathbf{x}, y) \leq t\}$ which is relevant for the definition of the Kendall distribution function given by

$$F_K^d(t) := \mathbb{P}(C(\mathbf{X}, Y) \leq t).$$

Theorem 3. *Consider $C \in \mathscr{C}_{ar}^d$ with generator φ and corresponding d-stochastic measure μ_C. Then for $t > 0$*

$$F_K^d(t) = (\varphi^{\leftarrow})_-^{(d-1)}(\varphi(t)) \frac{(-1)^{d-1}}{(d-1)!} \varphi(t)^{d-1} + \sum_{k=0}^{d-2} (\varphi^{\leftarrow})^{(k)}(\varphi(t)) \frac{(-1)^k}{k!} \varphi(t)^k. \quad (5)$$

For $t = 0$ and strict C we have $F_K^d(0) = 0$ and for non-strict C,

$$F_K^d(0) = (\varphi^{\leftarrow})_-^{(d-1)}(\varphi(0)) \cdot \frac{(-1)^{d-1}}{(d-1)!} \cdot \varphi(0)^{d-1}. \quad (6)$$

Proof. Applying disintegration and decomposing the integral yields

$$F_K^d(t) = \mu_C(E_t) = \int_{\mathbb{I}^{d-1}} K_C(\mathbf{x}, (E_t)_\mathbf{x}) \, \mathrm{d}\mu_{C^{1:d-1}}(\mathbf{x})$$

$$= \int_{\{\mathbf{x} \in \mathbb{I}^{d-1}: \sum_{i=1}^{d-1} \varphi(x_i) \leq \varphi(t)\}} K_C(\mathbf{x}, (E_t)_\mathbf{x}) \, \mathrm{d}\mu_{C^{1:d-1}}(\mathbf{x})$$

$$+ \int_{\{\mathbf{x} \in \mathbb{I}^{d-1}: \sum_{i=1}^{d-1} \varphi(x_i) > \varphi(t)\}} K_C(\mathbf{x}, (E_t)_\mathbf{x}) \, \mathrm{d}\mu_{C^{1:d-1}}(\mathbf{x})$$

Denote by (III) and (IV) the first and the second of the above summands, respectively. Analogously to the procedure in Theorem 2 we obtain

$$(III) = (\varphi^{\leftarrow})^{(d-1)}(\varphi(t)) \cdot \frac{(-1)^{d-1}}{(d-1)!}\varphi(t)^{d-1}.$$

For (IV), on the other hand, the condition $\sum_{i=1}^{d-1} \varphi(x_i) > \varphi(t)$ is equivalent to both, $(E_t)_{\mathbf{x}} = [0,1]$ and $C^{1:d-1}(\mathbf{x}) \leq t$. Therefore,

$$(IV) = \int\limits_{\{\mathbf{x}:\ C^{1:d-1}(\mathbf{x}) \leq t\}} 1 \,\mathrm{d}\mu_{C^{1:d-1}}(\mathbf{x}) = \int\limits_{\mathbb{I}^{d-1}} \mathbf{1}_{[0,t]}(C^{1:d-1}(\mathbf{x}))\,\mathrm{d}\mu_{C^{1:d-1}}(\mathbf{x}) = F_K^{d-1}(t)$$

and proceeding iteratively finally yields

$$F_K^d(t) = (\varphi^{\leftarrow})_-^{(d-1)}(\varphi(t)) \cdot \frac{(-1)^{d-1}}{(d-1)!} \cdot \varphi(t)^{d-1} + \sum_{k=1}^{d-2} (\varphi^{\leftarrow})^{(k)}(\varphi(t)) \frac{(-1)^k}{k!}\varphi(t)^k + t.$$

For $t = 0$ we have $F_K^d(t) = \int_{\{\mathbf{x} \in \mathbb{I}^{d-1}:\ \sum_{i=1}^{d-1} \varphi(x_i) \leq \varphi(0)\}} K_C(\mathbf{x}, (E_0)_{\mathbf{x}})\,\mathrm{d}\mu_{C^{1:d-1}}(\mathbf{x})$ and the result follows similarly. □

Acknowledgements. The first author gratefully acknowledges the financial support from Porsche Holding Austria and Land Salzburg within the WISS 2025 project 'KFZ' (P1900123). The second author gratefully acknowledges the support of the WISS 2025 project 'IDA-lab Salzburg' (20204-WISS/225/197-2019 and 20102-F1901166-KZP).

References

Charpentier, A., Segers, J.: Convergence of Archimedean copulas. Stat. Probab. Lett. **78**, 412–419 (2008)

Durante, F., Sempi, C.: Principles of Copula Theory. Taylor & Francis Group LLC, Boca Raton (2016)

Fernández-Sánchez, J., Trutschnig, W.: Singularity aspects of Archimedean copulas. J. Math. Anal. Appl. **432**, 103–113 (2015)

Genest, C., Rivest, L.: Statistical inference procedures for bivariate Archimedean copulas. J. Am. Stat. Assoc. **88**, 1034–1043 (1993)

Kallenberg, O.: Foundations of Modern Probability. Springer, New York (2002). https://doi.org/10.1007/978-3-030-61871-1

Kasper, T., Fuchs, S., Trutschnig, W.: On weak conditional convergence of bivariate Archimedean and extreme value copulas, and consequences to nonparametric estimation. Bernoulli **4**(27), 2217–2240 (2021)

Rüschendorf, L.: Wahrscheinlichkeitstheorie. Masterclass Series. Springer, Heidelberg (2008). https://doi.org/10.1007/978-3-662-48937-6

McNeil, A., Nešlehová, J.: Multivariate Archimedean copulas d-monotone functions and ℓ_1-norm symmetric distributions. Ann. Stat. **37**, 3059–3097 (2009)

Nelsen, R.: An Introduction to Copulas. Springer, Heidelberg (2006). https://doi.org/10.1007/0-387-28678-0

Using Fuzzy Cluster Analysis to Find Interesting Clusters

Frank Klawonn[1(✉)] and Georg Hoffmann[2]

[1] Helmholtz Centre for Infection Research, Biostatistics, Inhoffenstraße 7, 38124 Braunschweig, Germany
frank.klawonn@helmholtz-hzi.de

[2] Medizinischer Fachverlag Trillium GmbH, Jesenwanger Street 42b, 82284 Grafrath, Germany
georg.hoffmann@trillium.de

Abstract. Fuzzy clustering algorithms are not only popular within the field of fuzzy sets and systems, but are also often used in other areas. It is clear that membership degrees in fuzzy cluster analysis provide more information than crisp assignments of data objects to clusters. But fuzzy clustering has additional technical advantages compared to classical clustering approaches. Fuzzy clustering algorithms are less prone to converge to undesired results that correspond to local optima of the underlying objective function that is used for clustering. Furthermore, fuzzy clustering enables the definition of cluster validity measures that cannot be used in the context of crisp clustering. These purely technical advantages can be exploited to define fuzzy clustering algorithms that search for single clusters step-by-step. Finally, it is demonstrated how the approach of finding single clusters can be modified to estimate reference intervals or "normal ranges" in laboratory medicine.

1 Introduction: What Is Cluster Analysis?

In order to find an answer to the question what cluster analysis is, we take a look at definitions found in textbooks or at Wikipedia:

- "Cluster analysis or clustering is the task of grouping a set of objects in such a way that objects in the same group (called a cluster) are more similar (in some sense) to each other than to those in other groups (clusters)." Wikipedia (15/05/2022)
- "... This will lead us to various attempts to reformulate the problem as one of partitioning the data into subgroups or clusters." (Duda et al. 2021)
- "The goal of cluster analysis is to ascertain, on the basis of $x_1, \ldots, x_n$, whether the observations fall into relatively distinct groups." (James et al. 2001)
- "... cluster analysis comprises a large class of methods aiming at discovering a limited number of meaningful groups (or clusters) in

L. A. García-Escudero et al. (Eds.): SMPS 2022, AISC 1433, pp. 231–239, 2023.
https://doi.org/10.1007/978-3-031-15509-3_31

data. ... This defines a partition of n units in k clusters." (Giordani et al. 2020)

- "Cluster analysis ... looks for groups of similar data objects that can naturally be separated from other, dissimilar data objects." (Berthold et al. 2020)

A common ground of these definitions is that a single cluster is composed of similar data objects that are dissimilar to the other data. Except for the last definition, cluster analysis is supposed to partition the data into clusters. The concept of partition might not be meant in a strict set-theoretic sense but can also include probabilistic or fuzzy assignments of data objects to the clusters. Although in many applications of cluster analysis, the goal is to partition the data into clusters in order to identify a latent class variable, cluster analysis is often used as a flexible exploratory data analysis technique that also allows

- to find single interesting clusters that might only cover a smaller portion of the data or
- to identify outliers in terms of data objects that cannot be assigned to clusters even though the clusters might not satisfy the classical criterion of homogeneity inside each cluster and heterogeneity between clusters.

This paper reviews technical advantages of fuzzy clustering and focuses on methods for finding single clusters with an application to estimate reference intervals or "normal ranges" in laboratory medicine.

2 Fuzzy Cluster Analysis

The so-called fuzzy c-means algorithm (FCM) introduced by Dunn (1973) and Bezdek (1981) as a generalisation of k-means clustering is probably the most popular fuzzy clustering algorithm. It is based on the objective function

$$f = \sum_{i=1}^{c}\sum_{j=1}^{n} u_{ij}^{m} d_{ij} \tag{1}$$

that should be minimised under the constraints

$$\sum_{i=1}^{c} u_{ij} = 1 \qquad \text{for all } j = 1, \ldots, n \tag{2}$$

where $u_{ij} \in [0, 1]$ is the membership degree of a data vector x_j to cluster j. In the simplest case, the distance d_{ij} is the (squared) Euclidean distance of data vector x_j to the cluster centre v_i. For a predefined number c of clusters, the parameters u_{ij} and the cluster centres v_i need to be optimized, usually carried out by an alternating optimisation schemes that alternatingly updates the prototypes v_i and the membership degree u_{ij}. Without the so-called fuzzifier $m > 1$, the minimum of the objective function would always lead to crisp membership degrees $u_{ij} \in \{0, 1\}$.

The fundamental fuzzy c-means algorithm lead to a large variety of new algorithms. There are essentially to parts of the objective function (1) that can be modified: The way the membership degrees enter the objective function and the distance function d_{ij} allowing for more flexible and complex cluster shapes. It should be noted that a modification of the distance function has essentially nothing to do with fuzzy clustering because it could be applied in the same way in the crisp context with $u_{ij} \in \{0, 1\}$. Nevertheless, a large variety of algorithms have been introduced as fuzzy clustering algorithms although they only modified the distance function. Examples are the Gustafson-Kessel (GK) algorithm (Gustafson and Kessel 1979), the fuzzy c-varieties (FCV) algorithm by Bock (1979) and Bezdek (1981) for clusters in the form of linear manifolds or shell clustering algorithms as introduced by Krishnapuram et al. (1995) for image analysis to find lines, shells of circles or ellipses or more generally quadratic forms. Noise clustering introduced by Davé (1991) was designed to reduce the influence of outliers and noise by an additional noise cluster that does not have a prototype or cluster centre but only a fixed large distance to all data vectors, so that data vectors far away from all clusters are assigned to the noise cluster.

One could ask the question why all these algorithms where introduced in the context of fuzzy clustering and not in the context of ordinary crisp clustering because for all these algorithm one can directly implement a crisp version with a simpler formula for the (crisp) membership degree.

3 Technical Advantages of Fuzzy Clustering

A possible answer to this question might be that the fuzzy versions of these algorithm are less prone to local minima of the objective function (1) which is basically minimised by a heuristic greedy strategy with the danger of getting stuck in local minima leading to suboptimal or bad clustering results, especially when complex cluster shapes are considered. Indeed, it was shown empirically by Klawonn et al. (2015) and formally by Jayaram and Klawonn (2012) that fuzzy clustering can avoid local minima. This was, however, only demonstrated by specific examples. A general proof is still missing. Our conjecture is that the fuzzy clustering objective function can have fewer local optima because there is only one global optimum of the objective function when the fuzzifier m goes to infinity. This is however not the desired optimum because in this case all clusters collapse together. But on the way from $m = 1$ with many local optima to infinity, the objective function probably loses undesired local minima.

A reason why the fuzzy clustering objective function has fewer local optima is that the membership degrees can get close to zero but never zero except for some rare situations, so that all clusters take at least notice of all data vectors and can attract them if they "fit" to the corresponding cluster. This can also be a disadvantage when clusters are very imbalanced or for high-dimensional data where fuzzy clustering tends to collapse even for small values of the fuzzifier, according to Winkler et al. (2011a). A possible solution to this problem is to replace the fuzzifier function $g(u_{ij}) = u_{ij}^m$ by other suitable functions, e.g. by a convex combination of the crisp and the fuzzy objective function with m, the so-called polynomial fuzzifier, see Winkler et al. (2011b).

Another technical advantage of fuzzy clustering is that it allows for better adapted cluster validity measures taking the membership degrees into account. Cluster validity measures correspond to goodness of fit measures that can also help to determine the number of clusters. Examples for such validity measures are the partition coefficient (Bezdek 1981) (PC), the partition entropy (Bezdek 1981) (PE) or the Xie-Beni index (Xie and Beni 1991).

4 Identifying Clusters Step by Step

Although the above mentioned validity measures can be quite useful, they are often difficult to interpret when the assumption of more or less well-separated clusters in the data set is not satisfied. Subtractive clustering is one possible approach if there is only one or perhaps or few clusters in the data set, but a significant proportion of the data does not belong to any cluster. Subtractive clustering tries to identify a single "good" cluster in the data set, removes the cluster from the data set and continues with the remaining data until no more clusters can be found. Yager and Filev (1994) introduced with their mountain method an example for a subtractive fuzzy clustering algorithm.

The dynamic data assignment assessment (DDAA) algorithm by Georgieva and Klawonn (2006) also leads to a subtractive clustering approach. But what is more important is the additional visual information that the algorithm provides. The basic idea of DDAA is to use only one cluster plus a noise cluster, starting with a very large noise distance, so that in the beginning all data vectors are assigned to the actual cluster and the noise cluster is empty. Then the noise distance is decreased step by step until it finally reaches zero and all data vectors have moved from the actual cluster to the noise cluster. Observing the dynamics of what is happening when the noise distance is decreasing and the noise cluster absorbs more and more data can provide valuable information. For instance, one can look at the proportion of data that are assigned to the actual cluster.

Figure 1 shows a simple example data set and various measures depending on the noise distance when the data set is cluster with one cluster plus a noise cluster. The black solid line that indicates the proportion of data vectors assigned to the actual data cluster and not to the noise cluster. One should read the graph from left to right, i.e. starting with a large noise distance that decreases to zero from right to left. In the beginning at the very right, all data vectors are assigned to the actual cluster. When the noise distance is decreased – going to the left on the solid black line – the actual cluster loses data to the noise cluster. There are plateaus and steep slopes. A steep slope means that at least one data cluster is lost to the noise cluster. There are only two steep slopes on the solid black line although there are three data clusters because at one of the slopes two clusters are lost to the noise cluster simultaneously. The light blue dotted line corresponding to the partition entropy shows two local maxima that also indicate noise distance ranges where at least one complete cluster is lost to the

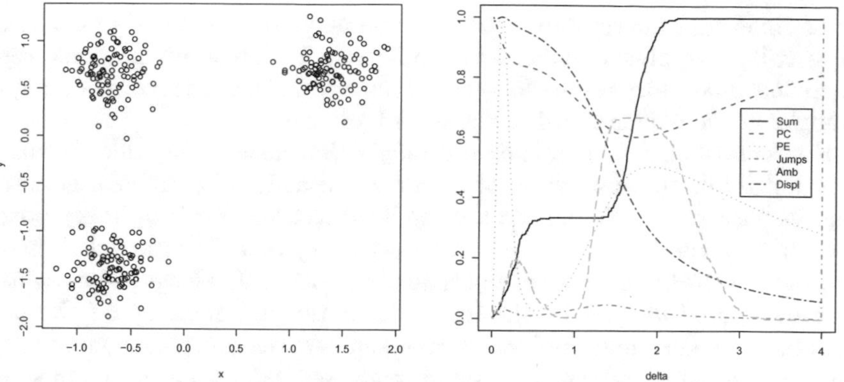

Fig. 1. A data set (left) and various measures depending on the noise distance (right)

noise cluster. For a detailed explanation of the other curves in Fig. 1 and how clusters can be identified based on such graphs see Klawonn (2016).

5 Application to Indirect Methods to Compute Reference Intervals in Laboratory Medicine

Laboratory medicine measures substances – very often in blood or urine – that can be considered as biomarkers that can serve as indicators of certain classes of diseases or pathological states. If a measured value is outside the "normal range" – the reference interval – this can be an indicator for a disease. Reference intervals are defined as the central 95% of values of a presumably healthy population (see Horowitz et al. 2010). Reference intervals often depend on sex and age, also on the measurement technique and device as well as on sample preparation. Therefore, laboratories are required to validate their reference intervals regularly. Taking into account the large number of analytes and the dependencies on sex and age, thousands of reference intervals need to be validated in each laboratory. Indirect methods provide a means to carry out this validation by using already measured values from all persons under the assumption that the majority of values lie in the "normal" range. In the view of cluster analysis, the data consist of a majority cluster of non-pathological values and one or more smaller clusters of pathological values. Concordet et al. (2009) follow this line and developed an algorithm based on a mixture of Box-Cox transformed normal distributions with the disadvantage that its computational costs are quite high.

The DDAA approach is very universally suited for single- and multidimensional data sets. A useful application in laboratory medicine could be isolating the proportion of non-pathological laboratory results as the main cluster from a mixed population of many healthy individuals and few sick individuals. We found that after successful removal of the "pathological noise", the main cluster can be used to determine reference intervals (so-called "normal ranges"), which are an important interpretation aid for laboratory results. In this final section, we describe a simplified mathematical approach for the one-dimensional case in which only laboratory results for a single analyte, such as the red blood pigment hemoglobin, were determined in the mixed population.

Klawonn et al. (2020) proposed an algorithm inspired by Tukey's boxplot. After a suitable transformation of the data – usually a log-transformation – it removes outliers that lie outside the modified whiskers of an ordinary boxplot. The whiskers are defined in such a way that they cover the range between the 2.5%- and the 97.5%-quantile of a normal distribution. The length of the whiskers is not computed directly by the interquartile range, but based on the ranges between the median and the first quartile and between the median and the third quartile. The larger of the two ranges might be under a stronger influence of pathological values on the corresponding side. Therefore, we use the shorter of the two ranges to obtain robust estimates for the 2.5%- and the 97.5%-quantile of a normal distribution. Let m, $q_{0.25}$ and $q_{0.75}$ denote the median, the first and the third quartile of the data, respectively. With

$$d = \min\{m - q_{0.25}, q_{0.75} - m\} \tag{3}$$

the estimates for the 2.5%- and the 97.5%-quantile are

$$\hat{q}_{0.025} = m - c_0 \cdot d, \tag{4}$$

$$\hat{q}_{0.975} = m + c_0 \cdot d \tag{5}$$

where c_0 is the adjusted whisker factor. In order to estimate the corresponding 2.5%- and the 97.5%-quantile, c_0 must be defined as

$$c_0 = \frac{\Phi^{-1}(0.025)}{\Phi^{-1}(0.25)} = 2.905847 \tag{6}$$

where Φ is the cumulative distribution function of the standard normal distribution and Φ^{-1} its inverse. After the whiskers – as initial estimates of the 2.5%- and the 97.5%-quantile of the healthy individuals – are calculated according to Eqs. (4) and (5), the data is truncated at these values. Then the same method is applied to the truncated data set and the procedure is iterated until no more data more are cut off. If we simply repeat this procedure, this would lead to a biased estimator if the data consists only of normally distributed values from healthy individuals because after the first truncation, the median and the other two quartiles required in Eqs. (4) and (5) are computed from a truncated distribution. So in case of a normal distribution, for estimating the 2.5%- and the 97.5%-quantile the factor c_0 must be adjusted taking into account that the estimates for the three quartiles come from a truncation to the central 95%. The factor c_0 must then be replaced by

$$c_{\text{trunc}} = \frac{\Phi^{-1}(0.025)}{\Phi^{-1}(0.25 \cdot 0.95 + 0.025)} = 3.083367. \tag{7}$$

We call this algorithm *iBoxplot95* where the letter *i* refers to "iterated". The algorithm is schematically described in Algorithm 1. This algorithm is extremely fast but can lead to slight biases when pathological values are concentrated on one side of the reference interal.

Algorithm 1 The iBoxplot95 algorithm.

```
procedure IBOXPLOT95(x)                          ▷ Input x: measured values
    c_0 ← 2.905847                               ▷ Eq. (6)
    c_trunc ← 3.083367                           ▷ Eq. (7)
    m ← median(x)                                ▷ Median of x
    q_1 ← q25(x)                                 ▷ 25% quantile of x
    q_3 ← q75(x)                                 ▷ 75% quantile of x
    d ← min{m − q_1, q_3 − m}
    a ← m − c_0 * d                              ▷ Initial lower truncation limit
    b ← m + c_0 * d                              ▷ Initial upper truncation limit
    y ← truncate(x, a, b)                        ▷ Remove all values <a or >b from x.
    y.old ← x
    while y ≠ y.old do                           ▷ Continue until no more values are truncated.
        y.old ← y
        m ← median(y)                            ▷ Median of y
        q_1 ← q25(y)                             ▷ 25% quantile of y
        q_3 ← q75(y)                             ▷ 75% quantile of y
        d ← min{m − q_1, q_3 − m}
        a ← m − c_trunc * d                      ▷ Lower truncation limit
        b ← m + c_trunc * d                      ▷ Upper truncation limit
        y ← truncate(y, a, b)                    ▷ Remove all values <a or >b from y.
    end while
    return a, b                                  ▷ The estimates for the 2.5%- and the 97.5%-quantile
end procedure
```

A compromise between the computationally expensive mixture model-based approach and the very fast iBoxplot95 algorithm with a possible bias for very asymmetric data can be based on the ideas described in Sect. 4. To make the algorithm more robust, we compute the centre of the actual cluster based on the median instead of the centre of gravity. To run the algorithm automatically, we do not use the graph as in Fig. 1. Instead we compute in each step of the decrease of the noise distance in addition to the cluster centre the standard deviation of the actual cluster. We then estimate the reference interval as the 2.5%- and the 97.5%-quantile of the normal distribution with the median of the cluster centres as expectation and the median of the standard deviations as standard deviation.

Experimental results show that this algorithm reduces the possible slight bias of iBoxplot95 for skewed data for the price of higher computational costs, but much lower computational costs than the mixture model-based approach.

References

Berthold, M., Borgelt, C., Höppner, F., Klawonn, F., Silipo, R.: Guide to Intelligent Data Science: How to Intelligently Make Use of Real Data, 2nd edn. Springer, Cham (2020). https://doi.org/10.1007/978-3-030-45574-3

Bezdek, J.: Pattern Recognition with Fuzzy Objective Function Algorithms. Plenum Press, New York (1981)

Bock, H.: Clusteranalyse mit unscharfen Partitionen. In: Bock, H. (ed.) Klassifikation und Erkenntnis. Numerische Klassifikation, vol. III, pp. 137–163. INDEKS, Frankfurt (1979)

Concordet, D., Geffré, A., Braun, J., Trumel, C.: A new approach for the determination of reference intervals from hospital-based data. Clinica Chimica Acta **405**(1), 43–48 (2009)

Davé, R.N.: Characterization and detection of noise in clustering. Pattern Recogn. Lett. **12**, 406–414 (1991)

Duda, R., Hart, P., Stork, D.: Pattern Classification, 2nd edn. Wiley, New York (2021)

Dunn, J.: A fuzzy relative of the isodata process and its use in detecting compact well-separated clusters. Cybern. Syst. **3**(3), 32–57 (1973)

Georgieva, O., Klawonn, F.: Cluster analysis via the dynamic data assigning assessment algorithm. Inf. Technol. Control **2**, 14–21 (2006)

Giordani, P., Ferraro, M.B., Martella, F.: An Introduction to Clustering with R. BQAHB, vol. 1. Springer, New York (2020). https://doi.org/10.1007/978-981-13-0553-5

Gustafson, D., Kessel, W.: Fuzzy clustering with a fuzzy covariance matrix. In: Proceedings of the IEEE Conference on Decision and Control and 17th Symposium on Adaptive Processes, San Diego, pp. 761–766 (1979)

Horowitz, G.L., et al.: Defining, establishing, and verifying reference intervals in the clinical laboratory. Approved guideline-third edition. Technical report, Clinical & Laboratory Standards Institute document EP28-A3c (2010). https://clsi.org/media/1421/ep28a3c_sample.pdf

James, G., Witten, D., Hastie, T., Tibshirani, R.: An Introduction to Statistical Learning: with Applications in R, 2nd edn. Springer, New York (2001). https://doi.org/10.1007/978-1-4614-7138-7

Jayaram, B., Klawonn, F.: Can fuzzy clustering avoid local minima and undesired partitions? In: Moewes, C., Nürnberger, A. (eds.) Computational Intelligence in Intelligent Data Analysis, pp. 31–44. Springer, Heidelberg (2012). https://doi.org/10.1007/978-3-642-32378-2_3

Klawonn, F.: Exploring data sets for clusters and validating single clusters. Procedia Comput. Sci. **19**, 1381–1390 (2016)

Klawonn, F., Kruse, R., Winkler, R.: Fuzzy clustering: more than just fuzzification. Fuzzy Set Syst. **281**, 272–279 (2015)

Klawonn, F., Orth, M., Hoffmann, G.: Quantitative laboratory results: normal or lognormal distribution? J. Lab. Med. **44**(3), 143–150 (2020)

Krishnapuram, R., Frigui, H., Nasraoui, O.: Possibilistic shell clustering algorithms and their application to boundary detection and surface approximation - Part 1 & 2. IEEE Trans. Fuzzy Syst. **1**, 29–60 (1995)

Winkler, R., Klawonn, F., Kruse, R.: Fuzzy c-means in high dimensional spaces. Fuzzy Syst. Appl. **1**, 1–17 (2011)

Winkler, R., Klawonn, F., Kruse, R.: Fuzzy clustering with polynomial fuzzifier in connection with M-estimators. Appl. Comput. Math. **10**, 146–163 (2011)

Xie, X., Beni, G.: A validity measure for fuzzy clustering. IEEE Trans. Pattern Anal. Mach. Intell. **13**, 841–847 (1991)

Yager, R.R., Filev, D.: Approximate clustering via the mountain method. IEEE Trans. Syst. Man Cybern. Syst. **24**, 1279–1284 (1994)

Learning Control Limits for Monitoring of Multiple Processes with Neural Network

Kamil Kmita(✉), Katarzyna Kaczmarek-Majer, and Olgierd Hryniewicz

Systems Research Institute, Polish Academy of Sciences,
Newelska 6, 01-447 Warsaw, Poland
{kmita,kaczmar,hryniewi}@ibspan.waw.pl

Abstract. In this work, inspired by the interpretability and usefulness of the statistical process control, we propose a novel procedure for simultaneous monitoring of multiple processes that is based on a neural network with learnable activation functions. The proposed procedure for learning control limits with neural network (CONNF) is aimed at scenarios where labeled data are available and makes use of these labels. CONNF can be particularly useful in monitoring processes when the amount of run-in data is insufficient, or the cost of obtaining such data is high. We illustrate the performance of CONNF method with a simulation study and preliminary results for real-life data collected from smartphones of patients with diagnosed bipolar disorder. These results show the potential of CONNF and indicate further research directions.

1 Introduction

Statistical process control (SPC) is a well-established field applied in various domains, and its primary goal is to monitor single or multiple processes to generate an alarm when a change is observed. One of the first SPC tools, a control chart, was introduced by W. Shewhart. It is assumed that a process is in control (stable) in its initial run-in stage. In this stage, mean value and standard deviation are recorded and used to design a control chart, known as the Shewhart control chart, which consists of control lines: (i) central line representing a target value of the process (usually its mean value), (ii) lower and/or upper lines located at 2 or 3 standard deviations from the central line. The process is considered stable when its future observations are located within control lines (limits). In Kaczmarek-Majer et al. (2019), the authors show the usefulness of the weighted model averaging in the residual control charts for early detection of change in the mental state of bipolar disorder patients. The proposed method uses behavioural data about smartphone usage with a limited amount of diagnostic data. Another tool of SPC that was shown to be effective, especially in detecting small shifts in the mean of a process, is the cumulative sum (CUSUM) control chart, see Hryniewicz and Kaczmarek-Majer (2022).

L. A. García-Escudero et al. (Eds.): SMPS 2022, AISC 1433, pp. 240–247, 2023.
https://doi.org/10.1007/978-3-031-15509-3_32

However, a problem often arises in practice if there is no initial stable phase that can be considered as in-control and used to learn control limits. Moreover, the collection of data from the initial stable period may be infeasible in practice. For example, in managing psychiatric disorders, patients consult psychiatrists usually when the disease episode has already started rather than in preceding periods of stable mental health.

In this work, inspired by the interpretability and usefulness of SPC, we propose a novel procedure for **learning control limits for monitoring of multiple processes with neural network** (CONNF) for simultaneous monitoring of multiple processes based on neural network (NN). The proposed method is aiming at resolving the aforementioned difficulties with the collection of data in the run-in period. We provide preliminary results that illustrate the effectiveness of the proposed approach. Our goal is to complement traditional statistical process control for situations of missing or imprecise information.

CONNF algorithm proposed in this work builds on the classical feedforward neural network architecture, which has proven successful in a wide range of statistical learning problems. Neural networks are based on the concept of modelling response variable as a non-linear function of linear combinations of inputs, the so called *activation functions* (*activations* in short). Activations have gained significant interest from researchers in recent years. An extensive overview of different activation functions can be found in Tavakoli et al. (2021). In an attempt to further increase the predictive power of NN, a field emerged exploring *learnable activations*, where activations become parametrized appropriately, and the neural network is learning optimal values of these parameters instead of treating them as fixed. Parametrizations proposed in the literature range from simple modifications of existing activations to more complex approaches (Tavakoli et al. 2021; Goyal et al. 2020). However, there is a price for complexity, and simplicity can be preferred as – inter alia – it reduces the hypothesis space. In this article, we propose a modification of the classical sigmoid activation that we parametrize to learn one-sided control limits for process monitoring.

2 Learning Control Limits with CONNF Neural Network

The goal of our method is to learn functions $\mathbb{R} \ni x \mapsto \{0, 1\}$ that could be used to generate alarms in joint monitoring of multiple processes of interest. To achieve that, we propose a special modification of the classical feedforward neural network. Let $\mathbf{x}^t_{i,j} = (x^t_{i,1}, x^t_{i,2}, \ldots, x^t_{i,p})$ be a vector representing p measurements for i–th observational unit, $i = 1, \ldots, n$, at time point $t = 1, \ldots, T$. Let $y^t_i \in \{0, 1\}$ be a response variable. Following Hryniewicz and Kaczmarek-Majer (2021), we consider processes consisting of segments and subsegments, and data from subsegments belonging to respective segments are aggregated using the possibilistic aggregation. In consequence, we assume that $\mathbf{x}^t_{i,j}$ contains characteristics that are already aggregated to a meaningful time interval (e.g. a day), and labels y^t_i are assumed to be known for this interval.

Let us now investigate one-sided control limits, either lower or upper. Specifically, we would like to use threshold functions $\text{th}(x; r, u), r \in \mathbb{R}, u \in \{0, 1\}$ such that $\text{th}(x; r, u = 1) = \mathbb{1}(x \geq r)$ and $\text{th}(x; r, u = 0) = \mathbb{1}(x \leq r)$, where r is a threshold value, and u decides on the direction of boundary. However, for the purpose of training NN, we need a differentiable function that approximates $\text{th}(x; r, u)$. A function we propose is *boosted sigmoid*:

$$s_j(x; r, u, b) = \frac{\left(e^{b\cdot(x-r)}\right)^u}{1 + e^{b\cdot(x-r)}}, \tag{1}$$

where parameter b is responsible for adjusting the shape of the nominal sigmoidal curve. Note that $s_j(x; r, u = 0) = 1 - s_j(x; r, u = 1)$.

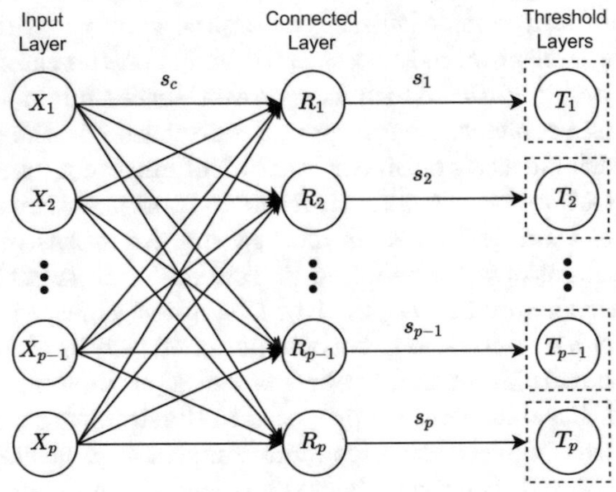

Fig. 1. Architecture of CONNF neural network

Learning r_j and u_j for each s_j can be achieved by fitting a special kind of neural network, which architecture is adjusted to the learnable activation problem. The architecture we propose is depicted in Fig. 1. First *Input Layer* consists of p features that are available before learning the network. The second layer *Connected Layer* is the dense layer that aims at finding linear combinations of p input features that may be more accurate for effective monitoring of the process. Note that all neurons R_j share a classical, non-learnable activation function s_c. In this work, s_c is simply the identity function instead of more complex activations, since our goal is to find possible interactions of features and ensure explainability. Thus,

$$R_j = \sum_{k=1}^{p} w_k \cdot X_k, j = 1, \ldots, p. \tag{2}$$

Next, an activation function s_j is applied to each neuron R_j separately, resulting in one-to-one connections between neurons R_j and T_j from *Threshold Layers*.

Note that each T_j is a separate layer in the network, leading to a total of $p+2$ layers. Threshold layers are the final outcome of the network. In our setting, each of them has its own loss function $\mathscr{L}_j$ assigned, and the algorithm is minimizing the sum of them $\mathscr{L} = \sum_{j=1}^{p} \mathscr{L}_j$. Since the response variable is assumed to be binary, we chose binary cross-entropy loss for every $\mathscr{L}_j$:

$$\mathscr{L}_j(x) = -\Big(y \cdot \log\big(s_j(x)\big) + (1-y) \cdot \log\big(1 - s_j(x)\big)\Big). \tag{3}$$

As much as insight into the parameters of each s_j may be of interest, we are finally interested in joint monitoring of multiple processes. To achieve this, we consider a sum of generated alarms for each data point $\mathbf{x}_{i,j}^t$:

$$A_i^t = \sum_{j=1}^{p} \mathbb{1}\big(s_j(x_{i,j}^t) > 0.5\big). \tag{4}$$

Ideally, we would like to obtain an alarms rate A_i^t/p close to 1 for data labeled with $y=1$, or close to 0 for data labeled with $y=0$. However, finding the optimal alarms rate threshold is beyond the scope of this article and requires further research.

3 Simulation Results

Let us consider a simulation experiment that shows the potential of CONNF. Figure 2 presents two processes observed over time, X_1 and X_2. It is assumed $y=1$ for $t \geq 30$, and $y=0$ for $t < 30$. Data were simulated in such a way, that $y=1$ only when two variables achieve values around 0 simultaneously. This setting is intended to resemble a scenario in which raising alarms correctly is possible only when the interaction of two variables is lower than a certain threshold.

Results of the proposed CONNF method are compared with two simple Shewhart control charts with 2σ control limits, as depicted in Fig. 3. We assumed $u=0$ when learning CONNF network according to the simulation setting. Also, we set $b=100$ as it modified the shape of the sigmoidal curve so that it approximates $\text{th}(x;r,u)$ closely. As a result, following equations were obtained: $R_1 = -0.22 \cdot X_1 + 0.89 \cdot X_2$ and $R_2 = 0.77 \cdot X_1 + 0.39 \cdot X_2$. The algorithm estimated control limits to be $r_1 = 0.046, r_2 = 0.162$. Figures 3c and 3d present estimated $\bar{X}_j \pm 2\sigma_j$ control limits. Alarms were generated when values were higher than upper limit, or lower than lower limit.

It is clearly visible that in this context CONNF outperformed two individual control charts by creating variable R_2 that was able to generate alarms accurately. There certainly exist more advanced SPC tools tailored for monitoring multiple processes that would be included as benchmarks in further research. Let us also note that variable R_1 created by CONNF did not perform as well as R_2, suggesting that the method could be enhanced with feature selection capabilities. Despite the limitations of this simulation experiment, it shows the potential of CONNF and presents the mechanism behind generating alarms for each monitored variable.

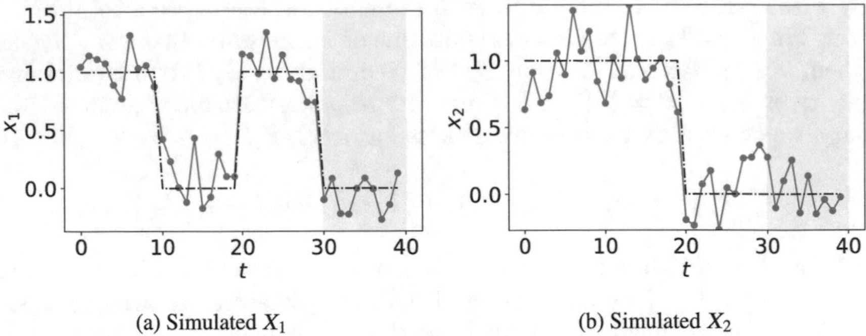

(a) Simulated X_1

(b) Simulated X_2

Fig. 2. Data for X_1 and X_2 were simulated around values of $\mu_1 = 0$ and $\mu_2 = 1$ plotted with black dashed lines. Numbers simulated from $\mathcal{N}(\mu_g, 0.1), g = 1, 2$, are plotted with blue dots connected by lines. Periods of t with $y = 1$ label are highlighted in grey

4 Experiment on Real-Life Data from Smartphone-Based Monitoring of Bipolar Disorder Patients

This work is motivated by a real-life application problem of smartphone-based monitoring of mental illnesses, where voice characteristics extracted from phone calls were observed to be non-stationary and subject to various uncertainties. We performed an experiment on such data collected in a prospective observational study[1] using $p = 10$ selected acoustic features describing patients voice following results of Kamińska et al. (2020).

The considered data streams were aggregated to daily series and summarised with center of gravity according to the possibilistic aggregation of Hryniewicz and Kaczmarek-Majer (2021). Such preprocessing resulted in $n = 28$ subjects with a number of daily summaries per patient ranging from 1 to 34. Each daily summary was labeled with $y_i^t \in \{0, 1\}$ representing either euthymic phase of disease ($y = 0$) or non-euthymic one ($y = 1$). Response y was extrapolated from the day of a psychiatric assessment onto the surrounding time period according to the so-called ground truth approach. It is important to note that CONNF is not patient-specific, thus we considered $s = 1, \ldots, N$ vectors $\mathbf{x}_s \in \mathbb{R}^p$ regardless of which subject generated given daily aggregation (and, consequently, y_s and A_s). In this experiment, $N = 365$.

Figure 4 presents a summary of the experiment's results. CONNF network was trained on all 365 data points. For each $\mathbf{x}_s$, the number of alarms generated was then predicted. Figure 4b shows that for 72% non-euthymic observations there were ≥8 alarms generated, which is a desirable outcome. For non-euthymic periods, we would like to generate a number of alarms close to the number of 10 variables monitored. Figure 4a presents a similar summary for

[1] The study was conducted in the Department of Affective Disorders, Institute of Psychiatry and Neurology in Warsaw, Poland within the project entitled "Smartphone-based diagnostics of phase changes in the course of bipolar disorder".

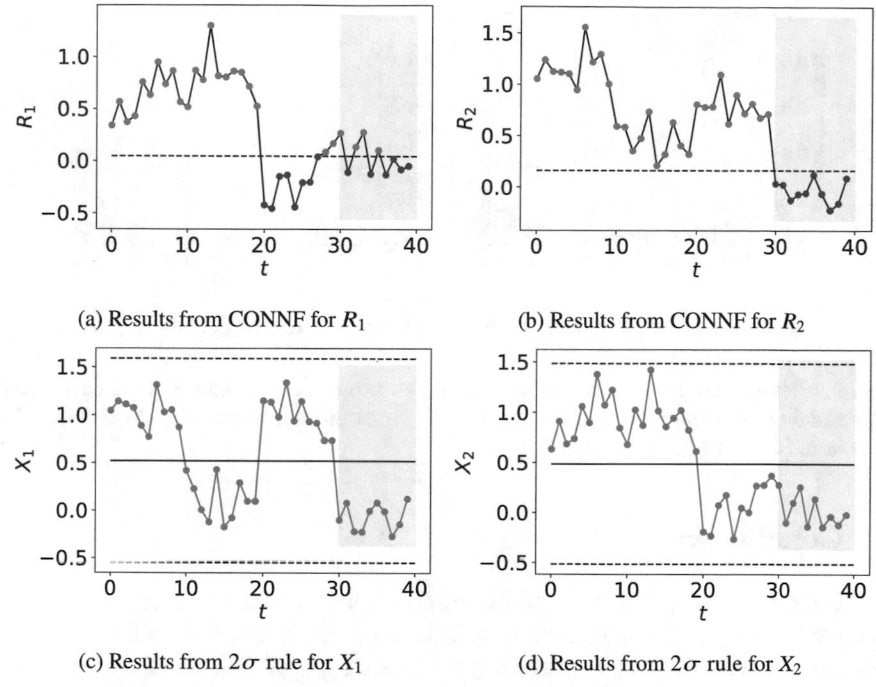

(a) Results from CONNF for R_1

(b) Results from CONNF for R_2

(c) Results from 2σ rule for X_1

(d) Results from 2σ rule for X_2

Fig. 3. Simulation results for CONNF approach and the classical 2σ control limits. In all graphs, data points were colored according to the results of the given approach such that green data points would not generate an alarm, and red data points would generate it. Figures 3a and 3b contain a dashed line representing corresponding r_1 and r_2. Figures 3c and 3d contain solid line representing $\bar{X}_j$, and dashed lines representing $\bar{X}_j \pm 2\sigma_{X_j}$

euthymic period. For 32% observations, there were ≤ 2 alarms generated. However, we also observed that for 52% observations with $y = 0$ label there were ≥ 8 alarms generated. These are false alarms that are undesirable in the context of BD monitoring. Taking these results into account, we conclude that CONNF approach must be further enhanced to achieve satisfactory in-training validation results before testing it on independent test data.

Note that in this experiment, as opposed to simulations in Sect. 3, binary parameter u was not known upfront. The classical feedforward network does not allow binary trainable parameters as it would result in the discontinuity that is not handled by standard optimization algorithms (Qin et al. 2020). We thus considered a two-step approach for learning u_j. In the first step, for each R_j we created two activations: (i) $T_j^1 = s_j(x; u = 1, r, b)$ and (ii) $T_j^0 = s_j(x; u = 0, r, b)$. CONNF was then trained and for each neuron R_j, u_j was extracted from the corresponding layer with lower loss. Parameters u_j selected this way were treated as non-trainable parameters in the second step, in which we trained the final model.

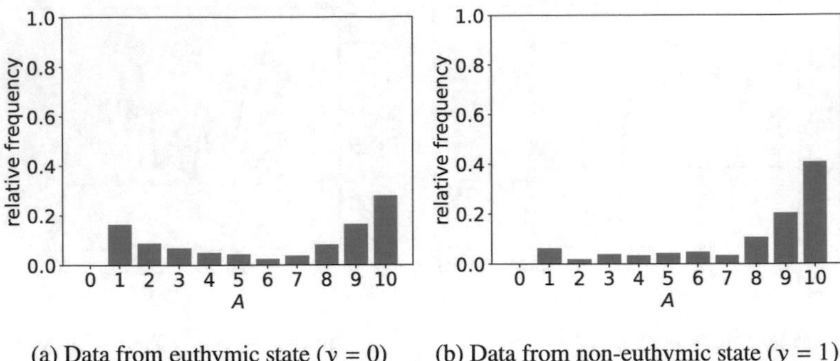

(a) Data from euthymic state ($y = 0$) (b) Data from non-euthymic state ($y = 1$)

Fig. 4. Histograms presenting validation of the proposed procedure on training data. The number of generated alarms A_s was aggregated separately for days labeled as euthymic ($y = 0$) or non-euthymic ($y = 1$)

5 Conclusions

In this work, we proposed a new data mining technique for learning control limits when monitoring complex processes. The main benefits of CONNF algorithm include inherent feature engineering and using all available information in a supervised learning paradigm. It must be noted that the presented results are of preliminary character, but they encourage further research. It is planned to: consider different activations for learning two-sided control limits, research on introducing feature selection capabilities, and test different loss functions that handle the total number of alarms generated directly.

Acknowledgements. This work is supported by the Small Grants Scheme (NOR/SGS/BIPOLAR/ 0239/2020-00) within the research project: Bipolar disorder prediction with sensor-based semi-supervised learning (BIPOLAR). The authors thank the researcher Gennaro Vessio for inspiring discussion and advice.

References

Goyal, M., Goyal, R., Venkatappa Reddy, P., Lall, B.: Activation functions. In: Pedrycz, W., Chen, S.-M. (eds.) Deep Learning: Algorithms and Applications. SCI, vol. 865, pp. 1–30. Springer, Cham (2020). https://doi.org/10.1007/978-3-030-31760-7_1

Hryniewicz, O., Kaczmarek-Majer, K.: Possibilistic aggregation of inhomogeneous streams of data. In: Proceedings of the 2021 IEEE International Conference on Fuzzy Systems (FUZZ-IEEE), pp. 1–6 (2021)

Hryniewicz, O., Kaczmarek-Majer, K.: Monitoring of possibilisticaly aggregated complex time series. In: García-Escudero, L.A., et al. (eds.) Building Bridges Between Soft and Statistical Methodologies for Data Science. Advances in Intelligent Systems and Computing. Springer, Cham (2022, forthcoming)

Kaczmarek-Majer, K., et al.: Control charts designed using model averaging approach for phase change detection in bipolar disorder. In: Destercke, S., Denoeux, T., Gil, M.Á., Grzegorzewski, P., Hryniewicz, O. (eds.) SMPS 2018. AISC, vol. 832, pp. 115–123. Springer, Cham (2019). https://doi.org/10.1007/978-3-319-97547-4_16

Kamińska, O., Kaczmarek-Majer, K., Hryniewicz, O.: Acoustic feature selection with fuzzy clustering, self organizing maps and psychiatric assessments. In: Lesot, M.J., et al. (eds.) IPMU 2020. CCIS, vol. 1237, pp. 342–355. Springer, Cham (2020). https://doi.org/10.1007/978-3-030-50146-4_26

Qin, H., Gong, R., Liu, X., Bai, X., Song, J., Sebe, N.: Binary neural networks: a survey. Pattern Recogn. **105**, 107281 (2020)

Tavakoli, M., Agostinelli, F., Baldi, P.: SPLASH: learnable activation functions for improving accuracy and adversarial robustness. Neural Netw. **140**, 1–12 (2021)

Testing the Homogeneity of Topic Distribution Between Documents of a Corpus

Louisa Kontoghiorghes(✉) and Ana Colubi

King's College London, London, UK
{louisa.1.kontoghiorghes,ana.colubi}@kcl.ac.uk

Abstract. A methodology is introduced to test the equality of topic distribution between documents of a corpus. This is achieved by using Latent Dirichlet Allocation (LDA) to estimate the topic distributions and the Kullback-Leibler divergence to measure the dissimilarity between the distributions. The testing approach combines Bayesian and frequentist statistics. Since the sampling distribution of the proposed statistics is unknown, a bootstrap test is suggested. The methodology is illustrated using scientific abstracts from the CMStatistics conference.

1 Introduction

The aim is to compare statistically the thematic content distribution of two documents picked from a given collection of documents. Documents have been compared in information retrieval from different viewpoints, e.g., by quantifying their word frequency divergence, keyword weights or terms relevance. Standard methods that have been used for this task are the cosine similarity, the Euclidean distance, the Jaccard coefficient, the Pearson correlation coefficient, or shift graphs (Gallagher et al. 2021; Gunawan et al. 2018; Huang 2008).

Alternatively, topic modelling can be used to identify the content efficiently first. In topic modeling, topics are latent variables modeled according to multinomial distributions, and the above methods do not consider these models. To take into account the topic distribution and test whether two documents of a given corpus are topic-wise homogeneous, the discrete Kullback-Leibler (KL) divergence can be used. An assumption is that both documents come from the same corpus, i.e., they cover the same set of thematic topics (although possibly with different proportions). The topics can be estimated with one of the most popular topic modeling approaches, Latent Dirichlet Allocation (LDA) (Blei et al. 2003). LDA is a Bayesian approach that allocates topic weights by considering Dirichlet distributions as priors.

Bayesian mixture models combined with the KL divergence have been widely used to handle latent variables, including clustering users' profiles by their historical clicked documents (Abri et al. 2020), extracting topic linkages (Xu et al. 2019) and observing topic evolution (Andrei and Arandjelović 2016). However,

L. A. García-Escudero et al. (Eds.): SMPS 2022, AISC 1433, pp. 248–254, 2023.
https://doi.org/10.1007/978-3-031-15509-3_33

(frequentist) hypothesis testing for the homogeneity of topic distributions has not been considered until now. Due to the unknown sampling distribution of the proposed statistic, a bootstrap statistical hypothesis test is suggested. The performance of the approach is shown with some simulations. The methodology is applied to test the topic distribution homogeneity of scientific abstracts from the CMStatistics conference.

The rest of the paper is organized as follows. Section 2 reviews some preliminaries on topic modeling and introduces the KL divergence in this context. Section 3 presents the methodology used to test the homogeneity of the topic distribution. Section 4 summarizes some simulations to assess the correctness of the method. In Sect. 5, the methodology is applied to abstracts from the CMStatistics conference. Finally, the paper concludes with some remarks in Sect. 6.

2 Topic Modelling

Let $C = \{d_1, ..., d_n\}$ be a corpus, that is, a set of n documents d_i, with $i \in \{1, ..., n\}$. Each document consists of a set of unique terms t_j from the population $t_1, \ldots, t_p$, where $j \in \{1, ..., p\}$. An n-gram is a combination of two or more words that can be considered as a term. A topic z_m, with $m \in \{1, ..., k\}$, is a (latent) mixture of terms identified with one of the k themes covered in C. Each document d_i, with $i \in \{1, ..., n\}$, is assumed to be a mixture of a predefined number of topics based on its content.

2.1 Pre-processing Step

Text documents usually contain unnecessary information for the pursued target that can blur and over-complicate the analysis. A typical step before applying text mining approaches is the so-called pre-processing step. When the focus is on topics, terms found with high frequency within most documents are labeled as stop-words and eliminated, as they cannot provide information to discern between topics. Punctuation is removed, n-grams are defined, and terms are converted to lowercase and steamed to their root. A non-normalized weight factor is allocated to each term to quantify the relevance of the terms within each document based on the term-frequency-inverse-document-frequency (tf-idf) Salton and Buckley (1988) defined as:

$$\omega_{ij} = tf_{ij} \times \log\left(\frac{n}{df_j}\right), \tag{1}$$

for all $i \in \{1, ..., n\}$ and $j \in \{1, ..., p\}$, where tf_{ij} is the relative frequency of the term t_j in the document d_i with respect to the total number of terms in d_i and df_j is the number of documents where the term t_j appeared. Based on these weights, a threshold is defined such that very rare and very common terms are removed. Finally, a matrix $dtm \in \mathbb{R}^{n \times p}$ is created to display in each of its entries the frequency of a term t_j in a document d_i.

2.2 Latent Dirichlet Allocation

Latent Dirichlet Allocation (LDA) (Blei et al. 2003) is an unsupervised generative probabilistic model that classifies the document's terms into topics by making use of the $dtm \in \mathbb{R}^{n\times p}$ with $i \in \{1,\ldots,n\}$ and $j \in \{1,\ldots,p\}$. The latent topics are identified based on the likelihood of word co-occurrence. From two Dirichlet priors, it generates probability distributions for the topics within the documents $\theta = (\boldsymbol{\theta}_1,\ldots,\boldsymbol{\theta}_n) \in \mathbb{R}^{n\times k}$ for $m \in \{1,\ldots,k\}$, and for the words within the topic $\phi = (\boldsymbol{\phi}_1,\ldots,\boldsymbol{\phi}_k) \in \mathbb{R}^{k\times p}$. Namely, LDA has the following generative model:

- $\theta \sim \text{Dir}_k(\alpha)$,
- $\phi \sim \text{Dir}_p(\beta)$,
- $z_m \sim \text{Multinomial}(\theta)$,
- $t_j \sim \text{Multinomial}(\phi)$,

where Dir(.) denotes a symmetric Dirichlet distribution and α and β are hyperparameters characterizing the topic-document and the term-topic sparseness respectively.

The estimates $\hat{\theta}$ and $\hat{\phi}$ are derived by approximating the posterior distribution

$$P(z_{d_i t_j}|\theta,\phi,z_{-(d_i t_j)},\mathbf{t},\alpha,\beta) \propto P(z,\theta,\phi,\mathbf{t}|\alpha,\beta), \tag{2}$$

where, $z_{d_i t_j}$ is the topic assignment of the term t_j in document d_i and $z_{-(d_i t_j)}$ is the vector containing all topics assignments to each term from each document except term t_j in document d_j and $\mathbf{t}$ is the set of categorical terms $t_1,\ldots,t_p$.

2.3 Kullback Leibler Divergence

The Kullback-Leibler (KL) divergence is a method used to measure the similarity of two distributions on the same probability space (Van Erven and Harremos 2014). If the distributions are discrete, e.g., $P = \{p_1,\ldots,p_k\}$ and $Q = \{q_1,\ldots,q_k\}$, then the KL divergence is defined as,

$$D_{KL}(P,Q) = \sum_{m=1}^{k} p_m log\left(\frac{p_m}{q_m}\right). \tag{3}$$

It is cleat that $D_{KL}(P,Q) \geq 0$, $D_{KL}(P,Q) = 0$ if and only if $P = Q$ and $D_{KL}(P,Q) \neq D_{KL}(Q,P)$. The less similar the distributions are the greater the KL divergence score will be.

3 Topic Distribution Equality Test

The goal is to test whether two documents of the same corpus that covers k predefined topics are homogeneous with respect to their topic distribution, that is,

$$\begin{matrix} H_0 : (\theta_{i_1 1},\ldots,\theta_{i_1 k}) = (\theta_{i_2 1},\ldots,\theta_{i_2 k}) \\ H_1 : (\theta_{i_1 1},\ldots,\theta_{i_1 k}) \neq (\theta_{i_2 1},\ldots,\theta_{i_2 k}) \end{matrix}.$$

The document's probability distribution over topics and the topic distribution over terms are estimated by using LDA, i.e., $\hat{\theta}$ and $\hat{\phi}$ are approximated either by Gibbs sampling or variational Bayesian inference (Blei et al. 2003; Koks 2019).

The dissimilarity of topic distributions between documents can be measured with the KL divergence (3), i.e., given $\hat{\boldsymbol{\theta}}_{\mathbf{i_1}} = (\hat{\theta}_{i_1 1}, \ldots, \hat{\theta}_{i_1 k})$ and $\hat{\boldsymbol{\theta}}_{\mathbf{i_2}} = (\hat{\theta}_{i_2 1}, \ldots, \hat{\theta}_{i_2 k})$. The divergence between the distributions is as follows,

$$D_{KL}(\hat{\boldsymbol{\theta}}_{\mathbf{i_1}}, \hat{\boldsymbol{\theta}}_{\mathbf{i_2}}) = \sum_{m=1}^{k} \hat{\theta}_{i_1 m} log\left(\frac{\hat{\theta}_{i_1 m}}{\hat{\theta}_{i_2 m}}\right), \tag{4}$$

where $i_1 \neq i_2$ and $\hat{\boldsymbol{\theta}}_{\mathbf{i}}$ with $i \in \{1, \ldots, n\}$, belongs to the $(k-1)$-dimensional closed simplex, such that $\theta_{im} \geq 0$ and $\sum_{m=1}^{k} \theta_{im} = 1$ (Blei et al. 2003).

The sampling distribution of the proposed (standardized) statistics is unknown; hence a (frequentist) "parametric" bootstrap test is suggested instead.

In order to set the bootstrap populations under the null hypothesis of equality of distributions as usual for bootstrap tests, a synthetic population merging the two considered samples is proposed in Step 3. Then, for each iteration, two bootstrap samples will be obtained from this population taking into account the number of terms of each document (see p. 21 in Efron and Tibshirani 1994). The bootstrap resampling procedure can be summarized as follows:

Step 1 Set the LDA model in C and obtain the estimates $\hat{\theta} \sim Dir_k(\alpha)$ and $\hat{\phi} \sim Dir_p(\beta)$ matrices.

Step 2 Compute the value of the (non-standardized) KL divergence statistic on the two document of interest, d_{i_1} and d_{i_2} where $i_1 \neq i_2$, with corresponding topic distribution $\hat{\boldsymbol{\theta}}_{\mathbf{i_1}} = \{\hat{\theta}_{i_1 1}, \ldots, \hat{\theta}_{i_1 k}\}$ and $\hat{\boldsymbol{\theta}}_{\mathbf{i_2}} = \{\hat{\theta}_{i_2 1}, \ldots, \hat{\theta}_{i_2 k}\}$

$$T = \sum_{m=1}^{k} \hat{\theta}_{i_1 m} log\Big(\frac{\hat{\theta}_{i_1 m}}{\hat{\theta}_{i_2 m}}\Big).$$

Step 3 Merge d_{i_1} and d_{i_2} to obtain $d_{i_1.i_2.}$ and compute the LDA model to obtain the estimates $\hat{\theta}_{merged} \sim Dir_k(\alpha)$ and $\hat{\phi}_{merged} \sim Dir_p(\beta)$ matrices.

Step 4 From the estimated distributions $\hat{\theta}_{merged}$ and $\hat{\phi}_{merge}$, resample according to the LDA model to obtain new documents $d^*_{i_1}$ and $d^*_{i_2}$. Then, apply again the LDA estimation to obtain $\hat{\boldsymbol{\theta}}^*_{\mathbf{i_1}} = \{\hat{\theta}^*_{i_1 1}, \ldots, \hat{\theta}^*_{i_1 k}\}$ and $\hat{\boldsymbol{\theta}}^*_{\mathbf{i_2}} = \{\hat{\theta}^*_{i_2 1}, \ldots, \hat{\theta}^*_{i_2 k}\}$ and compute

$$T^* = \sum_{m=1}^{k} \hat{\theta}^*_{i_1 m} log\Big(\frac{\hat{\theta}^*_{i_1 m}}{\hat{\theta}^*_{i_2 m}}\Big).$$

Step 5 Repeat Step 4 a large number of times B to obtain an approximation of the sampling distribution by Monte Carlo: $\{T^*_1, \ldots, T^*_B\}$.

Step 6 Compute the proportion of times that $T^* > T$ and reject the null hypothesis if this proportion is less than the pre-fixed significance level α.

Given the Bayesian nature of the underlying approach and the approximations made during the estimation process, the results vary depending on the selected seeds and hyperparameters. However, the bootstrap procedure achieves good empirical results.

4 Simulations

The correctness of the proposed method is tested by Monte Carlo simulations. The type I errors were calculated based on 169 bootstraps resamples with replacement at a significance level of 5%. This process is carried out 1000 times.

Each document d_i with $i \in \{1,..,n\}$ is composed of a vocabulary of 100 unique terms t_j, where $j \in \{1,..,100\}$. Each corpus C, covers 20 topics z_m, where $m \in \{1,..,20\}$ and $\alpha = 0.1$ and $\beta = 0.01$ where used to compute each LDA iteration. The simulation results show that the rejection percentage tends to 0.05 as the number of documents increases under the null hypothesis, H_0 : $(\theta_{i_1 1}, \ldots, \theta_{i_1 k}) = (\theta_{i_2 1}, \ldots, \theta_{i_2 k})$, (see Table 1).

5 Application

The methodology is applied to a series of Books of Abstracts (BoAs) of the CMStatistics conference. The prepossessing step was performed to obtain the *dtm*. LDA was run with different hyperparameters and topic numbers to determine the neatest topic classification, which was found with hyperparameters $\alpha = 0.1$, $\beta = 0.125$ and $k = 30$ topics. The estimated topic distribution matrix $\hat{\theta}$ was obtained. Table 2 illustrates five weights, $\hat{\theta}_{doc1}$, $\hat{\theta}_{doc2}$ and $\hat{\theta}_{doc3}$ corresponding to five topic proportion of three different abstracts with titles:

Table 1. Bootstrap rejection percentage under H_0 with $p = 100$, $k = 20$, $\alpha = 1/10$, $\beta = 1/100$ and different values of documents, n

Number of documents	Rejection percentage
50	0.02
100	0.03
200	0.04
400	0.05

doc1: Bayesian two level model for partially ordered repeated responses
doc2: A unified proposal for modelling ordinal data
doc3: Large sample behaviour of high dimensional autocovariance matrices.

Topics are manually labeled for further analysis, e.g., "Bayesian statistics", "Regression", "Functional Data Analysis (FDA)". The higher the weight, the more related the abstract is to the topic.

Table 2. Dirichlet weights of 5 out of 30 topics for each document

Document	Bayesian statistics	Stochastic	FDA	Clustering	Regression
$\hat{\theta}_{doc1}$	0.578	0.007	0.007	0.007	0.079
$\hat{\theta}_{doc2}$	0.157	0.014	0.014	0.157	0.014
$\hat{\theta}_{doc3}$	0.006	0.006	0.3	0.006	0.006

The KL divergences are computed on both pairs to evaluate the difference between the two probability distributions indicating that doc1 and doc2 are more similar distributed than doc1 and doc3. Their homogeneity was tested using the bootstrap equality test, where each bootstrap hypothesis test was performed under 1000 resamples with replacement. The results denote that doc1 and doc2 cannot be considered non-homogeneous contrary to doc1 and doc3 (see the p-values in Table 3).

Table 3. CMStatistics application: KLD scores and p-values for the different tests

Null hypothesis	KLD score	p-value
$\theta_{doc1} = \theta_{doc2}$	0.989	0.744
$\theta_{doc1} = \theta_{doc3}$	2.938	0.017

A drawback is that different seeds used for LDA may lead to different topic distributions for each document. The sensitivity of the methods should be analyzed in more detail.

6 Conclusion

A new frequentist procedure is introduced to test the topic equality of two documents of the same corpus. This is achieved by implementing a Bayesian approach, LDA, to estimate each document's topic distribution and using the KL divergence to measure the homogeneity of two distributions with a bootstrap test. The frequentist consistency of variational Bayes for mixture models has been proved (Chérief-Abdellatif and Alquier 2018) which can be the base to prove the correctness of the test.

References

Abri, S., Abri, R., Çetin, S.: Group-based personalization using topical user profile. In: Adjunct Publication of the 28th ACM Conference on User Modeling, Adaptation and Personalization (UMAP 2020), pp. 181–186. Association for Computing Machinery, New York, USA (2020)

Andrei, V., Arandjelović, O.: Complex temporal topic evolution modelling using the Kullback-Leibler divergence and the Bhattacharyya distance. EURASIP J. Bioinf. Syst. Biol. **1**, 1–11 (2016)

Blei, D.M., Ng, A.Y., Jordan, M.I.: Latent Dirichlet allocation. J. Mach. Learn. Res. **3**, 993–1022 (2003)

Chérief-Abdellatif, B.E., Alquier, P.: Consistency of variational Bayes inference for estimation and model selection in mixtures. Electron. J. Stat. **12**(2), 2995–3035 (2018)

Efron, B., Tibshirani, R.J.: An introduction to the Bootstrap. CRC Press, Boca Raton (1994)

Gallagher, R.J., et al.: Generalized word shift graphs: a method for visualizing and explaining pairwise comparisons between texts. EPJ Data Sci. **10**(1), 1–29 (2021). https://doi.org/10.1140/epjds/s13688-021-00260-3

Gunawan, D., Sembiring, C.A., Budiman, M.A.: The implementation of cosine similarity to calculate text relevance between two documents. J. Phys. Conf. Ser. **978**(1), 012120 (2018)

Huang, A.: Similarity measures for text document clustering. In: Proceedings of the Sixth New Zealand Computer Science Research Student Conference, vol. 4 (NZCSRSC2008, Christchurch, New Zealand), pp. 9–56 (2008)

Koks, I.: Latent Dirichlet allocation: explained and improved upon for applications in marketing intelligence. Ph.D. thesis, Delft University of Technology, The Netherlands (2019). http://resolver.tudelft.nl/uuid:faa7cd3f-a946-4685-a36e-d01a15c4159e

Salton, G., Buckley, C.: Term-weighting approaches in automatic text retrieval. Inf. Process. Manage. **24**(5), 513–523 (1988)

Van Erven, T., Harremos, P.: Rényi divergence and Kullback-Leibler divergence. IEEE Trans. Inf. Theory **60**(7), 3797–3820 (2014)

Xu, S., Zhai, D., Wang, F., An, X., Pang, H., Sun, Y.: A novel method for topic linkages between scientific publications and patents. J. Am. Soc. Inf. Sci. **70**(9), 1026–1042 (2019)

Circular Ordering Methods for Timing and Visualization of Oscillatory Signals

Yolanda Larriba(✉), Alejandro Rodríguez-Collado, and Cristina Rueda

IMUVa, University of Valladolid, Valladolid, Spain
{yolanda.larriba,alejandro.rodriguez.collado,cristina.rueda}@uva.es

Abstract. This paper is focused on the problem of inferring a circular order to solve two different issues arising in the analysis of real data in genomics: the development of a human atlas of circadian gene expressions and a taxonomy of neuronal mouse brain cells. The solutions are derived using different approaches to ordering in a circle the sampling points, the cells or the genes.

1 Introduction

Oscillatory systems govern many biological processes, for example, the molecular clock networks that drive tissue-specific circadian gene expressions (Larriba et al. 2022) or the neuronal dynamics orchestrated by action potential (AP) curves, that measure the fluctuation of the potential of a neuron (Rodríguez-Collado and Rueda 2021). Both circadian gene expressions and APs display up-down-up patterns periodically. The analysis of signals associated to oscillatory process generally involves a large amount of data, several noise sources and very heterogeneous patterns, and hence its study is a challenge. The inherent circular nature of oscillatory signals enables suitable mathematical formulations to efficiently deal with data analysis problems. In this paper, we focus on the problem of inferring a circular order to solve two different issues arising in the analysis of real data: the development of a human atlas of circadian gene expressions and a neuronal cell-type (Cre-lines) taxonomy in the mouse brain.

In practice, the interest in the identification of a circular order arises in quite different contexts. In some cases, as is the circadian atlas application, is the order among the sampling times what needs to be estimated because it is unknown; being the time of day a critical variable in circadian medicine, with implications in the effectiveness of therapies. This happens in most human postmortem gene studies, which are often conducted to reduce the experiment cost or the risk for the health (Mavroudis and Jusko 2021). In many other cases, the ordering of individuals is the focus, and eventually these are arranged in a circle. In particular, circadian genes have a pattern of expression with a peak attained just before their function. Therefore, these phase angles, that are circular parameters, are expected to be ordered on the circle according to the biological functions of these genes. Moreover, the organization of individuals into groups or types is the aim of multiple data analysis and the visualization using trees, matrices or

L. A. García-Escudero et al. (Eds.): SMPS 2022, AISC 1433, pp. 255–262, 2023.
https://doi.org/10.1007/978-3-031-15509-3_34

circles is quite useful, this is the case of our second application, where a circular taxonomy reveals interesting relations among the Cre-lines.

The problem of finding a circular ordering in the data is less known in the literature than that of finding a linear order. Some recent references on the subject are Cai and Ma (2022) and Armstrong et al. (2021). There are algorithms in the literature specifically designed to solve temporal estimation problem on genomic, among those, the most widely used is CYCLOPS (Anafi et al. 2017). Besides, Larriba et al. (2020) proposes a solution based on CIE (circular isotonic estimators), that compared with CYCLOPS is more simple to formulate and gives biologically interpretable solutions. The main drawback of the CIE solution is that it assumes equally spaced time samples. In addition, there is a simpler approach that is known as Circular Principal Component Analysis (CPCA), which given a circular arrangement uses a simple transformation of the two Principal Components (Scholz 2007). Finally, the most simple approach to derive a circular ordering or arrangement is the one that directly uses a circular measure. In particular, the parameters associated to the Frequency Modulated Möbius (FMM) model are suitable to analyse oscillatory signals as was shown in Rueda et al. (2021a).

In this paper we present two applications that illustrate the effectiveness of different approaches to ordering the sampling points and the individuals (genes or cells), giving answer to relevant biological questions.

2 Methods

Statistical tools for the analysis of oscillatory signal have been developed by our research group in the last few years. A review is Rueda et al. (2021b). The question of finding a circular ordering is specifically dealt in Barragán et al. (2017) and Larriba et al. (2020). The Frequency Modulated Möbius (FMM) model and the CPCA are briefly explained below, as they are the basic tools in both applications.

2.1 The FMM Model

Let us assume that the time points are in $[0, 2\pi)$. In other case, transform the time points $t' \in [t_0, T + t_0]$ by $t = \frac{(t'-t_0)2\pi}{T}$. In the following, oscillations are also referred to as waves. The single FMM signal is defined as the following wave: $W(t, A, \alpha, \beta, \omega) = A\cos(\phi(t, \alpha, \beta, \omega))$, where A is the wave amplitude and, $\phi(t, \alpha, \beta, \omega) = \beta + 2\arctan(\omega\tan(\frac{t-\alpha}{2}))$ is the wave phase.

$W(t, A, \alpha, \beta, \omega)$ is suitable for describing oscillatory patterns, as is well justified in Rueda et al. (2019). The parameters ($A \in \mathfrak{R}^+; \alpha, \beta \in [0, 2\pi); \omega \in [0, 1]$) characterize various aspects of an oscillatory pattern. A is measuring the signal's amplitude, α is a phase location parameter, while ω and β are parameters describing the shape. ω measures the sharpness, and β measures skewness and indicates upward and/or downward peak direction. Specifically, a sinusoidal curve corresponds to $\omega = 1$.

Let assume that we have observations in n time points $t_1 < ... < t_n$, the FMM$_m$ model is defined as a parametric m-component signal plus error model, as follows:

$$X(t_i) = M + \sum_{J=1}^{m} W(t_i, A_J, \alpha_J, \beta_J, \omega_J) + e(t_i); \quad i = 1, ..., n, \tag{1}$$

where, M is an intercept term with values in $\Re$ and, $(e(t_1), \ldots, e(t_n))' \sim N_n(0, \sigma^2 \boldsymbol{I})$. Furthermore, it is assumed that $\alpha_1 \leq \ldots \leq \alpha_m \leq \alpha_1$ and $A_1 = \max_{1 \leq J \leq m} A_J$. These restrictions guarantee the identifiability of the parameters and a backfiting algorithm is designed to efficiently derive the MLE (Rueda et al. 2021a).

The number of components depends on the application, being equal to 1 for the gene expression data and 3 for analysing APs.

2.2 The CPCA

Let $\boldsymbol{X}$ be a data matrix with r rows. The rows may represent temporal points or individuals while the columns are variables. Specifically, temporal points (rows) and gene expression data (columns) in the first application, and Cre-lines (rows) and morphological features (columns) in the second one. Let $\boldsymbol{E}_1$ and $\boldsymbol{E}_2$ be the two first eigenvectors of $\boldsymbol{X}$. When the columns of the data matrix are the gene expression of different genes and the columns are the values of the expression in different time points, the eigenvectors are often called eigengenes. The mapping of $\boldsymbol{E}_1$ against $\boldsymbol{E}_2$ may reveal an underlying circular structure.

The CPCA works using the transformation in which the eigenvectors are projected onto the unit circle as follows:

$$(e_{1,i}, e_{2,i}) = (\frac{E_{1,i}}{\sqrt{E_{1,i}^2 + E_{2,i}^2}}, \frac{E_{2,i}}{\sqrt{E_{1,i}^2 + E_{2,i}^2}}), i = 1, \ldots, r$$

Now, let us define:

$$\theta_i = \arctan(\frac{e_{1,i}}{e_{2,i}}), i = 1, \ldots, r$$

as the first CPC, a vector of angular phases giving a circular arrangement for the rows of $\boldsymbol{X}$. The CPCA also allows for outlier detection, identifying as such the samples for which the pair $(E_{1,i}, E_{2,i})$, $i = 1, \ldots, r$, is close to the origin $(0, 0)$ given a distance, see Larriba et al. (2022) for details.

3 Application 1: Human Circadian Atlas

This application is part of the circadian atlas described from the human post-mortem Genotype-Tissue Expression (GTEx) data collection in Larriba et al. (2022). For each tissue the atlas details: the circadian (rhythmic) genes, together with their peak times and rhythmicity measures, being both interesting circadian markers to describe human molecular clock networks.

GTEx comprises unordered gene expressions samples from postmortem individuals in 54 human tissues. In particular, GTEx consists of 17,382 RNA-seq samples from 948 donors. In the absence of biological times, GTEx provides the time of death (TOD) estimates to order samples across tissues. However, the annotated TODs may provide inaccurate ordering estimates, and unlike expected, the well-known circadian genes do not display rhythmic patterns accordingly to TOD order. For instance, see the pattern of *Per1* in Fig. 1 (left panel), where no rhythmicity is appreciated. More interesting estimates of the timing are given using CPCA, see Fig. 1 (right panel), where the rhythmicity of *Per1* is now discovered. CPCA is chosen as it efficiently works even in the case of high levels of noise, as happens in the GTEx database. Given the order, FMM predictions provide rhythmicity features. The FMM predicted curve for *Per1* is shown in Fig. 1 (right panel, blue line).

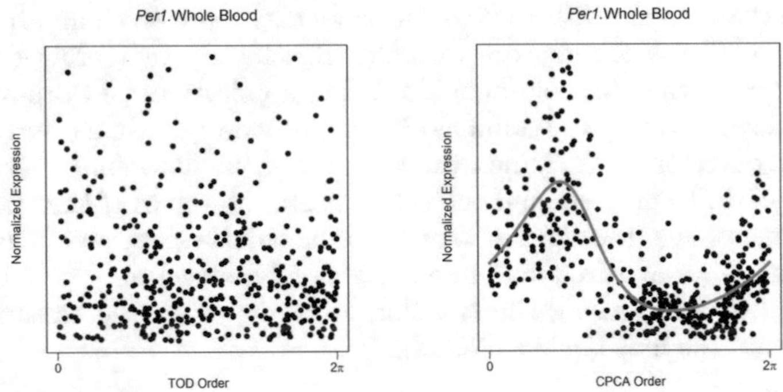

Fig. 1. *Per1* expression data in Whole Blood ordered according to TOD (left panel) and CPCA (right panel). Blue line is the FMM predictions

The main circadian markers are the peaks of the predicted curves as they reveal the activation times among the genes giving insights into the biological function relationships. Under the FMM model assumption, the peak (t_U) is given by: $t_U = \alpha + 2\arctan(\frac{1}{\omega}\tan(\frac{-\beta}{2}))$, see Rueda et al. (2019) for details. A goodness of fit measure R^2, computed from the FMM predictions, is also derived as a gene rhythmicity measure. We label a gene as circadian in a given tissue when the R^2 is high (> 0.5). The number of circadian genes, their peaks' arrangement, and even the number of peak modes are tissue-specific. In order to compare the activation times between tissues, a set of core genes is defined, which are highly rhythmic across tissues and species. Even for these genes, the peak arrangement changes from one tissue to another. These results are illustrated in Figs. 2 and 3 for two relevant tissues in circadian biology. Figure 2 shows circadian genes' peak distribution while Fig. 3 displays core clock genes' activation times arrangement.

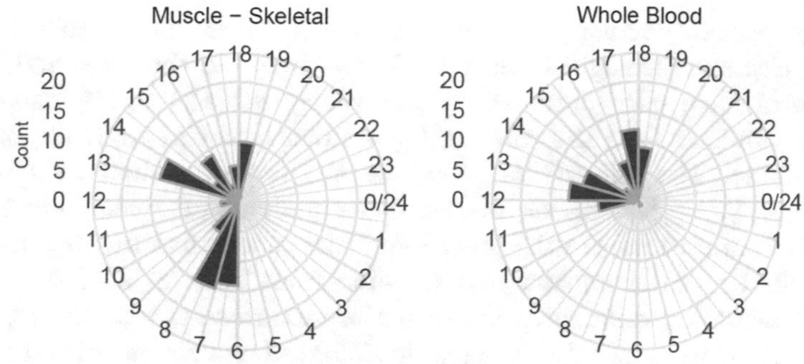

Fig. 2. Peak time distributions of the circadian genes along the 24-hour day. Left: in Muscle-Skeletal. Right: Whole Blood

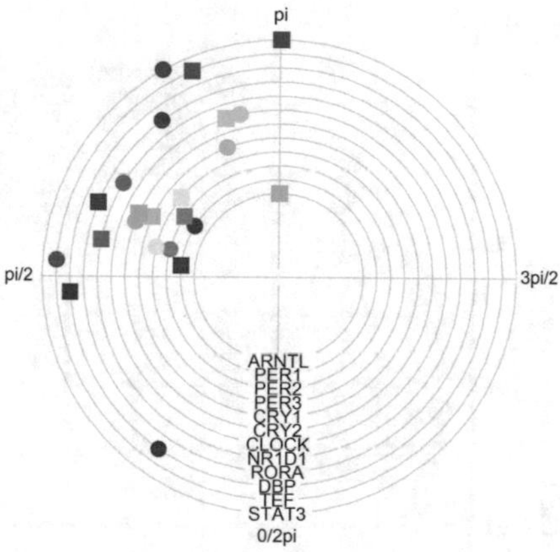

Fig. 3. Biological function times arrangement in Muscle-Skeletal (circles) and Whole Blood (squares) for twelve core clock genes. Colors denote distinct core clock genes

The human circadian expression atlas given in Larriba et al. (2022) is the study that involves, until now, the largest amount of analyzed tissues which allows the identification of new circadian genes.

4 Application 2: Neuronal Cell-Type Taxonomy

This study is part of Rodríguez-Collado and Rueda (2021), where more details about the analysis can be found. The classification of neuronal cells into different types is one of the most challenging open problems in neuroscience. At class level, cells are classified into non-neuronal cells and neurons, which in turn can be classified into GABAergic and glutamatergic based on their neurotransmitter. Habitually, these are further divided into different neuron subclasses or Cre-lines based on the expression of Cre recombinases markers.

A hierarchical, circular taxonomy for visual cortex mouse cells is defined based on electrophysiological and genetic features. On the one hand, the former features have been obtained by analysing with the FMM_3 model 1,892 APs of 24 Cre-lines from the Allen cell types database (https://celltypes.brain-map.org/data). The selected features are the median values by Cre-line of $(A_J, \alpha_J, \beta_J, \omega_J), J = \{1, 2, 3\}$, as well as other measures derived from the basic set of FMM parameters. On the other hand, the genetic features are the number of core cells by genetic cluster and Cre-line from Tasic et al. (2016). Due to notable distribution differences between the feature sets, separate PCAs have been conducted to extract two electrophysiological and six genetic components. These components are used in a final ensemble CPCA that results in the proposed taxonomy, shown in Fig. 4.

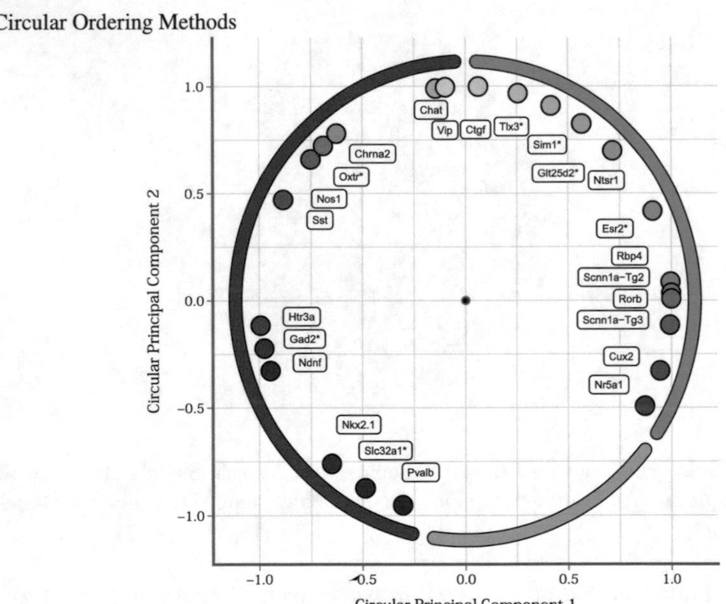

Fig. 4. First two circular principal components plot; included labels are the Cre-lines

The taxonomy is the first defined in the literature with a genuine circular topology. It combines a circular visualization tool (commonly used to represent genomic data) with an integrated clustering approach defined with CPCA. The proposal provides both an order and placement among the Cre-lines that mostly goes in agreement in several aspects with other recent mouse visual cortex taxonomies. It keeps together Cre-lines that have similar characteristics (Vip and Chat, Htr3a and Ndnf, Pvalb and Nkx2.1, among others). The blank space between Pvalb and Nr5a1 Cre-lines is assumed to correspond to non-neuronal cells, unavailable in the studied data. Moreover, the proposal locates certain Cre-lines for the first time in a taxonomy (those marked with an asterisk).

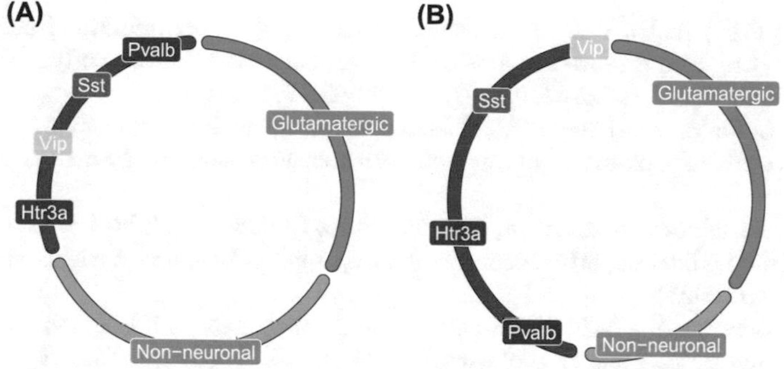

Fig. 5. Arrangement of the main neuronal cell-types in Tasic et al. (2016) (A), and in the novel taxonomy (B)

However, the differences existing between the taxonomy's circular arrangement and others, such as proposals Tasic et al. (2016), provide several interesting insights. This is briefly summarized in Fig. 5. First, the circular continuum between neuronal cell classes is not evenly distributed: GABAergic neurons display much more variation than other cells, specially non-neuronal cells. Secondly, significant differences in the order between the main GABAergic Cre-lines are found. Particularly, the GABAergic neurons with most similar characteristics to Glutamatergic neurons are the Vip neurons, whereas the Pvalb neurons are found to be the most alike to non-neuronal cells.

Acknowledgements. The authors thank the financial support given by the Spanish Ministry of Science, Innovation and Universities [PID2019-106363RB-I00].

References

Anafi, R.C., Francey, L.J., Hogenesch, J.B., Kim, J.: CYCLOPS reveals human transcriptional rhythms in health and disease. Proc. Natl. Acad. Sci. **114**(20), 5312–5317 (2017)

Armstrong, S., Guzmán, C., Sing Long, C.A.: An optimal algorithm for strict circular seriation. SIAM J. Math. Data Sci. **3**(4), 1223–1250 (2021)

Barragán, S., Rueda, C., Fernández, M.A.: Circular order aggregation and its application to cell-cycle genes expressions. IEEE/ACM Trans. Comput. Biol. Bioinf. **14**(4), 819–829 (2017)

Cai, T.T., Ma, R.: Matrix reordering for noisy disordered matrices: optimality and computationally efficient algorithms. arXiv preprint 10.48550/arXiv.2201.06438 (2022)

Downs, T.D., Mardia, K.V.: Circular regression. Biometrika **89**(3), 683–697 (2002)

Larriba, Y., Rueda, C., Fernández, M.A., Peddada, S.D.: Order restricted inference in chronobiology. Stat. Med. **39**, 265–278 (2020)

Larriba, Y., Rueda, C., Mason, I., Saxena, R., Scheer, F.: CIRCUST: a novel methodology for reconstruction of temporal order of molecular rhythms; validation and application towards a human circadian gene expression atlas (2022, preprint)

Mavroudis, P.D., Jusko, W.J.: Mathematical modeling of mammalian circadian clocks affecting drug and disease responses. J. Pharmacokinet Pharmacodyn. **48**(3), 375–386 (2021). https://doi.org/10.1007/s10928-021-09746-z

Rodríguez-Collado, A., Rueda, C.: Electrophysiological and transcriptomic features reveal a circular taxonomy of cortical neurons. Front. Hum. Neurosci. **15**, 684950 (2021)

Rueda, C., Fernández, I., Larriba, Y., Rodríguez-Collado, A.: The FMM approach to analyze biomedical signals: theory, software, applications and future. Mathematics **9**(10), 1145 (2021)

Rueda, C., Larriba, Y., Peddada, S.D.: Frequency modulated Möbius model accurately predicts rhythmic signals in biological and physical sciences. Sci. Rep. **9**, 18701 (2019)

Rueda, C., Rodríguez-Collado, A., Larriba, Y.: A novel wave decomposition for oscillatory signals. IEEE Trans. Signal Process. **69**, 960–972 (2021)

Scholz, M.: Analysing periodic phenomena by circular PCA. Proc. Conf. Bioinform. Res. Dev. **4414**, 38–47 (2007)

Tasic, B., Menon, V., Nguyen, T., et al.: Adult mouse cortical cell taxonomy revealed by single cell transcriptomics. Nat. Neurosci. **19**, 335–346 (2016)

The Extended Version of Cohen's *d* Index for Interval-Valued Data

M. Asunción Lubiano[1(✉)], José García-García[1], Antonio L. García-Izquierdo[2], and Ana M. Castaño[2]

[1] Department of Statistics, O.R. and D.M., University of Oviedo, C/Federico García Lorca 18, 33007 Oviedo, Spain
{lubiano,garciagarjose}@uniovi.es

[2] Department of Psychology, Universidad de Oviedo, Plaza Feijóo, s/n, 33003 Oviedo, Spain
{angarcia,castanoana}@uniovi.es

Abstract. Nowadays, new types of data are emerging from lots of distinct real-life experiments and statistical researchers need to develop new tools to deal with them. For instance, interval-valued responses arise as an alternative to Likert-type responses in questionnaires measuring people's behavior (their attitudes, opinions, perceptions, feelings, etc.). In order to facilitate the comparison of different analysis involving several rating scales and with the aim of studying the effect size measure for difference between two independent groups, in this paper we extend the concept of Cohen's *d* index established for real numbers to the interval-valued data context. Finally, a real-life example has been included to motivate and illustrate the problem.

1 Introduction

For many decades now, data analysts have been encouraging to enhance the presentation of research findings in the behavioral sciences by including an effect-size measure along with a statistical significance test (Cohen 1965; Hays 1963). Besides, the American Psychological Association (APA) Publication Manual stated "It is almost always necessary to include some measure of effect-size in the Result section" (APA 2010, p. 34) and Wilkinson et al. (1999) highlighted the importance of including the effect size for future systematic reviews and meta-analysis.

In Statistics, an effect-size is a quantitative measure that is independent of sample size and complements statistical hypothesis testing. This measure quantifies the magnitude of a phenomenon like the difference between populations or the relationship between explanatory and response variables and facilitates the statistical interpretation of the importance of a research result. Consequently, it allows the comparison of different results of a set of empirical studies carried out independently about a given research problem.

Nowadays, statistical data analysis methodology is constantly evolving due to the appearance of new types of real-life data that cannot be strictly classified as quantitative or qualitative ones.

L. A. García-Escudero et al. (Eds.): SMPS 2022, AISC 1433, pp. 263–270, 2023.
https://doi.org/10.1007/978-3-031-15509-3_35

In social and educational sciences and many other disciplines, *Likert-type Scales* (Likert 1932) are the most popular rating scales considered in the literature to rate evaluations, perceptions, judgments, classifications, etc., through questionnaires. This type of data cannot be numerically measured because they concern intrinsically imprecise valued attributes and Likert scales allow the respondent to choose among a small number of predetermined 'linguistic values' (discrete scale).

To overcome the limitations of Likert-type scales, since the individual differences are almost systematically overlooked, in the last years *Interval-Valued Scales* (IVSs) are gaining strength as an alternative to Likert-type scales by allowing respondents to select a range or interval of real data and not being constrained to choose among a few pre-specified responses (see, for instance, Ellerby et al. 2021; Wagner et al. 2015; Themistocleous et al. 2019).

In this paper, we will extend the definition of one of the most known standardized mean difference effect-size measures to fuzzy approach. We will analyze the results with a real life example where interval-valued data have been gathered.

2 Preliminary Concepts

Interval-valued scales make use of random intervals. In this section, we will recall the main concepts and the methodology used to analyze this type of data.

Let $\mathscr{K}_c(\mathbb{R})$ denote the class of nonempty compact intervals from $\mathbb{R}$. Each interval K in the space $\mathscr{K}_c(\mathbb{R})$ can be characterized in terms of either its infimum and supremum or its mid-point and spread or radius as follows:

$$K = [\inf K, \sup K] = [\operatorname{mid} K - \operatorname{spr} K, \operatorname{mid} K + \operatorname{spr} K].$$

When dealing with interval-valued data we use an arithmetic based on the *sum* and the *product by a scalar* operations defined as the corresponding image sets of the involved interval values (see Minkowski 1903) which are settled for $K, K' \in \mathscr{K}_c(\mathbb{R})$ and any $\lambda \in \mathbb{R}$ as follows

$$K + K' = [\inf K + \inf K', \sup K + \sup K'],$$

$$\lambda \cdot K = \begin{cases} [\lambda \cdot \inf K, \lambda \cdot \sup K] \text{ if } \lambda \geq 0 \\ [\lambda \cdot \sup K, \lambda \cdot \inf K] \text{ if } \lambda < 0. \end{cases}$$

In contrast to the real-valued case, the space $\mathscr{K}_c(\mathbb{R})$ is not linear with these two operations, but only semilinear with a conical structure, because of the lack of an opposite element for the Minkowski addition. For this reason, it is not possible to treat intervals directly as two-dimensional vectors. Therefore, distances play a crucial role in statistical developments.

To measure the distance between two interval-valued data, we will make use of a metric on $\mathscr{K}_c(\mathbb{R})$ extending the Euclidean one and being easy-to-use and interpret. More precisely, we will consider the *d_θ-metric* proposed by Gil et al. (2002) which is defined for two intervals $K, K' \in \mathscr{K}_c(\mathbb{R})$ as follows,

$$d_\theta(K, K') = \sqrt{(\text{mid}\, K - \text{mid}\, K')^2 + \theta \cdot (\text{spr}\, K - \text{spr}\, K')^2},$$

where $\theta \in (0, \infty)$ weighs the relative importance assessed to deviations in imprecision in contrast to deviations in trends. It is often imposed that $\theta \in (0, 1]$, in order to weigh the deviation in location no less than the deviation in imprecision, as well as to make d_θ coincide with the metric introduced by Bertoluzza et al. (1995). Actually, the d_1 metric coincides with the 2-norm metric between intervals which has been proposed by Vitale (1985).

Compact random intervals (see Matheron 1975) determine a well-stated and supported model for the random mechanisms generating interval-valued data within the probabilistic setting. They integrate both randomness and imprecision, so that the first one affects the generation of experimental data, whereas the second affects the nature of the experimental data which, for formal purposes, are assumed to be intrinsically interval-valued.

Following the general random set approach, given a probability space $(\Omega, \mathscr{A}, P)$, a mapping $X : \Omega \to \mathscr{K}_c(\mathbb{R})$ is said to be a *interval-valued random set* (IVRS for short) associated with it if X is measurable with respect to $\mathscr{A}$ and the Borel σ-algebra generated by the topology induced by the d_θ metric on $\mathscr{K}_c(\mathbb{R})$. Equivalently, X is a interval-valued random set if, and only if, both functions $\inf X$ and $\sup X$ (or alternatively, $\text{mid}\, X$ and $\text{spr}\, X$) are real-valued random variables.

As a consequence from the Borel measurability, crucial concepts in probabilistic and inferential developments, such as the (induced) distribution of a interval-valued random set or the stochastic independence of interval-valued random sets, are well-defined.

In performing inferential analysis about the distribution of interval-valued random sets, the best known involved parameters are the Aumann-type mean value (Aumann 1965) and the Fréchet-type variance (Korner 1997; Lubiano et al. 2000).

We are going to recall their sample version. Given a random sample $(X_1, \ldots, X_n)$ of size n from an IVRS X and a realization $\mathbf{x} = (x_1, \ldots, x_n)$,

The *sample Aumann mean* of $\mathbf{x}$ is given by the compact interval

$$\overline{\mathbf{x}} = \frac{1}{n} \cdot (x_1 + \cdots + x_n).$$

The *sample (d_θ Fréchet-type) variance* of $\mathbf{x}$ is given by the real number

$$s^2_{\mathbf{x}} = \frac{1}{n-1} \cdot \sum_{i=1}^{n} \left[d_\theta\left(x_i, \overline{\mathbf{x}}\right)\right]^2.$$

The above considered (sample) mean and variance preserve all the main properties from the numerical case. All of these properties allow us to consider the mean and the variance as suitable estimates of central tendency and dispersion, respectively.

In the next section, we are going to state an effect-size measure for interval-valued data considering the extension of the most common effect-size measure for real-valued data which is defined to compare the means of two groups.

3 Standardized Mean Difference for Interval-Valued Data

In research studies that involve the comparison of two groups, the standardized mean difference is one of the most frequently used effect-size measures.

Let $(\Omega, \mathscr{A}, P)$ be the probability space modeling a random experiment. Then, if X and Y are two independent IVRSs associated with $(\Omega, \mathscr{A}, P)$, we will consider the following effect size

$$\delta = \frac{d_\theta\left[E(X), E(Y)\right]}{SD},$$

where SD is the standard deviation of the population.

In the practical setting, population values are not typically known and must be estimated from sample statistics. Distinct versions of effect-sizes based on means proposed so far differ with respect to which statistics are used. The most known effect-size measure is the Cohen's d index suggested by Cohen (1969; 1988), see also Hedges (1981).

Definition 1. Let X and Y be two independent IVRSs associated with $(\Omega, \mathscr{A}, P)$ and consider a sample of independent observations from X, $\mathbf{x} = (x_1, \ldots, x_{n_1})$, and a sample of independent observations from Y, $\mathbf{y} = (y_1, \ldots, y_{n_2})$. The **extended Cohen's d index** of effect size is defined as the real number

$$d = \frac{d_\theta\left(\overline{\mathbf{x}}, \overline{\mathbf{y}}\right)}{SD_p} \quad \text{with} \quad SD_p = \sqrt{\frac{(n_1 - 1)s^2_{\mathbf{x}} + (n_2 - 1)s^2_{\mathbf{y}}}{n_1 + n_2 - 2}},$$

where SD_p is the pooled standard deviation for the two groups which is recommended if the standard deviations and sizes of the two groups differ (Cohen 1988, p. 67).

In the next section, we are going to apply the preceding measure on a dataset obtained from a real-life situation.

4 Real-Life Data Example

The COVID-19 pandemic has promoted a big change in the Higher Education due to the adjustment to a new scenario characterized by the need to quickly adapt from the face-to-face to the online distance modality.

An educational innovation project was carried out for the planning and improvement of teaching-learning processes of a subject of the Degree in Labor

Relations and Human Resources at the University of Oviedo (Spain) for the 2020/2021 academic year. In the study a blended learning method was considered for the teaching modality (face-to-face vs. online learning).

Specifically, we are going to examine the behavior of this effect-size measure in the example with respect to the influence of respondents' sex (men *vs.* women) for both teaching modalities.

A total of 50 participants have been requested to answer a questionnaire (available by means a custom web tool, see Fig. 1) by selecting the interval that best represents their level of agreement to the statements proposed in a interval-valued scale bounded between 1 and 7.

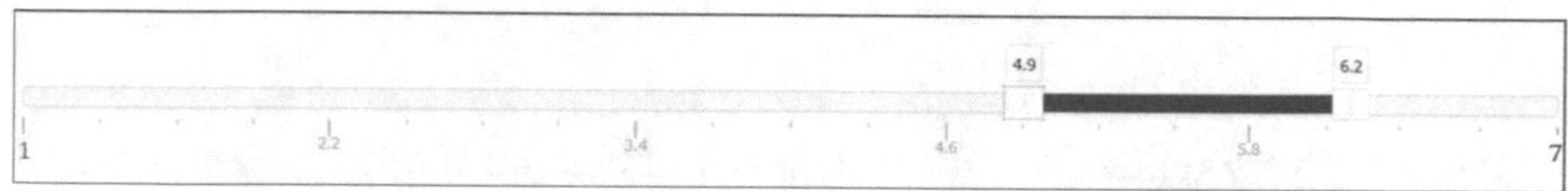

Fig. 1. Example of interval-valued based-responses to the online questionnaire

The online questionnaire was comprised of biographical information (i.e., age, gender, etc.) as well as 142 items that measured perception of lack of information and isolation (adapted from Weinert et al. 2015), perception of justice, the opportunity to carry out dishonest academic behavior, technical and contextual obstacles in monitoring of distance classes and satisfaction with the educational innovation project. We focus our attention on the seven items corresponding to the perception of lack of information and isolation displayed in Table 1.

Table 1. Constructs and measurement items

Information undersupply	
$I.1$	I receive too little information from my classmates
$I.2$	It is difficult to receive relevant information from my classmates
$I.3$	It is difficult to receive relevant information from the teacher
$I.4$	The amount of information I receive from my classmates is very low
$I.5$	The amount of information I receive from the teacher is very low
Isolation	
$I.6$	I feel less integrated in my team at class
$I.7$	I feel poorly informed about the relevant issues from my team at class

In Table 2 we show the results of the calculation of the Cohen's d for the seven items with this teaching method when we consider the interval-valued scale. Between parentheses, we show the approximate p-values obtained applying the bootstrapped two-sample test about means with fuzzy rating scale-based data

Table 2. Analyzing the influence of respondent's sex to the perception of Information undersupply (Items 1–5) and Isolation (Items 6–7)

Cohen's d (p-value)	Face-to-face	Online
$I.1$	.629 (*.027*)	.225 (*.502*)
$I.2$	.812 (*.003*)	.508 (*.082*)
$I.3$	.355 (*.196*)	.174 (*.594*)
$I.4$	.853 (*.002*)	.718 (*.012*)
$I.5$	.397 (*.134*)	.359 (*.192*)
$I.6$	.392 (*.160*)	.837 (*.002*)
$I.7$	.635 (*.021*)	.562 (*.030*)

for independent samples (see, for instance, Lubiano et al. 2016), since interval-valued data is a particular case of trapezoidal fuzzy data.

Figure 2 show the sample means for 18 men (black) and 32 women (gray).

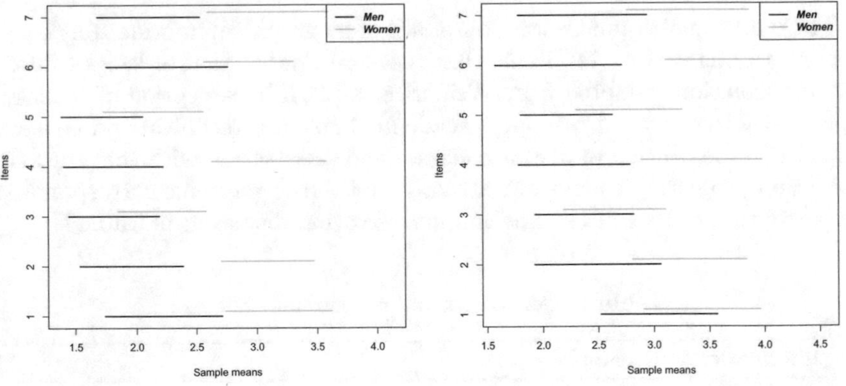

Fig. 2. Graphical sample means for men *vs.* women with face-to-face (left) and online (right) methodoly

According to Cohen (1988), values between .2 to .49, .50 to .79, and .80 and higher are considered small, medium and large, respectively.

We can observe that respondent's sex is a non-significant factor with a small-medium effect size both face-to-face and online modality for items $I.3$ and $I.5$ related to the information undersupply received by the teacher (values of d between .174 to .397).

On the other hand, the respondent's sex is a significant factor with a medium-large effect size with both teaching modality (values of d over than .562) for items $I.4$ and $I.7$.

When we analyze the influence of sex for question $I.1$, if responses have been obtained with the face-to-face teaching modality, it could be concluded that this factor is significant with a medium-effect size of $d = .629$, while it is not significant for online modality with a small-effect size of $d = .225$. Nevertheless, for question $I.6$, sex is significant when teaching modality is online and the effect size is large ($d = .837$) but it is not significant with the face-to-face modality with a small-effect size ($d = .392$).

5 Conclusions and Future Research

In this paper we have stated an extended version of Cohen's d index when the random experiment involve interval-valued data.

Besides, it would be desirable to study the properties of these sample measures estimating the corresponding population measure (like unbiasedness, consistency, and so on).

In a similar way, it is possible to calculate this index with fuzzy data obtained from the responses of fuzzy rating scales-based questions (Castaño et al. 2020). By extending the concept of effect size to more complex type of data, it will be possible to compare research results concerning both fuzzy or interval-valued data and real-valued data.

On the other hand, we are now studying other effect-size measures of difference on means and the extension of these concepts to other situations.

Acknowledgement. The research in this paper has been partially supported by from Principality of Asturias Grant AYUD/2021/50897, and the Spanish Ministry of Economy and Business Grant PID2019-104486GB-I00. Their financial support is gratefully acknowledged. The authors would like to the reviewers for valuable and helpful comments to improve the quality of this work.

References

American Psychological Association: Publication Manual of the American Psychological Association, 6th edn. APA, Washington, D.C. (2010)

Aumann, R.J.: Integrals of set-valued functions. J. Math. Anal. Appl. **12**, 1–12 (1965)

Bertoluzza, C., Corral, N., Salas, A.: On a new class of distances between fuzzy numbers. Mathware Soft Comput. **2**, 71–84 (1995)

Castaño, A.M., Lubiano, M.A., García-Izquierdo, A.L.: Gendered beliefs in STEM undergraduates: a comparative analysis of fuzzy rating versus likert scales. Sustainability **12**, 6227 (2020)

Cohen, J.: Some statistical issues in psychological research. In: Wolman, B.B. (ed.) Handbook of Clinical Psychology, pp. 95–121. Academic Press, New York (1965)

Cohen, J.: Statistical Power Analysis for the Behavioral Sciences, 1st edn. Academic Press, New York (1969)

Cohen, J.: Statistical Power Analysis for the Behavioral Sciences, 2nd edn. L. Erlbaum Associates, Hillsdale (1988)

Ellerby, Z., Wagner, C., Broomell, S.B.: Capturing richer information: on establishing the validity of an interval-valued survey response mode. Beha. Res. Methods **54**, 1240–1262 (2021). https://doi.org/10.3758/s13428-021-01635-0

Gil, M.Á., Lubiano, M.A., Montenegro, M., López, M.T.: Least squares fitting of an affine function and strength of association for interval-valued data. Metrika **56**, 97–111 (2002)

Hays, W.L.: Statistics for Psychologists. Holt Rinehart and Winston, New York (1963)

Hedges, L.V.: Distribution theory for Glass's estimator of effect size and related estimators. J. Educ. Stat. **6**(2), 106–128 (1981)

Körner, R.: On the variance of fuzzy random variables. Fuzzy Sets Syst. **92**(1), 83–93 (1997)

Likert, R.: A technique for the measurement of attitudes. Arch. Psychol. **22**, 140–155 (1932)

Lubiano, M.A., Gil, M.Á., López-Díaz, M., López, M.T.: The $\overrightarrow{\lambda}$-mean squared dispersion associated with a fuzzy random variable. Fuzzy Sets Syst. **111**, 307–317 (2000)

Lubiano, M.A., Montenegro, M., Sinova, B., de la Rosa de Sáa, S., Gil, M.Á.: Hypothesis testing for means in connection with fuzzy rating scale-based data: algorithms and applications. Eur. J. Oper. Res. **251**, 918–929 (2016)

Matheron, G.: Random Sets and Integral Geometry. Wiley, New York (1975)

Minkowski, H.: Vorlumen und Oberflache. Math. Ann. **57**, 447–495 (1903)

Themistocleous, C., Pagiaslis, A., Smith, A., Wagner, C.: A comparison of scale attributes between interval-valued and semantic differential scales. Int. J. Mark. Res. **61**(4), 394–407 (2019)

Vitale, R.A.: L_p metrics for compact, convex sets. J. Approx. Theory **45**(3), 280–287 (1985)

Wagner, C., Miller, S., Garibaldi, J.M., Anderson, D.T., Havens, T.C.: From interval-valued data to general type-2 fuzzy sets. IEEE Trans. Fuzzy Syst. **23**, 248–269 (2015)

Weinert, C., Maier, C., Laumer, S.: Why are teleworkers stressed? An empirical analysis of the causes of telework-enabled stress. In: Thomas, O., Teuteberg, F. (eds.) 12th International Conference on Wirtschaftsinformatik, pp. 1407–1421 (2015). aisel.aisnet.org/wi2015/94

Wilkinson, L.: Task force on statistical inference, American Psychological Association, Science Directorate: statistical methods in psychology journals: guidelines and explanations. Am. Psychol. **54**, 594–604 (1999)

Copulas, Lower Probabilities and Random Sets: How and When to Apply Them?

Roman Malinowski[1(✉)] and Sébastien Destercke[2]

[1] CNES/CS Group/UTC, 18 Avenue Edouard Belin, Toulouse, France
roman.malinowski@utc.fr

[2] Université de Technologie de Compiègne (UTC), Avenue de Landshut, Compiègne, France
sebastien.destercke@utc.fr

Abstract. A copula is an aggregation function, that can be used as a dependency model. Any multivariate distribution function to be characterized by its marginals and a copula. When introducing imprecision in the modelling of those distribution functions, different solutions are available to aggregate the univariate uncertainty representations into a multivariate one via the copula. We present some of those solutions, and discuss their respective inclusions for special cases: independence, belief functions, necessity functions and p-boxes.

1 Introduction

Credal sets, or convex sets of probability distributions, are useful tools to reason under uncertainty, especially in the presence of imprecision. They include many uncertainty models, such as belief functions, p-boxes, etc. (Destercke et al. 2008). How to combine such univariate models into multivariate ones remain a very active research question, with many researchers studying how tools used in the precise setting can be extended to the imprecise one, including in particular copulas(Gray et al. 2021; Montes et al. 2015).

When considering precise probabilities, copulas have been shown to be able to model any dependency structure between probability distributions. This is no longer true when adding imprecision to probabilities (Montes et al. 2015), for various reasons, such as the fact that imprecise cumulative distributions and credal sets are no longer in one-to-one correspondence.

In this paper, we first provide a short review of different models of uncertainty (Sect. 2), and then present multiple ways of combining creal sets or their lower envelopes with a copula (Sect. 3). In Sect. 4, we finally study their relationships for various specific cases of interest.

2 Preliminaries

In this section, we present the various uncertainty models we will consider, as well as copulas, which are aggregation functions typically used to model dependencies between random variables. We will work on finite spaces.

L. A. García-Escudero et al. (Eds.): SMPS 2022, AISC 1433, pp. 271–278, 2023.
https://doi.org/10.1007/978-3-031-15509-3_36

2.1 Imprecise Models

Definition 1. Given a credal set[1] $\mathscr{M}$ over a finite space $\mathscr{X}$, we define its lower probability $\underline{P}$ as its lower envelope on all events of the power set $\mathscr{P}(\mathscr{X})$:

$$\forall A \subseteq \mathscr{X}, \underline{P}(A) = \inf\{P(A)|P \in \mathscr{M}\} \tag{1}$$

Note that although lower probabilities cannot describe any credal set, we will mostly focus on them in this paper to privilege clarity of exposure. As credal sets are quite generic, it is useful to consider simpler, more practical models. We now define such models that we will consider later on.

Definition 2. Given a finite space $\mathscr{X}$ and its power set $\mathscr{P}(\mathscr{X})$, a probability mass function (Shafer (1976)) is a mapping $\mathscr{P}(\mathscr{X}) \to [0,1]$ satisfying:

$$m(\emptyset) = 0 \text{ and } \sum_{A \subseteq \mathscr{X}} m(A) = 1 \tag{2}$$

A set A of $\mathscr{X}$ is called a *focal set* if and only if $m(A) > 0$ and we will note $\mathscr{F}_m$ the set of all focal sets. Such a probability mass function defines a Belief function $Bel : \mathscr{P}(\mathscr{X}) \to [0,1]$ and a Plausibility function $Pl : \mathscr{P}(\mathscr{X}) \to [0,1]$:

$$\forall A \subseteq \mathscr{X}, Bel(A) = \sum_{B \subseteq A} m(B) \text{ and } Pl(A) = \sum_{B \cap A \neq \emptyset} m(B) \tag{3}$$

Those two functions are conjugate as $Bel(A) = 1 - Pl(A^c)$. $Bel(A)$ can be interpreted as our belief that the truth lies in A. A belief function is a lower probability (1) inducing a credal set $\mathscr{M}(Bel) = \{P \mid \forall A \subseteq \mathscr{X}, Bel(A) \leqslant P(A)\}$.

Definition 3. A necessity measure Nec is a minitive belief function:

$$\forall A, B \subseteq \mathscr{X}, Nec(A \cap B) = min(Nec(A), Nec(B)). \tag{4}$$

In a finite space, focal elements of a necessity function form a nested family of events: $\mathscr{F}_m = \{a_1 \subset ... \subset a_k\}$ (Dubois and Prade 2009). As a specific belief function, a necessity induces a credal set $\mathscr{M}(Nec) = \{P|Nec(A) \leqslant P(A), \forall A\}$.

Definition 4. A p-box (probability-box) is a pair of two cumulative distribution functions (CDFs) $[\underline{F}, \overline{F}]$ defined on the real line s.t. $\underline{F}(x) \leqslant \overline{F}(x)$, $\forall x \in \mathbb{R}$. A p-box is the extension of CDFs to imprecise probabilities. It induces a credal set, composed of all the CDFs dominating $\underline{F}$ and dominated by $\overline{F}$:

$$\mathscr{M}([\underline{F}, \overline{F}]) = \{P|\forall x \in \mathbb{R}, \underline{F}(x) \leqslant P(]-\infty, x]) \leqslant \overline{F}(x)\} \tag{5}$$

We can define 'α-levels' $\mathscr{C}^{\alpha}_{[\underline{F},\overline{F}]}$ ($\alpha \in [0,1]$) of a p-box as intervals $[\underline{x}, \overline{x}]$ whose lower bound (resp. upper bound) is the pseudo-inverse of $\underline{F}$ (resp. $\overline{F}$)

[1] A convex set of probability distributions.

at α (Fig. 1). It has been proven in Destercke et al. (2008) that to each p-box corresponds a belief function Bel s.t.:

$$\mathscr{M}([\underline{F}, \overline{F}]) = \mathscr{M}(Bel) \tag{6}$$

Additionally, the set of focal elements of a p-box is included in the α-levels $\mathscr{F}_m \subseteq \mathscr{C}^{\alpha}_{[\underline{F},\overline{F}]}$. We will note $\mathscr{F}_m = \{[\underline{x}_k, \overline{x}_k] \mid m([\underline{x}_k, \overline{x}_k]) > 0\}$. Due to the fact that both $\underline{F}$ and $\overline{F}$ are increasing mappings, the lower and upper bounds of those intervals are ordered. Thus all focal elements $[\underline{x}_k, \overline{x}_k]$ can be ordered using the natural order:

$$\forall i, j, [\underline{x}_i, \overline{x}_i] \leqslant_{nat} [\underline{x}_j, \overline{x}_j] \Leftrightarrow \underline{x}_i \leqslant \underline{x}_j \text{ and } \overline{x}_i \leqslant \overline{x}_j \tag{7}$$

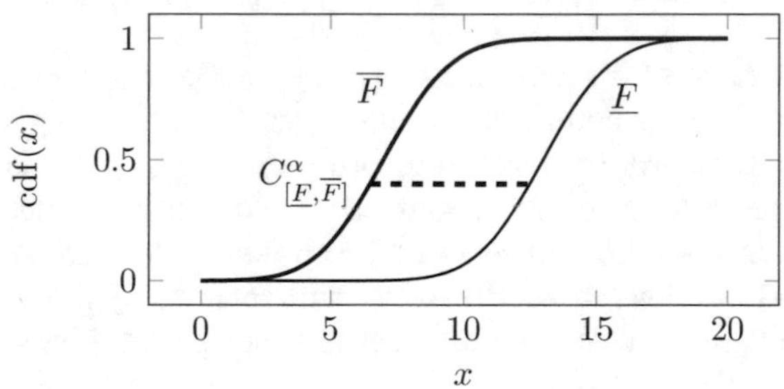

Fig. 1. p-box and one of its focal elements $[\underline{x}_i, \overline{x}_i]$

2.2 Copulas

Definition 5. A copula is a multivariate distribution function $C : [0,1]^N \to [0,1]$ whose marginals are uniform on $[0,1]$. It can be seen as a joint distribution function of N random variables. A copula verifies a number of properties, expressed here in the bivariate case. $\forall u, u', v, v' \in [0,1]^4$ s.t. $u \leqslant u', v \leqslant v'$:

$$C(u,0) = C(0,v) = 0,\ C(u,1) = u,\ C(1,v) = v \tag{8}$$

$$C(u',v') + C(u,v) - C(u,v') - C(u',v) \geqslant 0 \tag{9}$$

From (9) we have that C is a component-wise increasing mapping. There exists two copulas, the Lukasiewicz (C_L) and Minimum (C_M) copulas (also called lower and upper Fréchet-Hoeffding copulas in Nelsen (2006)), that bound all copulas C, i.e., $\forall u, v \in [0,1]$,

$$\max(0, u+v-1) \triangleq C_L(u,v) \leqslant C(u,v) \leqslant C_M(u,v) \triangleq \min(u,v) \tag{10}$$

The celebrated Sklar's theorem states that copulas, when applied to cumulative distributions of precise probabilities, can model any multivariate function, and that any multivariate function can be modelled by a copula applied to its marginals. As said earlier, this is however no longer true in the imprecise setting, hence the need to restudy how copulas can be applied to generic credal sets and lower probabilities (Montes et al. 2015).

3 Applying Copulas to Credal Sets

There are multiple ways to apply a copula to lower probabilities. We will describe some of them: a robust method on dominated probabilities, a method on mass distributions inducing belief functions, and an aggregation method.

3.1 Robust Method on Dominated Probabilities

Consider a copula C, and two credal sets $\mathscr{M}(\underline{P}_X)$ and $\mathscr{M}(\underline{P}_Y)$ defined over $\mathscr{P}(\mathscr{X})$ and $\mathscr{P}(\mathscr{Y})$ respectively. Applying C to every marginal $P_X \in \mathscr{M}(\underline{P}_X)$ and $P_Y \in \mathscr{M}(\underline{P}_Y)$ gives a joint lower probability such that for all $E \subseteq \mathscr{X} \times \mathscr{Y}$:

$$\underline{P}_{Robust}(E) = \inf\{P_{XY}(E)|F_{XY}(x,y) = C\left(F_X(x), F_Y(y)\right), \forall(x,y) \in \mathscr{X} \times \mathscr{Y}\} \tag{11}$$

with F_X, F_Y the CDF of P_X, P_Y and $F_{XY}(x,y) = P_{XY}(X \leqslant x, Y \leqslant y)$. Because F_{XY} is a precise CDF, it completely determines P_{XY}, allowing P_{XY} to be computed on events that are not Cartesian products. $\underline{P}_{Robust}(E)$ is then the infinimum of those probability distributions on E. Note that for defining the CDFs on finite spaces that are not subsets of $\mathbb{R}$, complete orderings on $\mathscr{X}$ and $\mathscr{Y}$ must be defined. In the following sections, we will refer to $\underline{P}_{Robust}$ as in (11) and its credal set generated with two univariate lower probabilities $\underline{P}_X$, $\underline{P}_Y$ and a copula C as:

$$\mathscr{M}_{Robust}(C, \underline{P}_X, \underline{P}_Y) = \{P_{XY} \mid \underline{P}_{Robust} \leqslant P_{XY}\} \tag{12}$$

or $\mathscr{M}_{Robust}$[2] for short when no ambiguity can arise.

3.2 Joint Masses from Copulas

As mass functions inducing belief functions can be seen as probabilities over sets, one could directly apply copulas to those masses. However, in general there is no natural order over the set of focal sets $\mathscr{F}_m$ (which then play the role of atoms). To apply a copula and define a joint mass, we have to chose an arbitrary ordering $\leqslant_{arb}$ that will determine the order inside the sets of focal sets $\mathscr{F}_{m_X} = \{a_1, a_2, ..., a_n\}$ and $\mathscr{F}_{m_Y} = \{b_1, b_2, ..., b_{n'}\}$ with $a_{i-1} \leqslant_{arb} a_i$ $\forall i \in [\![1, n]\!]$ and $b_{j-1} \leqslant_{arb} b_j$ $\forall j \in [\![1, n']\!]$ (with $a_0 = b_0 = \emptyset$). The bivariate mass associated to an element (a_i, b_j) of $\mathscr{F}_{m_X} \times \mathscr{F}_{m_Y}$ can be defined using the diagonal difference as in Ferson et al. (2004):

$$m^C_{XY}(a_i \times b_j) = C(A^i_X, B^j_Y) + C(A^{i-1}_X, B^{j-1}_Y) - C(A^i_X, B^{j-1}_Y) - C(A^{i-1}_X, B^j_Y) \tag{13}$$

with $A^i_X = \sum_{k \leqslant i} m_X(a_k)$ and $B^j_Y = \sum_{k \leqslant j} m_Y(b_k)$ being the cumulative masses over a_i and b_j (with $A^0_X = m_X(a_0) = B^0_Y = m_Y(b_0) = 0$). It is easy to check

[2] $\mathscr{M}_{Robust}$is the smallest credal set containing all probabilities from the marginal credal sets linked with C, but as it is convex, it can also contain probabilities linked to their marginals with a different copula C'.

that the joint mass defined in (13) is a mass distribution whose focal sets form a subset of the Cartesian product of the marginal focal sets $\mathscr{F}_{m_X} \times \mathscr{F}_{m_Y}$.

This mass induces a belief function Bel_{XY} from which can be generated a credal set. In the following sections, we will refer to this credal set as:

$$\mathscr{M}_{mass}(C, Bel_X, Bel_Y) = \{P_{XY} \mid Bel_{XY} = \sum m_{XY}^C \leqslant P_{XY}\} \quad (14)$$

or $\mathscr{M}_{mass}$ for short, where m_{XY}^C is defined as in (13). However, as shows the next example, the choice of $\leqslant_{arb}$ can strongly impact the joint model.

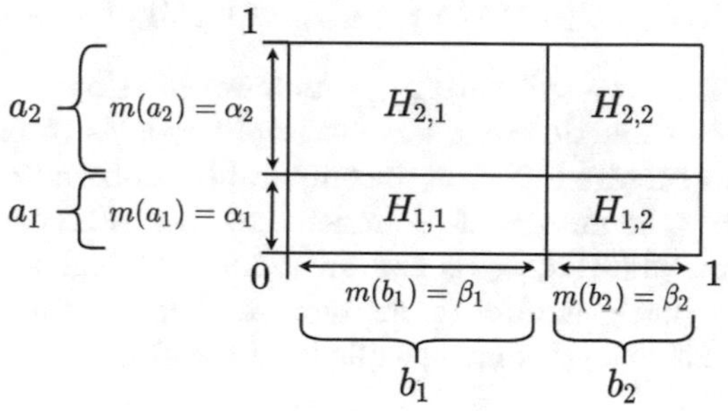

Fig. 2. Joint probability over events A and B

Example 1. Let m_X and m_Y be mass functions defined over $\mathscr{P}(\mathscr{X})$ and $\mathscr{P}(\mathscr{Y})$, and whose sets of focal sets are $\mathscr{F}_{m_X} = \{a_1, a_2\}$ and $\mathscr{F}_{m_Y} = \{b_1, b_2\}$ (Fig. 2). Assuming that $b_1 \leqslant_{arb} b_2$, consider the following $\leqslant_{arb}$ on $\mathscr{F}_{m_X}$:

- If "$a_1 \leqslant_{arb} a_2$": $m_{XY}(a_1 \times b_1) = C(\alpha_1, \beta_1)$, $m_{XY}(a_2 \times b_1) = \beta_1 - C(\alpha_1, \beta_1)$
- If "$a_1 \geqslant_{arb} a_2$": $m_{XY}(a_1 \times b_1) = \beta_1 - C(\alpha_2, \beta_1)$, $m_{XY}(a_2 \times b_1) = C(\alpha_2, \beta_1)$

Taking $C(u, v) = \min(u, v)$ and $\alpha_1 = \alpha_2 = \beta_1 = 0.5$, yields in the first case $m_{xy}(a_1 \times b_1) = 0.5$ and in the second case $m_{xy}(a_1 \times b_1) = 0$. There is, in general, no reason for two orders to have the same value of masses for the same Cartesian product of events. A notable exception is the product copula, explored below. Finally, note that such an approach is quite commonly encountered in the literature (Gray et al. 2021; Alvarez et al. 2018)

3.3 Copula Applied to the Lower Probabilities

Another way to aggregate two lower probabilities $\underline{P}_X, \underline{P}_Y$ with a copula C is by directly applying the copula to lower probabilities:

$$\forall A \in \mathscr{X}, B \in \mathscr{Y}, \underline{P}_{agg}(A \times B) = C(\underline{P}_X(A), \underline{P}_Y(B)) \quad (15)$$

Here, we use a copula only as an aggregation operator. Note that in general we cannot expect $\underline{P}_{agg}$ to induce a non-empty $\mathscr{M}(\underline{P}_{agg})$. We can nevertheless note that in the case of the product copula, the bivariate lower probability induced by (15) induces a non empty credal set if the marginals credal sets are not empty, i.e. $\exists P_X \geqslant \underline{P}_X, \exists P_Y \geqslant \underline{P}_Y \implies \exists P_{XY} = P_X \times P_Y \geqslant \underline{P}_{agg}$ due to the fact that it is true for any precise probability. It follows that for all copulas C such that $C_\Pi(u,v) = u \times v \geqslant C(u,v)$, we have

$$\underline{P}_X \underline{P}_Y \geqslant C(\underline{P}_X, \underline{P}_Y) = \underline{P}_{agg} \implies \mathscr{M}(\underline{P}_{agg}) \neq \emptyset \tag{16}$$

In the following sections, we will refer to the credal set generated by aggregating directly two univariate lower probabilities with a copula C as:

$$\mathscr{M}_{agg}(C, \underline{P}_X, \underline{P}_Y) = \{P_{XY} \mid \underline{P}_{agg} = C(\underline{P}_X, \underline{P}_Y) \leqslant P_{XY}\} \tag{17}$$

or $\mathscr{M}_{agg}$ for short, and we will note $\underline{P}_{agg}$ its lower envelope.

We have defined three different ways to apply copulas to probability sets. In general, they will generate different, incomparable probability sets. Also, while $\mathscr{M}_{Robust}$ is probably the most well-grounded way to extend copulas, it may generate a complex set, while $\mathscr{M}_{mass}$ is guaranteed to be a belief function and $\mathscr{M}_{agg}$ is easy to compute and generate. In the next section, we study different special cases where those sets enjoy some specific relationships.

4 Special Cases

This section explores what happens when considering the product copula, and when marginal models are both necessity measure or p-boxes.

Product Copula: In the case of the product copula, it is known (Couso et al. 2000) that all $\underline{P}_{robust}, \underline{P}_{mass}$ and $\underline{P}_{agg}$ factorize over Cartesian products, i.e., $\forall\ (A,B) \subseteq \mathscr{X} \times \mathscr{Y}, \underline{P}(A \times B) = \underline{P}_X(A) \times \underline{P}_Y(B)$. It is also known (Couso et al. 2000) that $\mathscr{M}_{Robust} \subsetneq \mathscr{M}_{mass}$, with the inclusion being sometimes strict. Also, as $\mathscr{M}_{agg}$ is the largest credal set enjoying this factorisation properties, we necessarily have

$$\mathscr{M}_{Robust}(C_\Pi, Bel_X, Bel_Y) \subseteq \mathscr{M}_{mass}(C_\Pi, Bel_X, Bel_Y) \subseteq \mathscr{M}_{agg}(C_\Pi, Bel_X, Bel_Y)$$

therefore having strong relationships in the case of the product copula[3].

Necessity Functions: Let us first compare $\mathscr{M}_{agg}$ and $\mathscr{M}_{mass}$. In the case of necessity functions, a natural ordering $\leqslant_{nat}$ between focal elements exists which corresponds to the inclusion ordering. Because of this, it holds that $\sum_{k \leqslant_{nat} i} \sum_{l \leqslant_{nat} j} m_{XY}(a_k \times b_l) = \sum_{a_k \subseteq a_i} \sum_{b_l \subseteq b_j} m_{XY}(a_k \times b_l)$ and thus computing the belief function Bel_{XY} defined in (14) on all focal elements a_i of Nec_X and b_j of Nec_Y yields:

$$Bel_{XY}(a_i \times b_j) = C\left(Nec_X(a_i), Nec_Y(b_j)\right)$$

[3] In this case, $\mathscr{M}_{mass}$ is insensitive to re-ordering, as C_Π corresponds to the uniform distribution over $[0,1]^2$.

Thus the lower envelope of $\mathscr{M}_{agg}$ and $\mathscr{M}_{mass}$ coincide on the Cartesian products of events. Since $\mathscr{M}_{agg}$ is again the largest set with this lower envelope, we do have $\mathscr{M}_{mass} \subseteq \mathscr{M}_{agg}$ in the case of necessity functions. However, as show the next example, there is no reason in general for

$$\mathscr{M}_{agg} \subseteq \mathscr{M}_{Robust} \text{ nor } \mathscr{M}_{agg} \supseteq \mathscr{M}_{Robust}$$

Example 2. Consider two necessity function Nec_X, Nec_Y defined over $\mathscr{X} = \{x_1, x_2\}$ and $\mathscr{Y} = \{y_1, y_2\}$ respectively, s.t $Nec_X(x_1) = Nec_Y(y_2) = 0$ and $Nec_X(x_2) = Nec_Y(y_1) = 0.9$. Let us first consider the Łukasiewicz copula $C(u, v) = max(0, u + v - 1)$ and events $x_2 \times y_1$. It is possible to show that on those events $\underline{P}_{Robust} = 0.9 > 0.8 = \underline{P}_{agg}$.

If we consider the Minimum copula $C(u, v) = \min(u, v)$, then taking any (P_X, P_Y) verifying $P_X(x_1) = 0.1$ and $P_Y(y_1) = 0.9$ and computing $\underline{P}_{agg}$ and $P_{XY} \in \mathscr{M}_{Robust}$ over event $x_2 \times y_1$ yields $\underline{P}_{agg}(x_2 \times y_1) = 0.9$ and $P_{XY}(x_2 \times y_1) = 0.8$. Thus it holds that on those events $\underline{P}_{agg} = 0.9 > 0.8 \geqslant \underline{P}_{Robust}$.

P-Boxes. When considering uncertainty represented by two p-boxes $[\underline{F}_X, \overline{F}_X]$, $[\underline{F}_Y, \overline{F}_Y]$, it does not hold in general that $\mathscr{M}_{Robust} \subseteq \mathscr{M}_{agg}$ nor $\mathscr{M}_{Robust} \supseteq \mathscr{M}_{agg}$. We refer to Example 2 when considering the p-boxes induced from necessity functions as in Baudrit and Dubois (2006). It however holds that:

$$\mathscr{M}_{Robust}(C, Bel_X, Bel_Y) \subseteq \mathscr{M}_{mass}(C, Bel_X, Bel_Y)$$

5 Conclusion

We presented different methods for joining uncertainty models with a copula: a robust method, a method based on cumulated masses, and a method using the copula as a direct aggregation operator. We showed that in the special case of the product copula, the lower probabilities coincide on Cartesian products. When using necessity functions, the aggregated lower probability coincides with the cumulated masses one on Cartesian products, and when using p-boxes we showed that the credal set obtained with the robust method is included in the credal set of the cumulated mass approach.

References

Alvarez, D.A., Uribe, F., Hurtado, J.E.: Estimation of the lower and upper bounds on the probability of failure using subset simulation and random set theory. Mech. Syst. Signal Process. **100**, 782–801 (2018)

Baudrit, C., Dubois, D.: Practical representations of incomplete probabilistic knowledge. Comput. Stat. Data Anal. **51**, 86–108 (2006)

Couso, I., Moral, S., Walley, P.: A survey of concepts of independence for imprecise probabilities. Risk Decis. Policy **5**, 165–181 (2000)

Destercke, S., Dubois, D., Chojnacki, E.: Unifying practical uncertainty representations – I: generalized p-boxes. Int. J. Approximate Reasoning **49**, 664–677 (2008)

Dubois, D., Prade, H.: Possibility Theory. Springer, New York (2009). https://doi.org/10.1007/978-0-387-30440-3_413

Ferson, S., et al.: Dependence in Probabilistic Modeling, Dempster-Shafer Theory, and Probability Bounds Analysis (2004)

Gray, A., Hose, D., De Angelis, M., Hanss, M., Ferson, S.: Dependent possibilistic arithmetic using copulas. In: Proceedings of the Twelveth International Symposium on Imprecise Probability: Theories and Applications. Proceedings of Machine Learning Research, vol. 147, pp. 169–179 (2021)

Montes, I., Miranda, E., Pelessoni, R., Vicig, P.: Sklar's theorem in an imprecise setting. Fuzzy Sets Syst. **278**, 48–66 (2015)

Nelsen, R.B.: An Introduction to Copulas. Springer, New York (2006). https://doi.org/10.1007/0-387-28678-0

Shafer, G.: A Mathematical Theory of Evidence. Princeton University Press, Princeton (1976)

Complex Dimensionality Reduction: Ultrametric Models for Mixed-Type Data

Marco Mingione[1(✉)], Maurizio Vichi[1], and Giorgia Zaccaria[2]

[1] University of Rome "La Sapienza", P.le A. Moro 5, 00185 Rome, Italy
{marco.mingione,maurizio.vichi}@uniroma1.it
[2] University of Rome Unitelma Sapienza, P.zza Sassari 4, 00161 Rome, Italy
giorgia.zaccaria@unitelmasapienza.it

Abstract. The factorial latent structure of variables, if present, can be complex and generally identified by nested latent concepts ordered in a hierarchy, from the most specific to the most general one. This corresponds to a tree structure, where the leaves represent the observed variables and the internal nodes coincide with latent concepts defining the general one (i.e., the root of the tree). Although several methodologies have been proposed in the literature to study hierarchical relationships among quantitative variables, very little has been done for more general mixed-type data sets. Hence, it is of the utmost importance to extend these methods and make them suitable to the even more frequent availability of mixed-type data matrices, as complex real phenomena are often described by both qualitative and quantitative variables. In this work, we propose a new exploratory model to study the hierarchical statistical relationships among variables of mixed-type nature by fitting an ultrametric matrix to the general dependence matrix, where the former is one-to-one associated with a hierarchical structure.

1 Introduction

In many statistical applications, the presence of mixed-type data brings an additional level of complexity to the analysis as they usually result in intricate dependence structures (De Leon and Chough 2013). When this is the case, the straightforward implementation of standard techniques is generally inappropriate, and specific solutions are required to deal with the mixed nature of the data depending on the final objective of the statistical analysis. For instance, the majority of conventional tools that are used both in the exploratory and the modeling phase relies on the assumption that the data, or at least a suitable transformation of them, follow a Normal distribution: an assumption that no longer applies in such contexts. For these reasons, several methods have been developed to deal with mixed-type data in many applications, especially concerning dimensionality reduction (McParland and Gormley 2016; Van de Velden et al. 2019; Vichi et al. 2019). These methodologies mostly rely on the computation of the general dependence matrix, say $\tilde{\mathbf{\Sigma}}$, which altogether includes the correct pairwise statistical relationship measures, namely: χ^2 in the case of qualitative pairs; ρ^2

L. A. García-Escudero et al. (Eds.): SMPS 2022, AISC 1433, pp. 279–286, 2023.
https://doi.org/10.1007/978-3-031-15509-3_37

between pairs of quantitative variables; η^2 in the mixed-type case. This general dependence matrix can be proven to be positive semi-definite and it can be paired with the matrix of the p-values assessing the statistical significance of the estimated relationship. Hence, while being useful as an exploratory tool, $\tilde{\mathbf{\Sigma}}$ has been already used to extend Factor Analysis (FA) and Principal Component Analysis (PCA) to the case of mixed-type data (Pagès 2004; Chavent et al. 2011) in order to detect latent structures underlying the data. However, if complex phenomena are considered, FA and PCA for mixed-type data are not able to pinpoint the hierarchical relationships among variables defining nested dimensions (concepts). In this paper, we introduce a new simultaneous, exploratory model for identifying a hierarchy of latent concepts. The proposal aims at reconstructing the general dependence matrix $\tilde{\mathbf{\Sigma}}$ via an ultrametric dependence matrix by extending the model proposed by Cavicchia et al. (2020b) to mixed-type data.

2 Background

Let $\mathbf{x}_i$ be a $(J \times 1)$ random vector corresponding to a generic multivariate observation i, where $i = 1, \ldots, N$ and $N > J$. Without loss of generality, $\mathbf{x}_i$ can be ordered such that the first P values are realizations of qualitative (categorical) variables, while the remaining $J - P$ values come from quantitative (numerical) variables. Hence, we can rewrite $\mathbf{x}_i = [{}_q\mathbf{x}_i{}', {}_n\mathbf{x}_i{}']'$, where ${}_q\mathbf{x}_i$ and ${}_n\mathbf{x}_i$ are the $(P \times 1)$ and the $((J-P) \times 1)$ vectors of the qualitative and quantitative variables, respectively. Specifically, ${}_q\mathbf{x}_i$ has elements ${}_qx_{ij} \in \{1, \ldots, c_j\}$, where $c_j \geq 2$ represents the number of distinct categories of variable j, for $j = 1, \ldots, P$. For the moment, we assume that the categories are not ordered, therefore defining the qualitative variables as *nominal* variables.

The sampled observations can be stacked together in the matrix $\mathbf{X} = [{}_q\mathbf{X}, {}_n\mathbf{X}]$, formed by two sub-matrices of dimensions $(N \times P)$ and $(N \times (J-P))$, having a qualitative and quantitative part.

The computation of $\tilde{\mathbf{\Sigma}}$ relies on the suitable standardization of matrix $\mathbf{X}$, which can be defined as in Vichi et al. (2019):

$$ {}_s\mathbf{X} = [{}_s\mathbf{G}, \frac{1}{\sqrt{N}}\mathbf{Z}] = [{}_s\mathbf{G}_1, \ldots, {}_s\mathbf{G}_P, \frac{1}{\sqrt{N}}\mathbf{z}_1, \ldots, \frac{1}{\sqrt{N}}\mathbf{z}_{J-P}], \tag{1} $$

where ${}_s\mathbf{G}$ is the $(N \times C)$ standardized matrix referring to the first P nominal variables, $C = \sum_{j=1}^{P} c_j$ and $\mathbf{Z} = \mathbf{J}_n\mathbf{X}\text{diag}(\text{diag}(\Sigma_n\mathbf{x}))^{-\frac{1}{2}}$ is the $(N \times (J-P))$ standardized matrix referring to the $J - P$ quantitative variables. ${}_s\mathbf{G}$ is obtained by appending together the single ${}_s\mathbf{G}_j = \mathbf{J}\mathbf{G}_j(\mathbf{G}_j'\mathbf{G}_j)^{-\frac{1}{2}}, j = 1, \ldots, P$, as in multiple correspondence analysis (Greenacre 2017), while $\mathbf{Z}$ has 0 mean and unit sum of square columnwise; $\mathbf{J} = \mathbf{I} - (\frac{1}{N})\mathbf{1}_N\mathbf{1}_N'$ is the centering idempotent matrix. Note that ${}_s\mathbf{X}$ has dimension $(N \times (C + J - P))$.

In the following section, we compute the general dependence matrix that includes the relationships among variables of different nature, qualitative and quantitative.

3 Statistical Relationships Between Mixed-Type Variables

Considering the matrix ${}_s\mathbf{X}$ in Eq. (1), we compute the matrix of the *correct* pairwise statistical relationships among mixed-type variables as follows

$$_e\boldsymbol{\Sigma}_{{}_s\mathbf{X}} = {}_s\mathbf{X}'{}_s\mathbf{X}.$$

$_e\boldsymbol{\Sigma}_{{}_s\mathbf{X}}$ is a square, positive semi-definite matrix of dimension $C + J - P$, and it can be divided into four blocks

$$_e\boldsymbol{\Sigma}_{{}_s\mathbf{X}} = \begin{pmatrix} \boldsymbol{\Phi}_{{}_s\mathbf{G}} & \mathbf{H}_{{}_s\mathbf{GZ}} \\ \mathbf{H}'_{{}_s\mathbf{GZ}} & \mathbf{R}_{{}_n\mathbf{X}} \end{pmatrix},$$

where

- $\boldsymbol{\Phi}_{{}_s\mathbf{G}} = [{}_s\mathbf{G}'_j \, {}_s\mathbf{G}_m : j, m = 1, \ldots, P]$ is a square matrix of order C, containing positive values if two modalities of ${}_q\mathbf{X}_j$ and ${}_q\mathbf{X}_m$, say l and h, are jointly assumed, negative otherwise;
- $\mathbf{R}_{{}_n\mathbf{X}} = [\frac{1}{N}\mathbf{z}'_k \, \mathbf{z}_r : k, r = P+1, \ldots, J]$ is the correlation matrix of order $J - P$ with elements $|\rho_{kr}| \leq 1$;
- $\mathbf{H}_{{}_s\mathbf{GZ}} = [{}_s\mathbf{G}_j{}' \mathbf{z}_k : j = 1, \ldots, P, k = P + 1, \ldots, J]$ is the matrix of order $(C \times (J - P))$ accounting for the association between the qualitative variable ${}_q\mathbf{X}_j$ and the quantitative variable ${}_n\mathbf{X}_k$.

The dependence between qualitative and quantitative variables can be defined by considering the matrix with elements equal to the square of elements of $_e\boldsymbol{\Sigma}_{{}_s\mathbf{X}} = {}_s\mathbf{X}'{}_s\mathbf{X}$, i.e.

$$_e\boldsymbol{\Sigma}^2_{{}_s\mathbf{X}} = \begin{pmatrix} \boldsymbol{\Phi}^2_{{}_s\mathbf{G}} & \mathbf{H}^2_{{}_s\mathbf{GZ}} \\ \\ \mathbf{H}^{2\prime}_{{}_s\mathbf{GZ}} & \mathbf{R}^2_{{}_n\mathbf{X}} \end{pmatrix}.$$

The latter is evidently in connection with well-known dependence/correlation measures.

- $\boldsymbol{\Phi}^2_{{}_s\mathbf{G}}$ can be exploited to compute

$$_r\chi^2_{jm} = tr({}_s\mathbf{G}_j \, {}_s\mathbf{G}'_j \, {}_s\mathbf{G}_m \, {}_s\mathbf{G}'_m)/min(c_j - 1, c_m - 1);$$

- $\mathbf{R}^2_{{}_n\mathbf{X}}$ is the matrix of the squared Pearson's correlation coefficients measuring the relationship between ${}_n\mathbf{X}_k$ and ${}_n\mathbf{X}_r$;
- $\mathbf{H}^2_{{}_s\mathbf{GZ}}$ is the matrix of the correlation ratios between the qualitative variable ${}_q\mathbf{X}_j$ and the quantitative variable ${}_n\mathbf{X}_k$.

It has to be noticed that all the three matrices have elements ranging between 0 and 1, where the former represents the case of independence and the latter that one of perfect dependence.

The matrix ${}_e\boldsymbol{\Sigma}^2_{s}{}_{\mathbf{X}}$ can be therefore synthesized into a $J \times J$ matrix of general dependence as follows

$$\tilde{\boldsymbol{\Sigma}} = \begin{pmatrix} {}_r\boldsymbol{\chi}^2_{s}{}_{\mathbf{G}} & \boldsymbol{\eta}^2_{s}{}_{\mathbf{GZ}} \\ \boldsymbol{\eta}^{2\prime}_{s}{}_{\mathbf{GZ}} & \mathbf{R}^2_{n}{}_{\mathbf{X}} \end{pmatrix}. \tag{2}$$

In the following section, we introduce a model for studying the relationships occurring among quantitative and quantitative variables into $\tilde{\boldsymbol{\Sigma}}$.

4 Methods

The ultrametricity notion is well-known in statistics for hierarchical clustering. Nevertheless, this notion has been introduced in mathematics with regard to the p-acid number theory (Dellacherie et al. 2014) associated with a generic matrix, which is not necessarily a distance matrix – specifically, a reverse relationship occurs with an ultrametric distance matrix. Hereinafter, we recall the definition of an *ultrametric* matrix.

Definition 1. A nonnegative[1] matrix $\mathbf{U}$ of order J is said to be ultrametric if

1. $u_{ij} = u_{ji}, i, j = 1, \ldots, J$ (symmetry);
2. $u_{jj} \geq \max\{u_{ij} : i = 1, \ldots, J\}, j = 1, \ldots, J$ (column pointwise diagonal dominance);
3. $u_{ij} \geq \min\{u_{il}, u_{jl}\}, i, j, l = 1, \ldots, J$ (ultrametricity).

Condition 3 can be rewritten as follows

- for each triplet i, j, l, there exists a reordering $\{i, j, l\}$ of the elements s.t. $u_{ij} \geq u_{il} = u_{jl}$,

by unraveling that an ultrametric matrix $\mathbf{U}$ is composed of a reduced number of distinct values subject to a specific order. This engenders an interesting feature of an ultrametric matrix, that is of being associated with a hierarchy over variables. Moreover, every ultrametric matrix is positive semi-definite (Dellacherie et al. 2014 pp. 60–61).

Dimensionality reduction for mixed-type data can be performed by Factor Analysis (Pagès 2004). However, FA is not able to reconstruct hierarchical relationships among variables (both qualitative and quantitative) and thus to unravel concepts of higher-order. Cavicchia et al. (2020b) introduced a model to study the hierarchical relationships among variables by reconstructing a nonnegative correlation matrix via an ultrametric correlation one. However, their proposal pertains *quantitative* variables only. In this paper, we extend the ultrametric model proposed by Cavicchia et al. (2020b) in order to detect hierarchical structures on variables of different nature (i.e., quantitative and qualitative).

[1] A nonnegative matrix $\mathbf{M} = [m_{ij}]$ is a matrix with nonnegative values, i.e., $m_{ij} \geq 0$.

Specifically, we fit an ultrametric matrix to the general dependence matrix $\tilde{\boldsymbol{\Sigma}}$ defined in Eq. (2) – that is nonnegative by definition – as follows:

$$\tilde{\boldsymbol{\Sigma}} = \tilde{\boldsymbol{\Sigma}}_{\mathrm{u}} + \mathbf{E}, \tag{3}$$

where $\tilde{\boldsymbol{\Sigma}}$ is an ultrametric dependence matrix of order P and $\mathbf{E}$ is a residual (error) matrix of the same order. $\tilde{\boldsymbol{\Sigma}}_{\mathrm{u}}$ pinpoints a hierarchical structure by defining a reduced number of variable groups associated with latent concepts and identifying broader concepts as bottom-up aggregations of the lower-order one. Model (3) can be estimated in the Least-Squares context or in the Maximum-Likelihood framework under suitable specific assumptions on the distribution of the error.

It is worthy to highlight that the proposal differs from hierarchical clustering methods applied on variables (see Cavicchia et al. 2020a, for the quantitative case). Indeed, even if the latter can be implemented on variables after a proper transformation of similarity measures into dissimilarity ones, they are sequential and greedy procedures affected by misclassification at the bottom of the hierarchy that can have an effect on higher levels. Other than detecting specific features related to variables, the proposed methodology overcomes the aforementioned drawbacks of hierarchical clustering procedures since this is a simultaneous and parsimonious model.

5 Application

We apply the ultrametric model described in Sect. 4 on the well-known benchmark data set `mtcars` (Henderson and Velleman 1981), available in the package `MASS` of `R` statistical software (Team et al. 2013). The data was extracted from the 1974 Motor Trend US magazine, and comprises fuel consumption and 10 aspects of design and performance for 32 cars. Specifically, the data set includes 6 quantitative variables:

- miles per gallon (`mpg`): a proxy for the fuel consumption efficiency;
- displacement (`disp`): the total volume (in cubic inches) of all the cylinders in an engine;
- gross horsepower (`hp`): a measurement of engine output, taken at the flywheel;
- rear axle ratio (`drat`): the number of revolutions the driveshaft must make to spin the axle one full turn;
- weight (`wt`): the weight of the vehicle (in 1000 lbs.);
- quarter per mile (`qsec`): the shortest time from a standing start to the end of a straight 1/4 mile track;

	cyl	vs	am	gear	carb	mpg	disp	hp	drat	wt	qsec
cyl	1	0.67	0.27	0.28	0.33	0.73	0.84	0.71	0.49	0.61	0.35
vs	0.67	1	0.03	0.38	0.47	0.44	0.5	0.52	0.19	0.31	0.55
am	0.27	0.03	1	0.65	0.09	0.36	0.35	0.06	0.51	0.48	0.05
gear	0.28	0.38	0.65	1	0.1	0.43	0.59	0.44	0.69	0.43	0.4
carb	0.33	0.47	0.09	0.1	1	0.43	0.26	0.48	0.11	0.34	0.38
mpg	0.73	0.44	0.36	0.43	0.43	1	0.72	0.6	0.46	0.75	0.18
disp	0.84	0.5	0.35	0.59	0.26	0.72	1	0.63	0.5	0.79	0.19
hp	0.71	0.52	0.06	0.44	0.48	0.6	0.63	1	0.2	0.43	0.5
drat	0.49	0.19	0.51	0.69	0.11	0.46	0.5	0.2	1	0.51	0.01
wt	0.61	0.31	0.48	0.43	0.34	0.75	0.79	0.43	0.51	1	0.03
qsec	0.35	0.55	0.05	0.4	0.38	0.18	0.19	0.5	0.01	0.03	1

Fig. 1. Observed $\tilde{\boldsymbol{\Sigma}}$ for the `mtcars` data set

and 5 qualitative variables:

- number of cylinders (`cyl`): low, medium, high;
- engine (`vs`): V-shaped or straight;
- transmission (`am`): automatic or manual;
- number of gears (`gear`): low, medium, high;
- number of carburetors (`carb`): single, low, medium, high.

We computed $\tilde{\boldsymbol{\Sigma}}$ as described in Sect. 3. From Fig. 1, we can clearly see an overall high dependence among the variables in the data set: the largest χ^2 is observed for the *number of cylinders* and the *engine shape*; the largest η^2 between the *number of cylinders* and the *displacement*; the largest ρ^2 between the *displacement* and the *weight*.

The results of the ultrametric model on the `mtcars` data set are provided in Fig. 2. The model applied herein considers a complete hierarchy over the 11 variables. Looking at Fig. 2b, we can highlight four main groups of variables, each one associated with a latent concept: the first composed of *displacement*, *number of cylinders*, *weight*, *miles per gallon* and *gross horsepower*, thus representing the *efficiency* of a car; the second group included *rear axle ration*, *number of gears* and *transmission*, thus identifying the *gearshift features*; the third group formed by *quarter per mile* and *engine*, hence depicting the *performances on the standing start*; the fourth group solely defined by *number of carburetors*, representing a singleton. Remark that we model the dependence between mixed-type data measured into $\tilde{\boldsymbol{\Sigma}}$, whose elements are non-negative and range between 0 and 1. In defining the hierarchical structure, the ultrametric model therefore considers the strength of the dependence among variables, but not its sign. Nonetheless, the direction of dependence can be taken into account for interpreting the results. For instance, in defining the first latent concept, i.e., *efficiency*, negative relationships manifest between *miles per gallon* and *weight*, and *miles per gallon* and *gross horsepower* since the lower the weight and the horsepower of a car are, the higher the performances of a car are and, consequently, its efficiency.

The existence of these four latent concepts, with highly dependent variables defining them, is clearly visible in Fig. 2a, as well as the order of their aggregations. Indeed, the first aggregation occurs between *efficiency* and *gearshift features*, the second one between the *performances on the standing start* and *number of carburetors* and the last one defines the aggregation of the aforementioned two broader groups.

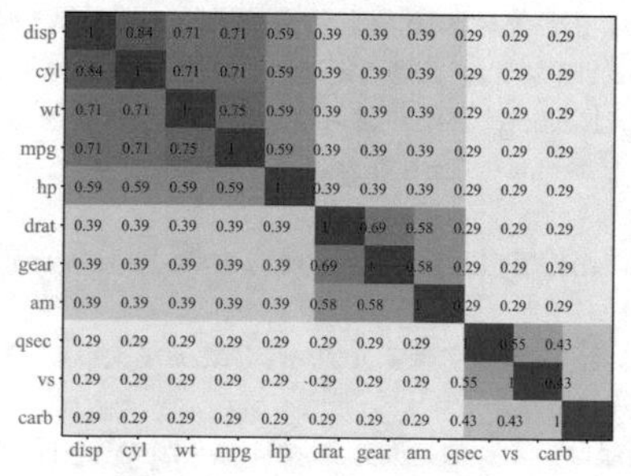

(a) Heatmap of the fitted $\tilde{\boldsymbol{\Sigma}}_u$ on the mtcars data set

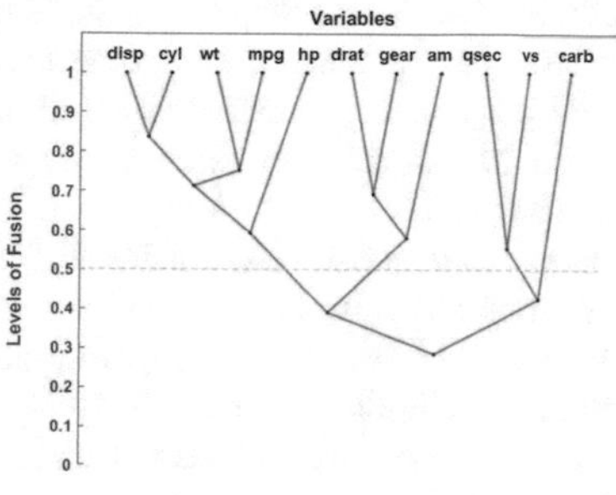

(b) Path diagram representation

Fig. 2. Results of the ultrametric model in Eq. (2).

6 Discussion

In this work, we firstly revised the importance of the general dependence matrix $\tilde{\boldsymbol{\Sigma}}$, which depicts pairwise relationships among variables of different nature. The latter is a fundamental tool for the application of methodologies for dimensionality reduction, such as Factor Analysis and Principal Component Analysis, to mixed-type data. Even if these methodologies are suitable to detect latent structures underlying multivariate data, they are not able to unravel hierarchical relationships among variables. In this paper, we extended the ultrametric model proposed by Cavicchia et al. (2020b) to mixed-type data with the aim of inspecting the relationships among dimensions with different levels of abstraction defining a complex phenomenon.

Further developments of this work can be explored. Indeed, it would be interesting to compare the performances of our proposal with the ones of traditional hierarchical clustering algorithms, i.e., to extended the proof in Cavicchia et al. (2020a) to the more general case of mixed-type data.

References

Cavicchia, C., Vichi, M., Zaccaria, G.: Exploring hierarchical concepts: theoretical and application comparisons. In: Imaizumi, T., Nakayama, A., Yokoyama, S. (eds.) Advanced Studies in Behaviormetrics and Data Science. BQAHB, vol. 5, pp. 315–328. Springer, Singapore (2020). https://doi.org/10.1007/978-981-15-2700-5_19

Cavicchia, C., Vichi, M., Zaccaria, G.: The ultrametric correlation matrix for modelling hierarchical latent concepts. Adv. Data Anal. Classif. **14**(4), 837–853 (2020). https://doi.org/10.1007/s11634-020-00400-z

Chavent, M., Kuentz, V., Liquet, B., Saracco, J.: Clustering of variables via the PCAMIX method. In: International Classification Conference, Saint Andrews, UK (2011). https://hal.archives-ouvertes.fr/hal-00646593

De Leon, A.R., Chough, K.C.: Analysis of Mixed Data: Methods & Applications. Chapman and Hall/CRC Press, New York (2013)

Dellacherie, C., Martinez, S., San Martin, J.: Hadamard functions of inverse M-matrices. In: Inverse M-Matrices and Ultrametric Matrices. LNM, vol. 2118, pp. 165–213. Springer, Cham (2014). https://doi.org/10.1007/978-3-319-10298-6_6

Greenacre, M.: Correspondence Analysis in Practice, 2nd edn. Chapman & Hall/CRC, New York (2017)

Henderson, H.V., Velleman, P.F.: Building multiple regression models interactively. Biometrics **37**, 391–411 (1981)

McParland, D., Gormley, I.C.: Model based clustering for mixed data: clustMD. Adv. Data Anal. Classif. **10**(2), 155–169 (2016)

Pagès, J.: Analyse factorielle de données mixtes. Revue de statistique appliquée **52**(4), 93–111 (2004)

Team RC, et al.: R: a language and environment for statistical computing (2013)

Van de Velden, M., Iodice D'Enza, A., Markos, A.: Distance-based clustering of mixed data. Wiley Interdiscip. Rev. Comput. Stat. **11**(3), e1456 (2019)

Vichi, M., Vicari, D., Kiers, H.A.L.: Clustering and dimension reduction for mixed variables. Behaviormetrika **46**(2), 243–269 (2019). https://doi.org/10.1007/s41237-018-0068-6

Partial Calibrated Multi-label Ranking

Serafín Moral-García[1] and Sébastien Destercke[2(✉)]

[1] University of Granada, Granada, Spain
seramoral@decsai.ugr.es
[2] University of Technology of Compiegne, Compiègne, France
sebastien.destercke@hts.utc.fr

Abstract. An approach to solve Multi-Label Classification (MLC) is to transform this task into a label ranking problem. An example of this is the Calibrated Label Ranking algorithm (CLR). In this research, we propose a new method, called Partial Calibrated Label Ranking (PCLR), that considers a binary classifier for each pair of labels. Such a classifier predicts, for a given instance, the relative relevance of a label versus the other one. Unlike CLR, PCLR predicts relevance through probability intervals. The predicted intervals are combined to obtain a partial order between the labels. Experimental results reveal that our proposed PCLR leads to incomplete but more robust predictions than CLR.

1 Introduction

The *Multi-Label Classification* task (MLC), in which multiple labels can simultaneously belong to a given instance, has attracted much attention in the past years. It generalizes traditional classification, where each instance is associated with a unique value of a class variable. The MLC problem arises in many situations: a music can stir several emotions (Trohidis et al. 2008), proteins can possess several functions (Yu et al. 2013), etc.

A common approach to solve MLC is to transform this problem into a *label ranking* task (Zhao et al. 2015). An example of this approach is the Calibrated Label Ranking method (Fürnkranz et al. 2008). This algorithm considers a binary classifier for each pair of labels which, for a new instance, outputs a score about the relative relevance of one of the labels versus the other one. This leads to a ranking of the labels for the instance. For separating relevant from irrelevant labels, CLR introduces a virtual label.

MLC is a much more complex learning task than standard multi-class classification, prone to make much more prediction mistakes. For this reason, it is interesting to investigate MLC methods that may provide *partial predictions* (Nguyen and Hullermeier 2020; Carranza Alarcón and Destercke 2021a,b). Partial predictions tend to be more robust as they are only made for the labels for which there is sufficient information. The methods developed so far to determine partial multi-label predictions only predict, for a given instance, a set of relevant labels and a set of labels that can be relevant. This may be restrictive because,

L. A. García-Escudero et al. (Eds.): SMPS 2022, AISC 1433, pp. 287–294, 2023.
https://doi.org/10.1007/978-3-031-15509-3_38

given two labels and a certain instance, we can have sufficient information about which label is more relevant for the instance even though there is no sufficient information available to predict whether such labels belong to the instance.

In this work, we propose a new version of the CLR method, called *Partial Calibrated Label Ranking* (PCLR) which, similarly to CLR, considers a binary classifier for each pair of labels. Unlike CLR, it outputs a probability interval about the relative relevance of the first label versus the second one. Then, for each label, the probability intervals of the relative relevances are combined into a unique interval. A dominance criterion is used for determining a partial order of relevance of the labels. To determine the partial multi-label prediction, PCLR uses a virtual label mechanism similar to the one utilized in CLR. We show, via an experimental study, that PCLR yields more robust predictions than CLR, whereas the proportion of incomplete predictions made by PCLR is low.

This paper is structured as follows: Sect. 2 describes the Multi-Label Classification paradigm and the Calibrated Label Ranking method. The Partial Calibrated Label Ranking algorithm is presented in Sect. 3. Section 4 details the experimental analysis carried out in this research.

2 Preliminaries

2.1 Multi-Label Classification

Let $\mathscr{X}$ be an attribute space and $\mathscr{Y} = \{y_1, y_2, \dots, y_q\}$ a set of labels, where $q > 1$. In traditional classification, each instance $\mathbf{x} \in \mathscr{X}$ has a single label $y \in \mathscr{Y}$. In contrast, in Multi-Label Classification (MLC), each instance $\mathbf{x} \in \mathscr{X}$ is associated with a set of labels $Y \subseteq \mathscr{Y}$, which is called the set of *relevant labels* of $\mathbf{x}$. The set $\mathscr{Y} \setminus Y$ is known as the set of *irrelevant labels* of $\mathbf{x}$.

Thus, MLC aims to learn a mapping $h : \mathscr{X} \to 2^{\mathscr{Y}}$ which assigns, to each instance $\mathbf{x} \in \mathscr{X}$, a set of labels $\mathbf{Y} \subseteq \mathscr{Y}$. For this purpose, a training set is employed, namely $\mathscr{D} = \{(\mathbf{x}^\mathbf{i}, \mathbf{Y}^\mathbf{i}), \quad \mathbf{x}^\mathbf{i} \in \mathscr{X}, \mathbf{Y}^\mathbf{i} \subseteq \mathscr{Y} \quad i = 1, 2, \dots, N\}$. We identify each $\mathbf{Y} \subseteq \mathscr{Y}$ with a binary vector $(y_1, y_2, \dots, y_q)$, where $y_j = 1$ if $y_j \in \mathbf{Y}$ and 0 otherwise, $\forall j = 1, 2, \dots, q$.

Ideally, the learned function h assigns to each instance a complete set of labels denoted by a complete vector $(y_1, y_2, \dots, y_q)$, where $y_j \in \{0, 1\} \quad \forall j = 1, 2, \dots, q$. Nonetheless, in many cases, it might be difficult to predict whether a label is relevant for an instance since the information available may not be sufficient. For this reason, we can allow partial predictions such as incomplete sets of relevant labels $\mathbf{Y}^*$ characterized by incomplete vectors $(y_1, y_2, \dots, y_q)$, where $y_j = 1$ ($y_j = 0$) means that y_j is predicted as relevant (irrelevant), and $y_j = *$ indicates that it is not possible to make a prediction about $y_j, \quad \forall j = 1, 2, \dots, q$.

2.2 Calibrated Label Ranking

The CLR method (Fürnkranz et al. 2008) learns a binary classifier for each pair of labels y_j, y_k, with $1 \le j < k \le q$. For this, it considers the training set $\mathcal{D}_{jk} = \left\{(\mathbf{x}^{\mathbf{i}}, y_j^i) \mid y_j^i \neq y_{k'}^i\right\}$. The learned classifier is often given by a function $f_{jk} : \mathcal{X} \to [0, 1]$, which, given an instance $\mathbf{x} \in \mathcal{X}$, predicts the probability $P(y_j \succ_{\mathbf{x}} y_k)$.

When a new instance $\mathbf{x}$ is required to be classified, for each label, the sum of the probabilities of that label being more relevant than each one of the remaining labels is considered:

$$P_j(\mathbf{x}) = \sum_{k=1}^{j-1} (1 - f_{kj}(\mathbf{x})) + \sum_{k=j+1}^{q} f_{jk}(\mathbf{x})), \quad \forall j = 1, 2, \ldots, q. \tag{1}$$

These scores yield a label ranking for $\mathbf{x}$. To separate relevant from irrelevant labels in the ranking, a virtual label y_0 is introduced. For each label, a binary classifier is learned for predicting the relative relevance of the label versus the virtual one, where the target value indicates whether the label is relevant for an instance. Such a classifier is often described via a function $f_j : \mathcal{X} \to [0, 1]$ which, for a given instance $\mathbf{x}$, returns the predicted probability that y_j is more relevant than the virtual one for $\mathbf{x}$.

Then, the final score of each label is computed by adding the predicted probability that the label is more relevant than the virtual one for $\mathbf{x}$:

$$S_j(\mathbf{x}) = P_j(\mathbf{x}) + f_j(\mathbf{x}), \quad \forall j = 1, 2, \ldots, q. \tag{2}$$

The score of the virtual label is computed by summing the probabilities that each label is irrelevant: $S_0(\mathbf{x}) = \sum_{j=1}^{q}(1 - f_j(\mathbf{x}))$.

These scores lead to the predicted complete label ranking for $\mathbf{x}$, $\succ_{\mathbf{x}}$. It is implicitly determined by the scores:

$$y_j \succ_{\mathbf{x}} y_k \Leftrightarrow S_j(\mathbf{x}) \geq S_k(\mathbf{x}), \quad \forall j, k \in \{0, 1, \ldots, q\}. \tag{3}$$

The set of labels predicted as relevant for $\mathbf{x}$ is composed of those labels ranked above the virtual one:

$$h(\mathbf{x}) = \left\{y_j \mid y_j \succ_{\mathbf{x}} y_0\right\}. \tag{4}$$

3 Partial Calibrated Label Ranking

The PCLR method, proposed here, takes the idea of pairwise decomposition used in label ranking methods (Hüllermeier et al. 2008) and the idea proposed in Destercke (2013) of predicting pairwise preferences in form of probability intervals.

For each pair of labels y_j, y_k, with $1 \leq j < k \leq q$, a binary classifier is learned similarly to CLR. Nevertheless, for a given instance x, the learned classifier outputs a probability interval $\left[\underline{f}_{jk}(\mathbf{x}), \overline{f}_{jk}(\mathbf{x})\right]$, where $\underline{f}_{jk}(\mathbf{x})$ and $\overline{f}_{jk}(\mathbf{x})$ indicate, respectively, the lower and upper probabilities that y_j is more relevant than y_k for $\mathbf{x}$. We will use the Imprecise Credal Decision Tree algorithm (ICDT) (Abellán and Masegosa 2012) to get them.

To classify a new instance $\mathbf{x}$, the probability intervals corresponding to the relative relevances are combined to provide an interval score for each label:

$$\begin{aligned}\underline{P}_j(\mathbf{x}) &= \sum_{k=1}^{j-1}\left(1-\overline{f}_{kj}(\mathbf{x})\right) + \sum_{k=j+1}^{q} \underline{f}_{jk}(\mathbf{x}),\\ \overline{P}_j(\mathbf{x}) &= \sum_{k=1}^{j-1}\left(1-\underline{f}_{kj}(\mathbf{x})\right) + \sum_{k=j+1}^{q} \overline{f}_{jk}(\mathbf{x}).\end{aligned} \tag{5}$$

The same procedure is applied to the virtual label y_0, using estimates $[\underline{f}_j(\mathbf{x}), \overline{f}_j(\mathbf{x})]$. The lower and upper scores for the virtual label are derived from these probability intervals:

$$\underline{S}_0(\mathbf{x}) = \sum_{j=1}^{k}\left(1-\overline{f}_j(\mathbf{x})\right), \quad \overline{S}_0(\mathbf{x}) = \sum_{j=1}^{k}\left(1-\underline{f}_j(\mathbf{x})\right). \tag{6}$$

Then, for each label, the final lower and upper scores are computed by adding the lower and upper probabilities that such a label is more relevant than the virtual label for $\mathbf{x}$, respectively:

$$\underline{S}_j(\mathbf{x}) = \underline{P}_j(\mathbf{x}) + \underline{f}_j(\mathbf{x}), \quad \overline{S}_j(\mathbf{x}) = \overline{P}_j(\mathbf{x}) + \overline{f}_j(\mathbf{x}). \tag{7}$$

For determining the partial ranking of the labels for $\mathbf{x}$, $\succ_{\mathbf{x}}$, in accordance with the method proposed in Destercke (2013) for label ranking through pairwise comparisons, we use the *strict dominance* criterion, according to which a label y_j precedes another label y_k if, and only if, the lower score of y_j is higher than the upper score of y_k:

$$y_j \succ_{\mathbf{x}} y_k \Leftrightarrow \underline{S}_j(\mathbf{x}) > \overline{S}_k(\mathbf{x}), \quad \forall j,k \in \{0,1,\ldots,q\}. \tag{8}$$

This criterion is coherent because the narrow the intervals predicted by the binary classifiers are, the more precise the predicted ranking is. Moreover, if the estimated ranking is consistent with the theoretical model, then the prediction can be extended to the optimal complete ranking (Destercke 2013).

In order to predict whether a label is relevant, irrelevant, or undetermined for $\mathbf{x}$, its partial order with regard to the virtual label is checked: if the label is ranked above than the virtual one ($y_j \succ_{\mathbf{x}} y_0$), then such a label is predicted as relevant; if the opposite happens ($y_0 \succ_{\mathbf{x}} y_j$), then the label is predicted as irrelevant; if there is no order between that label and the virtual one, then the label is predicted as undetermined for the instance.

3.1 Justification of Partial Calibrated Label Ranking

Our proposed PCLR method is based on the following observations:

- Estimating a unique probability value that a label is more relevant than another one for a given instance might not be very realistic since there might not be sufficient information for reliably making such a prediction. Thus, it may be more sensible to relax the assumption that this estimate is precise, and allow it to be an interval that reflects the lack of knowledge about which of the two labels is more relevant for the instance.
- In MLC, we can know whether a label is more relevant than another one for a given instance. Nonetheless, there might be pairs of labels for which it is difficult to predict which is more relevant for that instance. For this reason, it is probably more logical to predict the ranking in the form of a partial order rather than a total order.
- The methods proposed so far that allow partial predictions in MLC do not work for multi-label rankings. It might be difficult to predict whether a certain label belongs to a given instance. Nevertheless, even in these cases, it may be possible to predict whether that label is more or less relevant than another one for such an instance. Hence, we consider that the PCLR method presented in this work is more expressive than the algorithms for partial multi-label predictions proposed so far. As a simple example, consider the case of two labels $\{y_1, y_2\}$ and the partial ordering $y_1 \succ_{\mathbf{x}} y_2$. The set of possible multi-label predictions is $\{(0,0), (1,0), (1,1)\}$, which cannot be expressed by a partial vector.

4 Experiments

4.1 Experimental Settings

In this experimental study, seven datasets have been used, which can be found in the website of Mulan (Tsoumakas et al. 2011), a Java library for MLC[1]. Table 1 illustrates the main characteristics of each dataset: number of instances, number of attributes, number of labels, and label density, i.e. average proportion of labels per instance.

For the binary classifiers in CLR, the Credal Decision Tree method (Abellán and Moral 2003) has been used. CDT only differs from ICDT in that the former algorithm predicts precise probabilities.

To compare the performance of CLR and PCLR, we use the following metrics:

- **Completeness**: It indicates the average proportion of predicted pairwise orders between a relevant and an irrelevant label, also considering the pairwise preferences between a label and the virtual one.
- **Correctness_CLR** and **Correctness_PCLR** : They measure the average proportion of correct pairwise preferences among the ones predicted by CLR and PCLR, respectively.
- **Rate Error Covered (REC)**: It indicates, between the pairwise orders erroneously predicted by CLR, the proportion of them that correspond to an abstention in PCLR.

[1] http://mulan.sourceforge.net/datasets-mlc.html.

Table 1. Datasets employed in our experiments. N is the number of instances, N_A is the number of attributes, q is the number of labels, and L_D is the label density

Dataset	N	N_A	q	L_D
birds	645	260	19	0.053
emotions	593	72	6	0.311
flags	194	9	17	0.485
genbase	662	1186	27	0.046
medical	978	1449	45	0.028
scene	2407	294	6	0.179
yeast	2417	103	14	0.303

- **Rate Partial Incorrect (RPI)**: The proportion of abstentions made by PCLR associated with an incorrect pairwise order predicted by CLR.

For each algorithm and dataset considered in this experimental analysis, a cross-validation procedure of five folds has been carried out.

4.2 Results and Discussion

Table 2 presents, for each dataset, the results obtained PCLR in completeness and correctness, the results obtained by CLR in correctness, and the Rate Error Covered (REC) and Rate Partial Incorrect (RPI) results.

The following points should be noted about these results:

- The **Completeness** value is higher than 0.85 in most of the datasets. Consequently, the rate of abstentions of PCLR is generally very low.
- Concerning **Correctness**, PCLR always achieves a higher value than CLR. It means that the predictions made by PCLR are more robust.
- The average **Rate Error Covered** value is higher than 0.4 for all datasets and higher than 0.5 for most datasets. It implies that PCLR avoids many

Table 2. Results obtained over all data sets

Dataset	Completeness	Correctness CLR	Correctness PCLR	REC	RPI
birds	0.9176	0.9439	0.9618	0.7363	0.619
emotions	0.8538	0.7958	0.8354	0.4879	0.5938
flags	0.6825	0.7705	0.8475	0.6985	0.4143
genbase	0.9907	0.9913	0.9924	0.9723	0.7708
medical	0.9658	0.9531	0.9874	0.5878	0.5974
scene	0.9214	0.8875	0.9109	0.5984	0.7106
yeast	0.8635	0.8141	0.8479	0.4692	0.4607

pairwise preferences incorrectly predicted by CLR. Indeed, for most of the datasets, more than half of the prediction errors made by CLR correspond to an abstention with PCLR.

- Due to the results obtained in the **Rate Partial Incorrect** metric, we can state that many abstentions of PCLR are associated with a pairwise order incorrectly predicted by CLR. In fact, the average RPI value is higher than 0.5 for most datasets, which means that it is sensible to abstain from predicting a pairwise preference in CLR when there is not sufficient information available, as PCLR does.

To summarize, PCLR makes less erroneous predictions about pairwise preferences than CLR without the proportion of abstentions made by PCLR being very high. Many pairwise orders incorrectly predicted by CLR are avoided by PCLR by means of an abstention. Also, many of the abstentions made by PCLR avoid a prediction error with CLR.

5 Conclusions and Future Research

The Calibrated Label Ranking method (CLR) transforms the Multi-Label Classification problem into a label ranking task by considering pairwise label preferences and combining them. A new version of CLR, called Partial Calibrated Label Ranking (PCLR), has been presented in this research, which outputs the pairwise preferences in the form of probability intervals.

Unlike CLR, our proposed method allows partial orders, i.e., abstaining from predicting whether a label is more relevant than another one for a given instance. We have argued that relaxing the assumption that the label ranking is always total makes sense because, in Multi-Label Classification, the information available may not always be sufficient for predicting a pairwise preference. In consequence, it might be logical to output partial orders.

An experimental study carried out in this work has highlighted that PCLR outputs more robust label rankings than CLR, the proportion of abstentions not being very high; the abstentions of PCLR avoid many errors; most of the abstentions of PCLR correspond to an error of CLR. Therefore, it makes much sense to relax the assumption that there is always an order between a pair of labels by allowing abstentions.

As future work, other methods for partial label ranking in Multi-Label Classification could be developed. Moreover, it would be interesting to propose other metrics to evaluate the performance of algorithms for label ranking with abstentions in this filed.

Acknowledgements. This work has been supported by UGR-FEDER funds under Project A-TIC-344-UGR20, by the "FEDER/Junta de Andalucía-Consejería de Transformación Económica, Industria, Conocimiento y Universidades" under Project P20_00159, and by research scholarship FPU17/02685.

References

Abellán, J., Masegosa, A.R.: Imprecise classification with credal decision trees. Int. J. Uncertainty Fuzziness Knowl.-Based Syst. **20**(05), 763–787 (2012)

Abellán, J., Moral, S.: Building classification trees using the total uncertainty criterion. Int. J. Intell. Syst. **18**(12), 1215–1225 (2003). https://doi.org/10.1002/int.10143

Carranza Alarcón, Y.C., Destercke, S.: Distributionally robust, skeptical binary inferences in multi-label problems. In: Cano, A., De Bock, J., Miranda, E., Moral, S. (eds.) Proceedings of the Twelveth International Symposium on Imprecise Probability: Theories and Applications. PMLR, Proceedings of Machine Learning Research, vol. 147, pp. 51–60 (2021a)

Alarcón, Y.C.C., Destercke, S.: Multi-label chaining with imprecise probabilities. In: Vejnarová, J., Wilson, N. (eds.) ECSQARU 2021. LNCS (LNAI), vol. 12897, pp. 413–426. Springer, Cham (2021). https://doi.org/10.1007/978-3-030-86772-0_30

Destercke, S.: A pairwise label ranking method with imprecise scores and partial predictions. In: Blockeel, H., Kersting, K., Nijssen, S., Železný, F. (eds.) ECML PKDD 2013. LNCS (LNAI), vol. 8189, pp. 112–127. Springer, Heidelberg (2013). https://doi.org/10.1007/978-3-642-40991-2_8

Fürnkranz, J., Hüllermeier, E., Mencía, E.L., Brinker, K.: Multilabel classification via calibrated label ranking. Mach. Learn. **73**(2), 133–153 (2008)

Hüllermeier, E., Furnkranz, J., Cheng, W., Brinker, K.: Label ranking by learning pairwise preferences. Artif. Intell. **172**, 1897–1916 (2008)

Nguyen, V.L., Hullermeier, E.: Reliable multilabel classification: prediction with partial abstention. Proc. AAAI Conf. Artif. Intell. **34**(04), 5264–5271 (2020)

Trohidis, K., Tsoumakas, G., Kalliris, G., Vlahavas, I.P.: Multi-label classification of music into emotions. ISMIR **8**, 325–330 (2008)

Tsoumakas, G., Spyromitros-Xioufis, E., Vilcek, J., Vlahavas, I.: Mulan: a java library for multi-label learning. J. Mach. Learn. Res. **12**, 2411–2414 (2011)

Yu, G., Domeniconi, C., Rangwala, H., Zhang, G.: Protein function prediction using dependence maximization. In: Blockeel, H., Kersting, K., Nijssen, S., Železný, F. (eds.) ECML PKDD 2013. LNCS (LNAI), vol. 8188, pp. 574–589. Springer, Heidelberg (2013). https://doi.org/10.1007/978-3-642-40988-2_37

Zhao, F., Huang, Y., Wang, L., Tan, T.: Deep semantic ranking based hashing for multi-label image retrieval. In: Proceedings of the IEEE Conference on Computer Vision and Pattern Recognition, pp. 1556–1564 (2015)

Imprecise Learning from Misclassified and Incomplete Categorical Data with Unknown Error Structure

Aziz Omar[1,2,3(✉)] and Thomas Augustin[2]

[1] Department of Psychology, University of the German Federal Armed Forces in Munich, Munich, Germany
aziz.omar@stat.uni-muenchen.de
[2] Department of Statistics, LMU Munich, Munich, Germany
augustin@stat.uni-muenchen.de
[3] Department of Mathematics, Insurance and Applied Statistics, Helwan University, Cairo, Egypt

Abstract. This article addresses the problem of learning from potentially misclassified and incomplete categorical data when the error structure is unknown and no prior information about the distribution of the data is available. We propose to use the knowledge gained from the well-known practice of double sampling to accomplish two goals; First, we estimate the unknown error structure. Then, under the framework of imprecise probability, we derive a prior Dirichlet distribution that expresses a state of quasi-near-ignorance about the data. Updating this prior using sample data leads to a quasi-near-ignorance posterior distribution that produces non-trivial estimates.

1 Introduction

Analysis of categorical data is a common practice in many fields of applications such as health studies, social sciences and marketing research. A deeply rooted problem in analysing categorical data is misclassification because it can lead to severely biased estimates (see, e.g., Bross 1954; Küchenhoff et al. 2012, for demonstration of the resulting bias). Misclassification treatment depends, in general, on incorporating additional knowledge about misclassification probabilities (error structure) available through validation data or contextual assumptions about the error structure (see, e.g., the comprehensive reviews by Manski 2003; Frénay and Verleysen 2013; Hu 2021.

Under the Bayesian framework, different priors were used to express certain contextual assumptions about the error structure (see, e.g., Swartz et al. 2004; Bollinger and Van Hasselt 2017). The problem with most of the diffuse priors used to express prior ignorance about the probability distribution of possible categories is that they do not satisfy basic inference principles e.g., the symmetry principle, the representation invariance principle, the embedding principle, the likelihood principle, the coherence principle (see, e.g., Bernard 2005).

L. A. García-Escudero et al. (Eds.): SMPS 2022, AISC 1433, pp. 295–302, 2023.
https://doi.org/10.1007/978-3-031-15509-3_39

As an alternative, Walley (1996) introduced the *Imprecise Dirichlet Model* (IDM) and explained how it expresses a state of prior *near-ignorance*. The IDM gained a lot of attention and popularity among other models under the general framework of "Imprecise Probabilities" (see, e.g., Bernard and Ruggeri 2009; Augustin et al. 2014).

The employment of the IDM to learn from misclassified categorical data is, however, problematic as Piatti et al. (2009) showed that even when the error structure is known, the IDM cannot be used to learn from misclassified categorical data. As a solution, Omar et al. (2022) derived a *quasi-near-ignorance* prior corresponding to a condition that is easily met in most practical situations and showed that under this condition the IDM can be used to learn from misclassified categorical data with known error structure.

In this paper, we extend the forenamed work to the case of an unknown error structure. Specifically, we use the knowledge gained from applying a double sampling approach proposed by Tenenbein (1972) to estimate the unknown error structure and additionally, derive a quasi-near-ignorance prior under which the IDM can be used to learn from misclassified categorical data.

The rest of this paper is organized as follows. Section 2 presents the situation involving potentially misclassified and incomplete categorical data and the double sampling approach to estimate the unknown error structure. In Sect. 3, we utilize the knowledge obtained from the double sampling procedure to derive a quasi-near-ignorance prior and explain how to use it to learn from misclassified categorical data. In Sect. 4, we demonstrate our proposed approach using a simple numeric example. Finally, Sect. 5 discuses the result and outlines further possible extensions.

2 Misclassified and Incomplete Categorical Data

Consider a categorical (multinomial) random variable X that has $K \geq 2$ possible categories where population units belong to category j, $j = 1, 2, \cdots, K$, with probability θ_j, $0 \leq \theta_j$, $\sum_{j=1}^{K} \theta_j = 1$. It is of interest to infer $\boldsymbol{\theta} := (\theta_1, \theta_2, \cdots, \theta_K)$. Therefore, a sample of size n is randomly selected and exposed to a certain classifier in order to record the category of each sample unit. If the classifier is perfect, a dataset $\boldsymbol{x} := (x_1, x_2, \cdots, x_K)$, $0 \leq x_j$, $\sum_{j=1}^{K} x_j = n$ is obtained, where x_j is the true count in category j, $j = 1, 2, \cdots, K$.

If, however, the classifier is fallible such that it allows for misclassification and/ or missingness, we do not obtain the true dataset $\boldsymbol{x}$. Instead, we obtain the dataset $\boldsymbol{o} := (o_1, o_2, \cdots, o_K, o_{K+1})$, $0 \leq o_i$, $\sum_{i=1}^{K+1} o_i = n$, whose i-th element is the observed count in category i, $i = 1, 2, \cdots, K+1$ of an observed multinomial variable O that has $K+1$ possible categories. For simplicity, assume, without lose of generality, that the first K categories of O are equivalent to the K categories of X, while the $(K+1)$-th category of O represents unclassified sample units due to missingness.

The error structure is defined through the misclassification probabilities $\pi_{ij} := \mathrm{P}(O = i|X = j)$, $0 \leq \pi_{ij}$, $\sum_{i=1}^{K+1} \pi_{ij} = 1$, $i = 1, 2, \cdots, K+1$, $j =$

$1, 2, \cdots, K$. To estimate π_{ij} that are usually unknown, a double sampling scheme can be utilized where after observing $\boldsymbol{o}$, a sub-sample of m units where, $m < n$, is examined using a gold standard classifier that is believed to produce perfect classifications.[1] The result of the double sample procedure can is summarized in Table 1.

Table 1. Cross-classification of the m sub-sample units into their X and O categories

		X				
		1	2	$\cdots$	K	Σ
O	1	m_{11}	m_{12}	$\cdots$	m_{1K}	m_{1+}
	2	m_{21}	m_{22}	$\cdots$	m_{2K}	m_{2+}
	$\vdots$	$\vdots$	$\vdots$	$\ddots$	$\vdots$	$\vdots$
	K	m_{K1}	m_{K2}	$\cdots$	m_{KK}	m_{K+}
	$K+1$	$m_{(K+1)1}$	$m_{(K+1)2}$	$\cdots$	$m_{(K+1)K}$	$m_{(K+1)+}$
	Σ	m_{+1}	m_{+2}	$\cdots$	m_{+K}	m

Tenenbein (1972) derived the following maximum likelihood estimator of π_{ij} using the double sampling approach

$$\hat{\pi}_{ij} = \frac{o_i m_{ij}}{n\hat{\theta}_j m_{i+}}, \quad i = 1, 2, \cdots, K+1,\ j = 1, 2, \cdots, K. \tag{1}$$

An obvious problem in estimating π_{ij} using the estimator in (1) is the need for an estimate of the unknown θ_j. We discuss this issue in more details in Sect. 3.

3 Learning from Misclassified Categorical Data

In the IDM, prior near-ignorance about $\boldsymbol{\theta}$ is represented by defining the set $\mathscr{M}_s$ of prior densities such that it consists of all Dirichlet densities with an equivalent sample size s. That is, the general form of the elements of $\mathscr{M}_s$ is

$$dir(s, \boldsymbol{t})(\boldsymbol{\theta}) := \frac{\Gamma(s)}{\prod_{j=1}^{K} \Gamma(st_j)} \prod_{j=1}^{K} \theta_j^{st_j - 1}, \tag{2}$$

where $s > 0$, $\boldsymbol{t} = (t_1, t_2, \cdots, t_K)$, $0 < t_j < 1$, $\sum_{j=1}^{K} t_j = 1$. The parameter s represents the degree of prior near-ignorance about $\boldsymbol{\theta}$ (see, e.g., Bernard (2005) for a discussion of choice of s as a representative of the *total prior strength*). Assuming a perfect classifier, according to Walley's general coherence theory (Walley 1991, Chapter 6), each prior density in $\mathscr{M}_s$ is updated using observed

[1] Tenenbein (1972) discussed some rules to calculate m based on observation cost and desired precision.

true counts into a corresponding Dirichlet posterior density in the set $\mathscr{M}_{n+s}$ of posterior densities such that, the general form of the elements in $\mathscr{M}_{n+s}$ is

$$dir(n+s,\boldsymbol{x}+\boldsymbol{t})(\boldsymbol{\theta}|\boldsymbol{x}) := \frac{\Gamma(n+s)}{\prod_{j=1}^{K}\Gamma\left(x_j+st_j\right)}\prod_{j=1}^{K}\theta_j^{x_j+st_j-1}. \tag{3}$$

Consequently, estimates of functions of $\boldsymbol{\theta}$ are interval-valued and their upper and lower bounds are obtained by optimizing their classical single-valued counterparts with respect to $\boldsymbol{t}$. To illustrate, consider, for example, the classical single-valued posterior predictive probability used to learn about the chances of the next observation, which takes the form

$$P(X=j|\boldsymbol{x}) := \mathbb{E}(\theta_j|\boldsymbol{x}) = \frac{x_j+st_j}{n+s}, \qquad j=1,2,\cdots,K. \tag{4}$$

The equivalent interval-valued estimate for $P(X=j|\boldsymbol{x})$ is received as

$$\begin{aligned}\left(\underline{P}(X=j|\boldsymbol{x}),\overline{P}(X=j|\boldsymbol{x})\right) &:= \left(\lim_{t_j\to 0}P(X=j|\boldsymbol{x}),\lim_{t_j\to 1}P(X=j|\boldsymbol{x})\right)\\ &= \left(\frac{x_j}{n+s},\frac{x_j+s}{n+s}\right), \qquad j=1,2,\cdots,K.\end{aligned} \tag{5}$$

3.1 Incorporating the Knowledge Gained from Double Sampling

The knowledge obtained from the double sampling scheme described in Sect. 2 can be used to update our beliefs about $\boldsymbol{\theta}$. Consider the case where the following condition is satisfied

$$m_{+j} \geq 1, \qquad j=1,2,\cdots,K. \tag{6}$$

That is, the sub-sample of size m has at least one unit in each possible category $j=1,2,\cdots,K$. Using the corrected counts from the sub-sample, an *initial* maximum likelihood estimator of θ would be

$$\hat{\theta}_j = \frac{m_{+j}}{m}, \qquad j=1,2,\cdots,K. \tag{7}$$

Hence, the misclassification probabilities π_{ij} can be estimated using (1) such that

$$\hat{\pi}_{ij} = \frac{mo_i}{n}\frac{m_{ij}}{m_{+j}m_{i+}}, \quad i=1,2,\cdots,K+1,\ j=1,2,\cdots,K. \tag{8}$$

Moreover, assuming a state of prior near-ignorance represented by the set $\mathscr{M}_s$, the counts m_{+j}, $j=1,2,\cdots,K$ can used to update the elements in $\mathscr{M}_s$ into corresponding elements in a new set $\mathscr{M}_{m+s}$ where the general form of the elements in the later set is

$$dir\left(m+s,\boldsymbol{t}\right)(\boldsymbol{\theta}) := \frac{\Gamma\left(m+s\right)}{\prod_{j=1}^{K}\Gamma\left(m_{+j}+st_j\right)}\prod_{j=1}^{K}\theta_j^{m_{+j}+st_j-1}. \tag{9}$$

3.2 Learning Under the State of Quasi-Near-Ignorance

In the sense of Omar et al. (2022), the knowledge implied through condition (6) and represented through $\mathscr{M}_{m+s}$ defines a state of prior quasi-near-ignorance about $\boldsymbol{\theta}$ when the remaining potentially misclassified $n-m$ sample units are considered. Theorem 1 shows that it is possible to learn from misclassified categorical data under the state of prior quasi-near-ignorance.

Theorem 1. *Under the state of prior quasi-near-ignorance implied through condition* (6)*, the posterior predictive probability* $P_c(X = j|\boldsymbol{o})$ *has non-trivial boundaries such that*

$$0 < \underline{P_c}(X = j|\boldsymbol{o}) < \overline{P_c}(X = j|\boldsymbol{o}) < 1, \quad j = 1, 2, \cdots, K. \tag{10}$$

Proof. Omitted due to space limitations.[2] □

The binary version of the posterior predictive probability under condition (6) can be expressed as

$$P(X = j|\boldsymbol{o}) = \frac{m_{+j} + st_j}{n+s} + \frac{\sum_{h=0}^{\tilde{o}_1}\sum_{l=0}^{\tilde{o}_2}\sum_{r=0}^{\tilde{o}_3}(h+l+r)f_{\tilde{n}}}{(n+s)\sum_{h=0}^{\tilde{o}_1}\sum_{l=0}^{\tilde{o}_2}\sum_{r=0}^{\tilde{o}_3}f_{\tilde{n}}}, \quad j = 1, 2, \tag{11}$$

where

$$f_{\tilde{n}} := \binom{\tilde{o}_1}{h}\binom{\tilde{o}_2}{l}\binom{\tilde{o}_3}{r}\left(\frac{\pi_{11}}{\pi_{12}}\right)^h\left(\frac{\pi_{21}}{\pi_{22}}\right)^l\left(\frac{\pi_{31}}{\pi_{32}}\right)^r$$
$$\Gamma(h+l+r+st_1+m_{+1})\Gamma(n-m-h-l-r+st_2+m_{+2})$$

and $\tilde{o}_i := o_i - m_{i+}$, $i = 1, 2, 3$.

It is notable that the posterior predictive probability in (11) has non-trivial upper and lower boundaries when $t \to 0$ and $t \to 1$, respectively, which means that the IDM can be used to learn about $\boldsymbol{\theta}$ under the assumed setting.

4 Example

To infer about success probability θ of a binary variable X, a sample of size 20 is drawn and the observed dataset $\boldsymbol{o} = (8, 12, 0)$ is obtained. Out of the 20 sampled units, a sub-sample of size $m = 6$ is drawn and its units are exposed to a gold standard classifier. Table 2 displays the cross-classification of the units in the sub-sample.

[2] Interested readers are referred to Omar et al. (2022, Theorem 3).

Table 2. Cross-classification of sub-sample units into their X and O categories

		X		
		1	2	$\sum$
O	1	2	1	3
	2	2	1	3
	3	0	0	0
	$\sum$	4	2	6

As previously discussed, the result of the sub-sample is used to estimate π_{ij} as in (8). Then, these estimates are plugged in the binary version of the posterior predictive probability in (11). If we set $s = 1$, we receive the interval-valued estimate

$$\left(\underline{P_c}(X=1|\boldsymbol{o}), \overline{P_c}(X=1|\boldsymbol{o})\right) = (0.571, 0.714). \tag{12}$$

If, however, we ignore the possibility that the classifier is fallible and treat the observed dataset $\boldsymbol{o}$ as if it were the true dataset $\boldsymbol{x}$ (without double sampling) using (5) and setting $s = 1$, we obtain the interval-valued estimate

$$\left(\underline{P}(X=1|\boldsymbol{x}), \overline{P}(X=1|\boldsymbol{x})\right) = (0.381, 0.429). \tag{13}$$

The difference between the two resulting estimates indicates the effect of taking misclassification into account.

5 Discussion

In this article, we proposed a solution to the problem of learning from categorical data subject to misclassification with an unknown error structure under the state of prior near-ignorance using the IDM. Our solution utilizes the known procedure of double sampling and the resulting split of the data to estimate the unknown misclassification probabilities and update our prior beliefs about the distribution of the data. In doing so, we followed an approach that mixes both the frequentist and the Bayesian frameworks. This blending has quite a long history in the literature (see, e.g., Bayarri and Berger 2004; Raue et al. 2013; Bickel 2015.

Our proposed treatment is a step towards an alternative class of near-ignorance priors where the IDM can be utilized to learn in complex data situations. An immediate extension to the current work is to compare our results to those of the existing approaches to misclassification treatment, such as the employment of the Imprecise Sample Size Dirichlet Model suggested by Masegosa and Moral (2014). Another possible extension to the current work is, for example, to utilize the IDM to model the prior near-ignorance about the unknown error structure as well as the multinomial distribution of the data and devise alternatives if the classical application of the IDM leads to empty estimates.

Acknowledgements. The authors are grateful to an anonymous reviewer for their helpful comments and suggestions.

References

Augustin, T., Coolen, F.P.A., de Cooman, G., Troffaes, M.C.M. (eds.): Introduction to Imprecise Probabilities. Wiley Series in Probability and Statistics, Wiley, Chichester (2014)

Bayarri, M.J., Berger, J.O.: The interplay of Bayesian and frequentist analysis. Stat. Sci. **19**(1), 58–80 (2004)

Bernard, J.-M.: An introduction to the imprecise Dirichlet model for multinomial data. Int. J. Approximate Reasoning **39**(2–3), 123–150 (2005)

Bernard, J.-M., Ruggeri, F. (eds.): Special section on the imprecise Dirichlet model (Issues in imprecise probability). Int. J. Approximate Reasoning **50**(2), 201–268 (2009)

Bickel, D.R.: Blending Bayesian and frequentist methods according to the precision of prior information with applications to hypothesis testing. Stat. Methods Appl. **24**(4), 523–546 (2015). https://doi.org/10.1007/s10260-015-0299-6

Bollinger, C.-R., Van Hasselt, M.: A Bayesian analysis of binary misclassification. Econ. Lett. **156**, 68–73 (2017)

Bross, I.: Misclassification in 2 × 2 tables. Biometrics **10**(4), 478–486 (1954)

Frénay, B., Verleysen, M.: Classification in the presence of label noise: a survey. IEEE Trans. Neural Netw. Learn. Syst. **25**(5), 845–869 (2013)

Hu, Z.H.: Dirichlet process probit misclassification mixture model for misclassified binary data. Ph.D. thesis, UCL (2021). https://discovery.ucl.ac.uk/id/eprint/10140643/. Cited 15 May 2022

Küchenhoff, H., Augustin, T., Kunz, A.: Partially identified prevalence estimation under misclassification using the Kappa coefficient. Int. J. Approximate Reasoning **53**(8), 1168–1182 (2012)

Manski, C.F.: Partial Identification of Probability Distributions. Springer, New York (2003). https://doi.org/10.1007/b97478

Masegosa, A.R., Moral, S.: Imprecise probability models for learning multinomial distributions from data. Applications to learning credal networks. Int. J. Approximate Reasoning **55**(7), 1548–1569 (2014)

Omar, A., von Oertzen, T., Augustin, T.: Learning from categorical data subject to non-random misclassification and non-response under prior quasi-near-ignorance using an imprecise Dirichlet model. In: Ciucci, D., et al. (eds.) Proceedings of the 19th International Conference on Information Processing and Management of Uncertainty in Knowledge-Based Systems (IPMU 2022) (2022, forthcoming). https://link.springer.com/chapter/10.1007/978-3-031-08974-9_43

Piatti, A., Zaffalon, M., Trojani, F., Hutter, M.: Limits of learning about a categorical latent variable under prior near-ignorance. Int. J. Approximate Reasoning **50**(4), 597–611 (2009)

Raue, A., Kreutz, C., Theis, F.J., Timmer, J.: Joining forces of Bayesian and frequentist methodology: a study for inference in the presence of non-identifiability. Philos. Trans. R. Soc. A Math. Phys. Eng. Sci. **371**(1984), 20110544 (2013)

Swartz, T., Haitovsky, Y., Vexler, A., Yang, T.: Bayesian identifiability and misclassification in multinomial data. Can. J. Stat. **32**(3), 285–302 (2004)

Tenenbein, A.: A double sampling scheme for estimating from misclassified multinomial data with applications to sampling inspection. Technometrics **14**(1), 187–202 (1972)

Walley, P.: Statistical Reasoning with Imprecise Probabilities. Chapman and Hall, London (1991)

Walley, P.: Inferences from multinomial data: learning about a bag of marbles. J. Roy. Stat. Soc. **58**(1), 3–57 (1996)

Note on Efron's Monotonicity Property Under Given Copula Structures

Patricia Ortega-Jiménez[1], Franco Pellerey[2(✉)], Miguel A. Sordo[1], and Alfonso Suárez-Llorens[1]

[1] Universidad de Cádiz, 11002 Cádiz, Spain
{patricia.ortega,mangel.sordo,alfonso.suarez}@uca.es
[2] Politecnico di Torino, 10129 Turin, Italy
franco.pellerey@polito.it

Abstract. Given a multivariate random vector, Efron's marginal monotonicity (EMM) refers to the stochastic monotonicity of the variables given the value of their sum. Recently, based on the notion of total positivity of the joint density of the vector, Pellerey and Navarro (2021) obtained sufficient conditions for EMM when the monotonicity is in terms of the likelihood ratio order. We provide in this paper new sufficient conditions based on properties of the marginals and the copula. Moreover, parametric examples are provided for some of the results included in Pellerey and Navarro (2021) and in the present paper.

1 Introduction and Background

Given a random vector of independent continuous marginals with logconcave densities, Efron (1965) studied the stochastic monotonicity of the marginals given the value of their sum, obtaining the following result.

Proposition 1 (Efron 1965**).** *Let $X_1, X_2, \cdots, X_n$ be n independent random variables with ILR densities, let $S = \sum_{i=1}^{n} X_i$ be their sum, and let $\Phi(x_1, x_2, \cdots, x_n)$ be a real measurable function, increasing in each of its arguments. Then, the function $s \mapsto E(\Phi(X_1, X_2, \cdots, X_n) \mid S = s)$ is increasing.*

Throughout this paper, the term "increasing" is used for "non-decreasing" and "decreasing" is used for "non-increasing". Recall that a continuous random variable X having density f is said to have the *increasing likelihood ratio* (ILR) property if $\frac{f(x+y)}{f(x)}$ decreases in x for all $y \geq 0$, i.e., if $\log f(x)$ is concave. X is said to have the *increasing proportional likelihood ratio* ($IPLR$) property if $f(\lambda x)/f(x)$ is increasing in x for any positive constant $\lambda < 1$. Equivalently, a random variable X with density f is $IPLR$ if and only if $x\ \eta(x)$ increases in x, where $\eta(x) = -f'(x)/f(x)$ (see Oliveira and Torrado 2015). Note that ILR implies $IPLR$, but the reverse does not hold (Ramos and Sordo 2001). Let us also recall that, given X and Y two continuous random variables with respective distribution functions F, G and densities f, g, respectively, X is said to be smaller

L. A. García-Escudero et al. (Eds.): SMPS 2022, AISC 1433, pp. 303–310, 2023.
https://doi.org/10.1007/978-3-031-15509-3_40

than Y in the *usual stochastic order* ($X \leq_{st} Y$) if $F(x) \geq G(x)$ for all $x \in \mathbb{R}$. The order $X \leq_{st} Y$ holds if and only if, for all increasing functions $\phi : \mathbb{R} \to \mathbb{R}$, $\mathrm{E}[\phi(X)] \leq \mathrm{E}[\phi(Y)]$, provided that these expectations exist. Analogously, X is said to be smaller than Y in the *likelihood ratio order* ($X \leq_{lr} Y$) if $f(x)/g(x)$ is decreasing in the union of the supports (Shaked and Shanthikumar 2007).

From now on, based in Proposition 1, we will refer as "Efron's strong monotonicity" (ESM) to the monotonicity of $s \mapsto \{(X_1, X_2, \cdots, X_n) \mid S = s\}$ in terms of any stochastic order, and "Efron's marginal monotonicity" (EMM) to the monotonicity of $s \mapsto \{X_i \mid S = s\}$. Efron's monotonicity and its subsequent generalizations have been of great interest in different areas, as economics, combinatorial probability, dependence modeling and statistical theory. For a list of references on its applications, the interested reader may consult Saumard and Wellner (2018); Pellerey and Navarro (2021).

The results obtained in Efron (1965) have been extended in several ways. Lehmann (1966) showed that the conditions stated in Proposition 1 imply EMM in terms of the lr-order. More recently, Saumard and Wellner (2018) extended Efron's results to bivariate vectors with non-independent variables, providing conditions in terms of the second derivatives of $-logf(x, y)$, which imply ESM and EMM in the usual stochastic order. In the same framework, Oudghiri (2021) provided sufficient conditions for an stronger ESM and EMM assumption, considering not only that $\phi(s) = E\left(\Phi\left(X_1, X_2, \cdots, X_n\right) \mid S = s\right)$ increases in s, but also that $\alpha(s)\phi(s)$ increases in s, for some functions α. Also in the bivariate setting, for non-independent vectors, Pellerey and Navarro (2021) give sufficient conditions which imply ESM in the usual stochastic order. They also generalize the result in Lehmann (1966), connecting the EMM monotonicity in the lr-order to the notion of total positivity.

Let us recall that, in the bivariate framework, a function $p : \mathbb{R}^2 \to \mathbb{R}^+$ is said to be *totally positive of order 2* (TP_2) if, for $x_1 < x_2$ and $y_1 < y_2$, it is verified that $p(x_2, y_2)p(x_1, y_1) \geq p(x_1, y_2)p(x_2, y_1)$. Let us note that, for such p, if at each point (x, y), the second order partial derivative $\frac{\partial^2}{\partial x \partial y} \log(p(x, y))$ exists, then p is TP_2 if and only if $\frac{\partial^2}{\partial x \partial y} \log(p(x, y)) \geq 0$ (Karlin (1967)).

Pellerey and Navarro (2021) provide the following results.

Proposition 2 (Pellerey and Navarro 2021**).** *Let the vector (X_1, X_2) have a joint density f. Then, the following conditions are equivalent:*

1. *The function $f(x, s - x)$ is TP_2 in (x, s);*
2. *$[X_1 \mid S = s_1] \leq_{lr} [X_1 \mid S = s_2]$ whenever $s_1 \leq s_2$;*
3. *$[S \mid X_1 = x_1] \leq_{lr} [S \mid X_1 = x_2]$ whenever $x_1 \leq x_2$.*

Proposition 3 (Pellerey and Navarro 2021**).** *Let the vector (X_1, X_2) have a joint density f. If $f\left(x_1, x_2\right)$ is TP_2 in (x_1, x_2) and logconcave in x_2 (respectively, x_1) for every x_1 (respectively, x_2), then $f(x, s - x)$ $(f(s - x, x))$ is TP_2 in (x, s).*

Let $\mathbf{X} = (X_1, X_2)$ be a random vector with joint density f and survival copula $\hat{C}$. Let $\hat{c}$ the second mixed partial derivative of $\hat{C}$ (here $\hat{c}$ is referred to as the density of the survival copula $\hat{C}$). As pointed out in Example 2.4 in Pellerey and Navarro (2021), $f(x_1, x_2)$ is TP_2 in (x_1, x_2) if and only if $\hat{c}(u_1, u_2)$ is TP_2. However, the condition $\hat{c}(u, z-u)$ is TP_2 in (u, z) does not imply that $f(x, s-x)$ is TP_2 in (x, s). This suggests to find conditions on $\hat{c}$ and the marginals implying that $f(x_1, x_2)$ is TP_2 in (x_1, x_2). This is what we do in Sect. 2 below. In Sect. 3, we provide a list of copulas that, when joined to exponential or uniform marginals, imply that $f(x, s-x)$ is TP_2 in (x, s). The final Sect. 4, instead, is devoted to an application of Proposition 3 in the context of generalized order statistics (GOSs).

2 Efron's Marginal Monotonocity in Terms of the Copula

Given a random vector (X_1, X_2) with exponential marginals and joint density f, our first result provides a conditions on the density of the corresponding survival copula $\hat{c}$ that ensures that $f(x, y-x)$ is TP_2 in (x, y). Note that $f(x_1, x_2) = \hat{c}(\bar{F}_1(x_1), \bar{F}_2(x_2)) f_1(x_1) f_2(x_2)$.

Proposition 4. *Let the vector (X_1, X_2), with $X_1, X_2 \sim \exp(\lambda)$ have joint density f and density of the survival copula $\hat{c}$. If $\hat{c}(u, v/u)$ is TP_2 in (u, v), for all $0 < u < 1, 0 < v < u$, then $f(x, y-x)$ is TP_2 in (x, y).*

Proof. The function $f(x, s-x)$ is TP_2 in (x, s) if

$$\frac{f(x, s_2 - x)}{f(x, s_1 - x)} = \frac{\hat{c}(\bar{F}_1(x), \bar{F}_2(s_2 - x)) f_1(x) f_2(s_2 - x)}{\hat{c}(\bar{F}_1(x), \bar{F}_2(s_1 - x)) f_1(x) f_2(s_1 - x)} = \frac{\hat{c}(\bar{F}_1(x), \bar{F}_2(s_2)/\bar{F}_2(x))}{\hat{c}(\bar{F}_1(x), \bar{F}_2(s_1)/\bar{F}_2(x))}$$

is increasing in x for all $s_1 < s_2$. The monotonicity follows from the fact that $\bar{F}_2(s-x) = \frac{\bar{F}_2(s)}{\bar{F}_2(x)}$, $\frac{f_2(s_2-x)}{f_2(s_1-x)}$ is constant in x and $\frac{\hat{c}(u, v_1/u)}{\hat{c}(u, v_2/u)}$ is decreasing in u for $v_1 < v_2$. This last condition is equivalent to say that $\hat{c}(u, v/u)$ is TP_2 in (u, v), for all $0 < u < 1, 0 < v < u$.

In Proposition 3, the joint density function $f(x_1, x_2)$ is required to be logconcave in x_2 for every x_1, which is equivalent to say that $\{X_2 | X_1 = x_1\}$ is ILR for all x_1. This condition can be weakened when it is expressed in terms of the density of the survival copula whenever the marginal X_2 is exponential.

Proposition 5. *Let (X_1, X_2) be a random vector with joint density f and density of the survival copula $\hat{c}$. Let $X_2 \sim \exp(\lambda)$. If $\hat{c}(u, v)$ is TP_2 in (u, v) and $\{U_2 | U_1 = u_1\}$ is $IPLR$, where $U_i \sim \bar{F}_i(X_i)$ for $i = 1, 2$ for all u_1, then $f(x, y-x)$ is TP_2 in (x, y).*

Proof. Since $\hat{c}(u, v)$ is TP_2 in (x_1, x_2), then $f(x_1, x_2)$ is TP_2 in (x_1, x_2). By Proposition 3, it remains to see that $f(x_1, x_2)$ is logconcave in x_2 for all x_1. This is the same as proving that

$$\frac{f(x_1, x_2 + y)}{f(x_1, x_2)} = \frac{\hat{c}(\bar{F}_1(x_1), \bar{F}_2(x_2 + y)) f_2(x_2 + y)}{\hat{c}(\bar{F}_1(x_1), \bar{F}_2(x_2)) f_2(x_2)} = \frac{\hat{c}(\bar{F}_1(x_1), e^{-\lambda y} \bar{F}_2(x_2))}{\hat{c}(\bar{F}_1(x_1), \bar{F}_2(x_2))}$$

is decreasing in x_2 for all x_1 and $y \geq 0$, where we have used that $X_2 \sim \exp(\lambda)$. Taking into account that $f_{\{U_2|U_1=v\}}(u) = \hat{c}(v,u)$, this follows from the fact that $\{U_2|U_1 = u_1\}$ is $IPLR$ for all u_1, which is equivalent to say that $f_{\{U_2|U_1=v\}}(\alpha u)/f_{\{U_2|U_1=v\}}(u)$ increases in u for all v and any $0 < \alpha \leq 1$.

It should be noted that the sufficient conditions in Proposition 4 neither imply, nor are implied, by the conditions in Proposition 5. For example, if $X_1, X_2 \sim \exp(\lambda)$ and $\hat{C}$ is the Ali-Mikhail-Haq copula with parameter $\theta \leq 1$ (see Sect. 3.3) then (X_1, X_2) satisfies the conditions on Proposition 4 but not those in Proposition 5. Similarly, if $X_2 \sim \exp(\lambda)$, $X_1 \nsim \exp(\lambda)$ and $\hat{C}$ is a Clayton copula (see Sect. 3.2) then (X_1, X_2) satisfies the conditions on Proposition 5 but not those in Proposition 4.

3 Examples of Copulas

Given a copula C (joint distribution function of (U_1, U_2)) with density c, we consider the following properties:

(P1) $c(u, s-u)$ is TP_2 in (u, s) for $0 < u < 1, u < s < 1+u$.
(P2) $c(u, v)$ is TP_2 in (u, v) for $0 < u < 1, 0 < v < 1$.
(P3) $c(u, v)$ is logconcave in v for all u.
(P4) $c(u, v/u)$ is TP_2 in (u, v) for $0 < u < 1, 0 < v < u$.
(P5) $v\eta_{\{U_2|U_1=u\}}(v) = -(v\frac{\partial}{\partial v}c(u,v))/c(u,v)$ increases in v for all u.

Given $\mathbf{X} = (X_1, X_2)$ with copula C and survival copula $\hat{C}$, if any of these conditions hold:

- $X_1, X_2 \sim U(0,1)$ and C verifies (P1) (Proposition 2),
- $X_1, X_2 \sim U(0,1)$ and C verifies (P2) and (P3) (Proposition 3),
- $X_1, X_2 \sim \exp(\lambda)$ and $\hat{C}$ verifies (P4) (Proposition 4),
- $X_1 \sim \exp(\lambda)$ and $\hat{C}$ verifies (P2) and (P5) (Proposition 5),

then $f(x, y-x)$ is TP_2 in (x, y). Next, we provide examples of parametric families of copulas satisfying the property (P2) that also satisfy some of the other properties. Note that this is not the general case, for example, copulas (4.1.4) (Gumbel-Hougaard copula) and (4.1.2) in Nelsen (2007) satisfy (P2) but not (P1), (P3), (P4) or (P5).

3.1 Farlie-Gumbel-Morgenstern Copula

Let consider the copula given by $C_\theta(u,v) = uv(1+\theta(1-u)(1-v))$, for $\theta \in [-1,1]$, $u, v \in [0,1]$. The copula density $c_\theta(u,v) = 1 + \theta(1-2v)(1-2u)$ is TP_2 for $\theta \in [0,1]$, thus (P2) holds and, for such values, $\frac{d}{ds}\frac{d}{du}(\log(c_\theta(u, s-u)) \geq 0$ for all $0 < u < 1, u < s < u+1$, and (P1) holds. Moreover, $\frac{d^2}{dv^2}(\log(c_\theta(u,v)) \leq 0$, therefore c_θ verifies (P3). Analogously, as $\frac{d}{dv}\frac{d}{du}(\log(c_\theta(u, v/u))) \geq 0$ for all $\theta \in [0,1]$, (P4) is verified for such values. Finally, as $\frac{d}{dv}\left(\frac{-v\frac{\partial}{\partial v}c(u,v)}{c(u,v)}\right) \geq 0$ if and only if $(1-2u)\theta \geq 0$, (P5) does not hold for all u and $\theta \neq 0$.

3.2 Clayton Copula

Let consider the copula 1 in Table 4.1 in Nelsen (2007), given by $C_\theta(u,v) = (u^{-\theta}+v^{-\theta}-1)^{-1/\theta}$ for $\theta > -1$ with $\theta \neq 0$ (note that, when θ tends to zero, this is the independence copula, which trivially verifies (P1)-(P5), as the density is 1). The density is given by $c_\theta(u,v) = u^{-1-\theta}v^{-1-\theta}\left(u^{-\theta}+v^{-\theta}-1\right)^{-2-\frac{1}{\theta}}(1+\theta)$. It is easy to see that, for $\theta > 0$, (P2) holds, but (P1) does not. Consequently, (P3) also fails to be satisfied (because (P2) and (P3) imply (P1)). It can be easily computed that both $\frac{d}{dv}\frac{d}{du}(\log(c_\theta(u,v/u)))$ and $\frac{d}{dv}\left(\frac{-v\frac{\partial}{\partial v}c(u,v)}{c(u,v)}\right)$ are positive for all $\theta > 0$, so, for such values, (P4) and (P5) holds. Here, we can see an example of the fact that $IPLR$ does not imply ILR. Considering a vector (U_1, U_2) with a Clayton copula, there are values of u_1 for which $\{U_2|U_1 = u_1\}$ is $IPLR$ but it is not ILR.

3.3 Ali-Mikhail-Haq Copula

Let us now consider the copula 3 in Table 4.1 in Nelsen (2007), given by $C_\theta(u,v) = \frac{uv}{(1-\theta(1-u)(1-v))}$ for $\theta \in [-1,1)$. The density of the copula, given by

$$c_\theta(u,v) = \frac{1+\theta(u+v+uv-2+(1-u)(1-v)\theta)}{(1-(1-u)(1-v)\theta)^3}$$

is TP_2 for $\theta \in [0,1)$. Computing $\frac{d}{ds}\frac{d}{du}(\log(c_\theta(u,s-u)))$ and $\frac{d^2}{dv^2}(\log(c_\theta(u,v)))$, we see that, for all $\theta \in [0,1)$, $c(u,s-u)$ is TP_2 in (u,s) but $c(u,v)$ is not logconcave in v for all u. It can be also shown that $\frac{d}{dv}\frac{d}{du}(\log(c_\theta(u,v/u)))$ is positive for all $0 < u \leq 1$ and $0 \leq v \leq u$ if and only if $\theta \in [0,1/2]$ (which means that (P4) holds whenever $\theta \in [0,1/2]$). Finally, (P5) does not hold.

3.4 Frank Copula

Given the copula $C_\theta(u,v) = -\frac{1}{\theta}\log\left(1+\frac{(e^{-u\theta}-1)(e^{-v\theta}-1)}{e^{-\theta}-1}\right)$, for $\theta \in \mathbb{R}\setminus\{0\}$ (copula 5 in Table 4.1 in Nelsen, 2007), the density is given by

$$c_\theta(u,v) = \frac{e^{(1+u+v)\theta}\left(e^\theta-1\right)\theta}{\left(e^{(u+v)\theta}-e^\theta\left(e^{u\theta}+e^{v\theta}-1\right)\right)^2}.$$

This function is TP_2 in (u,v) if $\theta \geq 0$ and, since $\frac{d^2}{dv^2}(\log(c_\theta(u,v))) < 0$ for all $\theta \neq 0$, (P3) and (P1) hold for $\theta \geq 0$. It can be verified that $\frac{d}{dv}\frac{d}{du}(\log(c_\theta(u,v/u)))$ is positive (and therefore (P4) holds) for all $0 < u \leq 1$ and $0 \leq v \leq u$ if and only if $\theta \in [0,1]$. Property (P5) does not hold whatever θ.

The results of this section can be summarized in Table 1:

Table 1. Values of the parameters under which properties (P1) to (P5) are satisfied

	P1	P2	P3	P4	P5
FGM copula	$\theta \in [0,1]$	$\theta \in [0,1]$	$\theta \in [-1,1]$	$\theta \in [0,1]$	
Clayton copula		$\theta > 0$		$\theta > 0$	$\theta > 0$
AMH copula	$\theta \in [0,1)$	$\theta \in [0,1)$		$\theta \in [0, \frac{1}{2}]$	
Frank copula	$\theta > 0$	$\theta > 0$	$\theta \in \mathbb{R} \setminus \{0\}$	$\theta \in (0,1]$	

4 Other Examples: Generalized Order Statistics (GOSs)

In this section, we aim to provide further examples of parametric families where Proposition 3 can be applied. With this purpose, we introduce the notion of generalized order statistics (GOSs) (see Kamps 1995).

Definition 1. *Let $n \in \mathbb{N}, k \geq 1, m_1, \ldots, m_{n-1} \in \mathbb{R}, M_r = \sum_{j=r}^{n-1} m_j, 1 \leq r \leq n-1$, be parameters such that $\gamma_r = k+n-r+M_r \geq 1$ for all $r \in \{1, \ldots, n-1\}$, and let $\tilde{m} = (m_1, \ldots, m_{n-1})$ if $n \geq 2$ ($\tilde{m} \in \mathbb{R}$ arbitrary, if $n = 1$). If the random variables $U_{(r,n,\tilde{m},k)}$, $r = 1, \ldots, n$, possess a joint density of the form*

$$h(u_1, \ldots, u_n) = k \left(\prod_{j=1}^{n-1} \gamma_j \right) \left(\prod_{j=1}^{n-1} (1-u_j)^{m_j} \right) (1-u_n)^{k-1}$$

defined on $0 \leq u_1 \leq \ldots \leq u_n \leq 1$, then they are called GOSs. For a given distribution function F, the random variables $X_{(r,n,\tilde{m},k)} = F^{-1}\left(U_{(r,n,\tilde{m},k)}\right)$, for $r = 1, ..n$, are called the GOSs based on F.

Several models of ordered random variables are included in this model:

- Considering $m_i = 0$ for all $i = 1, \ldots, n-1$ and $k = 1$, we get the order statistics from a distribution F.
- Taking $m_i = -1$ for all $i = 1, \ldots, n-1$ and $k = 1$, we get the first n record values from a sequence of random variables with distribution F (or the first n epoch times of a nonhomogeneous Poisson process).
- A generalization of the previous model is the case in which $k \in N$, resulting in the so-called k-records.
- GOS also includes some other models of interest such as sequential order statistics and progressively type-II censored order statistics.

Proposition 6. *Let F be an absolutely continuous distribution function with logconcave density f. Let $(X_{(i,n,\tilde{m},k)}, X_{(j,n,\tilde{m},k)})$, $1 \leq i < j \leq n$, a bivariate random vector of GOSs from F. Assume that any of the following conditions is satisfied:*

(a) $k \geq 1, m_i \geq 0$.
(b) $k > 0, m_i \geq -1$ *and the failure rate function* $\lambda(\cdot)$ *of* F *is log-concave.*

Then, $\{X_{(i,n,\tilde{m},k)}|X_{(i,n,\tilde{m},k)} + X_{(j,n,\tilde{m},k)} = t\}$ *increases in* t *in the likelihood ratio order.*

Proof. It is well-known that any random vector of GOS is MTP_2 (Belzunce et al. 2005). Then, the bivariate vector $(X_{(i,n,\tilde{m},k)}, X_{(j,n,\tilde{m},k)}), 1 \leq i \leq j \leq n$, is TP_2. In order to obtain the result, by Proposition 2.4 in Pellerey and Navarro (2021), it is sufficient to prove that $\{X_{(j,n,\tilde{m},k)}|X_{(i,n,\tilde{m},k)} = t\}$ has a logconcave density function (or it is ILR). It is known that

$$\{X_{(j,n,\tilde{m},k)} \mid X_{(i,n,\tilde{m},k)} = t\} \simeq_{\text{st}} \{X_{(j-i,n-i,\tilde{m}',k)} \mid X_{(1,n-r+1,\tilde{m}',k)} > t\},$$

where $\tilde{m}' = (m'_1, \ldots, m'_{n-r})$ is such that $m'_j = m_{n-j}$ for $j = 1, \ldots, n-r$. If condition (a) or (b) holds, then $X_{(j-i,n-i,\tilde{m}',k)}$ has a logconcave density (Chen et al. (2009)) and it is easy to see that logconcavity is preserved by right truncations, that is, $\{X_{(j-i,n-i,\tilde{m}',k)} \mid X_{(1,n-r+1,\tilde{m}',k)} > t\}$ has also a logconcave density and the assertion follows.

In particular:

- If f is logconcave and $X_{i:n}, X_{j:n}$ are two order statistics with $1 \leq i < j \leq n$ then, $\{X_{i:n}|X_{i:n} + X_{j:n} = s\}$ increases in s in the lr-ratio order. In particular, it holds for $\{\min(X_1, X_2)|X_1 + X_2 = s\}$. It can be also checked that, when $n = 2$, $\{\max(X_1, X_2)|X_1 + X_2 = s\}$ increases in s in the lr-ratio order.
- If X_{L_n}, X_{L_m}, $n < m$ are two record values of F and f and the failure rate function $\lambda(t)$ are both logconcave, then $\{X_{L_n}|X_{L_n} + X_{L_m} = s\}$ increases in s in the likelihood ratio order.
- If $X_{L_n^k}, X_{L_m^k}$, $n < m$ are two k-record values of F, and $f(.)$ and $\lambda(.)$ are both logconcave, then $\{X_{L_n^k}|X_{L_n^k} + X_{L_m^k} = s\}$ increases in s in the lr-ratio order.

References

Belzunce, F., Mercader, J.A., Ruiz, J.M.: Stochastic comparisons of generalized order statistics. Probab. Eng. Inf. Sci. **19**(1), 99–120 (2005)

Chen, H., Xie, M., Hu, T.: Log-concavity of generalized order statistics. Stat. Probab. Lett. **79**, 396–399 (2009)

Efron, B.: Increasing properties of Polya frequency function. Ann. Math. Stat. **36**(1), 272–279 (1965)

Kamps, U.: A concept of generalized order statistics. J. Stat. Plann. Inference **48**(1), 1–23 (1995)

Karlin, S.: Total Positivity. Stanford University Press, Stanford (1967)

Lehmann, E.L.: Some concepts of dependence. Ann. Math. Stat. **37**(5), 1137–1153 (1966)

Nelsen, R.B.: An Introduction to Copulas. Springer, New York (2007). https://doi.org/10.1007/0-387-28678-0

Oliveira, P.E., Torrado, N.: On proportional reversed failure rate class. Stat. Pap. **56**(4), 999–1013 (2015). https://doi.org/10.1007/s00362-014-0620-8

Oudghiri, Y.: Generalizations of Efron's theorem. Stat. Probab. Lett. **177**, 109158 (2021)

Pellerey, F., Navarro, J.: Stochastic monotonicity of dependent variables given their sum. TEST **31**, 543–561 (2021). https://doi.org/10.1007/s11749-021-00789-5

Ramos, H.M., Sordo, M.A.: The proportional likelihood ratio order and applications. Qüestiió **25**(2), 211–223 (2001)

Saumard, A., Wellner, J.A.: Efron's monotonicity property for measures on R2. J. Multivar. Anal. **166**, 212–224 (2018)

Shaked, M., Shanthikumar, J.G.: Stochastic Orders. Springer, New York (2007). https://doi.org/10.1007/978-0-387-34675-5

A Minimizing Problem of Distances Between Random Variables with Proportional Reversed Hazard Rate Functions

Patricia Ortega-Jiménez[1(✉)], Franco Pellerey[2], Miguel A. Sordo[1], and Alfonso Suárez-Llorens[1]

[1] Universidad de Cádiz, 11002 Cádiz, Spain
{patricia.ortega,mangel.sordo,alfonso.suarez}@uca.es
[2] Politecnico di Torino, 10129 Turin, Italy
franco.pellerey@polito.it

Abstract. Let X be a random variable with distribution function F and let $\mathscr{F}_X$ be the family of proportional reversed hazard rate distribution functions associated to F. Given the random vector (X, Y) with copula C and respective marginal distribution functions F and $G \in \mathscr{F}_X$, we obtain sufficient conditions for the existence of $G \in \mathscr{F}_X$ that minimizes $E_C|X - Y|$.

1 Introduction

Given a random variable X, several location measures can be defined as the argument that minimizes a variability functional of X. Examples of such measures are:

- The expectation, μ_X, which can be defined as the value that minimizes the mean square error of X,

$$\mu_X = \operatorname{argmin}_{t\in\mathbb{R}} E[(X - t)^2].$$

- The median, Me_X, defined as the value that minimizes the mean absolute deviation:

$$Me_X = \operatorname{argmin}_{t\in\mathbb{R}} E|X - t|$$

- The α-expectile, $\rho_\alpha(X)$, which can be defined as the value that minimizes the following linear combination of expected square excesses (Krätschmer and Zähle 2017):

$$\rho_\alpha(X) = \operatorname{argmin}_{t\in\mathbb{R}} \left\{\alpha E\left[\left((X - t)^+\right)^2\right] + (1 - \alpha)E\left[\left((t - X)^+\right)^2\right]\right\}$$

- Analogously, the α-quantiles, $q_\alpha(X)$, that can be defined in a similar way:

$$q_\alpha(X) = \operatorname{argmin}_{t\in\mathbb{R}} \left\{\alpha E\left[(X - t)^+\right] + (1 - \alpha)E\left[(t - X)^+\right]\right\}$$

L. A. García-Escudero et al. (Eds.): SMPS 2022, AISC 1433, pp. 311–318, 2023.
https://doi.org/10.1007/978-3-031-15509-3_41

Therefore, it is natural to wonder whether we can proceed analogously with a different variability functional of X, that is, if the functional can be minimized in order to have a measure that gives information about X.

Let us consider a random variable X with strictly increasing distribution function F and $h : [0,1] \to [0,1]$, a strictly increasing distortion function, that is, a strictly increasing function such that $h(0) = 0$ and $h(1) = 1$. Throughout the paper, we will consider that all variables are absolutely continuous and that all distribution functions and copulas are continuously differentiable. If we now consider (X, Y), a random vector with copula C and marginal distribution functions F and $G = h(F)$ respectively, in Ortega-Jiménez et al. (2021) was shown that $\nu(X) = E_C\,|X - Y|$, where

$$E_C\,|X - Y| = \int_{-\infty}^{\infty} \left(F(x) + G(x) - 2\,C\left(F(x), G(x)\right)\right) dx, \tag{1}$$

is a comonotonic additive measure of variability in the sense of Bickel and Lehmann (1979), that is, it is a measure that satisfies the following properties:

(P0) Law invariance: if X and Y have the same distribution, then $\nu(X) = \nu(Y)$.
(P1) Translation invariance: $\nu(X + k) = \nu(X)$ for all X and all constant k.
(P2) Positive homogeneity: $\nu(0) = 0$ and $\nu(\lambda X) = \lambda\nu(X)$ for all X and all $\lambda > 0$.
(P3) Non-negativity: $\nu(X) \geq 0$ for all X, with $\nu(X) = 0$ if X is degenerated at $c \in \mathbf{R}$.
(P4) Consistency with dispersive order: if $X \leq_{disp} Y$, then $\nu(X) \leq \nu(Y)$.
(P5) Comonotonic additivity: if X and Y are comonotonic, then $\nu(X + Y) = \nu(X) + \nu(Y)$.

Recall that, given two random variables X and Y with distribution functions F and G, respectively, we say that X is smaller than Y in the dispersive order ($X \leq_{disp} Y$) if $F^{-1}(p) - F^{-1}(q) \leq G^{-1}(p) - G^{-1}(q)$ for all $0 \leq q < p \leq 1$.

Let us consider the distortion given by the power function $h(t) = t^{\alpha}, \alpha > 0$. This distortion function characterizes the proportional reversed hazard rate (PRHR) model, which has interesting applications in insurance risk (see Psarrakos and Sordo 2019). A variable satisfies such a model if its reversed hazard rate function ($\tilde{r}(t) = \frac{f(t)}{F(t)}$) is proportional to the baseline reversed hazard rate function. Given a random variable X, we will denote as $\mathscr{F}_X = \{X_{\alpha} : F_{X_{\alpha}}(x) = F(x)^{\alpha}, \alpha > 0\}$ the family of all random variables that satisfy the PRHR model. The interest on considering such distortion function lies in the importance of the PRHR model, which has applications in various areas, such as statistics, reliability engineering, demography, physics or forensic science. The interested reader may consult Gupta and Gupta (2007) for an extensive list of further applications.

Following the approach that initialized the paper, we can now consider the following problem. Fixed the copula C and the marginal distribution function F, we study sufficient conditions for the existence of a distribution function G of $Y \in \mathscr{F}_X$ such that the distance (1) is minimal. The first false intuition may suggest, at least when X and Y have the same support, that the smallest value of $E_C|X-Y|$ is reached when $F = G$, that is, when $Y =_{st} X$. This is not necessarily true, as we can see in the following counterexample. Considering C the independence copula and $X \sim U(0,1)$, it is easy to see that, considering any $\alpha \in (1,2)$, if Y_1 has a distribution function $G(u) = u^\alpha$ for $u \in [0,1]$ and $Y_2 \sim U(0,1)$, then $E_I|X-Y_1| < E_I|X-Y_2|$. The minimum is reached when $\alpha = \sqrt{2}$. It may even happen that there is not a minimizer α for the function. If we consider the independence copula and $X \sim Weibull(k,1)$, we can see that, for some values of k, the minimizer exists and for some others it does not. For $k=1$ the minimum is reached in $\alpha = 0.390$ and for $k = 1.3$ is reached in $\alpha = 0.714$. Although, it can be checked that, for example, for $k = 0.7$ there is not $\alpha > 0$ that minimizes such functional (Fig. 1).

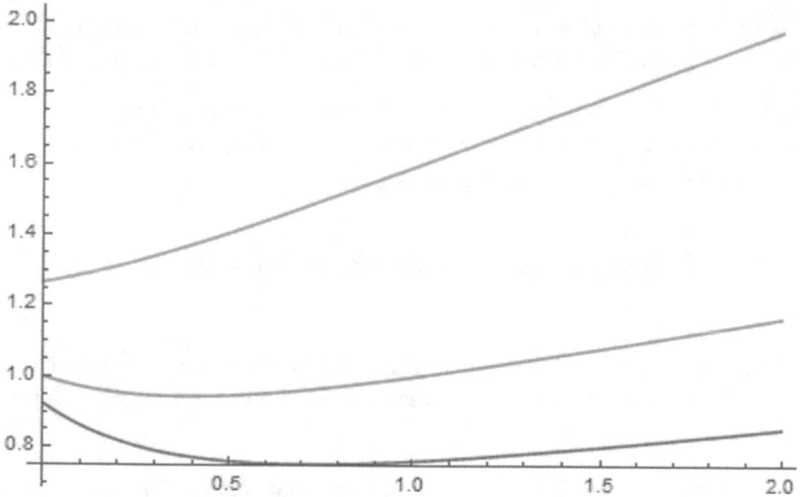

Fig. 1. $E_I|X-X_\alpha|$ in terms of α considering $X \sim Weibull(k,1)$ with $k = 0.7$ (green), $k = 1$ (orange) and $k = 1.3$ (blue)

The rest of the work is organized as follows. Section 2 contains preliminaries. In Sect. 3, given a random variable X, a copula C and the family of variables that satisfy the PRHR model, $\mathscr{F}_X$, we obtain sufficient conditions for the existence of $Y^* \in \mathscr{F}_X$ that minimizes $E_C|X-Y|$ within all $Y \in \mathscr{F}_X$. Finally, Sect. 4 includes examples of the result, both for existence and nonexistence of such minimizer. A second step in this research would be to obtain analytically the minimizer α and to provide a plausible interpretation. Such work remains for future research.

2 Preliminaries

Let us consider a random vector $\boldsymbol{X} = (X, Y)$ with marginal distribution functions F and G and joint distribution function K. Throughout the paper, as mentioned in the introduction, we will consider K, F, G continuously differentiable. By the Sklar theorem, the joint distribution K can be written as $K(x,y) = C(F(x), G(y))$, where C is the joint distribution function of the vector-copula $(F(X), G(Y))$. Such C is the copula (Nelsen 2007) of the vector $\boldsymbol{X}$ and, under the given assumptions, it is unique and continuously differentiable.

We need the following definitions. Here and throughout the paper, the term "increasing" is used for "non-decreasing" and "decreasing" is used for "non-increasing".

Definition 1. *Let $\boldsymbol{X} = (X, Y)$ be a random vector with copula C and marginal distribution functions F, G. X is stochastically increasing in Y ($X \uparrow_{SI} Y$), if $P[X > x \mid Y = y]$ increases in y for all x. A vector is positively dependent through stochastic ordering (PDS) if $X \uparrow_{SI} Y$ and $Y \uparrow_{SI} X$.*

As $P[X > x \mid Y = y] = 1 - \partial_2 C(F(x), G(y))$, $X \uparrow_{SI} Y$ iff $\partial_2 C(u,v)$ decreases in v for all u. We will say that the copula C is PDS if the vector is PDS.

Definition 2. *A random variable X with distribution function F and density function f is said to have the increasing failure rate property (IFR) if its hazard rate function $r(x) = \frac{f(x)}{1-F(x)}$, is increasing.*

3 Determining $\min_{\{Y \in \mathscr{F}_X\}} E_C|X - Y|$ with X and C Fixed

We will see that, under some conditions on the fixed copula, if the variable X is IFR, then $\min_{\{Y \in \mathscr{F}_X\}} E_C|X - Y|$ is reached. We need some results to prove it. From now on, we will denote

$$\varphi_0(u) = \lim_{v \to 0} \partial_2 C(u,v) \quad \text{and} \quad \varphi_1(u) = \lim_{v \to 1} \partial_2 C(u,v).$$

Such limits exist for all $u \in (0,1)$. When the copula is PDS, $\partial_2 C(u,v)$ decreases in v. Since $\int_0^1 \partial_2 C(u,v)dv = u$, for all $u \in (0,1)$, it follows that $\varphi_0(u) \geq u \geq \varphi_1(u)$.

Lemma 1. *Let C be an absolutely continuous copula, and let X be an IFR random variable with F, f and r its respective distribution, density and hazard rate functions. Assume that $r(F^{-1}(0)) \neq 0$. Then:*

$$\lim_{\alpha \to 0} \int_0^1 \frac{1 - 2\,\partial_2 C(u, u^\alpha)}{f(F^{-1}(u))} u^\alpha \log(u) du = \int_0^1 \frac{1 - 2\varphi_1(u)}{f(F^{-1}(u))} \log(u) du, \qquad (2)$$

$$\lim_{\alpha \to \infty} \int_0^1 \frac{1 - 2\,\partial_2 C(u, u^\alpha)}{f(F^{-1}(u))} \log(u) du = \int_0^1 \frac{1 - 2\varphi_0(u)}{f(F^{-1}(u))} \log(u) du. \qquad (3)$$

Proof. If we define, for all $u \in (0,1), n \in \mathbb{N}$, $g_n(u) = \frac{1-2\,\partial_2 C(u, u^{1/n})}{f(F^{-1}(u))} u^{1/n} \log(u)$, $g(u) = \frac{1-2\varphi_1(u)}{f(F^{-1}(u))} \log(u)$, $h_n(u) = \frac{1-2\,\partial_2 C(u, u^n)}{f(F^{-1}(u))} \log(u)$, $h(u) = \frac{1-2\varphi_0(u)}{f(F^{-1}(u))} \log(u)$, then, for all $u \in (0,1)$ $\lim_{n\to\infty} g_n(u) = g(u)$ and $\lim_{n\to\infty} h_n(u) = h(u)$. As $|g_n(u)|$ and $|h_n(u)|$ are bounded by $\frac{-\log(u)}{f(F^{-1}(u))}$, by the Dominated Convergence Theorem, if $\frac{-\log(u)}{f(F^{-1}(u))}$ is a positive integrable function, (2) and (3) hold. By the Cauchy-Schwarz inequality and considering the hazard rate function r:

$$\int_0^1 \frac{-\log(u)}{f(F^{-1}(u))}du = \int_0^1 \frac{-\log(u)}{1-u}\frac{1}{r(F^{-1}(u))}du$$
$$\leq \left(\int_0^1 \left(\frac{-\log(u)}{1-u}\right)^2 du\right)^{1/2} \left(\int_0^1 \left(\frac{1}{r(F^{-1}(u))}\right)^2 du\right)^{1/2} \quad (4)$$

Note that $\int_0^1 \left(\frac{-\log(u)}{1-u}\right)^2 du = \frac{\pi^2}{3}$ and, since r increases and $r(F^{-1}(0)) \neq 0$, $\frac{1}{r(F^{-1}(u))}$ is bounded. Therefore, (4) is finite and the assertion follows. □

Lemma 2. *Let X be an IFR random variable with F, f and r its respective distribution, density and hazard rate functions. Then:*

$$\int_0^1 \frac{1-2u}{f(F^{-1}(u))}(-\log(u))du \quad (5)$$

is strictly positive and finite.

Proof. First we will show that

$$b(s) = \int_0^s \left(\frac{1-2u}{1-u}(-\log(u))\right) du > 0 \text{ for all } s \in [0,1].$$

Since $b'(s) = \frac{1-2s}{1-s}(-\log(s))$, $b(s)$ increases if $s < 1/2$ and decreases if $s > 1/2$. Moreover, $b(0) = 0$ and $b(1) = 2 - \pi^2/6 > 0$, therefore $b(s) > 0$ for all $s \in (0,1]$. We can rewrite (5) in the following form:

$$\int_0^1 \left(\frac{1-2u}{1-u}(-\log(u))\right) \frac{1}{r(F^{-1}(u))}du. \quad (6)$$

As X is IFR, $1/r(F^{-1}(u))$ decreases in u. By Lemma 4.7.1 in Barlow and Proschan (1975), if, for all $s \in [0,1]$, $b(s) = \int_0^s \left(\frac{1-2u}{1-u}(-\log(u))\right) du$ is positive, then (6) is also positive. $b'(s) = \frac{1-2s}{1-s}(-\log(s))$, so $b(s)$ increases if $s < 1/2$ and decreases if $s > 1/2$. As $b(0) = 0$ and $b(1) = 2 - \pi^2/6 > 0$, $b(s) > 0$ for all $s \in (0,1]$ and (6) is strictly positive. Also, by Cauchy-Schwarz inequality, (6) is smaller or equal than:

$$\left(\int_0^1 \left(\frac{1-2u}{1-u}(-\log(u))\right)^2 du\right)^{\frac{1}{2}} \left(\int_0^1 \frac{1}{r(F^{-1}(u))^2}du\right)^{\frac{1}{2}}.$$

Integrating by parts and considering the Spence's function, given by $Li_2(u) = \int_0^u \frac{-\log(t)}{1-t}dt$, we can see that $\int_0^1 \left(\frac{-\log(u)(1-2u)}{1-u}\right)^2 du \leq 1 + Li_2(1) < 3$ (we can obtain, computationally, that the value is approximately 1.2936). As we already saw in Lemma 1 that the second element is finite, (5) is finite. □

We can now move on to the main result:

Proposition 1. *Let X be a random variable with F and f its respective distribution and density function. Let us consider the family of random variables $\mathscr{F}_X$ described above. Let us consider the random vector $\mathbf{X} = (X, Y)$ with PDS copula C, such that $\partial_2 C(u, u^n)$ increases in u for all $n \in \mathbb{N}$ and $\lim_{u\to 1}\varphi_1(u) = 1$. If X is IFR, there exists $\alpha_0 > 0$ such that $X_{\alpha_0} \in \mathscr{F}_X$ and:*

$$E_C|X - X_{\alpha_0}| \leq E_C|X - Y| \text{ for all } Y \in \mathscr{F}_X.$$

Proof. Given X, we can consider the function $\alpha \mapsto E_C|X - X_\alpha|$, given by:

$$\int_{-\infty}^{\infty} (F(x) + F(x)^\alpha - 2\,C(F(x), F(x)^\alpha))\,dx = \int_0^1 \frac{u + u^\alpha - 2C(u, u^\alpha)}{f(F^{-1}(u))} du$$

This is a continuous and derivable function for $\alpha > 0$, and

$$\partial_\alpha E_C|X - X_\alpha| = \int_0^1 \frac{1 - 2\,\partial_2 C(u, u^\alpha)}{f(F^{-1}(u))} u^\alpha \log(u) du$$

In order to see that there exists α_0 that minimizes $E_C|X - X_\alpha|$, it would be enough to see that

1. $\lim_{\alpha_1\to 0}(\partial_\alpha E_C|X - X_\alpha|_{\alpha=\alpha_1}) < 0$, and
2. $\lim_{\alpha_2\to +\infty}(\partial_\alpha E_C|X - X_\alpha|_{\alpha=\alpha_2}) \geq 0$.

It would mean that there exists at least one value $\alpha_0 > 0$ where $E_C|X - X_\alpha|$ attains a local minimum, and necessarily one of them will be the global one.

By Lemma 1, $\lim_{\alpha_1\to 0}(\partial_\alpha E_C|X - X_\alpha|_{\alpha=\alpha_1}) = -\left(\int_0^1 \frac{1-2\varphi_1(u)}{f(F^{-1}(u))}(-\log(u))du\right)$ and, as $u \geq \varphi_1(u)$, by Lemma 3,

$$\lim_{\alpha_1\to 0}(\partial_\alpha E_C|X - X_\alpha|_{\alpha=\alpha_1}) \leq -\left(\int_0^1 \frac{1-2u}{f(F^{-1}(u))}(-\log(u))du\right) < 0.$$

In order to study $\lim_{n\to\infty}\partial_\alpha E_C|X - X_\alpha|_{\alpha=n}$, let us note that, as $\partial_2 C(u, u^n)$ increases in u, for each n there exists $c_n \in [0, 1)$ such that, for $u < c_n$, $\partial_2 C(u, u^n) \leq \frac{1}{2}$ and for $u > c_n$, $\partial_2 C(u, u^n) \geq \frac{1}{2}$. Note also that $\lim_{n\to\infty} c_n < 1$ (such limit exists due to the smoothness of C); otherwise, if $\lim_{n\to\infty} c_n = 1$, then $\varphi_1(u) \leq \frac{1}{2}$ for all $u \in (0, 1)$, which contradicts the fact that $\lim_{u\to 1}\varphi_1(u) = 1$. Taking this into consideration, we have that, for all $n \in \mathbb{N}$:

$$\partial_\alpha E_C|X-X_\alpha|_{\alpha=n} = \int_0^1 \frac{1-2\,\partial_2 C(u,u^n)}{f(F^{-1}(u))} u^n \log(u)du =$$

$$\int_0^{c_n} \frac{1-2\,\partial_2 C(u,u^n)}{f(F^{-1}(u))} u^n \log(u)du + \int_{c_n}^1 \frac{1-2\,\partial_2 C(u,u^n)}{f(F^{-1}(u))} u^n \log(u)du >$$

$$c_n^n \int_0^{c_n} \frac{1-2\,\partial_2 C(u,u^n)}{f(F^{-1}(u))} \log(u)du + c_n^n \int_{c_n}^1 \frac{1-2\,\partial_2 C(u,u^n)}{f(F^{-1}(u))} \log(u)du$$

$$> (c_n)^n \int_0^1 \frac{1-2\partial_2 C(u,u^n)}{f(F^{-1}(u))} \log(u)du \qquad (7)$$

Taking limits in (7), $\lim_{n\to\infty}(c_n)^n = 0$ and by Lemma 2,

$$\lim_{\alpha\to\infty} \int_0^1 \frac{1-2\,\partial_2 C(u,u^\alpha)}{f(F^{-1}(u))} \log(u)du = \int_0^1 \frac{1-2\varphi_0(u)}{f(F^{-1}(u))} \log(u)du$$

Note that, as $u \le \varphi_0(u) \le 1$,

$$\int_0^1 \frac{1-2u}{f(F^{-1}(u))} \log(u)du \le \int_0^1 \frac{1-2\varphi_0(u)}{f(F^{-1}(u))} \log(u)du \le \int_0^1 \frac{-\log(u)}{f(F^{-1}(u))} du$$

In Lemmas 2 and 3, we saw that both the bounds are finite, so we can conclude that $\lim_{n\to\infty}\left((c_n)^n \int_0^1 \frac{1-2\partial_2 C(u,u^n)}{f(F^{-1}(u))}(\log(u))du\right) = 0$, and, therefore

$$\lim_{\alpha_2\to+\infty} (\partial_\alpha E_C|X-X_\alpha|_{\alpha=\alpha_2}) \ge 0.$$

This concludes the proof. □

4 Examples

Let us give some examples of *PDS* copulas that satisfy both that $\partial_2 C(u,u^n)$ increases in u for all $n \in \mathbb{N}$ and that $\lim_{u\to 1} \varphi_1(u) = 1$:

1. Independence, $C(u,v) = uv$.
2. Farlie-Gumbel-Morgenstern copula, $C_\theta(u,v) = uv(1+\theta(1-u)(1-v))$, for $\theta \in [0,1]$.
3. Frank copula, $C_\theta(u,v) = -\frac{1}{\theta}\log\left(1+\frac{(e^{-u\theta}-1)(e^{-v\theta}-1)}{e^{-\theta}-1}\right)$, for $\theta > 0$.
4. Copula 17 in Table 4.1 in Nelsen (2007), for $\theta > 1$,

$$C_\theta(u,v) = \left(1+\frac{\left[(1+u)^{-\theta}-1\right]\left[(1+v)^{-\theta}-1\right]}{2^{-\theta}-1}\right)^{-1/\theta} - 1$$

Note that there are many *PDS* copulas that do not verify such conditions. For example, there are copulas that satisfy $\lim_{u\to 1}\varphi_1(u) = 1$ but do not verify that $\partial_2 C(u,u^n)$ increases in u. Examples of this are the following copulas in Table 4.1 in Nelsen (2007): 1 (Clayton), 3 (Ali-Mikhail-Haq), 13 and 19. Also, it can be checked that, for some PDS copulas, $\lim_{u\to 1} \varphi_1(u) = 0$ and therefore, do not satisfy conditions on Proposition 1. Examples of this are the Gaussian copula for $\rho > 0$ or the following copulas in Table 4.1 in Nelsen (2007): 2, 4 (Gumbel-Hougard), 6, 12 and 14.

Let us also recall the importance of the property IFR in X. If (X, Y) is independent and $X \sim Weibull(k, 1)$, the IFR assumption fails for $k < 1$. Figure 1 shows that for $k = 0.7$, $E_I|X - X_\alpha|$ increases in α and the minimizer does not exists.

References

Barlow, R.E., Proschan, F.: Statistical Theory of Reliability and Life Testing. Holt, Rinehart and Winston, New York (1975)

Bickel, P.J., Lehmann, E.L.: Descriptive statistics for nonparametric models, IV. Spread. In: Jureckova, J. (ed.) Contributions to Statistics, Jaroslaw Hajek Memorial Volume, pp. 33–40. Reidel, Dordrecht (1979)

Gupta, R.C., Gupta, R.D.: Proportional reversed hazard rate model and its applications. J. Stat. Plann. Inference **137**(11), 3525–3536 (2007)

Krätschmer, V., Zähle, H.: Statistical inference for expectile-based risk measures. Scand. J. Stat. **44**(2), 425–454 (2017)

Nelsen, R.B.: An Introduction to Copulas. Springer, New York (2007). https://doi.org/10.1007/0-387-28678-0

Ortega-Jiménez, P., Sordo, M.A., Suárez-Llorens, A.: Stochastic comparisons of some distances between random variables. Mathematics **9**(9), 981 (2021)

Psarrakos, G., Sordo, M.A.: On a family of risk measures based on proportional hazards models and tail probabilities. Insur. Math. Econ. **86**, 232–240 (2019)

Polytopes of Discrete Copulas and Applications

Elisa Perrone(✉)

Eindhoven University of Technology, Groene Loper 5,
5612AZ Eindhoven, The Netherlands
e.perrone@tue.nl

Abstract. Discrete copulas are flexible tools in statistics to describe the joint distribution of discrete random vectors. In addition, they are interesting mathematical objects that can be represented as convex polytopes. In this work, we summarize the most important results related to polytopes of discrete copulas and discuss their use in applications. Along the way, we also highlight some differences between the space of bivariate and d-dimensional discrete copulas (with $d > 2$), thereby raising interesting questions on their geometric properties and statistical interpretation.

1 Introduction

Copula functions are flexible tools in dependence modeling to represent multivariate distributions (see, e.g., Durante and Sempi 2015; Joe 2014; Nelsen 2006). The popularity of copulas in probability and statistics is due to *Sklar's Theorem* (Sklar 1959), which affirms that, for every $(x_1, \ldots, x_d) \in \mathbb{R}^d$, the joint distribution function $F_{\mathbf{X}}$ of any d-dimensional random vector $\mathbf{X} = (X_1, \ldots, X_d)$ can be written as

$$F_{\mathbf{X}}(x_1, \ldots, x_d) = C(F_{X_1}(x_1), \ldots, F_{X_d}(x_d)), \qquad (1)$$

where the function $C : [0,1]^d \to [0,1]$ is a *d-dimensional copula* and $F_{X_1}, \ldots, F_{X_d}$ are univariate marginal distributions.

The function C in Eq. (1) is uniquely identified on the set $\text{Range}(F_{X_1}) \times \cdots \times \text{Range}(F_{X_d})$. The special case of $\text{Range}(F_{X_1}) \times \cdots \times \text{Range}(F_{X_d})$ resulting in a grid domain has received special attention. In such a case, Sklar's theorem identifies the so-called *discrete copulas*, i.e., restrictions of copulas on grid domains of the unit hyper-cube (see, e.g., Mesiar 2005; Mayor et al. 2005; Kolesárová et al. 2006). Discrete copulas can be transformed into full-domain copulas by simply spreading the probability mass on each hyper-rectangle of their grid domain. However, the extension is generally not unique (de Amo et al. 2017). Nevertheless, the mathematical features of discrete copulas together with their probabilistic interpretation make them useful tools to tackle applied problems in different domains.

In the following, we summarize the most interesting mathematical properties of discrete copulas and relate them to several application domains.

L. A. García-Escudero et al. (Eds.): SMPS 2022, AISC 1433, pp. 319–325, 2023.
https://doi.org/10.1007/978-3-031-15509-3_42

2 Polytopes of Bivariate Discrete Copulas

We now discuss the representations of discrete copulas as convex polytopes, i.e., bounded convex spaces consisting of all points satisfying a finite list of inequalities (Ziegler 1995). The matrix representation of bivariate discrete copulas has been extensively studied in the literature by several authors (see, for example, Aguiló et al. 2010; Fernández-Sánchez et al. 2021; Kolesárová et al. 2006; Mayor et al. 2005; Mesiar 2005; Mordelová and Kolesárová 2007; Perrone et al. 2019; Perrone and Durante 2021). Here, we use the same notation as in Perrone et al. (2019), where the most general case of arbitrary grid partitions was presented.

2.1 Discrete Copulas on Arbitrary Grid Domains

We consider $p \in \mathbb{Z}_{>0}$ and denote $[p] = \{1, \ldots, p\}$, $\langle p \rangle = \{0, \ldots, p\}$, and $I_p = \{0, 1/p, \ldots, (p-1)/p, 1\}$. For p and q in $\mathbb{Z}_{>0}$, we define U_p and V_q as two finite (not necessarily identical) grid partitions of the unit interval. The most general discrete copulas C_{U_p,V_q} are those defined on $U_p \times V_q$ as functions that satisfy the properties of a copula function on the grid domain $U_p \times V_q$.

As noted in Perrone et al. (2019), the space of discrete copulas C_{U_p,V_q} corresponds to special convex polytopes called *transportation polytopes* (De Loera and Kim 2014). Starting with two vectors $\tilde{u} = (\tilde{u}_1, \ldots, \tilde{u}_p) \in \mathbb{R}^p_{>0}$ and $\tilde{v} = (\tilde{v}_1, \ldots, \tilde{v}_q) \in \mathbb{R}^q_{>0}$, we can define the transportation polytope $\mathcal{T}(\tilde{u}, \tilde{v})$ as the convex polytope in the pq variables $x_{i,j}$ satisfying, for all $i \in [p]$ and $j \in [q]$, the following conditions:

$$x_{i,j} \geq 0, \quad \sum_{h=1}^{q} x_{i,h} = \tilde{u}_i, \quad \sum_{\ell=1}^{p} x_{\ell,j} = \tilde{v}_j.$$

We refer to the two vectors $\tilde{u}$ and $\tilde{v}$ as the margins of $\mathcal{T}(\tilde{u}, \tilde{v})$. As showed in Perrone et al. (2019), any discrete copula C_{U_p,V_q} corresponds to a matrix within a transportation polytope $\mathcal{T}(\tilde{u}, \tilde{v})$, and viceversa. For any discrete copula $C_{U_p,V_q} : U_p \times V_q \to [0, 1]$, there is a $(p \times q)$ transportation matrix $(x_{i,j})$ in $\mathcal{T}(\tilde{u}, \tilde{v})$, with $\sum_{h=1}^{q} \tilde{v}_h = \sum_{\ell=1}^{p} \tilde{u}_\ell = pq$, such that for every $i \in \langle p \rangle$, $j \in \langle q \rangle$

$$c_{i,j} = C_{U_p,V_q}(u_i/(pq), v_j/(pq)) = \frac{1}{pq} \sum_{\ell=1}^{i} \sum_{h=1}^{j} x_{\ell,h}.$$

Conversely, any discrete copula C_{U_p,V_q} can be transformed into a transportation matrix in $\{1/(pq)\}\mathcal{T}(\tilde{u}, \tilde{v})$, with margins $\tilde{u} = (u_1, u_2 - u_1, \ldots, u_k - u_{k-1}, \ldots, u_p - u_{p-1})$ and $\tilde{v} = (v_1, v_2 - v_1, \ldots, v_k - v_{k-1}, \ldots, v_q - v_{q-1})$, through the linear map $T : \mathbb{R}^{(p+1)\times(q+1)} \longrightarrow \mathbb{R}^{p \times q}$ defined as $T(c_{i,j}) = c_{i,j} + c_{i-1,j-1} - c_{i,j-1} - c_{i-1,j}$ for all $i \in [p]$ and $j \in [q]$.

Intuitively, the transportation matrix corresponding to a discrete copula indicates how the probability mass is distributed in the rectangles of the unit squares identified by the partition $U_p \times V_q$. Moreover, the zero-pattern of such a transportation matrix indicates whether a discrete copula is *irreducible*, i.e., an extreme point of the polytope of discrete copulas. Clearly, a discrete copula is irreducible (or extremal) if and only if its corresponding transportation matrix $T_{p,q}$ is also extremal. As discussed in Perrone and Durante (2021), $T_{p,q}$ is extremal if and only if it does not contain any square sub-matrix with at least two positive entries in every row and column. This characterization makes it convenient to look at $T_{p,q}$ to assess whether a discrete copula is an extreme point or not.

We now report an example of an irreducible discrete copula that was originally presented in Perrone and Durante (2021). We assume $p = 3$, $q = 4$ and consider the two grid partitions $U_3 \times V_4 = \{0, \ 2/12, \ 7/12, \ 1\} \times \{0, 3/12, \ 6/12, \ 9/12, \ 1\}$. The following matrix C_1 is a discrete copula defined on $U_3 \times V_4$, while T_1 is its associated (3×4)-transportation matrix

$$C_1 = \begin{pmatrix} 0 & 0 & 0 & 0 & 0 \\ 0 & 2/12 & 2/12 & 2/12 & 2/12 \\ 0 & 3/12 & 6/12 & 7/12 & 7/12 \\ 0 & 3/12 & 6/12 & 9/12 & 1 \end{pmatrix}, \quad T_1 = \begin{pmatrix} 2 & 0 & 0 & 0 \\ 1 & 3 & 1 & 0 \\ 0 & 0 & 2 & 3 \end{pmatrix}.$$

T_1 belongs to the transportation polytope $\mathscr{T}([2, 5, 5], [3, 3, 3, 3])$ and is one of its extremal points since it does not contain any square sub-matrix with at least two positive entries. As a consequence, C_1 is extremal as well.

2.2 Empirical Copulas as Extreme Points

We now consider the case of discrete copulas defined on I_p^2, originally analyzed in Kolesárová et al. (2006). In their paper, the authors show a one-to-one correspondence between the discrete copulas on I_p^2 and the doubly stochastic matrices, which are square matrices with non-negative entries, and row and column sums equal to one. In addition, they prove that the extreme points of the class of discrete copulas on I_p^2 are those of minimal range, i.e., the discrete copulas that only take values in I_p. This result highlights a beautiful connection between the geometric properties of discrete copulas and their probabilistic interpretation. Indeed, as described by Mesiar (2005), the discrete copulas on I_p^2 of minimal range correspond to the so-called *empirical copulas*, which are the foundation of rank-based (non-parametric) copula approaches (Rüschendorf 2009; Segers 2012). In terms of polytopes, we can say that the space of discrete copulas on I_p^2 corresponds to the popular Birkhoff polytope (Ziegler 1995), and the empirical copulas to the permutation matrices. In the next section, we see that this elegant interpretation is not entirely preserved in a higher dimensional setting.

3 Polytopes of d-Dimensional Discrete Copulas, $d > 2$

Despite the several studies of the matrix representations of bivariate discrete copulas, the d-dimensional discrete copulas (with $d > 2$) are still quite unexplored. To the best of the author's knowledge, there is only one paper where the (hyper-) matrix representation of multivariate discrete copulas has been analyzed, namely, Schefzik (2015). The results presented in Schefzik (2015) are limited to the simple case of $d-$dimensional discrete copulas defined on uniform grid partitions I_p^d, which is a natural extension of the setting discussed in Sect. 2.2. We believe that a more general framework similar to the one defined in Perrone et al. (2019) is also possible for d-dimensional discrete copulas. However, we do not explore this further in this work, and we focus on highlighting differences between the bivariate and higher dimensional case for uniform grid partitions I_p^d.

We define $d-$dimensional discrete copulas as functions that satisfy the properties of a $d-$dimensional copula function on the uniform grid domain I_p^d. As proved in Schefzik (2015), such discrete copulas are in one-to-one correspondence with special d-dimensional matrices (also called *stochastic arrays*) that have (1) non-negative entries, and (2) every coordinate hyperplane sum equal to one. Interestingly, the space of such d-dimensional matrices is one of the many possible extensions of the Birkhoff polytope in higher dimensions, i.e., the so-called *d-index axial assignment polytope* (see Linial and Luria 2014, and references therein). Therefore, the correspondence proved by Schefzik (2015) is the multidimensional analogous of the relationship between bivariate discrete copulas and doubly stochastic matrices discussed in the previous section. Schefzik (2015) also shows that d-dimensional discrete copulas with minimal range I_p correspond to the empirical copulas (and the $d-$dimensional permutation arrays, respectively), in a similar fashion as Mesiar (2005).

Although everything seems to be working exactly as in the bivariate case, there is a crucial difference in the geometry of the d-index axial assignment polytope if compared to the classic Birkhoff polytope. Indeed, the $d-$dimensional permutation arrays are not the only extreme points of the polytope as there are other noninteger extreme points. To give a simple example, we consider the case when $d = 3$ and $p = 2$. As noted in Linial and Luria (2014), the 3$-$dimensional matrix T defined by the two following slices

$$T(i, j, 0) = T_1 = \begin{pmatrix} \frac{1}{2} & 0 \\ 0 & \frac{1}{2} \end{pmatrix}, \text{ and } \quad T(i, j, 1) = T_2 = \begin{pmatrix} 0 & \frac{1}{2} \\ \frac{1}{2} & 0 \end{pmatrix}$$

is an extreme point of the 3-index axial assignment polytope. This simple example can be generalized to show that the $d-$dimensional permutation arrays are a vanishingly small subset of the total extreme points of the space (Linial and Luria 2014). This difference has implications in our interpretation of the extreme points of the polytope of the $d-$dimensional discrete copulas as the empirical copulas, i.e., those discrete copulas one can construct from the data, are not the only extreme points. Specifically, there are many more extreme points which do not correspond to empirical copulas and, as such, do not have a clear statistical interpretation.

Understanding the role of the noninteger extreme points of the space of d-dimensional discrete copulas remains an interesting direction for future work.

4 Discussion

In the previous sections, we summarized the most important theoretical results on bivariate and $d-$dimensional discrete copulas available in the literature. We now briefly discuss how the theoretical framework described above can be useful in applications.

The main advantage of considering discrete copulas as convex polytopes is the possibility to exploit their geometric structure to solve convex optimization problems for model selection. For example, Piantadosi et al. (2007, 2012); Radi et al. (2017) exploit the polytopal representation of discrete copulas to derive flexible copula models of maximum entropy. They use these models in hydrology to generate synthetic data of rainfall totals. Recently, Kuzmenko et al. (2020) propose another application to copula selection in portfolio optimization. In Perrone et al. (2019), the authors adopted a discrete geometry approach to study convex sub-families of discrete copulas with a negative stochastic property. Their findings have been used to design a hypothesis test for desirable stochastic properties (Durante and Perrone 2020).

Besides, the theoretical framework defined in Schefzik (2015) constitutes the base of rank-based approaches in weather forecasting applications. In weather forecasting, accounting for more uncertainty is key to produce accurate predictions. To do so, field experts often use probabilistic numerical weather prediction (NWP) models and construct an ensemble of possible forecasts instead of a single one. Then, they use statistical methods to reduce errors and biases and produce accurate predictions by modeling the forecast distribution. In such a context, non-parametric (rank-based) approaches based on empirical copulas represent an efficient alternative to parametric methods to account for a dependence structure in the weather forecasting problem without making any restrictive assumption (Schefzik et al. 2013; Schefzik 2015; Perrone et al. 2020). For example, assuming we are interested in multi-site temperature forecasts, we could follow a commonly used 2-step approach. First, we obtain a univariate corrected (parametric) distribution for temperature at each location. Then, we construct a new sample from each of the corrected univariate marginal distributions, and we glue them together according to a suitable empirical copula which can come from past observations or from the numerical weather prediction models, directly.

All these ideas and works collectively demonstrate the benefits of studying discrete copulas and their mathematical properties. In the future, more research could be done to shed light on the polytopal representation of d-dimensional discrete copulas on arbitrary grid domains in connection with multi-way transportation polytopes (De Loera and Kim 2014).

References

Aguiló, I., Suñer, J., Torrens, J.: Matrix representation of copulas and quasi-copulas defined on non-square grids of the unit square. Fuzzy Sets Syst. **161**(2), 254–268 (2010) de Amo, E., Díaz Carrillo, M., Durante, F., Fernández Sánchez, J.: Extensions of subcopulas. J. Math. Anal. Appl. **452**(1), 1–15 (2017)

De Loera, J.A., Kim, E.D.: Combinatorics and geometry of transportation polytopes: an update. In: Discrete Geometry and Algebraic Combinatorics, Contemporary Mathematics, vol. 625, pp. 37–76. American Mathematical Society, Providence, RI (2014)

Durante, F., Perrone, E.: Stochastic dependence with discrete copulas. In: Pollice, A., Salvati, N., Schirripa Spagnolo, F. (eds.) Book of Short Papers SIS 2020, pp. 1344–1349. Pearson (2020)

Durante, F., Sempi, C.: Principles of Copula Theory. CRC/Chapman & Hall, Boca Raton (2015)

Fernández-Sánchez, J., Quesada-Molina, J.J., Úbeda Flores, M.: New results on discrete copulas and quasi-copulas. Fuzzy Sets Syst. **415**, 89–98 (2021)

Joe, H.: Dependence Modeling with Copulas, 2nd edn. Chapman and Hall/CRC, Boca Raton (2014)

Kolesárová, A., Mesiar, R., Mordelová, J., Sempi, C.: Discrete Copulas. IEEE Trans. Fuzzy Syst. **14**(5), 698–705 (2006)

Kuzmenko, V., Salam, R., Uryasev, S.: Checkerboard copula defined by sums of random variables. Depend. Model. **8**(1), 70–92 (2020)

Linial, N., Luria, Z.: On the vertices of the d-dimensional Birkhoff polytope. Discrete Comput. Geom. **51**, 161–170 (2014)

Mayor, G., Suñer, J., Torrens, J.: Copula-like operations on finite settings. IEEE Trans. Fuzzy Syst. **13**(4), 468–477 (2005)

Mesiar, R.: Discrete copulas - what they are. In: Montseny, E., Sobrevilla, P. (eds.) Proceedings of EUSFLAT-LFA Conference, pp. 927–930. Universitat Politecnica de Catalunya, Barcelona (2005)

Mordelová, J., Kolesárová, A.: Some results on discrete copulas. In: Proceedings of the 4th International Summer School on Aggregation operators (AGOP, Ghent, Belgium), pp. 145–150 (2007)

Nelsen, R.B.: An Introduction to Copulas. Springer Series in Statistics, 2nd edn. Springer, New York (2006). https://doi.org/10.1007/0-387-28678-0

Perrone, E., Durante, F.: Extreme points of polytopes of discrete copulas. In: Joint Proceedings of the 19th World Congress of the International Fuzzy Systems Association (IFSA), the 12th Conference of the European Society for Fuzzy Logic and Technology (EUSFLAT), and the 11th International Summer School on Aggregation Operators (AGOP), pp. 596–601 (2021)

Perrone, E., Solus, L., Uhler, C.: Geometry of discrete copulas. J. Multivar. Anal. **172**, 162–179 (2019)

Perrone, E., Schicker, I., Lang, M.N.: A case study of empirical copula methods for the statistical correction of forecasts of the ALADIN-LAEF system. Meteorol. Z. **29**, 277–288 (2020)

Piantadosi, J., Howlett, P., Boland, J.: Matching the grade correlation coefficient using a copula with maximum disorder. J. Ind. Manag. Optim. **3**(1), 305–312 (2007)

Piantadosi, J., Howlett, P., Borwein, J.: Copulas with maximum entropy. Optim. Lett. **6**(1), 99–125 (2012). https://doi.org/10.1007/s11590-010-0254-2

Ahmad Radi, N.F., et al.: Generating synthetic rainfall total using multivariate skew-t and checkerboard copula of maximum entropy. Water Resour. Manag. **31**(5), 1729–1744 (2017). https://doi.org/10.1007/s11269-017-1597-6

Rüschendorf, L.: On the distributional transform, Sklar's theorem, and the empirical copula process. J. Stat. Plan. Inference **139**(11), 3921–3927 (2009)

Schefzik, R.: Multivariate discrete copulas, with applications in probabilistic weather forecasting. Publications de l'Institut de Statistique de l'Université de Paris **116**, 59–87 (2015)

Schefzik, R., Thorarinsdottir, T.L., Gneiting, T.: Uncertainty quantification in complex simulation models using ensemble copula coupling. Stat. Sci. **28**(4), 616–640 (2013)

Segers, J.: Asymptotics of empirical copula processes under non-restrictive smoothness assumptions. Bernoulli **18**(3), 764–782 (2012)

Sklar, A.: Fonctions de répartition à n dimensions et leurs marges. Publications de l'Institut de Statistique de Paris **8**, 229–231 (1959)

Ziegler, G.M.: Lectures on Polytopes. Springer, New York (1995). https://doi.org/10.1007/978-1-4613-8431-1

Penalized Estimation of a Finite Mixture of Linear Regression Models

Roberto Rocci[1(✉)], Roberto Di Mari[2], and Stefano Antonio Gattone[3]

[1] Sapienza University of Rome, Rome, Italy
roberto.rocci@uniroma1.it
[2] University of Catania, Catania, Italy
roberto.dimari@unict.it
[3] University G. d'Annunzio, Chieti-Pescara, Italy
gattone@unich.it

Abstract. Finite mixtures of linear regressions are often used in practice in order to classify a set of observations and/or explain an unobserved heterogeneity. Their application poses two major challenges. The first is about the maximum likelihood estimation, which is, in theory, impossible in case of Gaussian errors with component specific variances because the likelihood is unbounded. The second is about covariate selection. As in every regression model, there are several candidate predictors and we have to choose the best subset among them. These two problems can share a similar solution, and here lies the motivation of the present paper. The pathway is to add an appropriate penalty to the likelihood. We review possible approaches, discussing and comparing their main features.

1 Introduction

In many applications, investigating the relationship between a response variable and a set of explanatory variables is of central interest. Nonetheless, in case of heterogeneous populations, estimating a single set of regression coefficients is often inadequate. To the purpose, finite mixture of linear regressions (FMLR), i.e. finite mixture of conditional normal distributions, can be used to estimate clusterwise regression parameters in a maximum likelihood context. Clusterwise linear regression is also known under the names of finite mixture of linear regressions or switching regressions (McLachlan and Peel 2000).

Let $\mathrm{y}_1, \ldots, \mathrm{y}_n$ be a sample of independent observations drawn from the response random variable Y_i, each respectively observed conditionally on a vector of J regressors $\mathbf{x}_1, \ldots, \mathbf{x}_n$. Let us assume $\mathrm{Y}_i|\mathbf{x}_i$ to be distributed as a finite mixture of linear regression models, that is

$$f(\mathrm{y}_i|\mathbf{x}_i;\boldsymbol{\psi}) = \sum_{g=1}^{G} p_g \phi_g(\mathrm{y}_i|\mathbf{x}_i;\sigma_g^2,\boldsymbol{\beta}_g) = \sum_{g=1}^{G} p_g \frac{1}{\sqrt{2\pi\sigma_g^2}} \exp\left[-\frac{(\mathrm{y}_i - \mathbf{x}_i'\boldsymbol{\beta}_g)^2}{2\sigma_g^2}\right],$$

L. A. García-Escudero et al. (Eds.): SMPS 2022, AISC 1433, pp. 326–333, 2023.
https://doi.org/10.1007/978-3-031-15509-3_43

where G is the number of clusters and p_g, $\boldsymbol{\beta}_g$, and σ_g^2 are respectively the mixing proportion, the vector of $J+1$ regression coefficients, including an intercept, and the variance term for the g-th cluster. The vector of model parameters is $\boldsymbol{\psi} = (p_1, \ldots, p_G, \boldsymbol{\beta}'_1, \ldots, \boldsymbol{\beta}'_G, \sigma_1^2, \ldots, \sigma_G^2)'$. It belongs to the parameter space $\boldsymbol{\Psi} = \{\boldsymbol{\psi} \in \mathbb{R}^{G+(J+1)G+G} : \sum_g p_g = 1, p_g > 0, \sigma_g^2 > 0 \text{ for } g = 1, \ldots, G\}$.

The log-likelihood function can be formulated as

$$\ell(\boldsymbol{\psi}) = \sum_{i=1}^{n} \log \left(\sum_{g=1}^{G} p_g \frac{1}{\sqrt{2\pi\sigma_g{}^2}} \exp\left[- \frac{(y_i - \mathbf{x}'_i\boldsymbol{\beta}_g)^2}{2\sigma_g^2} \right] \right), \tag{1}$$

which is maximized in order to estimate $\boldsymbol{\psi}$. Numerical computations are done by using direct maximization and/or the EM algorithm (Dempster et al. 1977).

The maximum likelihood (ML) estimation of this model can be very challenging because:

- the log-likelihood (1) is unbounded;
- the number of candidate predictors could be very high.

It is interesting to note that the two aforementioned issues can be solved in a similar manner, i.e. by adding a penalty to the log-likelihood. The aim of this paper is to review the main contributions that we had on this approach. They can be classified as regularization methods to:

- make the likelihood bounded;
- select the best (sparse) model.

In what follows, in particular, in the first two sections, we will review some proposals representative of each class. We complete our review with a section presenting some contributions where the authors tried to achive both objectives with a unique penalty. A final section with some concluding remarks ends the paper.

2 Likelihood Unboundedness

It is easy to show that when an observation has a zero residual on a particular component then the likelihood increases without bound if the corresponding variance decreases to zero. However, ML estimation is still feasible. Kiefer (1978) has shown that there exists a root of the likelihood equations leading to a consistent estimate of model parameters. Nevertheless, some authors prefer to introduce constraints or penalties on the log-likelihood in order to regularize it. This is so because the log-likelihood presents several local maxima that correspond to bad estimates. In practice, the optimization routine is often trapped into a degenerate solution, corresponding to a point where the log-likelihood is unbounded, or a spurious maximizer, i.e. a local maxima that does not correspond to the consistent root aforementioned.

The constrained estimation is based on the seminal work of Hathaway (1985) which, for univariate mixtures of normals, suggested to impose a lower bound to the ratios of the scale parameters in the maximization step. The method is equivariant under linear affine transformations of the data. That is, if the data are linearly transformed, the estimated posterior probabilities do not change and the clustering remains unaltered. Recently, in the multivariate case, Rocci et al. (2018) incorporated constraints on the eigenvalues of the component covariances matrices of Gaussian mixtures that are tuned on the data based on a cross–validation strategy. These constraints are built upon Ingrassia (2004)'s reformulation and provide an equivariant sufficient condition for Hathaway's constraints. Estimation is done in a familiar ML environment (Ingrassia and Rocci 2007), with data–driven selection of the scale balance. Di Mari et al. (2017) formulated the following constraints

$$\sqrt{c} \leq \frac{\sigma_g^2}{\bar{\sigma}^2} \leq \frac{1}{\sqrt{c}},$$

adapting Rocci et al. (2018)'s method to clusterwise linear regression and investigating its properties. The above constraints have the effect of shrinking the variances to a suitably chosen $\bar{\sigma}^2$, the *target* variance representing our prior information about the scale structure, and the level of shrinkage is given by the value of $c \in (0, 1]$.

An alternative to the constrained estimator is the penalized approach, in which a penalty $p(\sigma_1^2, \ldots, \sigma_G^2; \lambda)$ is imposed on the component variances and it is added to the log-likelihood. Under certain conditions on the penalty function, the penalized estimator is know to be consistent (Chen and Tan 2009). A function that satisfies these conditions is

$$p(\sigma_1^2, \ldots, \sigma_G^2; \lambda) = -\lambda \sum_{g=1}^{G} \left(\frac{\bar{\sigma}^2}{\sigma_g^2} + \log(\sigma_g^2) \right).$$

Similarly to the constrained approach, the component variances are shrunk to the target variance with a strength proportional to $\lambda \in [0, +\infty)$. Thus, the penalized log-likelihood can be written as

$$p\ell(\boldsymbol{\psi}; \lambda) = \ell(\boldsymbol{\psi}) + p(\sigma_1^2, \ldots, \sigma_G^2; \lambda)$$

and the set of unknown parameters is found by ML with computation done by means of an EM algorithm that is available in closed-form. As well as with the constrained approach, the penalized approach is equivariant with respect to linear transformations in the response provided that the target variance is consequently changed.

A crucial step in practical implementations of the constrained and penalized estimators is the calibration of the optimal amount of shrinkage. Technically, the choice of the values c and λ. Di Mari et al. (2021) considered the following two strategies.

i. **Cross-Validation**

It finds the value of the tuning parameter which maximizes the cross-validated log-likelihood. For a given c or λ, this is computed as follows.

1. Temporary estimates for the model parameters are obtained from the entire sample, and these are used as starting values to initialize the cross-validation procedure.
2. The data set is partitioned into training and test set.
3. Parameters are estimated on the training set and the contribution to the log-likelihood of the test set is computed.
4. Steps 2–3 are repeated M times and the M contributions to the log-likelihood of the test sets are added up computing the so-called cross-validated log-likelihood.

ii. ***k*-Deleted**

It is based on a modification of the k-deleted method (Seo and Lindsay 2010; Seo and Kim 2012) which finds the tuning parameter value maximizing the (modified) k-deleted log-likelihood. For a given c or λ, it is computed by taking out the k units with the largest log-likelihood.

On an extensive simulation study, Di Mari et al. (2021) evaluated the performance of the four estimators obtained by combining the two regularization methods with the two calibration strategies. The constrained and the penalized approaches performed almost equally well in all simulation conditions. That is, the simulation study showed no significant evidence in favor of one of the two approaches. By contrast, the cross-validation tuning performed slightly better than the k-deleted method. This, however, at the cost of a higher computation time.

3 Covariate Selection

In the application of a regression model, quite often several variables are available as possible explanatory variables and/or predictors, while, probably, only a few of them have a predictive and/or explanatory ability over the response. Being able to identify such covariates is therefore crucial.

The LASSO, proposed by Tibshirani (1996), has become very popular for the estimation of high-dimensional linear regression models where a vector of regression coefficients with a relatively small number of non-zero elements is desirable - this is typically referred to as *sparsity* of the regression coefficients. It consists in adding to the log-likelihood the penalty

$$p(\boldsymbol{\beta};\lambda) = -\lambda \sum_{j=1}^{J} |\beta_j| .$$

It reduces the variability of the estimator by shrinking the regression coefficients to zero. Its appeal as a covariate selection method derives from the fact that by increasing the value of λ some coefficient estimates become exactly equal to zero. Several algorithms have been proposed for solving the LASSO problem such as the active set method, the LARS (Efron et al. 2004), and the greedy coordinate descent method (Friedman et al. 2007).

Recent research has paid attention to LASSO-type penalties for sparse clusterwise regression modelling. Within this line, Khalili and Chen (2007) were the first presenting a penalty for FMLR models, which was specified as follows

$$p(\psi; \lambda_1, \ldots, \lambda_G) = -\sum_{g=1}^{G} \lambda_g p_g \sum_{j=1}^{J} |\beta_{jg}|, \tag{2}$$

where the tuning parameters are component-wise and weighted by p_g, i.e. the importance that a component has into the mixture. This approach poses two main challenges. The first is about the iterative algorithm. Differently from the traditional model, the EM algorithm does not admit analytical updates of model parameters. The second is about the calibration of the tuning parameters. A search over a multidimensional grid could be rather cumbersome or even impossible when the number of components/predictors increases. Khalili and Chen (2007) solve the first problem by developing an EM algorithm where the penalty function is locally approximated by a quadratic function and the prior probabilities $p_1, \ldots, p_G$ are updated by using the traditional, non optimal in this context, update. The calibration of the tuning parameters is done by minimizing the GCV computed for each component. Lloyd-Jones et al. (2018) adopted the same penalty but developed a minorization-maximization algorithm for fitting the penalized FMLR model. Even in this case the LASSO penalty is approximated by a derivable function. The calibration is instead performed by minimizing the BIC. In order to reduce the computational burden the minimization is done by using an optimization routine. An adaptive version of (2) has been proposed by Mortiera et al. (2015). The penalty is

$$p(\psi; \lambda_1, \ldots, \lambda_G) = -\sum_{g=1}^{G} \lambda_g p_g \sum_{j=1}^{J} \frac{|\beta_{jg}|}{|\hat{\beta}_{jg}|},$$

where $\hat{\beta}_{jg}$ is the unpenalized maximum likelihood estimate of β_{jg}. Even in this case, they use the standard, non optimal, update for the prior probabilities. The calibration of the tuning parameters is done by using the cross validation during the EM iterations.

4 Bounded Covariate Selection

Another L_1-penalization was proposed by Städler et al. (2010), who integrate a coordinate descent algorithm in a block-wise EM algorithm to update the regression coefficient parameters. They considered the following penalty

$$p(\boldsymbol{\psi};\lambda) = -\lambda \sum_{g=1}^{G} \frac{p_g^{\gamma}}{\sigma_g^{\delta}} \sum_{j=1}^{J} |\beta_{jg}|, \tag{3}$$

with $\gamma, \delta \in \{0, 1/2, 1\}$. Note that it penalizes values of the component variances that are too low preventing the degeneracy of the algorithm. In the case where γ is equal to zero, the Authors have shown that the penalized likelihood is bounded and conjectured that this should be hold true even when γ is different from zero. However, there is not a rigorous proof excluding pathological cases of unboundedness. The update of the mixing proportions is done by using a line search over a grid along the direction given by the traditional update. Calibration is greatly simplified by having a single λ: the optimal value is chosen over a unidimensional grid by minimizing an ad *hoc* BIC, adjusted to take into account of the effective number of degrees of freedom.

Di Mari et al. (2022) proposed a new and more flexible LASSO type penalty as follows

$$p(\boldsymbol{\psi};\lambda_1,\ldots,\lambda_G) = -\sum_{g=1}^{G} \lambda_g \frac{p_g^{\gamma}}{2\sigma_g^{2\delta}} \sum_{j=1}^{J} |\beta_{jg}|$$

with $\gamma, \delta \in \{0, 1\}$. It is more flexible than penalty (3) since each component has its own tuning parameter. The mixing proportions are updated by doing one step of the Gauss-Newton algorithm along with a line search to guarantee the monotonicity of the algorithm.

An issue common to all the aforementioned approaches is that, for tuning, the algorithm fitting the LASSO penalized FMLR must be run for all plausible values of the penalty parameter(s). This results in a computationally intensive and possibly time consuming, task. Di Mari et al. (2022) address this issue by generalizing the use of the LARS algorithm in combination with BIC to the specific case of LASSO penalized FMLR models: this allows performing the calibration in a single run. In LASSO penalized linear modelling, one of the most attractive properties of the LARS algorithm is that of solving the LASSO problem over the whole domain of the penalty parameter so to obtain the entire regularization path - from an empty set of predictors until the full set of predictors is reached. The appropriate level of sparsity is then chosen by BIC minimization. The idea is to first fit a FMLR with given $\lambda_1, \ldots, \lambda_G$. Then perform on each estimated component a covariate selection by using the BIC on the LARS path, weighted by the corresponding posterior probabilities. A FMLR is then fitted by retaining on each component only the explanatory variables selected in the previous step. Di Mari et al. (2022) have shown, in a large simulation study, that the proposed technique massively reduces the computation time, with an irrelevant loss in efficiency.

5 Concluding Remarks

In this short review we have shown how penalties, or constraints, help in the estimation and/or covariate selection of a finite mixture of linear regression models. A lot of work has been done. Still, there are several potential lines of future research that are worthwhile developing. Khalili and Chen (2007) have already identified some of them: to consider other finite mixture of generalized linear regression models, and other penalties different from LASSO. Anyhow, we believe that all proposals should leverage on computationally efficient methods to calibrate the strength of the penalty.

References

Chen, J., Tan, X.: Inference for multivariate normal mixtures. J. Multivar. Anal. **100**(7), 1367–1383 (2009)

Dempster, A., Laird, N., Rubin, D.: Maximum likelihood from incomplete data via the EM algorithm. J. Roy. Stat. Soc.: Ser. B (Methodol.) **39**(1), 1–22 (1977)

Di Mari, R., Gattone, S.A., Rocci, R.: Penalized versus constrained approaches for clusterwise linear regression modeling. In: Balzano, S., Porzio, G.C., Salvatore, R., Vistocco, D., Vichi, M. (eds.) CLADAG 2019. SCDAKO, pp. 89–95. Springer, Cham (2021). https://doi.org/10.1007/978-3-030-69944-4_10

Di Mari, R., Rocci, R., Gattone, S.: Clusterwise linear regression modeling with soft scale constraints. Int. J. Approx. Reason. **91**, 160–178 (2017)

Di Mari, R., Rocci, R., Gattone, S.A.: A two-step local least angle regression algorithm for lasso-penalized clusterwise linear regression modeling (2022, Submitted)

Efron, B., Hastie, T., Johnstone, I., Tibshirani, R.: Least angle regression. Ann. Stat. **32**(2), 407–499 (2004)

Friedman, J., Hastie, T., Holger, H., Tibshirani, R.: Pathwise coordinate optimization. Ann. Appl. Stat. **1**(2), 302–332 (2007)

Hathaway, R.: A constrained formulation of maximum-likelihood estimation for normal mixture distributions. Ann. Stat. **13**(2), 795–800 (1985)

Ingrassia, S.: A likelihood-based constrained algorithm for multivariate normal mixture models. Stat. Methods Appl. **13**(2), 151–166 (2004)

Ingrassia, S., Rocci, R.: Constrained monotone EM algorithms for finite mixture of multivariate Gaussians. Comput. Stat. Data Anal. **51**(11), 5339–5351 (2007)

Khalili, A., Chen, J.: Variable selection in finite mixture of regression models. J. Am. Stat. Assoc. **102**(479), 1025–1038 (2007)

Kiefer, N.: Discrete parameter variation: efficient estimation of a switching regression model. Econometrica **46**(2), 427–434 (1978)

Lloyd-Jones, L.R., Nguyen, H.D., McLachlan, G.J.: A globally convergent algorithm for lasso-penalized mixture of linear regression models. Comput. Stat. Data Anal. **119**, 19–38 (2018)

McLachlan, G., Peel, D.: Finite Mixture Models. Wiley, New York (2000)

Mortiera, F., et al.: Mixture of inhomogeneous matrix models for species-rich ecosystems. Environmetrics **26**, 39–51 (2015)

Rocci, R., Gattone, S., Di Mari, R.: A data driven equivariant approach to constrained Gaussian mixture modeling. Adv. Data Anal. Classif. **12**(2), 235–260 (2018)

Seo, B., Kim, D.: Root selection in normal mixture models. Comput. Stat. Data Anal. **56**(8), 2454–2470 (2012)

Seo, B., Lindsay, B.G.: A computational strategy for doubly smoothed MLE exemplified in the normal mixture model. Comput. Stat. Data Anal. **54**(8), 1930–1941 (2010)
Städler, N., Bühlmann, P., Van De Geer, S.: ℓ_1-penalization for mixture regression models. TEST **19**(2), 209–256 (2010)
Tibshirani, R.: Regression shrinkage and selection via the lasso. J. Roy. Stat. Soc. B **58**(1), 267–288 (1996)

Robust Bayesian Regression for Mislabeled Binary Outcomes

Massimiliano Russo[1(✉)] and Luca Greco[2]

[1] Division of Pharmacoepidemiology and Pharmacoeconomics, Harvard Medical School, Boston, USA
mrusso@bwh.harvard.edu

[2] University Giustino Fortunato Benevento, Benevento, Italy
l.greco@unifortunato.eu

Abstract. Logistic regression is an ubiquitous model for prediction and inference with binary outcomes. In some applications, part of the observed binary outcomes can be mislabeled, leading to severely biased estimates for the coefficients of the standard logistic regression model, and invalid subsequent inference. Taking a Bayesian approach, we propose a latent-logistic regression model that directly accounts for subject-specific misclassification probabilities. The true unknown correctly labeled outcomes are considered missing data, and their value is imputed from the model conditional posterior. A Bayesian logistic regression is then applied to the imputed outcomes, accounting for uncertainty in the imputation process. We study the posterior distribution of the proposed model under different prior choices, and provide a formal framework to identify misclassified outcomes with explicit control of the false discovery rate, comparing our proposed approach with robust frequentist alternatives in a comprehensive simulation study.

1 Introduction

We consider the problem of modeling mislabeled binary outcomes in a regression context. In particular, we focus on non differential misclassification, that occurs when the probability of reporting a 1 as a 0 is the same of reporting a 0 as a 1. Let $Y_i^* \in \{0, 1\}$ be the true unknown binary outcomes, for $i = 1, \dots n$, and $\mathbf{X}_i = (X_{i1}, \dots, X_{ip}\} \in \mathbb{R}^p$ a set of covariates, we assume that the outcomes are generated from the model $P(Y_i^* = 1 \mid \mathbf{X}_i = x_i) = \pi_0(x_i)$, without loss of generality we let $\pi_0(x_i) = \text{expit}\{x_i^\intercal \boldsymbol{\beta}_0\}$, where $\text{expit}\{a\} = \exp\{a\}/[1 + \exp\{a\}]$ is the cumulative distribution function of a standard logistic distribution, and $\boldsymbol{\beta}_0 \in \mathbb{R}^p$. The outcome variables Y_i^* can be considered latent, and we just observe a noisy version Y_i. Since the outcome is binary, the observed variable Y_i can either coincide with Y_i^* or with $(1 - Y_i^*)$, and their generative mechanism can be written as:

$$P(Y_i = 1 \mid X_i = x_i) = (1 - \epsilon_i)\pi_0(x_i) + \epsilon_i(1 - \pi_0(x_i)), \quad (1)$$

L. A. García-Escudero et al. (Eds.): SMPS 2022, AISC 1433, pp. 334–342, 2023.
https://doi.org/10.1007/978-3-031-15509-3_44

where $\epsilon_i = P(Y_i^* = 1 - Y_i)$. Model (1) can be re-written in terms of Y_i^* imputing the latent outcomes via

$$Y_i^* = (1 - \delta_i)Y_i + \delta_i(1 - Y_i), \tag{2}$$

where $\delta_i \in \{0, 1\}$ is a 'misclassification' indicator with $P(\delta_i = 1) = \epsilon_i$.

If we model the data with a standard binary regression model—ignoring the presence of misclassification—our estimate $\pi(x)$ of $\pi_0(x)$ can be severely biased. Moreover, in some cases it is of interest to determine the set $\mathscr{I} \subset \{1, \ldots, n\}$ of misclassified observations. In the following sections, we introduce a Bayesian model for misclassified binary outcomes, and study its posterior distribution under different prior specifications. In addition, we provide a principled way to estimate the set $\mathscr{I}$ explicitly controlling the False Discovery Rate (FDR).

The problem of mislabeled outcomes has been considered by several authors. In the frequentist setting various alternatives to the maximum likelihood estimator have been proposed, for example in Cantoni and Ronchetti (2001), Croux and Haesbroeck (2003) and Hung et al. (2018)—among others. In the Bayesian setting a robust nonparameteric regression is presented in Wood and Kohn (1998), while an explicit formulation of the logistic regression with misclassified outcome is presented in Verdinelli and Wasserman (1991). We discuss this formulation briefly in Sect. 3.1.

2 Model Formulation

2.1 A Bayesian Model for Misclassified Binary Data

We focus on Bayesian logistic regression, mimicking the data generative mechanism introduced in (2) via the following hierarchical model

$$\begin{aligned} Y_i^* &= (1 - \delta_i)Y_i + \delta_i(1 - Y_i), \\ Y_i^* \mid \mathbf{X}_i = x_i &\sim \text{Be}(\pi(x_i)), \\ \pi(x_i) &= \text{expit}\{x_i^\intercal \boldsymbol{\beta}\}, \\ \boldsymbol{\beta} &\sim \mathscr{N}(m, V), \\ \delta_i &\sim \text{Be}(\epsilon_i), \\ \epsilon_i &\sim P_\epsilon. \end{aligned} \tag{3}$$

From model (3) the posterior distribution of ϵ_i, conditional to the observed outcome and the linear predictor, can be expressed as a mixture:

$$P(\epsilon_i \mid Y_i = y, \pi(x_i)) \propto \pi(x_i)(1 - \epsilon_i)P_\epsilon + (1 - \pi(x_i))\epsilon_i P_\epsilon, \tag{4}$$

and its expectation

$$\mathbb{E}(\epsilon_i \mid Y_i = y, \pi(x_i)) = \pi(x_i)\mathbb{E}[(1 - \epsilon_i)P_\epsilon] + (1 - \pi(x_i))\mathbb{E}[\epsilon_i P_\epsilon]. \tag{5}$$

In the following section we describe the functional form of (5) for some choices of the prior distribution.

3 Misclassification Probability

3.1 Common Misclassification Probability

A model for misclassified binary outcomes with common misclassification probability has been introduced in Sect. 5 of Verdinelli and Wasserman (1991). This model can be expressed as in (3), under the constraint that $\epsilon_i = \epsilon$ for $i = 1, \ldots, n$. The authors use a Beta(2, 18) for the common misclassification probability parameter ϵ, and show that the model estimates are robust to outcome misclassification. However, the model itself it is not suitable to study individual-specific misclassification probabilities, that can only be approximated averaging the latent indicators δ_i across MCMC iterations. We will show in simulations (Sect. 6) that when the goal is to characterize the set $\mathscr{I}$, using the posterior estimates of δ_i can be sub-optimal compared to studying the posterior distribution of the individual misclassification parameters ϵ_i using the more general model formulation in (3).

3.2 Subject-Specific Misclassification Probabilities

In model (3) we ideally want the switching probability ϵ_i to be large when $\pi(x_i)$ is close to 1 but $Y_i = 0$, and close to 0 when $\pi(x_i)$ is small but $Y_i = 1$. Assuming the each subject has a different probability of being misclassified with common prior $\epsilon_i \sim \text{Beta}(a, b)$, (4) becomes

$$P(\epsilon_i \mid Y_i = y, \pi(x_i)) = \pi(x_i)\text{Beta}(a + (1 - y), b + y) + (1 - \pi(x_i))\text{Beta}(a + y, b + (1 - y)), \quad (6)$$

and (5)

$$\mathbb{E}(\epsilon_i \mid Y_i = y, \pi(x_i)) = \pi(x_i)\frac{a + (1 - y)}{a + b + 1} + (1 - \pi(x_i))\frac{a + y}{a + b + 1}, \quad (7)$$

which is a linear function of $\pi(x_i)$ A sensible prior would put a lot of mass near values close to 0 (e.g., 10%) with a relative small variance. For example, setting $a = 2$ and $b = 18$ as in Verdinelli and Wasserman (1991), we have a mean of 0.1 and a variance of close to 0.004.

The model behaves as expected: the posterior probability of switching is (linearly) increasing with $\pi(x_i)$ if $Y_i = 0$, and (linearly) decreasing with $\pi(x_i)$ if $Y_i = 1$, in a sensible ranges of values $(0.09, 0.15)$. However, when $\pi(x_i) = 0.5$ we have a relatively high probability of switching, which is not necessarily desirable. Changing the values of the hyperparameters of the Beta prior has the impact of modifying the range of the posterior, but the linear relationship remain. The Beta prior offer computational advantages but its effect on the posterior does not reflect the intuition that "well behaved observations" should have a posterior probability of switching which is essentially 0. We provide a possible solution in the following section (Fig. 1).

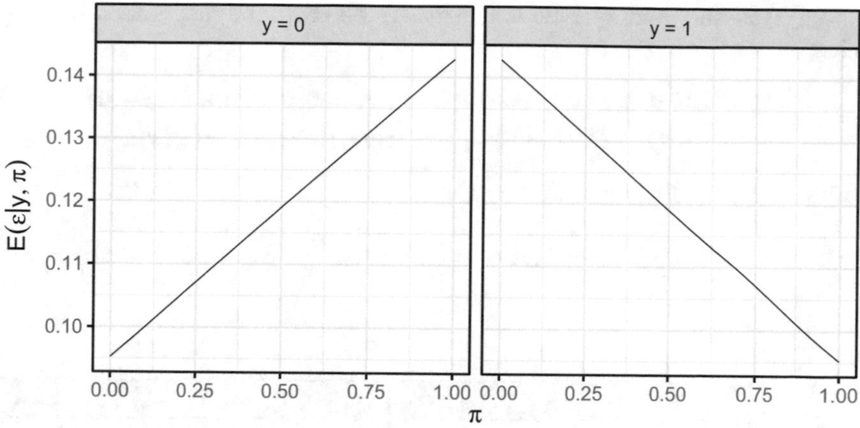

Fig. 1. Representation of the conditional posterior expectation in (7) with $a = 2$, $b = 18$

3.3 Spike-and-Slab Prior for Misclassification Probabilities

A potential problem with model (6) is that the probability of switching is relatively high also for 'small' (or 'high') values of the $\pi(x_i)$. A possible solution consist in shrinking the misclassification probability toward 0 for most of $\pi(x_i)$ via a spike-and-slab prior. Let's assume that *a priori* we have

$$\begin{aligned} \epsilon_i &\sim T_i \times 1\{\epsilon_i = 0\} + (1 - T_i)\text{Beta}(a, b), \\ T_i &\sim \text{Be}(\gamma), \end{aligned} \tag{8}$$

for a $\gamma \in [0, 1]$. Note that when $T_i = 0$, we are back to the previous specification, while when $T_i = 1$ we are considering a standard logistic regression model, without mislabeled outcomes. Using prior (8), we obtain the conditional posterior

$$P_s(\epsilon_i \mid Y_i = y, \pi(x_i), T_i) = T_i \times \pi(x)^y(1 - \pi(x))^{1-y}1\{\epsilon_i = 0\} + (1 - T_i) \times P(\epsilon_i \mid Y_i = y, \pi(x_i)), \tag{9}$$

with

$$P(T_i = 1 \mid Y_i = y, \pi(x_i)) = \frac{\gamma\pi(x_i)^y(1 - \pi(x_i))^{1-y}}{\gamma\pi(x_i)^y(1 - \pi(x_i))^{1-y} + (1 - \gamma)\int L_i(\beta, \epsilon_i)P(\epsilon_i)d\epsilon_i}, \tag{10}$$

where $P(\epsilon_i)$ is the probability mass function of a Beta(a,b), and

$$L_i(\beta, \epsilon_i) = [(1 - \epsilon)\pi(x_i) + \epsilon_i(1 - \pi(x_i))]^{Y_i} [(1 - \epsilon)(1 - \pi(x_i)) + \epsilon_i\pi(x_i)]^{1-Y_i},$$

is the i-th contribution to the likelihood function. The integral is available in closed form

$$\int L_i(\beta, \epsilon_i)P(\epsilon_i)d\epsilon_i = \frac{\{\pi(x_i)\Gamma(a + (1 - Y_i))\Gamma(b + Y_i) + (1 - \pi(x_i))\Gamma(a + Y_i)\Gamma(b + (1 - Y_i))\}}{(a + b)\Gamma(a)\Gamma(b)}, \tag{11}$$

where $\Gamma(\cdot)$ is the gamma function. Finally, we can integrate out T_i from (9), having

$$\begin{aligned} P_s(\epsilon_i \mid Y_i = y, \pi(x_i)) &= P(T_i = 1 \mid Y_i = y, \pi(x_i)) \times 1\{\epsilon_i = 0\} \\ &+ (1 - P(T_i = 1 \mid Y_i = y, \pi(x_i)))P(\epsilon_i \mid Y_i = y, \pi(x_i)), \end{aligned} \tag{12}$$

and its conditional expectation

$$\mathbb{E}(\epsilon_i \mid Y_i = y, \pi(x_i)) = (1 - P(T = 1 \mid Y_i = y, \pi(x_i))) \left\{ \pi(x_i) \frac{a + (1 - y)}{a + b + 1} + (1 - \pi(x_i)) \frac{a + y}{a + b + 1} \right\}. \tag{13}$$

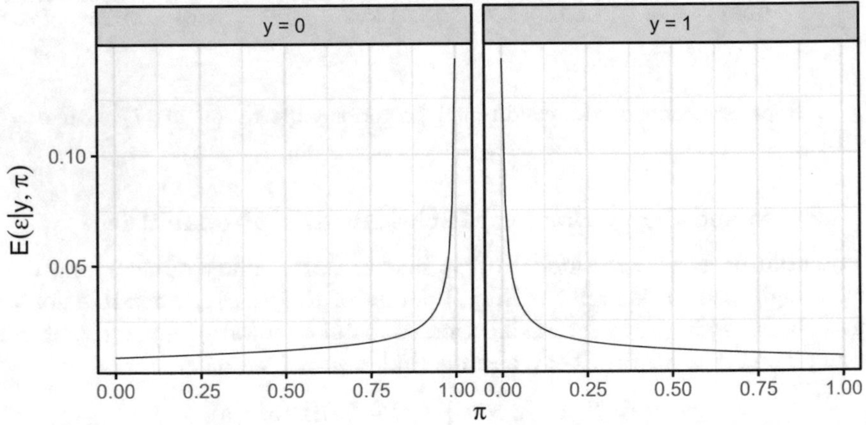

Fig. 2. Representation of the conditional posterior expectation (13) with $a = 2$ and $b = 18$, and $\gamma = 0.90$

Figure 2 shows that in average, prior (8) induces a negligible switching probability for most of the $\pi(x_i)$ space.

4 Posterior Computation

The posterior distribution of the three models presented in Sect. 2 can be efficiently approximated via a Gibbs sampler, exploiting the data augmentation scheme introduced in Polson et al. (2013). We report the details of the Gibbs sampler for the model with the spike-and-slab prior considered in Sect. 3.3. The corresponding algorithms for the models presented in Sects. 3.1 and 3.2 can be derived with minimal variations. The algorithm iterates through the following steps:

1. For $i = 1, \ldots, n$ generate auxiliary variables for a Pólya-gamma random variable

$$\omega_i \mid \boldsymbol{\beta} \sim \text{PG}(1, x_i^\intercal \boldsymbol{\beta}).$$

2. Let $\mathbf{k} = (Y_1^* - 1/2, \ldots, Y_n^* - 1/2)$, and $\Omega = \text{diag}(\omega_1, \ldots, \omega_n)$, update the regression coefficients via

$$\boldsymbol{\beta} \mid \mathbf{k}, \boldsymbol{\omega} \sim \mathcal{N}(m_\omega, V_\omega),$$

with $V_\omega = (X^\intercal \Omega X + S^{-1})^{-1}$ and $m_\omega = V_\omega(X^\intercal \mathbf{k} + S^{-1}m)^{-1}$.
3. For $i = 1, \ldots, n$ sample the indicator T_i from

$$T_i \mid Y_i = y, x_i, \boldsymbol{\beta} \sim \text{Be}(P(T = 1 \mid Y_i = y, \pi(x_i))),$$

where $P(T = 1 \mid Y_i = y, \pi(x_i)))$ is defined in (10).
4. For $i = 1, \ldots, n$ sample the switching probabilities from their conditional posterior. If $T_i = 1$ set $\epsilon_i = 0$, else sample ϵ_i from

$$\epsilon_i \mid Y_i = y, \pi(x_i), T_i = 0 \sim P(\epsilon_i \mid Y_i = y, \pi(x_i))$$

where $P(\epsilon_i \mid Y_i, \pi(x_i))$ is defined in (6).
5. For $i = 1, \ldots, n$ sample the misclassification indicators δ_i from a Bernoulli variable with probability of success

$$\frac{\epsilon_i(1 - \pi(x_i))^{Y_i}\pi(x_i)^{1-Y_i}}{\epsilon_i(1 - \pi(x_i))^{Y_i}\pi(x_i)^{1-Y_i} + (1 - \epsilon_i)\pi(x_i)^{Y_i}(1 - \pi(x_i))^{1-Y_i}},$$

and set $Y_i^* = (1 - \delta_i)Y_i + (1 - \delta_i)(1 - Y_i)$.
6. Finally, assuming a conjugate $\text{Beta}(a_\gamma, b_\gamma)$ for the parameter γ, update

$$\gamma \mid \boldsymbol{\delta} \sim \text{Beta}\left(a_\gamma + \sum_{i=1}^{n} \delta_i, b_\gamma + n - \sum_{i=1}^{n} \delta_i\right).$$

5 Outlier Detection

Beside obtaining reliable estimates of the logistic regression coefficients, it is sometime of interest to understand which data-points are misclassified. For example, if Y_i is the result of a lab test, it might be useful to repeat the test for the subjects whose outcome has been flagged as false positive or false negative. We can estimate the set $\mathcal{I}$ via the estimator $\widehat{\mathcal{I}} = \{i \in \{1, \ldots, n\} : \bar{\epsilon}_i > t\}$, where $\bar{\epsilon}_i$ is the posterior mean of ϵ_i and t is a suitable threshold chosen to control false positive results. An optimal threshold t that explicitly control the (Bayesian) FDR, can be found using the method proposed by Müller et al. (2004).

6 Simulation Study

We consider the same simulation settings presented in Croux and Haesbroeck (2003), focusing on three scenarios with $p = 3$ covariates and a sample size $n = 100$. We set $x_i = (1, x_{i1}, x_{i2})$, with x_{i1} and x_{i2} generated from a standard normal distribution, and the dependent variables generated according to

$Y_i = \text{Be}(\text{expit}\{x_i^\intercal \boldsymbol{\beta}_0\})$ with $\boldsymbol{\beta}_0 = (0, 2, 2)$. We propose three simulation scenarios: 1) in the first simulation scenarios we consider no contamination; 2) in the second scenario we add 5 outliers on a hyperplane parallel to the true discriminating hyperplane, the shift between the two hyperplanes being equal to $1.5\sqrt{2}$; 3) finally, the third scenario considers bad leverage points with contaminating hyperplane shifted $5\sqrt{2}$. We replicate each simulation scenario $R = 100$ times and compute the bias and mean square error as

$$\text{bias} = \left\| R^{-1} \sum_{r=1}^{R} \widehat{\boldsymbol{\beta}_r} - \boldsymbol{\beta}_0 \right\|, \quad \text{MSE} = R^{-1} \sum_{r=1}^{R} \left\| \widehat{\boldsymbol{\beta}_r} - \boldsymbol{\beta}_0 \right\|^2,$$

$\|\cdot\|$ indicates the euclidean distance, and r indexes the simulation replicates. We consider six estimation methods: 1) a standard logistic regression (MLE); 2) the Bianco and Yohai (BY) estimator proposed in Croux and Haesbroeck (2003) (BY); 3) the Huber type robust estimators described in Cantoni and Ronchetti (2001) (HUBER), downweighting the points in the covariate-space using the elements of the hat matrix; 4) the Bayesian robust logistic regression proposed in Verdinelli and Wasserman (1991) (VW); 5) the Bayesian robust estimator with subject-specific Beta priors introduced in Sect. 3.2 (BETA); 6) the Bayesian robust estimator with spike-and-slab priors introduced in Sect. 3.3 (SS-BETA). In particular, for the three considered Bayesian methods we use the posterior mean as an estimate of the regression coefficients, with standard normal priors for the regression coefficients, and $a = 2$ and $b = 18$ as prior for the misclassification probability. For the SS-BETA, we set $a_\gamma = 2$ and $b_\gamma = 18$.

Table 1 summarizes the simulation results in terms of bias and MSE. In scenario 1—where there are no outliers—the MLE is the method that presents less bias. The BY, HUBER, VW, BETA and SS-BETA estimators share a very similar performance. In scenario 2—with a moderate contamination—the MLE shows poor performance. The BY has a slightly lower bias compared to the MLE, while HUBER, VW, BETA and SS-BETA present substantially lower bias than the MLE. Similarly, in scenario 3—with high contamination—MLE performs poorly in term of bias, and the BY slightly improves its performance. The three Bayesian methods perform better in terms of bias reduction compared to MLE and BY, while the HUBER estimator has the lowest bias. In terms of MSE, the Bayesian methods perform consistently better than the other methods across all the three scenarios. The BY has a lower MSE compared to the MLE in scenario 2 and 3, but not substantially. The HUBER estimator performs better than the BY, but still presenting substantially higher MSE compared the Bayesian methods.

Table 1. Bias and MSE for the three simulation scenarios described in Sect. 6

	Bias			MSE		
	Scenario 1	Scenario 2	Scenario 3	Scenario 1	Scenario 2	Scenario 3
MLE	0.165	1.280	2.140	0.595	1.824	4.732
BY	0.195	1.119	1.628	0.768	1.487	3.470
HUBER	0.276	0.847	0.219	1.180	1.334	1.759
VW	0.290	0.899	0.334	0.310	1.050	0.475
BETA	0.166	0.729	0.320	0.268	0.810	0.447
SS-BETA	0.206	0.831	0.370	0.279	0.958	0.513

Table 2. Percentage of times a point is incorrectly flagged as an outlier (FDR) and percentage of times a true outlier is flagged as such (TPR), using the method described in Sect. 5. The symbol '–' indicates that the result does not exist for the particular method/scenario

	ϵ							δ					
	Scenario 1		Scenario 2		Scenario 3			Scenario 1		Scenario 2		Scenario 3	
	FDR	TPR	FDR	TPR	FDR	TPR		FDR	TPR	FDR	TPR	FDR	TPR
VW	–	–	–	–	–	–	VW	5	–	3	94	6	98
BETA	3	–	0	60	0	96	BETA	7	–	4	97	6	98
SS-BETA	3	–	0	61	0	93	SS-BETA	6	–	3	95	6	98

From Table 2 we see that applying the method sketched in Sect. 5 we have a strong control of FDR. In particular, in scenario 2 and 3—where there are outliers—we never flag any points that is not outlier as such, still preserving a good power to detect true outliers (about 60% in scenario 2 and 95% in scenario 3 for both BETA and SS-BETA). When applying the same method to the posterior means of the δ_is, we observe a slightly higher proportion of false rejection in all scenarios, and consequently also an higher power in detecting true outliers (grater or equal to 95% for all scenarios and methods).

7 Concluding Remarks

We discussed a Bayesian model for binary data that is robust to misclassification of the outcomes, and compared its performances, under three different prior specifications, with popular frequentist alternatives. We showed in simulations that the proposed Bayesian approach outperforms the frequentist alternatives in terms of MSE, still providing reasonably low bias for the estimate of the regression coefficients. Additionally, we provided a framework to identify outliers with explicit control the false discovery rate.

References

Cantoni, E., Ronchetti, E.: Robust inference for generalized linear models. J. Am. Stat. Assoc. **96**(455), 1022–1030 (2001)

Croux, C., Haesbroeck, G.: Implementing the Bianco and Yohai estimator for logistic regression. Comput. Stat. Data Anal. **44**(1–2), 273–295 (2003)

Hung, H., Jou, Z.Y., Huang, S.Y.: Robust mislabel logistic regression without modeling mislabel probabilities. Biometrics **74**(1), 145–154 (2018)

Müller, P., Parmigiani, G., Robert, C., Rousseau, J.: Optimal sample size for multiple testing. J. Am. Stat. Assoc. **99**(468), 990–1001 (2004)

Polson, N.G., Scott, J.G., Windle, J.: Bayesian inference for logistic models using Pólya-gamma latent variables. J. Am. Stat. Assoc. **108**(504), 1339–1349 (2013)

Verdinelli, I., Wasserman, L.A.: Bayesian analysis of outlier problems using the Gibbs sampler. Stat. Comput. **1**, 105–117 (1991)

Wood, S., Kohn, R.: A Bayesian approach to robust binary nonparametric regression. J. Am. Stat. Assoc. **93**(441), 203–213 (1998)

Case-Wise and Cell-Wise Outliers Detection Based on Statistical Depth Filters

Giovanni Saraceno(✉) and Claudio Agostinelli

Department of Mathematics, University of Trento, Trento, Italy
{giovanni.saraceno,claudio.agostinelli}@unitn.it

Abstract. According to the classical case-wise contamination model, observations are considered as the units to be identified as outliers. Alqallaf et al. (2009) showed the limits of this approach, especially for a larger number of variables, and introduced the Independent contamination model, or cell-wise contamination, where the cells are the units to be identified as outliers. For the estimation problem, one approach to deal, at the same time, with both type of contamination is filter out the contaminated cells from the data set and then apply a robust procedure able to handle case-wise outliers and missing values. In this work we deal with the outliers detection task, taking into account both types of contamination. We propose to use the depth filters introduced by Agostinelli (2021) as detection procedure which is able to identify both case-wise and cell-wise outliers. We investigated the finite sample performance by a small simulation study, comparing the depth filters with the detection rules available in literature.

1 Introduction

It is well-known that one of the common problem in real data is the possible presence of outliers. According to the situation, they may be errors that affect the data analysis or can suggest unexpected information. In both cases, it can be crucial to detect such observations. Indeed, the investigation of their source could reveal hidden random mechanisms. Furthermore, the outliers are model dependent then an effective detection rule is a successful strategy to improve the model estimation. In statistics, the outliers are typically referred to the observations, the rows of a data matrix. All the methods developed since 1960s s in the field of robust statistics had the objective to be less sensitive to the presence of these row-wise outliers. For example, outlying observations can be identified by their large residuals from the fit or by large Mahalanobis distances. See Maronna et al. (2006) for a complete description of the developments in robust statistics. Among others, it is worth to cite the Forward Search (FS), a powerful general model for detecting multiple masked outliers in continuous

L. A. García-Escudero et al. (Eds.): SMPS 2022, AISC 1433, pp. 343–349, 2023.
https://doi.org/10.1007/978-3-031-15509-3_45

multivariate data (Cerioli et al. 2018). The idea is that the search starts by fitting a multivariate normal model starting from a subset of size d. Then, all the n observations are ordered by their Mahalanobis distance and the subset size is updated to $d+1$ by taking the $d+1$ observations with the smallest Mahalanobis distance until all the observations are included. This approach not only identifies the outlying observations but also provide a monitoring tool of the main statistics during the search.

The case-wise contamination paradigm can be not sufficient in modern applications with high-dimensional data, where only some of the entries, or cells, of a row can be contaminated. Alqallaf et al. (2009) firstly formulated this cell-wise contamination scheme and they noticed how they propagate, i.e. given a proportion ε of contaminated cells, the expected proportion of rows that contain at least one outlying cell is

$$1-(1-\varepsilon)^p$$

where p denotes the number of variables. This proportion easily exceed the 50% breakdown point for increasing contamination level ε and dimension p. For this reason, the existing methods may fail under the cell-wise contamination scheme. Finally, Agostinelli et al. (2015) showed that case-wise and cell-wise outliers can occur at the same time. One of the first and successful method proposed to cope with the cell-wise contamination is called *Detecting Deviating Cells* (DDC), introduced by Rousseeuw and Van den Bossche (2018), which takes into account the correlations between variables and provides predicted values of the outlying cells.

2 Half-Space Depth-Filter

Let $\boldsymbol{X}$ be a $\mathbb{R}^d$-valued random variable and F a continuous distribution function.

Definition 1 (Half-space depth). For a point $\boldsymbol{x} \in \mathbb{R}^d$, the half-space depth of $\boldsymbol{x}$ with respect to F is defined as the minimum probability of all closed half-spaces including $\boldsymbol{x}$:

$$d_{HS}(\boldsymbol{x};F) = \min_{H \in \mathscr{H}(\boldsymbol{x})} P_F(\boldsymbol{X} \in H),$$

where $\mathscr{H}(\boldsymbol{x})$ indicates the set of all half-spaces in $\mathbb{R}^d$ containing $\boldsymbol{x} \in \mathbb{R}^d$.

Given an independent and identically distributed sample $\boldsymbol{X}_1, \ldots, \boldsymbol{X}_n$ of size n, we denote by $\hat{F}_n(\cdot)$ its empirical distribution function and by $d_{HS}(\boldsymbol{x};\hat{F}_n)$ the sample half-space depth. We have that $d_{HS}(\boldsymbol{x};\hat{F}_n)$ is a uniform consistent estimator of $d_{HS}(\boldsymbol{x};F)$ (Donoho and Gasko 1992), that is,

$$\sup_{\boldsymbol{x}} |d_{HS}(\boldsymbol{x};\hat{F}_n) - d_{HS}(\boldsymbol{x};F)| \overset{a.s.}{\to} 0 \qquad n \to \infty.$$

Given a statistical depth function, it is possible to define the α-depth trimmed region $R_\alpha(F)$, with $\alpha \in [0, m]$ and m denotes the maximum value obtained by the chosen depth. For the half-space depth it is given by

$$R_\alpha(F) = \{\boldsymbol{x} \in \mathbb{R}^d : d_{HS}(\boldsymbol{x}; F) \geq \alpha\},$$

and $\alpha \in [0, \frac{1}{2}]$. For any $\beta \in [0, 1]$, $R^\beta(F)$ will denote the smallest region $R_\alpha(F)$ that has probability larger than or equal to β according to F. Let $C^\beta(F)$ be the complement in $\mathbb{R}^d$ of the set $R^\beta(F)$.

Given a high order probability β, the half-space depth-filter of general dimension d is defined by (Saraceno and Agostinelli 2021).

$$d_n = \sup_{\boldsymbol{x} \in C^\beta(F)} \{d_{HS}(\boldsymbol{x}; \hat{F}_n) - d_{HS}(\boldsymbol{x}; F)\}^+, \tag{1}$$

where $\{a\}^+$ represents the positive part of a. Then, the $n_0 = \lfloor n d_n \rfloor$ d-variate observations with the smallest population half-space depth are marked as outliers, where $\lfloor a \rfloor$ is the largest integer less then or equal to a. The half-space depth-filter satisfies the desired property of a consistent filter, that is $\frac{n_0}{n} \to 0$ as $n \to \infty$.

3 Outlier Detection Based on Half-Space Depth-Filters

Consider a sample $\boldsymbol{X}_1, \ldots, \boldsymbol{X}_n$ where $\boldsymbol{X}_i \in \mathbb{R}^p, i = 1, \ldots, n$. The definition of the d-dimensional half-space depth-filter given in Eq. (1) allows to use the filtering procedure in a versatile way. In principle, it is possible to apply the d-variate depth-filter for all the values in the set $\{1, \ldots, p\}$. Each filter identifies the d-dimensional outliers that can be used to study possible substructures in the data. In practical situations, the choice of d can be dictated by previous knowledge about the phenomenon under investigation.

An efficient way to detect the outlying cells is to combine the output of the univariate and bivariate depth-filter, for $d = 1$ and $d = 2$, respectively. The procedure used for this purpose is described in Leung et al. (2017) and summarized here.

We first apply the univariate filter to each variable separately. Let $\boldsymbol{X}^{(j)} = \{X_{1j}, \ldots, X_{nj}\}$, $j = 1, \ldots, p$, be a single variable. The univariate filter will flag $\lfloor n d_{nj} \rfloor$ observations as outliers, where d_{nj} is as in Eq. (1). Filtered data are indicated through an auxiliary matrix $\boldsymbol{U}$ of zeros and ones, with zero corresponding to an outlying cell. Next, we identify the bivariate outliers by iterating the filter over all possible pairs of variables. Consider a pair of variables $\boldsymbol{X}^{(jk)} = \{(X_{ij}, X_{ik}), i = 1, \ldots, n\}$. For bivariate points with no flagged components by the univariate filter, we apply the bivariate filter. Given the pair of variables $\boldsymbol{X}^{(jk)}$, $1 \leq j < k \leq p$, we compute the value $d_n^{(jk)}$. Then, $n_0^{(jk)}$ couples will be identified as bivariate outliers. Finally, in order to identify the cells (i, j) which have to be flagged as cell-wise outliers, let

$$J = \{(i, j, k) : (X_{ij}, X_{ik}) \text{ is flagged as bivariate outlier}\}$$

be the set of triplets which identifies the pairs of cells flagged by the bivariate filter where $i = 1, \ldots, n$ indicates the row. For each cell (i, j) in the data, we count the number of flagged pairs in the i-th row in which the considered cell is involved:

$$m_{ij} = \#\{k : (i, j, k) \in J\}.$$

In absence of contamination, m_{ij} follows approximately a binomial distribution $Bin(\sum_{k \neq j} \boldsymbol{U}_{jk}, \delta)$ where δ represents the overall proportion of cell-wise outliers undetected by the univariate filter. Hence, we flag the cell (i, j) if $m_{ij} > c_{ij}$, where c_{ij} is the 0.99-quantile of $Bin(\sum_{k \neq j} \boldsymbol{U}_{jk}, 0.1)$.

Finally, in order to detect both case-wise and cell-wise outliers simultaneously we can consider the filtering procedure which consists in applying the d-dimensional depth-filter three times in sequence, using $d = 1$, $d = 2$ and $d = p$. In practice, after applying the univariate and bivariate filters as described above, the p-variate filter is performed to the full data matrix. Detected observations (rows) are directly flagged as p-variate (case-wise) outliers. This procedure based on univariate, bivariate and p-variate filters has been denoted as HS-UBPF.

4 Simulation Study

In order to illustrate the behaviour of the depth-filter HS-UBPF as detection rule, we consider a small simulation study where their performance is compared with the *Detecting Deviating Cells* algorithm, as implemented in the `R` (R Core Team 2019) package `cellWise`, and the detection rule based on the Minimum Covariance Determinant (MCD).

We considered samples from a $N_p(\mathbf{0}, \boldsymbol{\Sigma}_0)$, where all values in $diag(\boldsymbol{\Sigma}_0)$ are equal to 1, $p = 20$ and the sample size is $n = 200$. We consider the following scenarios:

- Clean data: data without changes.
- Cell-Wise contamination: a proportion ε of cells in the data is replaced by $X_{ij} \sim N(k, 0.1^2)$, where $k = 1, \ldots, 10$.
- Case-Wise contamination: a proportion ε of cases in the data is replaced by $\boldsymbol{X}_i \sim 0.5N(c\boldsymbol{v}, 0.1^2\boldsymbol{I}) + 0.5N(-c\boldsymbol{v}, 0.1^2\boldsymbol{I})$, where $c = \sqrt{k(\chi_p^2)^{-1}(0.99)}$, $k = 2, 4, \ldots, 20$ and $\boldsymbol{v}$ is the eigenvector corresponding to the smallest eigenvalue of $\boldsymbol{\Sigma}_0$ with length such that $(\boldsymbol{v} - \boldsymbol{\mu}_0)^\top \boldsymbol{\Sigma}_0^{-1} (\boldsymbol{v} - \boldsymbol{\mu}_0) = 1$.

The proportions of contaminated rows chosen for case-wise contamination are $\varepsilon = 0.1, 0.2$, and $\varepsilon = 0.02, 0.05, 0.1$ for cell-wise contamination. The number of replicates in our simulation study is $N = 50$.

We measure the performance of the considered rules by computing the values of accuracy and precision across the simulation parameters.

Figures 1 and 2 show the average accuracy and precision for different contamination values k and in case of case-wise and cell-wise contamination, respectively. The half-space depth-filter shows competitive performances with respect to DDC in terms of both accuracy and precision in case of cell-wise contamination. When the case-wise contamination is considered, the depth-filter has poorer results, while DDC confirms its high-quality results.

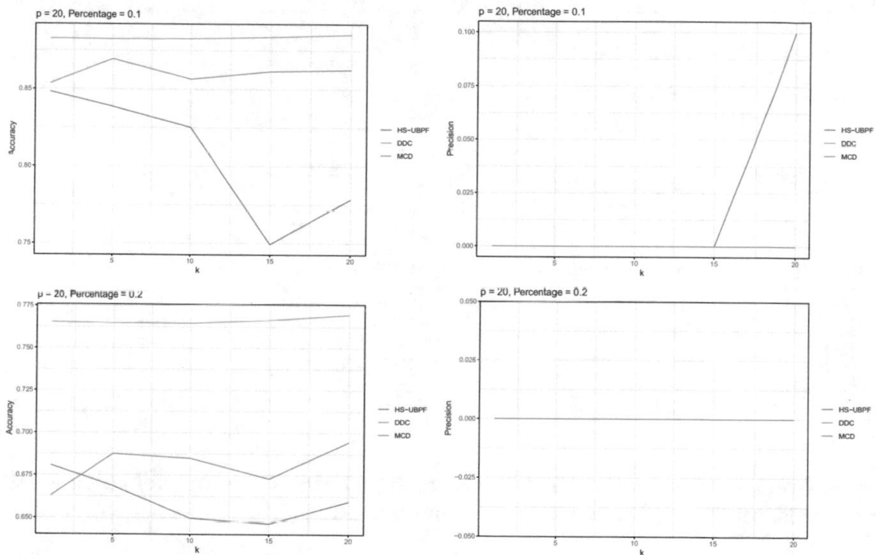

Fig. 1. Average accuracy (left) and precision (right) versus the contamination value k, for different case-wise contamination level and $p = 20$

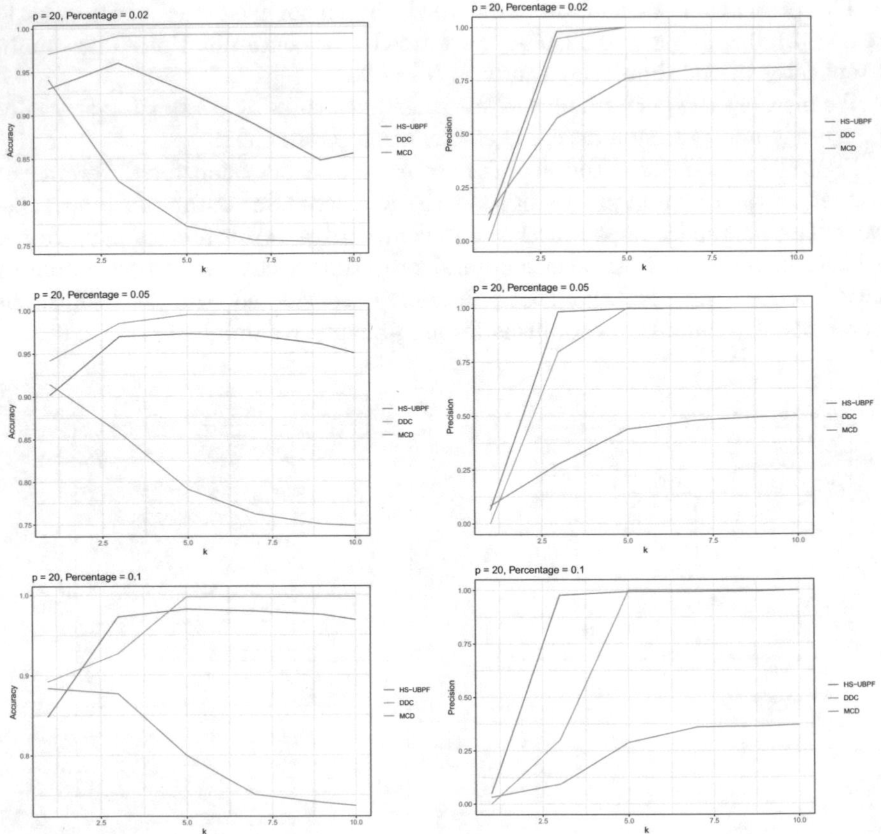

Fig. 2. Average accuracy (left) and precision (right) versus the contamination value k, for different cell-wise contamination level and $p = 20$

References

Agostinelli, C., Leung, A., Yohai, V.J., Zamar, R.H.: Robust estimation of multivariate location and scatter in the presence of cellwise and casewise contamination. TEST **24**(3), 441–461 (2015). https://doi.org/10.1007/s11749-015-0450-6

Alqallaf, F., Van Aelst, S., Zamar, R.H., Yohai, V.J.: Propagation of outliers in multivariate data. Ann. Stat. **37**(1), 311–331 (2009)

Cerioli, A., Riani, M., Atkinson, A.C., Corbellini, A.: The power of monitoring: how to make the most of a contaminated multivariate sample. Stat. Meth. Appl. **27**(4), 559–587 (2018)

R Core Team: R: A Language and Environment for Statistical Computing (2019). https://www.R-project.org/

Donoho, D.L., Gasko, M.: Breakdown properties of location estimates based on halfspace depth and projected outlyingness. Ann. Stat. **20**(4), 1803–1827 (1992)

Leung, A., Yohai, V.J., Zamar, R.H.: Multivariate location and scatter matrix estimation under cellwise and casewise contamination. Comput. Stat. Data Anal. **111**, 59–76 (2017)

Maronna, R.A., Martin, R.D., Yohai, V.J.: Robust Statistic: Theory and Methods. Wiley, Chichester (2006)

Rousseeuw, P.J., Van den Bossche, W.: Detecting deviating data cells. Technometrics **60**(2), 135–145 (2018)

Saraceno, G., Agostinelli, C.: Robust multivariate estimation based on statistical depth filters. TEST **30**(4), 935–959 (2021). https://doi.org/10.1007/s11749-021-00757-z

The d_θ-Depth-Based Interval Trimmed Mean

Beatriz Sinova(✉)

Departamento de Estadística e I.O. y D.M., Universidad de Oviedo, Oviedo, Spain
sinovabeatriz@uniovi.es

Abstract. Central tendency of interval-valued random elements has been mainly described in terms of different notions of medians and location M-estimators in the literature, whereas the approach consisting of medians and trimmed means based on a depth function has been rarely considered. Recently, depth-based trimmed means have been adapted to the more general framework of fuzzy number-valued data in terms of the so-called D_θ-depth. The aim of this work is to study the empirical behaviour of the particularization of such a location measure when data are interval-valued.

1 Introduction

Interval-valued data arise in numerous real-life experiments when imprecise information is handled. For instance, we could refer to surveys that collect opinions, judgements or perceptions; fluctuations or ranges of a characteristic along certain period of time; imprecise observations of a real-valued random variable due to measurement errors; interval-type censoring data; aggregated information, etc. Different statistical techniques have already been adapted to deal with interval-valued data, such as regression analysis, hypotheses testing procedures, clustering... Concerning central tendency measures, the Aumann mean is the most frequently used in these techniques, despite its excessive sensitivity to outliers or data changes. Other alternatives from the literature present a more robust behaviour, which makes them more suitable for summarizing datasets that contain some contaminated data: the median based on the generalized Hausdorff metric (Sinova et al. 2010), the 1-norm median (Sinova and Van Aelst 2013), the spatial-type interval-valued median (Sinova and Van Aelst 2018) and interval-valued M-estimators of location (Sinova 2016). Apart from these measures, which were specifically proposed for the interval-valued setting, medians and trimmed means based on the halfspace and simplicial depths have also been considered in Sinova (2016) by identifying each interval-valued datum with the point whose coordinates are the mid-point and the spread of the interval.

Recently, Sinova (2021) studied trimmed means based on a depth function in the more general framework of fuzzy number-valued data. Since they have shown very promising empirical results, this work is focused on their particularization to the interval-valued setting (the d_θ-depth-based interval trimmed

L. A. García-Escudero et al. (Eds.): SMPS 2022, AISC 1433, pp. 350–357, 2023.
https://doi.org/10.1007/978-3-031-15509-3_46

mean) and the empirical comparison with the central tendency measures mentioned before. Sections 2 and 3 recall the preliminaries related to the space of (compact) intervals and interval-valued central tendency measures, respectively. Section 4 presents the d_θ-depth-based interval trimmed mean and the comparative simulation study carried out to analyze its empirical performance. Finally, Sect. 5 contains some concluding remarks.

2 The Space $\mathscr{K}_c(\mathbb{R})$

Let $\mathscr{K}_c(\mathbb{R})$ denote the class of nonempty compact intervals of $\mathbb{R}$. Any interval $K \in \mathscr{K}_c(\mathbb{R})$ can be characterized in terms of either its extremes, $K = [\inf K, \sup K]$, or its mid-point (centre) $\text{mid}\, K = (\inf K + \sup K)/2$ and spread (radius) $\text{spr}\, K = (\sup K - \inf K)/2$, $K = [\text{mid}\, K - \text{spr}\, K, \text{mid}\, K + \text{spr}\, K]$.

For the statistical analysis of interval-valued data, the usual interval arithmetic will be considered. Given two intervals $K, K' \in \mathscr{K}_c(\mathbb{R})$, their *Minkowski sum* is defined as $K + K' = [\inf K + \inf K', \sup K + \sup K']$ or, in terms of the second characterization, $\text{mid}\,(K + K') = \text{mid}\, K + \text{mid}\, K'$ and $\text{spr}\,(K + K') = \text{spr}\, K + \text{spr}\, K'$. Analogously, the *product of an interval* $K \in \mathscr{K}_c(\mathbb{R})$ *by a scalar* $\gamma \in \mathbb{R}$ can be introduced in terms of any of the two characterizations:

$$\gamma \cdot K = \begin{cases} [\gamma \inf K, \gamma \sup K] & \text{if } \gamma \geq 0 \\ [\gamma \sup K, \gamma \inf K] & \text{if } \gamma < 0 \end{cases} = [\gamma \text{mid}\, K - |\gamma| \text{spr}\, K, \gamma \text{mid}\, K + |\gamma| \text{spr}\, K].$$

Metrics play a relevant role in the development of statistical techniques for this kind of data due to the lack of linearity of the space $(\mathscr{K}_c(\mathbb{R}), +, \cdot)$. The location measures in Sect. 3 will be based on the following metrics:

- Given $\theta \in (0, \infty)$, the *generalized Hausdorff metric* (Sinova et al. 2010) between two intervals $K, K' \in \mathscr{K}_c(\mathbb{R})$ is defined as

$$d_{H,\theta}(K, K') = |\text{mid}\, K - \text{mid}\, K'| + \theta |\text{spr}\, K - \text{spr}\, K'|.$$

- The *1-norm metric* (Vitale 1985) between any two intervals $K, K' \in \mathscr{K}_c(\mathbb{R})$ is given by

$$\rho_1(K, K') = \frac{1}{2} |\inf K - \inf K'| + \frac{1}{2} |\sup K - \sup K'|.$$

- Given $\theta \in (0, \infty)$, the d_θ *metric* (Gil et al. 2002) between any two intervals $K, K' \in \mathscr{K}_c(\mathbb{R})$ is defined as

$$d_\theta(K, K') = \sqrt{(\text{mid}\, K - \text{mid}\, K')^2 + \theta(\text{spr}\, K - \text{spr}\, K')^2}.$$

The value of the parameter is usually $\theta \in (0, 1]$ not to weigh the deviation in location less than the deviation in shape/imprecision. All the metrics $d_{H,\theta}$, ρ_1 and d_θ are strongly equivalent.

Interval-valued data are assumed to come from a (compact) random interval, that is, a Borel measurable mapping $X : \Omega \to \mathscr{K}_c(\mathbb{R})$, where $(\Omega, \mathscr{A}, P)$ is a probability space and the Borel σ-field on $\mathscr{K}_c(\mathbb{R})$ is generated by the topology induced by any of the previous metrics. Therefore, this notion models the random mechanism that repeatedly produces interval-valued observations. It holds that X is a random interval if, and only if, $\inf X$ and $\sup X$ are real-valued random variables such that $\inf X \leq \sup X$ (or, alternatively, $\operatorname{mid} X$ and $\operatorname{spr} X \geq 0$ are real-valued random variables).

3 Central Tendency Measures for Random Intervals

One of the best known central tendency measures for random intervals is the Aumann mean (Aumann 1965), which generalizes the mean of a real-valued random variable as follows.

Definition 1. The *Aumann mean* of a random interval X is the interval $E[X] = [E(\inf X), E(\sup X)]$ (whenever these expectations exist) or, equivalently, $E[X] = [E(\operatorname{mid} X) - E(\operatorname{spr} X), E(\operatorname{mid} X) + E(\operatorname{spr} X)]$.

The Aumann mean is the Fréchet expectation with respect to the d_θ metric, that is, $E[X]$ is the unique solution of $\min\limits_{K \in \mathscr{K}_c(\mathbb{R})} E[(d_\theta(X, K))^2]$.

It also fulfills very convenient statistical and probabilistic properties, but, unfortunately, it is highly sensitive to outliers and data changes. The following concepts from the literature (see Sinova et al. 2010; Sinova and Van Aelst 2013; Sinova 2016; Sinova and Van Aelst 2018, for more details) provide us with more robust central tendency measures for random intervals.

Definition 2. The *Hausdorff-type median* of a random interval X is the interval(s) $\operatorname{Med}[X] \in \mathscr{K}_c(\mathbb{R})$ such that minimizes $E[d_{H,\theta}(X, K)]$ over $K \in \mathscr{K}_c(\mathbb{R})$ (whenever this expectation exists).

In particular, any interval $[\operatorname{Me}(\operatorname{mid} X) - \operatorname{Me}(\operatorname{spr} X), \operatorname{Me}(\operatorname{mid} X) + \operatorname{Me}(\operatorname{spr} X)]$ is a Hausdorff-type median (and it does not depend on the value θ). In case the medians of real-valued random variables $\operatorname{Me}(\operatorname{mid} X)$ and/or $\operatorname{Me}(\operatorname{spr} X)$ are not unique, a convention such as choosing the mid-point of the interval of possible medians makes the Hausdorff-type median become unique.

Definition 3. The *1-norm median* of a random interval X is the interval(s) $\operatorname{Me}[X] \in \mathscr{K}_c(\mathbb{R})$ such that minimizes $E[\rho_1(X, K)]$ over $K \in \mathscr{K}_c(\mathbb{R})$ (whenever this expectation exists).

In particular, the interval $[\operatorname{Me}(\inf X), \operatorname{Me}(\sup X)]$ is a 1-norm median. In this case, the convention of choosing the mid-point of the corresponding interval of medians is adopted to guarantee that the 1-norm median is not empty even when $\operatorname{Me}(\inf X)$ or $\operatorname{Me}(\inf X)$ may not be unique.

Definition 4. The *d_θ-median* of a random interval X is the interval(s) $\mathrm{M}_\theta[X] \in \mathscr{K}_c(\mathbb{R})$ such that minimizes $E[d_\theta(X, K)]$ over $K \in \mathscr{K}_c(\mathbb{R})$ (whenever this expectation exists).

In contrast to what happens with the Hausdorff-type median, the d_θ-median depends on the value θ.

Finally, M-estimators could be understood as "intermediaries" between the Aumann mean and interval-valued medians because they weigh distances by means of a loss function that is generally less rapidly increasing than the square function.

Definition 5. Given a continuous and non-decreasing loss function $\rho : \mathbb{R}^+ \to \mathbb{R}$ that vanishes at 0, the *M-location measure* of a random interval X is the interval(s) $\mathrm{K}_\rho^M[X] \in \mathscr{K}_c(\mathbb{R})$ which minimizes $E[\rho(d_\theta(X, K))]$ over $K \in \mathscr{K}_c(\mathbb{R})$ (whenever this expectation exists).

A simple random sample $(X_1, \ldots, X_n)$ from a random interval X consists of n independent random intervals X_i, $i = 1, ..., n$, that are identically distributed as X. In this context,

- the sample Hausdorff-type median is given by
$$\widehat{\mathrm{Med}[X]}_n = [\mathrm{Me}\widehat{(\mathrm{mid}\, X)}_n - \mathrm{Me}\widehat{(\mathrm{spr}\, X)}_n, \mathrm{Me}\widehat{(\mathrm{mid}\, X)}_n + \mathrm{Me}\widehat{(\mathrm{spr}\, X)}_n];$$
- the sample 1-norm median is $\widehat{\mathrm{Me}[X]}_n = [\mathrm{Me}\widehat{(\inf X)}_n, \mathrm{Me}\widehat{(\sup X)}_n]$;
- the sample d_θ-median is the random interval $\widehat{\mathrm{M}_\theta[X]}_n$ that takes, for each realization $\mathbf{x}_n = (x_1, \ldots x_n)$, the value $\widehat{\mathrm{M}_\theta[\mathbf{x}_n]}$ that minimizes $\dfrac{1}{n}\displaystyle\sum_{i=1}^{n} d_\theta(x_i, K)$ over $K \in \mathscr{K}_c(\mathbb{R})$;
- the M-estimator of location is the random interval $\widehat{\mathrm{K}_\rho^M[X]}_n$ that takes, for each realization $\mathbf{x}_n$, the value $\widehat{\mathrm{K}_\rho^M[\mathbf{x}_n]}$ that minimizes $\dfrac{1}{n}\displaystyle\sum_{i=1}^{n} \rho(d_\theta(x_i, K))$ over $K \in \mathscr{K}_c(\mathbb{R})$.

The sample d_θ-median always exists and is unique whenever the points $\{(\mathrm{mid}\, X_i,\ \mathrm{spr}\, X_i)\}_{i=1}^n$ are not all collinear. On the other hand, the Representer Theorem (see Sinova 2016) provides necessary and sufficient conditions to express M-estimators of location as a weighted mean of the interval-valued observations, which hold for some well-known loss functions, such as the function by Hampel (1974) under some additional conditions. Unfortunately, neither of these two measures admits an explicit expression. For details about their practical computation, see Sinova (2016); Sinova and Van Aelst (2018).

4 The d_θ-Depth-Based Interval Trimmed Mean

Sinova (2021) has followed a different approach for summarizing the central tendency by studying trimmed means based on a depth function in the fuzzy number-valued setting (one-dimensional fuzzy sets). The main novelty of the work by Sinova (2021) consists in introducing a depth function for fuzzy-number valued data for the first time and applying it to adapt depth-based trimmed means to the fuzzy framework. Indeed, such depth-based trimmed means have shown very promising empirical results and, for this reason, it would be interesting to analyze whether their advantages remain when data are interval-valued.

The particularization of the depth-based trimmed means by Sinova (2021) to the interval-valued setting would be as follows.

Definition 6. Given a random interval X and a fixed value $\theta \in (0, \infty)$, the *d_θ-depth* of $K \in \mathscr{K}_c(\mathbb{R})$ with respect to the distribution of X is given by

$$DD_\theta(K; X) = \frac{1}{1 + E[d_\theta(X, K)]}.$$

Given a simple random sample from X, $(X_1, \ldots, X_n)$, the *empirical d_θ-depth* of K is given by

$$DD_{\theta,n}(K; (X_1, \ldots, X_n)) = \frac{1}{1 + \frac{1}{n}\sum_{i=1}^{n} d_\theta(X_i, K)}.$$

Notice that the centremost element with respect to the d_θ-depth is the d_θ-median.

Definition 7. Given a simple random sample $(X_1, \ldots, X_n)$ from a random interval X and a trimming proportion $\beta \in (0, 1)$, the *d_θ-depth-based interval trimmed mean* estimator is defined as

$$DD_\theta\text{-}\overline{X}_{n,\beta} = \frac{\sum_{i=1}^{n} I_{[\gamma,\infty)}(DD_{\theta,n}(X_i)) \cdot X_i}{\sum_{i=1}^{n} I_{[\gamma,\infty)}(DD_{\theta,n}(X_i))},$$

with $\frac{1}{n}\sum_{i=1}^{n} I_{[\gamma,\infty)}(DD_{\theta,n}(X_i)) \simeq 1 - \beta$.

As mentioned in Sinova (2021), the d_θ-depth is not affine invariant, but it is enough to replace the distances $d_\theta(X, K)$ by the standardized distances $d_\theta(X, K)/\sigma_X$ (where $\sigma_X = \sqrt{E((d_\theta(X, E[X]))^2)}$ denotes the standard deviation of the random interval X) to get an affine invariant version.

4.1 Comparative Simulation Study

The empirical behaviour of the d_θ-depth-based trimmed mean has been compared to that of the central tendency measures recalled in Sect. 3. The common choice $\theta = 1/3$ has been considered for the computation of the location measures (whenever the d_θ distance is involved), and the estimates of their bias, variance and mean square error (MSE): Bias $= d_{1/3}(1/N\sum_{i=1}^{N}\widehat{T}_i, T)$, Var $= 1/N\sum_{i=1}^{N}(d_{1/3}^2(\widehat{T}_i, \sum_{i=1}^{N}\widehat{T}_i/N))$ and MSE $= 1/N\sum_{i=1}^{N}(d_{1/3}^2(\widehat{T}_i, T))$, with T the population value of a location measure, N the number of samples and $\widehat{T}_i$ the estimate of T for the ith generated sample.

Regarding the M-estimator of location, the Hampel loss function with tuning parameters the median, the 75th and the 85th percentiles of the distribution of sample distances (see Sinova 2016, for details) has been considered due to its flexibility and good empirical performance in previous studies. Simulations have been designed as follows, inspired by the simulation strategy proposed in De la Rosa de Sáa et al. (2015).

1. A random sample of 100 interval-valued observations is generated from a random interval $X = [X_1 - X_2, X_1 + X_2]$, where
$$X_1 \rightsquigarrow \beta(6,1),\ X_2 \rightsquigarrow \begin{cases} \exp\,(100 + 4X_1) & \text{if } X_1 < .25, \\ \exp\,(200) & \text{if } .25 \le X_1 \le .75, \\ \exp\,(500 - 4X_1) & \text{if } X_1 > .75. \end{cases}$$
2. A proportion $c_p = 0, .1, .2, .3$ or $.4$ of the observations is then contaminated in both location and spread with
$$X_1 \rightsquigarrow \beta(1,6),\ X_2 \rightsquigarrow \begin{cases} \exp\,(100 + 4X_1)\,/(C_D^2 + 1) & \text{if } X_1 < .25, \\ \exp\,(200)\,/(C_D^2 + 1) & \text{if } .25 \le X_1 \le .75, \\ \exp\,(500 - 4X_1)\,/(C_D^2 + 1) & \text{if } X_1 > .75, \end{cases}$$ where
$C_D = 0, 1, 5, 10$ or 100 measures the relative distance between the distributions of the regular and contaminated observations.
3. The population parameters T are approximated by Monte Carlo simulation using $N = 10000$ replications of *Step 1*.
4. For each (c_p, C_D), *Steps 1–2* are repeated $N = 10000$ times and, for each of these contaminated samples, the location estimates $\widehat{T}_i$ are calculated.

Table 1 shows the outputs and the smallest value of bias, variance and MSE for each choice of c_p and C_D has been highlighted in bold. Bias, variance and MSE have also been computed in terms of the Hausdorff and 1-norm distances, but conclusions do not differ.

5 Concluding Remarks

Table 1 shows that there is no uniformly most appropriate location estimate. However, the performance of the d_θ-depth-based trimmed mean (with an appropriate choice of the trimming parameter) is among the best: it can outperform the behaviour of the other measures or, at least, become the second best option in terms of bias and MSE.

Table 1. Bias, variance and MSE (all multiplied by a factor 1000) of the Aumann mean, the 1-norm, Hausdorff-type and $d_{1/3}$- medians, the Hampel M-estimator of location and the $d_{1/3}$-depth-based trimmed means (DTM) with $\beta = .2$ and $\beta = .45$

c_p	C_D		$E[X]$	Me$[X]$	Med$[X]$	M$_{1/3}[X]$	K$^M_{Hampel}$	DTM$_{\beta=.2}$	DTM$_{\beta=.45}$
.0	0	Bias	.0	**.0**	.1	.1	.0	.1	.2
		Var	.2	.2	.2	.2	**.1**	.2	.2
		MSE	.2	.2	.2	.2	**.1**	.2	.2
.1	0	Bias	71.4	17.3	17.3	17.3	**2.5**	14.3	14.2
		Var	**.2**	.3	.3	.3	.2	.2	.3
		MSE	5.3	.6	.6	.6	**.2**	.4	.5
.1	1	Bias	71.3	17.3	17.3	17.3	**2.5**	14.2	14.2
		Var	**.2**	.3	.3	.3	.2	.2	.3
		MSE	5.2	.6	.6	.6	**.2**	.4	.5
.1	5	Bias	71.3	17.3	17.3	17.3	**2.7**	14.3	14.2
		Var	**.2**	.3	.3	.3	.2	.2	.3
		MSE	5.2	.6	.6	.6	**.2**	.4	.5
.1	10	Bias	71.6	17.5	17.6	17.6	**2.9**	14.6	14.3
		Var	**.2**	.3	.3	.3	.2	.2	.3
		MSE	5.3	.6	.6	.6	**.2**	.4	.5
.1	100	Bias	71.6	17.4	17.4	17.4	**2.8**	14.4	14.2
		Var	**.2**	.3	.3	.3	.2	.2	.3
		MSE	5.3	.6	.6	.6	**.2**	.4	.5
.2	0	Bias	142.9	41.7	41.7	41.6	**20.6**	44.3	32.1
		Var	**.2**	.4	.4	.4	.2	.2	.4
		MSE	20.6	2.1	2.2	2.1	**.7**	2.2	1.4
.2	1	Bias	142.6	41.4	41.4	41.4	**21.0**	44.1	31.5
		Var	**.2**	.4	.4	.4	.2	.2	.4
		MSE	20.5	2.1	2.1	2.1	**.7**	2.1	1.4
.2	5	Bias	142.8	41.5	41.5	41.5	**21.7**	44.2	31.4
		Var	**.2**	.4	.4	.4	.2	.2	.4
		MSE	20.5	2.1	2.1	2.1	**.7**	2.1	1.4
.2	10	Bias	142.8	41.8	41.9	41.9	**21.9**	44.4	32.0
		Var	**.2**	.4	.4	.4	.2	.2	.4
		MSE	20.5	2.2	2.2	2.2	**.7**	2.2	1.4
.2	100	Bias	142.9	41.8	41.9	41.9	**21.8**	44.4	32.0
		Var	**.2**	.4	.4	.4	.2	.2	.4
		MSE	20.6	2.2	2.2	2.2	**.7**	2.2	1.4
.3	0	Bias	214.3	79.2	79.2	78.9	127.3	116.9	**52.0**
		Var	**.2**	.6	.6	.6	1.5	.2	.5
		MSE	46.1	6.9	6.9	6.9	17.7	13.9	**3.2**
.3	1	Bias	214.0	78.6	78.7	78.5	133.7	116.6	**51.6**
		Var	**.1**	.6	.6	.6	1.4	.2	.5
		MSE	45.9	6.8	6.8	6.8	19.2	13.8	**3.1**
.3	5	Bias	214.3	79.5	79.5	79.3	138.3	117.0	**52.2**
		Var	**.2**	.6	.6	.6	1.3	.2	.5
		MSE	46.1	7.0	7.0	6.9	20.4	13.9	**3.2**
.3	10	Bias	214.4	79.5	79.6	79.3	138.4	117.0	**51.9**
		Var	**.2**	.6	.6	.6	1.3	.2	.5
		MSE	46.1	7.0	7.0	6.9	20.5	13.9	**3.2**
.3	100	Bias	214.3	79.5	79.5	79.3	138.1	117.0	**52.2**
		Var	**.2**	.6	.6	.6	1.3	.2	.5
		MSE	46.1	7.0	7.0	6.9	20.4	13.9	**3.2**
.4	0	Bias	285.6	147.9	148.0	147.1	241.6	200.5	**96.3**
		Var	**.2**	1.2	1.2	1.2	2.0	.2	1.5
		MSE	81.7	23.1	23.1	22.9	60.4	40.4	**10.8**
.4	1	Bias	285.5	148.0	148.1	147.3	246.5	200.5	**96.1**
		Var	**.2**	1.2	1.2	1.2	1.6	.2	1.5
		MSE	81.7	23.1	23.1	22.9	62.3	40.4	**10.7**
.4	5	Bias	285.6	148.4	148.5	147.5	250.3	200.6	**96.8**
		Var	**.2**	1.2	1.2	1.2	1.2	.2	1.5
		MSE	81.7	23.2	23.3	23.0	63.9	40.5	**10.9**
.4	10	Bias	285.8	148.5	148.6	147.8	249.7	200.8	**96.9**
		Var	**.2**	1.2	1.2	1.2	1.3	.2	1.5
		MSE	81.8	23.3	23.3	23.1	63.6	40.5	**10.9**
.4	100	Bias	285.8	148.4	148.5	147.6	250.3	200.8	**96.6**
		Var	**.2**	1.2	1.2	1.2	1.2	.2	1.4
		MSE	81.8	23.2	23.2	23.0	63.9	40.5	**10.8**

Acknowledgement. This research has been partially supported by the Spanish Ministry of Science, Innovation and Universities (Grant MTM-PID2019-104486GB-I00) and by Principality of Asturias/FEDER Grants (SV-PA-21-AYUD/2021/50897).

References

Aumann, R.J.: Integrals of set-valued functions. J. Math. Anal. Appl. **12**, 1–12 (1965)

de la Rosa de Sáa, S., Gil, M.A., González-Rodríguez, G., López, M.T., Lubiano, M.A.: Fuzzy rating scale-based questionnaires and their statistical analysis. IEEE Trans. Fuzzy Syst. **23**(1), 111–126 (2015)

Gil, M.A., Lubiano, M.A., Montenegro, M., López-García, M.T.: Least squares fitting of an affine function and strength of association for interval data. Metrika **56**, 97–111 (2002)

Hampel, F.R.: The influence curve and its role in robust estimation. J. Am. Stat. Assoc. **69**(346), 383–393 (1974)

Sinova, B.: M-estimators of location for interval-valued random elements. Chemom. Intell. Lab. Syst. **156**, 115–127 (2016)

Sinova, B.: On depth-based fuzzy trimmed means and a notion of depth specifically defined for fuzzy numbers. Fuzzy Sets Syst. **443**, 87–105 (2022)

Sinova, B., Van Aelst, S.: Comparing the medians of a random interval defined by means of two different L^1 metrics. In: Borgelt, C., Gil, M.A., Sousa, J., Verleysen, M. (eds.) Towards Advanced Data Analysis by Combining Soft Computing and Statistics. Studies in Fuzziness and Soft Computing, vol. 285, pp. 75–86. Springer, Heidelberg (2013). https://doi.org/10.1007/978-3-642-30278-7_7

Sinova, B., Van Aelst, S.: A spatial-type interval-valued median for random intervals. Statistics **52**(3), 479–502 (2018)

Sinova, B., Casals, M.R., Colubi, A., Gil, M.A.: The median of a random interval. In: Borgelt, C., et al. (eds.) Combining Soft Computing and Statistical Methods in Data Analysis. Advances in Intelligent and Soft Computing, vol. 77, pp 575–583. Springer, Heidelberg (2010). https://doi.org/10.1007/978-3-642-14746-3_71

Vitale, R.A.: L_p metrics for compact, convex sets. J. Approx. Theory **45**, 280–287 (1985)

Characterization of Extreme Points of p-Boxes via Their Normal Cones

Damjan Škulj(✉)

Faculty of Social Sciences, University of Ljubljana, Kardeljeva pl. 5, 1000 Ljubljana, Slovenia
damjan.skulj@fdv.uni-lj.si

Abstract. Probability boxes, also known as p-boxes, correspond to sets of probability distributions bounded by a pair of distribution functions, inducing sets of expectation functionals, commonly referred as credal sets. These sets are convex, and in the case of a finite domain, they form convex polyhedra. Characterizing extreme points of credal sets is of significant interest, especially in relation with the corresponding numerical analysis. In this article we provide a complete characterization of extreme points corresponding to p-boxes on finite domains via their normal cones. In our context, a normal cone is a set of random variables whose expectations with respect to a credal set are minimized in a common extreme point. We show that the basic building blocks of normal cones can be completely characterized in terms of subsets of the domain, independently of the particular realization of a p-box. Moreover, we then demonstrate how the building blocks can then be related to extreme points corresponding to a particular realization.

1 Introduction

Probability boxes or p-boxes (Ferson et al. 2003) for short, belong to a wider family of imprecise probabilistic models. Thus they are capable of modelling situations that no single probabilistic model can adequately describe (see e.g. Augustin et al. (2014) and references therein for further sources on general models of imprecise probabilities). A p-box is given as a pair of distribution functions $(\underline{F}, \overline{F})$ giving rise to a set of distribution functions lying between the bounds:

$$\mathscr{D} = \{F | F \text{ is a distribution function}, \underline{F} \leqslant F \leqslant \overline{F}\}, \tag{1}$$

usually interpreted as the set of models compatible with the available information. We restrict to p-boxes on finite domains $\mathscr{X} \subset \mathbb{R}$. Thus let $\mathscr{X} = \{x_1, \dots, x_n\} \subset \mathbb{R}$, where the ordering $x_1 < x_2 < \cdots < x_n$ is assumed. Denote also $A_i = \{x_1, \dots, x_i\}$ for every $i = 1, \dots, n$. The lower and upper bounds are then two non-decreasing functions $\underline{F}, \overline{F} \colon \mathscr{X} \to [0, 1]$, such that $\underline{F} \leq \overline{F}$ and $\underline{F}(x_n) = \overline{F}(x_n) = 1$. Note that every distribution function F on $\mathscr{X}$ corresponds to a probability mass function p_F, such that $p_F(x_i) = F(x_i) - F(x_{i-1})$, for every $i = 1, \dots, n$, where we set $x_0 = -\infty$ and $F(x_0) = 0$. Consequently, we let P_F

L. A. García-Escudero et al. (Eds.): SMPS 2022, AISC 1433, pp. 358–365, 2023.
https://doi.org/10.1007/978-3-031-15509-3_47

be the corresponding *linear prevision* or *expectation* with respect to p_F such that $P_F(h) = \sum_{i=1}^{n} h(x_i)p_F(x_i)$, for every real valued function h on $\mathscr{X}$ (the real valued functions on $\mathscr{X}$ can also be viewed as $|\mathscr{X}|$-dimensional real vectors). The induced set of all linear previsions corresponding to $\mathscr{D}$ is a convex set too, and is denoted as *credal set* $\mathscr{M}(\underline{F}, \overline{F})$.

The primary goal of this paper is the analysis of the credal sets from the viewpoint of convex analysis. A credal set $\mathscr{M}(\underline{F}, \overline{F})$ is described through a set of constraints induced by (1) and the standard monotonicity requirements for distribution functions. We give the constraints in the unified form $P_F(\mathbb{I}_A) \geq L(A)$ for sets A in the corresponding domain.

$$\mathscr{E} = \{A_i | i = 1, \ldots, n\} \cup \{A_i^c | i = 1, \ldots, n\} \cup \{\{x_i\} | i = 1, \ldots, n\},$$

where L is a suitable mapping (also called a *lower probability*) and $\mathbb{I}_A$ is the indicator function such that $\mathbb{I}_A(x) = 1$ exactly if $x \in A$ and 0 elsewhere:

(P1) $P_F(\mathbb{I}_{A_i}) \geq \underline{F}(x_i)$ for every $i = 1, \ldots, n$, following from $F(x_i) \geq \underline{F}(x_i)$;
(P2) $P_F(\mathbb{I}_{A_i^c}) \geq 1 - \overline{F}(x_i)$ for every $i = 1, \ldots, n$, following from $F(x_i) \leq \overline{F}(x_i)$;
(P3) $P_F(\mathbb{I}_{\{x_i\}}) \geq 0$ for every $i = 1, \ldots, n$, following from the monotonicity of distribution functions, or equivalently, the fact that P_F must be a positive functional.

(See also Troffaes and Destercke 2011). In addition, we assign to every p-box $(\underline{F}, \overline{F})$ the lower expectation functional by the means of the *natural extension*:

$$\underline{P}(h) = \min_{F \in \mathscr{D}} P_F(h)$$

for every real valued function on $\mathscr{X}$.

The importance of identification of extreme points (see also Montes and Destercke (2017) and Montes and Miranda (2018)) follows from the optimization problems of finding the minimal or the maximal expectation of a real-valued function h on $\mathscr{X}$ with respect to the credal set, which are known to be always attained in extreme points. A related problem of calculating extremal expectations with respect to p-boxes was addressed by Utkin and Destercke (2009).

Our approach to characterization of extreme points of p-boxes can be considered as an alternative to the existing one of Montes and Destercke (2017), who provide a characterization by means of belief functions. We build on the idea from Škulj (2022b), where extreme points are described through the corresponding normal cones. We skip the general theoretical definitions that can be found in the mentioned references and move directly to the description of normal cones for the specific case of p-boxes. A normal cone in an extreme point $P_F \in \mathscr{M}(\underline{F}, \overline{F})$ is the cone of all real valued functions on $\mathscr{X}$ minimized by P_F:

$$N\left(\mathscr{M}(\underline{F}, \overline{F}), F\right) = \{h \colon \mathscr{X} \to \mathbb{R} | P_F(h) = \underline{P}(h)\}.$$

It can be shown that normal cones of extreme points in $\mathscr{M}(\underline{F}, \overline{F})$ are cones generated by the indicator functions of sets forming the constraints given by

(P1)–(P3). Let $\mathscr{A}$ be a collection of sets containing $\mathscr{X}$. Then we write

$$\begin{aligned}\operatorname{cone}(\mathscr{A}) &:= \operatorname{posi}\{\mathbb{I}_A | A \in \mathscr{A}\backslash\{\mathscr{X}\}\} + \operatorname{lin}\{\mathbb{I}_{\mathscr{X}}\} \\ &= \left\{\sum_{A\in\mathscr{A}} \alpha_A \mathbb{I}_A + \alpha_0 \mathbb{I}_{\mathscr{X}} | \alpha_A \geq 0\ \forall A \in \mathscr{A}\backslash\{\mathscr{X}\}, \alpha_0 \in \mathbb{R}\right\}.\end{aligned}$$

The reason $\mathbb{I}_{\mathscr{X}}$ appears with any coefficient is that it denotes the equality constraint $P_F(\mathbb{I}_{\mathscr{X}}) = 1$.

In general we have that

$$N\left(\mathscr{M}(\underline{F}, \overline{F}), F\right) = \operatorname{cone}(\mathscr{A}),$$

where $\mathscr{A} = \{A | A \in \mathscr{E}, P_F(A) = \underline{P}(\mathbb{I}_A)\}$. Every normal cone can be triangulated into a collection of cones of the form cone($\mathscr{A}$), such that the indicator functions of $\mathscr{A}$ form a basis of the vector space $\mathbb{R}^{|\mathscr{X}|}$. Such cones are called *maximal elementary simplicial cones (MESC)*, and are characterized as follows. A cone($\mathscr{A}$) is a MESC if it satisfies the following conditions:

(M1) The vectors $\{\mathbb{I}_A | A \in \mathscr{A}\}$ are linearly independent.
(M2) If $B \in \mathscr{E}\backslash\mathscr{A}$, then $\mathbb{I}_B \notin \operatorname{cone}(\mathscr{A})$.

MESCs are of our primary interest because they are exactly the smallest possible normal cones corresponding to extreme points of $\mathscr{M}(\underline{F}, \overline{F})$. Their characterization will therefore allow us to induce the characterization of the corresponding extreme points.

In Škulj (2022a) detailed derivations and proofs of the results presented here can be found.

The paper is structured as follows. In the consequent section we give the general structure of the maximal elementary simplicial cones corresponding to extreme points of p-boxes. Whether a MESC corresponds to an extreme point and to which one is analyzed in Sect. 3, where some examples are also given.

2 General Structure of Maximal Elementary Simplicial Cones of p-Boxes

As explained in the previous section, every MESC corresponding to a p-box is of the form cone($\mathscr{B}$), where $\mathscr{B}$ consists of sets in $\mathscr{E}$. We first list some necessary conditions following from (M1) and (M2) for $\mathscr{B}$ to form a maximal elementary simplicial cone. Let $\mathscr{B}$ be a collection of sets such that cone($\mathscr{B}$) is a MESC. Then:

(C1) $\mathscr{X} = A_n \in \mathscr{B}$;
(C2) $A_i \in \mathscr{B} \implies A_i^c \notin \mathscr{B}$;
(C3) $A_i \in \mathscr{B} \implies \{x_{i+1}\} \notin \mathscr{B}$;
(C4) $A_i^c \in \mathscr{B} \implies \{x_i\} \notin \mathscr{B}$.

Let $\mathscr{A}$ be a collection of sets A_i such that either $A_i \in \mathscr{B}$ or $A_i^c \in \mathscr{B}$. By (C2), at most one of the inclusions is possible. We denote $\mathscr{A} = \{A_{i_1}, \ldots, A_{i_k}\}$, where $\{i_j\}$ is an increasing sequence in j and

$$B_j = \begin{cases} A_{i_j}, & \text{if } A_{i_j} \in \mathscr{B}, \\ A_{i_j}^c, & \text{if } A_{i_j}^c \in \mathscr{B} \end{cases}$$

for every $j = 1, \ldots, k$. In addition to sets B_j, singleton sets $\{x_i\}$ need to complete $\mathscr{A}$ to a basis. Before proceeding to details on the possible configurations, we describe the cones, assuming $\mathscr{B}$ given. That is, we have a collection of sets $B_1, \ldots, B_k$, and a collection of singletons $\{x_{j_1}\}, \ldots, \{x_{j_r}\}$. Let us also define a sign function for the sets B_j:

$$s(j) = \begin{cases} 1, & \text{if } B_j = A_{i_j}, \\ -1, & \text{if } B_j = A_{i_j}^c. \end{cases}$$

The cone can then be decomposed as a Minkowski sum

$$\text{cone}(\mathscr{B}) = \text{cone}(\{B_1, \ldots, B_k\}) + \text{cone}(\{x_{j_1}\}, \ldots, \{x_{j_r}\}).$$

While the second part is simply the set of all non-negative functions with support $\{x_{j_1}, \ldots, x_{j_r}\}$, a more detailed description of the first part follows.

Recall that the cone is insensitive to adding a multiple of a constant, which is $\mathbb{I}_{\mathscr{X}}$, and therefore the functions $\mathbb{I}_{A^c}$ can be replaced by $-\mathbb{I}_A = \mathbb{I}_{A^c} - \mathbb{I}_{\mathscr{X}}$. Any element $h \in \text{cone}(\{B_1, \ldots, B_k\})$ is then of the form

$$h = \sum_{j=1}^{k} s(j)\alpha_j \mathbb{I}_{A_j},$$

where α_j are non-negative constants. Moreover, $h \in \text{ri}(\text{cone}(\{B_1, \ldots, B_k\}))$ if all $\alpha_j > 0$ ($\text{ri}(C)$ denotes the relative interior of C, which contains all elements of C not belonging to its proper faces). The following lemma characterizes the elements of this cone, or more precisely, its relative interior.

Lemma 1. *Let a function $h\colon \mathscr{X} \to \mathbb{R}$ be given. The following propositions are equivalent:*

(i) $h \in \text{ri}(\text{cone}(\{B_1, \ldots, B_k\}))$;
(ii) h *is $\mathscr{A}$-measurable and* $\text{sign}(h(x') - h(x)) = s(j)$ *whenever* $x' \in A_{j+1}$ *and* $x \in A_j$.

(Function sign *assigns value* 1 *to positive,* -1 *to negative numbers and 0 to 0 – this case is impossible here if we restrict to the relative interior).*

It remains to give detailed relation between the two cone components. We proceed as follows. With respect to the ordering for the sets in $\mathscr{A}$ according to set inclusion, we take two adjacent sets, say $A \subset A'$. They correspond to some B_j and B_{j+1} in one of the four possible ways: $\{B_j, B_{j+1}\} = \{A, A'\}, \{B_j, B_{j+1}\} = \{A^c, A'\}$, $\{B_j, B_{j+1}\} = \{A, A'^c\}$ or $\{B_j, B_{j+1}\} = \{A^c, A'^c\}$. Additionally, we may add $A_0 = \emptyset$ to $\mathscr{A}$. The classification will rely on the following lemma.

Lemma 2. *Let $A \subset A'$ be a pair of sets as described above. Then $\mathscr{B}$ must contain exactly $|A' \backslash A| - 1$ singletons of the form $\{x\}$ for $x \in A' \backslash A$.*

Let us denote the elements of $A' \backslash A$ with $x_1 \leq x_2 \leq \ldots \leq x_m$. We consider each of the four cases separately:

Case 1 $\{B_j, B_{j+1}\} = \{A, A'\}$. By (C3), $\{x_1\}$ cannot be in $\mathscr{B}$. Thus exactly $m-1$ singletons $\{x_2\}, \ldots \{x_m\}$ must belong to $\mathscr{B}$.

Case 2 $\{B_j, B_{j+1}\} = \{A, A'^c\}$. By (C3) and (C4), neither $\{x_1\}$ nor $\{x_m\}$ can appear in $\mathscr{B}$. The remaining singletons cannot ensure the elements forming a basis. The conclusion is that such a pair can only appear in the case where $|A' \backslash A| = 1$.

Case 3 $\{B_j, B_{j+1}\} = \{A^c, A'\}$. In this case, all singletons $\{x_1\}, \ldots, \{x_m\}$ are allowed. Yet only $m-1$ are needed, which means that this case induces m different possible configurations.

Case 4 $\{B_j, B_{j+1}\} = \{A^c, A'^c\}$. A similar reasoning as above shows that exactly the singletons $\{x_1\}, \ldots, \{x_{m-1}\}$ must belong to $\mathscr{B}$ in this case.

The following example demonstrates why $|A' \backslash A| = 1$ is necessary in Case 2.

Example 1. Let $\mathscr{X} = \{1, \ldots, 5\}$ and denote $A_i = \{1, \ldots, i\}$. Consider a situation where $\mathscr{B}$ contains sets $A_1, A_3^c, A_4, A_5 = \mathscr{X}$. It corresponds to Case 2, where $B_j = A_1$ and $B_{j+1} = A_3^c$. Because of $|A_3 \backslash A_1| > 1$, according to our analysis, this collection cannot be completed to a MESC by adding only a singleton.

Let us demonstrate that indeed no singleton can be added to $\mathscr{B}$ to complete it to a MESC. As $\{1\} = A_1$ and $A_1 \cup \{2\} = A_2$ and $A_4 \cup \{5\} = \mathscr{X}$, the only candidates to be considered are $\{3\}$ and $\{4\}$. Yet, $A_3^c \cup \{3\} = A_2^c$, which is not allowed by (C3). It remains to analyze adding $\{4\}$. In this case, however, $\mathbb{I}_{A_4} + \mathbb{I}_{A_3^c} - \mathbb{I}_{\{4\}} = \mathbb{I}_{A_5}$, whence the set is linearly dependent.

3 Relating Extreme Points to Their Normal Cones

Classification of cases related to normal cones in Sect. 2 allows us to characterize all possible extreme points of p-boxes according to their global and local behaviour. Recall that every normal cone is characterized by a chain $\mathscr{A}$ and a sign function s that specifies whether a set $A \in \mathscr{A}$ or its complement A^c belongs to the set of positive generators of the cone. The missing generators are singletons, whose possible configurations are analyzed at the end of that section. The collection of all generator sets is denoted by $\mathscr{B}$.

Let F be an extreme point whose corresponding normal cone is generated by a collection $\mathscr{B}$, and P_F be the corresponding linear prevision. The lower expectation functional obtained via the natural extension is denoted by $\underline{P}$. Take any pair (B_j, B_{j+1}) as described in Sect. 2, and denote the elements of $A' \backslash A$ with $x_1 \leq x_2 \leq \ldots \leq x_m$ and set $x_0 = \max A$.

Case 1 $\{B_j, B_{j+1}\} = \{A, A'\}$. Given that A and A' are among the generators of the cone, $P_F(A) = \underline{P}(A) = \underline{F}(x_0)$ and $P_F(A') = \underline{P}(A') = \underline{F}(x_m)$ must hold. In addition, singletons $\{x_2\}, \ldots, \{x_m\}$ belong to $\mathscr{B}$. The constraints on singletons are all of the form $P(x_i) \geq 0$, hence, in the extreme point we have $0 = P_F(x_i) = F(x_i) - F(x_{i-1})$ for every $i = 2, \ldots, m$. Thus

$$F(x_i) = \begin{cases} \underline{F}(x_0), & i = 0; \\ \underline{F}(x_m), & 1 \leq i \leq m. \end{cases}$$

This case is possible exactly if $\overline{F}(x_1) \geq \underline{F}(x_m)$, which is needed to ensure $F(x_1) \leq \overline{F}(x_1)$.

Case 2 $\{B_j, B_{j+1}\} = \{A, A'^c\}$. This case is only possible if $|A' \backslash A| = 1$, and in that case no other restrictions are needed. Hence,

$$F(x_0) = P_F(A) = \underline{P}(A) = \underline{F}(x_0)$$

and $P_F(A'^c) = \underline{P}(A'^c) = 1 - \overline{P}(A')$, whence

$$F(x_1) = P_F(A') = \overline{P}(A') = \overline{F}(x_1).$$

Case 3 $\{B_j, B_{j+1}\} = \{A^c, A'\}$. In this case, m different extreme points are possible, as any collection of $m - 1$ singletons out of m can be used as generators. Let $\{x_k\}$ be the missing one of the singletons. First we have that $P_F(A^c) = \underline{P}(A^c) = 1 - \overline{P}(A) = 1 - \overline{F}(x_0) = 1 - P_F(A)$, and therefore $F(x_0) = \overline{F}(x_0)$. On the other hand, we have that $F(x_m) = \underline{F}(x_m)$. The constraints on the singletons imply that

$$F(x_i) = \begin{cases} \overline{F}(x_0), & 0 \leq i \leq k-1; \\ \underline{F}(x_m), & k \leq i \leq m. \end{cases}$$

This case is possible whenever $\overline{F}(x_0) \geq \underline{F}(x_{k-1})$ and $\underline{F}(x_m) \leq \overline{F}(x_k)$.

Case 4 $\{B_j, B_{j+1}\} = \{A^c, A'^c\}$. This remaining case is symmetric to the first one. Reasoning similar as above gives that $F(x_0) = \overline{F}(x_0)$ and $F(x_m) = \overline{F}(x_m)$. The fact that $P_F(x_i) = 0$ for $i = 1, \ldots m-1$ implies that $F(x_i) = F(x_{i-1})$ for those indices. Thus,we have

$$F(x_i) = \begin{cases} \overline{F}(x_0), & 0 \leq i \leq m-1; \\ \overline{F}(x_m), & i = m. \end{cases}$$

This case is possible whenever $\overline{F}(x_0) \geq \underline{F}(x_{m-1})$.

Example 2. We illustrate the above cases with the following examples. Let $\mathscr{X} = \{1, \ldots, 5\}$ and denote $A_i = \{1, \ldots i\}$ for $i \in \mathscr{X}$.

1. Let p-box $(\underline{F}, \overline{F})$ be given with $\underline{F}(i) = \frac{i}{5}$ and $\overline{F}(i) = 1$ for every $i \in \mathscr{X}$. Consider the cone generated by $\mathscr{B} = \{A_1, A_4, A_5, \{3\}, \{4\}\}$. This example corresponds to Case 1 with $B_j = A_1$ and $B_{j+1} = A_4$. The corresponding extreme distribution is depicted with blue line in Fig. 1(left). Its values are $F(1) = \underline{F}(1), F(2) = F(3) = F(4) = \underline{F}(4)$ and $F(5) = \underline{F}(5)$.

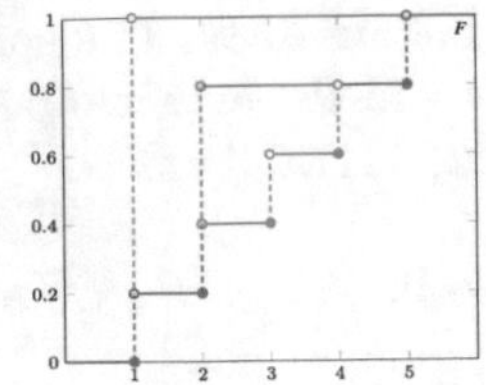

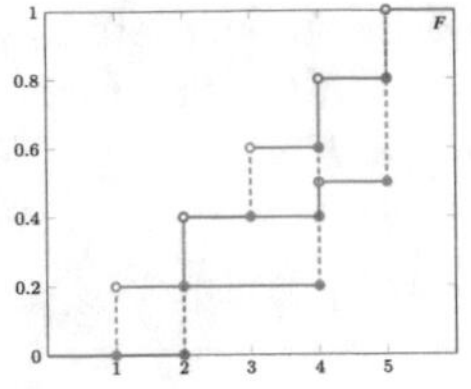

Fig. 1. Extreme points corresponding to Example 2–1 (left) and –2 (right)

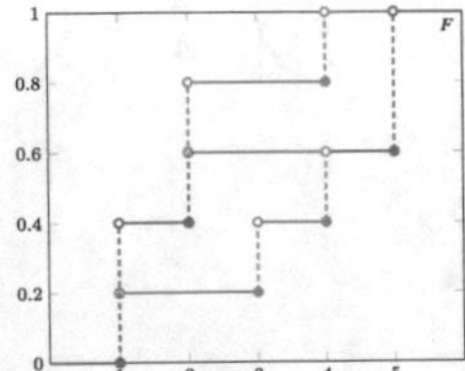

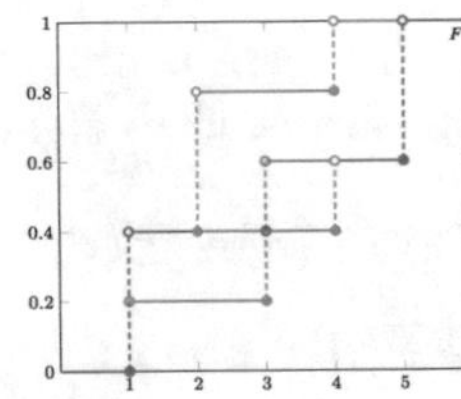

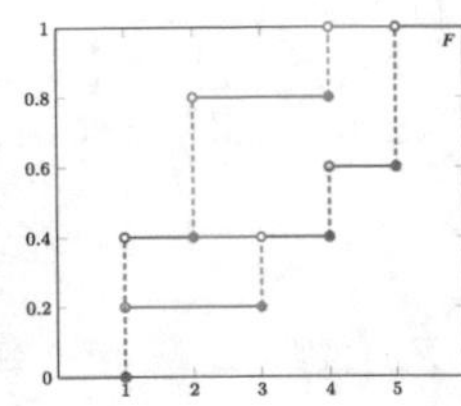

Fig. 2. Extreme points corresponding to Example 2–3, with the excluded singletons $\{2\}$ (left), $\{3\}$ (middle) and $\{4\}$ (right)

2. Let this time $\underline{F} = (0, 0.2, 0.2, 0.5, 1)$, given as a vector of values $\underline{F}(i)_{i=1,\dots,5}$, and similarly, $\overline{F} = (0.2, 0.4, 0.6, 0.8, 1)$. Let the cone generated by $\mathscr{B} = \{A_1, A_2^c, A_4^c, A_5, \{3\}\}$ be given. We thus have a combination of Cases 2 and 4. The corresponding extreme distribution is $F = (0, 0, 0.4, 0.4, 0.8, 1)$. The depiction of this case is given in Fig. 1 (right).
3. Finally, we give an example illustrating Case 3. Let $\underline{F} = (0.2, 0.2, 0.4, 0.6, 1)$ and $\overline{F} = (0.4, 0.8, 0.8, 1, 1)$, and let $\mathscr{B}$ contain A_1^c, A_4 and A_5. According to Case 3, it must additionally contain exactly two of the three singletons $\{2\}, \{3\}$ and $\{4\}$, yielding three different cones, whose corresponding extreme points are depicted in Fig. 2.
   The probability mass functions corresponding to the three cases are respectively $p_1 = (0.4, 0.2, 0, 0, 0.4), p_2 = (0.4, 0, 0.2, 0, 0.4)$ and $p_3 = (0.4, 0, 0, 0.2, 0.4)$. Representatives of each cone are for instance the sums of the indicator functions of the generator sets (except $A_5 = \mathscr{X}$), giving $h_1 = (1, 2, 3, 3, 1), h_2 = (1, 3, 2, 3, 1)$ and $h_3 = (1, 3, 3, 2, 1)$. Let P_i be the linear previsions corresponding to the above respective probability mass functions p_i. Then we have that $P_i(h_i) = 1.2$ and $P_j(h_i) = 1.4$ for all $i \neq j$, which clearly confirms the p_i are extreme distributions corresponding to the respective normal cones.

Acknowledgement. The author acknowledges the financial support from the Slovenian Research Agency (research core funding No. P5-0168).

References

Augustin, T., Coolen, F.P., de Cooman, G., Troffaes, M.C.: Introduction to Imprecise Probabilities. Wiley, Chichester (2014)

Ferson, S., Kreinovich, V., Ginzburg, L., Myers, D., Sentz, K.: Constructing probability boxes and Dempster-Shafer structures. Technical report (2003). https://doi.org/10.2172/809606

Montes, I., Destercke, S.: On extreme points of p-boxes and belief functions. Ann. Math. Artif. Intell. **81**(3), 405–428 (2017)

Montes, I., Miranda, E.: Extreme points of the core of possibility measures and maxitive p-boxes. Internat. J. Uncertain. Fuzz. Knowl.-Based Syst. **26**(06), 1017–1051 (2018)

Škulj, D.: A complete characterization of normal cones and extreme points for p-boxes. arXiv preprint arXiv:2203.11634 (2022a)

Škulj, D.: Normal cones corresponding to credal sets of lower probabilities. arXiv preprint arXiv:2201.10161 (2022b)

Troffaes, M., Destercke, S.: Probability boxes on totally preordered spaces for multivariate modelling. Int. J. Approx. Reason. **52**(6), 767–791 (2011)

Utkin, L., Destercke, S.: Computing expectations with continuous p-boxes: univariate case. Int. J. Approx. Reason. **50**(5), 778–798 (2009)

Measurability and Products of Random Sets

Pedro Terán[(✉)]

Universidad de Oviedo, Gijón, Spain
teranpedro@uniovi.es

Abstract. Some nice properties of a possible definition of random sets in certain non-metrizable spaces are studied. These concern specially applications to products of random sets and to random functions.

1 Introduction

The standard framework for the study of random sets since Matheron (1975) includes $\mathbb{R}^d$ as a possible ambient space but applies, more generally, to any topological space which is locally compact, second countable, and Hausdorff (often called an LCSH space in the literature). A random set is defined as a set-valued mapping for which all events of the form $\{X \cap K \neq \emptyset\}$, for compact K, are measurable.

Some spaces of interest are not LCSH, e.g., the space of all real continuous functions on $[0, 1]$. When the family of LCSH spaces is enlarged to consider all complete separable metric spaces, a set-valued mapping is usually called a random set if the events $\{X \cap G \neq \emptyset\}$ are measurable for all open sets G (Effros measurability). This defining property comes from previous research in areas like descriptive set theory and set-valued analysis. In LCSH spaces, it is equivalent to the ordinary definition. It also has a number of positive consequences, including a Choquet theorem characterizing the probability distributions of random sets via capacities (Terán 2019). However, some of those advantages are tied to the metric and the definition becomes less forceful once we move into non-metric spaces.

Terán (2014) established a Choquet theorem in locally compact, σ-compact, Hausdorff spaces (a different generalization of LCSH spaces). This theorem is naturally associated to a definition of measurability via the events $\{X \cap G \neq \emptyset\}$ for a subfamily of open sets but not all of them. Moreover, in LCSH spaces it coincides with Effros measurability and the ordinary definition of random sets.

A hypothetical Choquet theorem for Effros measurable mappings in those spaces is unknown. It is worth pointing out that the theorem in Terán (2014) characterizes the distribution of *a superset* of the Effros measurable mappings by using information from *less* events than those involved in that hypothetical theorem. That suggests that the measurability in Terán (2014) deserves further study in order to better understand its properties and whether it has advantages over Effros measurability in those spaces.

L. A. García-Escudero et al. (Eds.): SMPS 2022, AISC 1433, pp. 366–373, 2023.
https://doi.org/10.1007/978-3-031-15509-3_48

In this contribution, it will be shown that this framework accomodates some situations which do not fit in the usual one. First, it is a weaker measurability requirement than some alternatives. Second, it works well with taking arbitrary Cartesian products of random sets. That is, a product is measurable if and only if each factor is measurable, whereas the classes of LCSH or complete separable metric spaces are not closed under uncountable products (which prevents even speaking of arbitrary products of random sets). Third, that opens a way to regard (point-valued or set-valued) stochastic processes as random points or sets in a product space. Fourth, we show that a concrete example of a random set in a space of functions, the *bands* used in non-parametric functional data analysis, make rigorous sense even if no regularity assumptions are placed on the functions.

2 Preliminaries

Let $(\mathbb{E}, \tau)$ be a topological space. Let $\mathscr{F}(\mathbb{E})$ be the class of all non-empty closed subsets of $\mathbb{E}$. We will denote by $\mathscr{C}(\mathbb{E})$ and $\mathscr{C}_0(\mathbb{E})$ the sets of all real functions on $\mathbb{E}$ which are continuous, respectively continuous and vanishing at infinity (f vanishes at infinity if, for each $\varepsilon > 0$, there exists a compact set out of which $|f|$ is smaller than ε). Clearly, in a compact space $\mathscr{C}(\mathbb{E}) = \mathscr{C}_0(\mathbb{E})$.

Let $\mathscr{F}_f(\mathbb{E}) \subseteq \mathscr{F}(\mathbb{E})$ be the class of all sets of the form $\{f \geq 0\}$ (equivalently $\{f = 0\}$) for some $f \in \mathscr{C}_0(\mathbb{E})$. Let $\mathscr{G}_f(\mathbb{E}) \subseteq \tau$ be the class of all sets of the form $\{f > 0\}$ for some $f \in \mathscr{C}_0(\mathbb{E})$ (the complements of those in $\mathscr{F}_f(\mathbb{E})$). Please note that $\mathscr{F}_f(\mathbb{E})$ is called $\mathscr{F}_{f0}$ in Terán (2014).

The *Fell topology* in $\mathscr{F}_\emptyset(\mathbb{E}) = \mathscr{F}(\mathbb{E}) \cup \{\emptyset\}$ is defined by its subbase formed by all sets $\{A \in \mathscr{F}_\emptyset(\mathbb{E}) \mid A \cap G \neq \emptyset\}$ for G open, and $\{A \in \mathscr{F}_\emptyset(\mathbb{E}) \mid A \cap K = \emptyset\}$ for K compact. It is denoted by τ_F. The relative Fell topology in $\mathscr{F}(\mathbb{E})$ will be just called the Fell topology of $\mathscr{F}(\mathbb{E})$. It will occasionally be necessary to indicate in brackets the carrier space of τ_F. The Fell topology is traditionally called the *hit-and-miss topology* in the random sets literature, and is generally weaker than other possible topologies like the Wijsman topology, the Hausdorff metric topology and the Vietoris topology.

A *directed set* is a non-empty set endowed with a preorder $\leq$ such that every $s, t \in T$ have a common upper bound. A property holds *cofinally* in a directed set T if, for every $t_0 \in T$, it holds for some $t \geq t_0$; and *eventually* if, for some $t_0 \in T$, it holds for all $t \geq t_0$. A *net* is a generalization of a sequence, indexed by a directed set T instead of $\mathbb{N}$. Nets will be needed to carry out some proofs via convergence arguments.

A net of sets $\{F_t\}_{t \in T}$ *converges in the Painlevé–Kuratowski sense* to a limit F if

$$\begin{aligned} F &= \{x \in \mathbb{E} \mid N \cap F_t \neq \emptyset \text{ cofinally for every neighbourhood } N \text{ of } x\} \\ &= \{x \in \mathbb{E} \mid N \cap F_t \neq \emptyset \text{ eventually for every neighbourhood } N \text{ of } x\}. \end{aligned}$$

In a locally compact Hausdorff space, Fell convergence of closed sets is equivalent to their Painlevé–Kuratowski convergence (Beer 1993, Theorem 5.2.6).

Let $(\Omega, \mathscr{A})$ be a measurable space. A mapping $X : \Omega \to \mathscr{F}(\mathbb{E})$ is called *Effros measurable* if $\{X \cap G \neq \emptyset\} \in \mathscr{A}$ whenever G is open in $\mathbb{E}$. If $\{X \cap G \neq \emptyset\} \in \mathscr{A}$ just for $G \in \mathscr{G}_f(\mathbb{E})$, it will be called Σ-*measurable* (Terán 2014). This corresponds to ordinary measurability with respect to the σ-algebra Σ generated by all families of sets of the form $\{A \in \mathscr{F}(\mathbb{E}) \mid A \cap G \neq \emptyset\}$ for $G \in \mathscr{G}_f(\mathbb{E})$, or (more concisely) all $\mathscr{F}(F)$ for $F \in \mathscr{F}_f(\mathbb{E})$. The name *random closed set* will be reserved for the case in which $\mathbb{E}$ is an LCSH space.

The *Borel* σ-*algebra* in $\mathbb{E}$ is the smallest one containing the open sets, while the *Baire* σ-*algebra* is the smallest one which makes all real continuous functions on $\mathbb{E}$ measurable. These notions are equivalent in metric spaces but the Baire σ-algebra is smaller in general.

Proposition 1 (Terán 2014, **Proposition 3.1).** *Let $\mathbb{E}$ be a locally compact Hausdorff space, and let X be a function from a measurable space to $\mathscr{F}(\mathbb{E})$. Then,*

(a) X is Σ-measurable if and only if $\inf f(X)$ *is a random variable for each $f \in \mathscr{C}_0(\mathbb{E})$.*
(b) If $\mathbb{E}$ is second countable, then X is Σ-measurable if and only if X is Effros measurable.

3 Comparison of Measurability Conditions

In this section, we show that Σ-measurability is a weaker requirement than some reasonable alternatives. The following lemma is surely known.

Lemma 1. *Let $\mathbb{E}$ be a compact Hausdorff space. Then the Fell topology in $\mathscr{F}(\mathbb{E})$ is compact Hausdorff.*

Proof. By (Beer 1993, Corollary 5.1.4), since $\mathbb{E}$ is locally compact Hausdorff, $\mathscr{F}_\emptyset(\mathbb{E})$ is τ_F-compact and Hausdorff. Let us show that $\mathscr{F}(\mathbb{E})$ is closed. Indeed, Fell convergence to $\emptyset$ of a net of sets would imply, by its definition, that some of them are disjoint from the compact set $\mathbb{E}$, i.e., empty. Thus $\emptyset$ is not in the closure of $\mathscr{F}(\mathbb{E})$, and the latter is closed. □

The key step in this section is as follows.

Theorem 1. *Let $\mathbb{E}$ be a compact Hausdorff space, and let X be a mapping from a measurable space to $\mathscr{F}(\mathbb{E})$. Then,*

(a) If X is Baire measurable with respect to the Fell topology, then X is Σ-measurable.
(b) The converse holds if $\mathbb{E}$ is metrizable.

Proof. *Part (a).* Fix $f \in \mathscr{C}(\mathbb{E})$, and let $\tilde{f} : \mathscr{F}(\mathbb{E}) \to \mathbb{R}$ be given by

$$\tilde{f}(K) = \inf f(K).$$

From Proposition 1, it suffices to show that $\tilde{f}(X)$ is a random variable for each $f \in \mathscr{C}(\mathbb{E})$. That will follow from the Baire measurability of X if we prove that each $\tilde{f}$ is τ_F-continuous.

Consider a net $\{K_t\}_{t\in T} \subseteq \mathscr{F}(\mathbb{E})$ converging to some $K \in \mathscr{F}(\mathbb{E})$ in the Fell topology. Let us show that $\tilde{f}(K_t) \to \tilde{f}(K)$.

First, since K is compact, $\tilde{f}(K) = f(x)$ for some $x \in K$. Fix $\varepsilon > 0$ and take a neighbourhood N_ε of x for which $\sup f(N_\varepsilon) \le f(x) + \varepsilon$. By the equivalence between Fell and Painlevé–Kuratowski convergence, there is some $t_0 \in T$ such that $K_t \cap N_\varepsilon \neq \emptyset$ whenever $t \ge t_0$. Take $x_t \in K_t \cap N_\varepsilon$, then

$$\tilde{f}(K_t) = \inf f(K_t) \le f(x_t) \le \sup f(N_\varepsilon) \le f(x) + \varepsilon = \tilde{f}(K) + \varepsilon.$$

Now assume, reasoning by contradiction, that, for some $\varepsilon > 0$, $\tilde{f}(K_t) \ge \tilde{f}(K) - \varepsilon$ does not hold eventually. Then $\{K_t\}_t$ has a subnet $\{K_t\}_{t\in T'}$ such that $\tilde{f}(K_t) < \tilde{f}(K) - \varepsilon$ for every t. Like before, there exist $x_t \in K_t$ with $f(x_t) = \tilde{f}(K_t)$. Since $\mathbb{E}$ is compact, $\{x_t\}_{t\in T'}$ has a further subnet converging to some $y \in \mathbb{E}$. Accordingly, every neighbourhood of y is intersected by a cofinal subfamily of the K_t. Since K is the Painlevé-Kuratowski limit of the K_t, we have $y \in K$. The continuity of f yields

$$\tilde{f}(K) - \varepsilon > \tilde{f}(K_t) = f(x_t) \to f(y) \ge \inf f(K) = \tilde{f}(K),$$

a contradiction. Thus, for each $\varepsilon > 0$,

$$\tilde{f}(K) - \varepsilon \le \tilde{f}(K_t) \le \tilde{f}(K) + \varepsilon$$

for some $t_0 \in T$ and all $t \ge t_0$, i.e. $\tilde{f}$ is indeed continuous.

Part (b). Assume now that X is Σ-measurable. Since $\mathbb{E}$ is additionally taken metrizable, then also $(\mathscr{F}(\mathbb{E}), \tau_F)$ is metrizable (e.g., Beer 1993, Theorem 5.1.5). And, by Lemma 1, $\mathscr{F}(\mathbb{E})$ is τ_F-compact. Therefore,

(1) $\mathbb{E}$ is compact Hausdorff second countable, so X is Effros measurable by Lemma 1.
(2) $\mathbb{E}$ is compact metric, so Effros measurable implies Borel measurable with respect to τ_F (Srivastava 1998, Exercise 3.3.12, p. 98).
(3) $\mathbb{E}$ is metric, so Borel measurable is the same thing as Baire measurable. □

Corollary 1. *Let $\mathbb{E}$ be a compact Hausdorff space. The requirement of Σ-measurability is weaker than Effros measurability, Baire and Borel measurability with respect to the Fell topology, and hence also weaker than Baire and Borel measurability with respect to any finer topology.*

Proof. It stems from the definitions that Σ-measurability is weaker than Effros measurability, Baire measurability is weaker than Borel measurability, and Baire/Borel measurability with respect to a coarser topology is weaker than Baire/Borel measurability with respect to a finer topology. Thus the statement follows from Theorem 1.(a). □

4 Measurability of Products of Random Sets

We will show now that arbitrary products of random closed sets in compact metric spaces are Σ-measurable. Notice that uncountable products of compact metric spaces are compact but they are neither metric nor second countable, whence they do not fit in the frameworks of LCSH spaces or complete separable metric spaces.

Theorem 2. *Let $\{\mathbb{E}_i\}_{i\in I}$ be a family of compact metric spaces, and let X be a mapping from a measurable space Ω to $\prod_{i\in I}\mathscr{F}(\mathbb{E}_i)$ such that $X = \prod_{i\in I} X_i$ for some $X_i : \Omega \to \mathbb{E}_i$. Then, the following are equivalent:*

(a) X is Σ-measurable,
(b) Each X_i is a random closed set,
(c) X is Baire measurable with respect to the Fell topology of $\mathscr{F}(\prod_{i\in I}\mathbb{E}_i)$.

Proof. *Part (a)⟹(b).* For each $i \in I$, let $\pi_i : \prod_{i\in I}\mathbb{E}_i \to \mathbb{E}_i$ be the coordinate projection onto $\mathbb{E}_i$. Fix $g \in \mathscr{C}(\mathbb{E}_i)$. Since X is Σ-measurable and the composition $g \circ \pi_i$ is continuous on $\prod_{i\in I}\mathbb{E}_i$, the mapping $\inf g(X_i) = \inf(g\circ\pi_i)(X)$ is a random variable. By Proposition 1.(a) and the arbitrariness of g, each X_i is a random closed set.

Part (b)⟹(c). By Lemma 1, the Fell topology of each $\mathscr{F}(\mathbb{E}_i)$ is compact and Hausdorff. By Theorem 1.(b), each X_i is Baire measurable with respect to the Fell topology of $\mathscr{F}(\mathbb{E}_i)$. Thus, X is measurable with respect to the smallest σ-algebra σ_{proj} which makes all the π_i measurable.

Since the product topology makes all projections continuous, σ_{proj} is contained in the Baire σ-algebra. Let us show that they are actually equal. Using the Stone–Weierstrass theorem, the subalgebra of $\mathscr{C}(\prod_{i\in I}\mathbb{E}_i)$ generated by the continuous functions which depend on a single coordinate (all of which are σ_{proj}-measurable) is dense in $\mathscr{C}(\prod_{i\in I}\mathbb{E}_i)$. For any $F = \{f = 0\} \in \mathscr{F}_f(\mathbb{E})$, taking a sequence $\{f_n\}_n$ from the subalgebra such that $f_n \to f$ uniformly, we obtain

$$F = \bigcap_{k\in\mathbb{N}}\bigcup_{n_0\in\mathbb{N}}\bigcap_{n\ge n_0}\{-k^{-1} \le f_n \le k^{-1}\} \in \sigma_{\text{proj}}.$$

Since $\mathscr{F}_f(\mathbb{E})$ generates the Baire σ-algebra because $\mathscr{C}(\mathbb{E}) = \mathscr{C}_0(\mathbb{E})$, the latter is contained in (and thus equal to) σ_{proj}.

Accordingly, X is Baire measurable with respect to the product topology $\bigotimes_{i\in I}\tau_F(\mathscr{F}(\mathbb{E}_i))$. There remains to check that this topology is the same as the Fell topology of $\mathscr{F}(\prod_{i\in I}\mathbb{E}_i)$.

By considering in the definition of $\tau_F(\mathscr{F}(\prod_{i\in I}\mathbb{E}_i))$ the open and compact sets that depend on a single coordinate, it is readily shown that the Fell topology in the product space is finer than the product of the Fell topologies. But both topologies are compact Hausdorff: the former from Lemma 1 and the latter by the Tikhonov theorem and the fact that an arbitrary product of Hausdorff spaces is Hausdorff. Thus they must be equal.

Part (c)⟹(a). It follows from Theorem 1.(a). □

5 Random Functions

A natural setting for arbitrary products of spaces is that of random functions. As observed by Dellacherie and Meyer (1978, p. 89), *'A path governed by chance –*

that of a particle getting hit from all sides, for example – has no reason for being very regular.' Very often, however, it is assumed that the paths of a stochastic process enjoy at least some regularity, such as being continuous or being a càdlàg function. The framework described in this paper can accommodate stochastic processes with arbitrary trajectories (provided these are confined between two pre-specified bounds or functions) and even set-valued stochastic processes. Let T be an index set. For clarity, we will denote the elements of $\mathbb{R}^T$ as points in a product space, with coordinates x_t, rather than as functions with values $x(t)$.

Corollary 2. *Let $f, g : T \to \mathbb{R}$ be such that $f \leq g$. Let $(\Omega, \mathscr{A}, P)$ be a probability space and $X_t : \Omega \to \mathbb{R}$ such that $X_t \in [f(t), g(t)]$ for each $t \in T$.*

Then $\{X_t\}_{t\in T}$ is a stochastic process if and only if $X : \Omega \to \mathscr{F}(\mathbb{R}^T)$ given by

$$(X(\omega))_t = \{X_t(\omega)\}$$

is Σ-measurable.

Proof. By the assumption, X lives in the space

$$\{x \in \mathbb{R}^T \mid x_t \in [f(t), g(t)] \ \forall t \in T\} = \prod_{t\in T}[f(t), g(t)].$$

If $\{X_t\}_{t\in T}$ is a stochastic process, each X_t is a random variable with $X_t \in [f(t), g(t)]$. Each projection $\pi_t \circ X$ is exactly $\{X_t\}$, which is then a random closed set in $[f(t), g(t)]$. Theorem 2 implies X is Σ-measurable. The converse is analogous. □

Remark 1. This corollary extends to the case of set-valued stochastic processes since the fact that each $(X(\omega))_t$ is a singleton is nowhere used in the proof.

The following example shows how band depth, a notion from non-parametric statistics with functional data, makes sense without any assumption on the regularity of the functions.

Example 1. Functional data are continuously observed random variables. If the time interval is $T = [0, M]$ and the values are in an interval $[a, b]$, each datum is an element of $[a, b]^{[0,M]}$, a compact Hausdorff space when endowed with the pointwise convergence topology. The aim of this example is to show that the bands used to define band depth are Σ-measurable in this general setting in which no regularity assumptions about the functions are made.

Statistical depth measures try to provide a center–outward ordering of the potential observations. In the setting of functional data, a well-known example is *band depth* introduced by López-Pintado and Romo (2006). It involves a fixed value $J \geq 2$, taking a finite i.i.d. sequence of $n \geq J$ random functions $\{\xi_i\}_{i=1}^n$, and considering their j*-bands*

$$X_j = \{f \in [a, b]^{[0,M]} \mid \min_{1\leq i\leq j} \xi_i \leq f \leq \max_{1\leq i\leq j} \xi_i\}$$

for $j \in \{2, \ldots, J\}$.

That each j-band is Σ-measurable follows from Theorem 2, since

$$X_j = \prod_{x\in[0,M]} [\min_{1\le i\le j} \xi_i(x), \max_{1\le i\le j} \xi_i(x)]$$

and each interval is a random closed set in $[a,b]$ because its end-points are random variables. Also by Theorem 2, X_j is Baire measurable with respect to the Fell topology of $[a,b]^{[0,M]}$.

The depth value of a function f is calculated, in its original definition, as

$$BD_J(f) = \sum_{j=2}^{J} P(f \in X_j).$$

Notice that

$$\{f \in X_j\} = \{X_j \cap \{f\} \neq \emptyset\} \tag{1}$$

but singletons of $[a,b]^{[0,M]}$ are in general not in $\mathscr{F}_f(\mathbb{E})$, whence $P(f \in X_j)$ is not guaranteed to be well-defined. To overcome this situation, we will replace the induced distribution P_{X_j} by its unique extension $\widehat{P_{X_j}}$ to a Radon measure (Vakhania *et al.*, 1987, Theorem I.3.3.(b)) on the Borel σ-algebra of $\mathscr{F}([a,b]^{[0,M]})$. The assumptions in Vakhania *et al.* are indeed satisfied: since $(\mathscr{F}([a,b]^{[0,M]}), \tau_F)$ is compact by Lemma 1, it is a completely regular space and P_{X_j} is trivially tight.

Moreover, the event 'containing f' or 'intersecting $\{f\}$' in (1) is Borel measurable as the Borel σ-algebra of the Fell topology includes the event of intersecting any compact set. Denoting by Cont(f) the family of all closed sets that contain f, the final definition in this more general setting is

$$BD_J(f) = \sum_{j=2}^{J} \widehat{P_{X_j}}(\mathrm{Cont}(f)).$$

Acknowledgements. Research partially supported by Spain's Ministerio de Economía y Competitividad (MTM2015–63971–P) and Ministerio de Ciencia, Innovación y Universidades (PID2019–104486GB–100), and Asturias' Consejería de Empleo, Industria y Turismo (GRUPIN–IDI2018–000132).

References

Beer, G.: Topologies for Closed and Closed Convex Sets. Kluwer, Dordrecht (1993)

Dellacherie, C., Meyer, P.-A.: Probabilities and Potential. North-Holland, Amsterdam (1978)

López-Pintado, S., Romo, J.: Depth-based classification for functional data. In: Liu, R.Y., Serfling, R., Souvaine, D.L. (eds.) Data Depth: Robust Multivariate Analysis, Computational Geometry and Applications. DIMACS Series, vol. 72, pp. 103–119. American Mathematical Society, Providence (2006)

Matheron, G.: Random Sets and Integral Geometry. Wiley, New York (1975)

Srivastava, S.M.: A Course on Borel Sets. Springer, New York (1998). https://doi.org/10.1007/b98956

Terán, P.: Distributions of random closed sets via containment functionals. J. Nonlinear Convex Anal. **15**, 907–917 (2014)

Terán, P.: Choquet theorem for random sets in polish spaces and beyond. In: Destercke, S., Denoeux, T., Gil, M.Á., Grzegorzewski, P., Hryniewicz, O. (eds.) SMPS 2018. AISC, vol. 832, pp. 208–215. Springer, Cham (2019). https://doi.org/10.1007/978-3-319-97547-4_27

Vakhania, N.N., Tarieladze, V.I., Chobanyan, S.A.: Probability Distributions on Banach Spaces. Springer, Dordrecht (1987). https://doi.org/10.1007/978-94-009-3873-1

Fast CP Model Fitting with Integrated ASD-ALS Procedure

Valentin Todorov[1], Violetta Simonacci[2], Michele Gallo[3(✉)], and Nickolay Trendafilov[3]

[1] United Nations Industrial Development Organization (UNIDO), Vienna, Austria
valentin@todorov.at

[2] University of Naples Federico II, Naples, Italy
violetta.simonacci@unina.it

[3] University of Naples-L'Orientale, Naples, Italy
mgallo@unior.it

Abstract. The CP decomposition is the most appropriate tool for modeling data arrays with a trilinear structure. Model fitting can be hindered by several issues, including computational inefficiency, bad initialization, excessive modeled noise, sensitivity to over-factoring and collinearity. Many algorithms have been proposed for parameter estimation, each with specific strengths and weaknesses. Fast procedures tend to be less stable and vice-versa. Stability is usually prioritized by preferring the least-square approach ALS, albeit slow and sensitive to excess factors. As a solution integrated methods have been proposed in the literature. First, estimation is initialized with a fast procedure to ensure competitive speed then results are refined with ALS to improve precision. In this work, we implement a novel integrated algorithm called INT-3 where ASD steps are concatenated with ALS. ASD was selected because of its remarkable speed and low memory consumption requirements. INT-3 performance is tested against ALS on artificial data.

1 Introduction

The CANDECOMP/PARAFAC (CP) model is an extension of PCA to higher-order data and decomposes the systematic part of a three-way array intocfactors (Carroll and Chang 1970; Harshman 1970). Model fitting aims to find a unique solution where extracted components represent the same constructs throughout modes (Cattell 1944). By imposing component uniqueness, the CP model yields an easily interpretable latent structure, which can be also assessed with a confirmatory outlook.

The main limitation of the CP model concerns the estimation process of trilinear parameters. Computations are generally carried out with the PARAFAC-ALS (ALS) procedure. This algorithm is selected thanks to the in-built capability of minimizing modeled noise and well-defined properties of its loss function.

L. A. García-Escudero et al. (Eds.): SMPS 2022, AISC 1433, pp. 374–381, 2023.
https://doi.org/10.1007/978-3-031-15509-3_49

The main issue with ALS estimation relates to computational efficiency. Processing time and memory requirements may become prohibitive for large data. Additionally, degenerate solutions can occur, especially under challenging data conditions such as bad initialization, factor collinearity, and over-factoring (Yu et al. 2011).

Other estimating procedures have been proposed in the literature. Some of the fastest alternatives include the Alternating TriLinear Decomposition (Wu et al. 1998) (ATLD), the Self-Weighted Alternating TriLinear Decomposition (Chen et al. 2000) (SWATLD), and the Alternating Slice-wise Diagonalization (Jiang et al. 2000) (ASD). These procedures are also known to be insensitive to over-factoring. Competitive speed is granted to ATLD and SWATLD by the use of three loss functions, while ASD employs an SVD-based compression procedure that allows a significant efficiency improvement.

The somewhat unclear property of their convergence, instability, and excess modeled noise are the main reasons such fast alternaitves were never used on a large scale. The general recommendation in comprehensive comparative studies is to prefer ALS in terms of quality of the solution (Tomasi and Bro 2006; Faber et al. 2003).

A recent research thread has attempted to find a compromise by proposing an integrated system. The logic of this approach is quite simple. Estimation is split into two stages. In the first stage, a fast procedure that ensures quick leaps in the convergence process is used to boost efficiency and ensure over-factoring resistance. Successively, a clean, least-squares solution is reached by performing ALS iterations in a second and conclusive stage.

Two integrated algorithms have been proposed so far. INT was the first introduced variant, which uses SWATLD in the first stage (Simonacci and Gallo 2019). INT-2 was then implemented where ATLD was employed (Simonacci and Gallo 2020). ASD, albeit faster and more efficient from a memory usage standpoint, was not considered yet, due to the indication provided in (2006, p. 1728) where ASD is defined as consistently inferior.

The performance of INT and INT-2, however, demonstrates that none of the concerning issues of SWATLD and ATLD were passed on to their integrated counterpart. On the contrary, throughout simulations, the integrated algorithms are fast, stable, and more reliable than ALS when over-factoring.

In this perspective, we implemented a novel integrated procedure, INT-3, where ASD is used instead. Based on the previous considerations, INT-3 is expected to perform even better than INT and INT-2, as it should not inherit ASD's unpredictability. A simulation study has been implemented to assess INT-3 performance against ALS under different data conditions. Both efficiency and accuracy diagnostics confirm the anticipated favorable results.

In Sect. 2 the CP model is introduced with ALS estimation; Sect. 3 details the ASD and the novel INT-3 approach. Section 4 presents the simulation study and, lastly Sect. 5 concludes with some remarks.

2 The CP Model and Its ALS Estimation

The CP model (Carroll and Chang 1970; Harshman 1970) decomposes the three-way data array $\underline{\mathbf{X}}(I \times J \times K)$ with a generic element x_{ijk} into three loading matrices $\mathbf{A}$ $(I \times F)$, $\mathbf{B}$ $(J \times F)$, $\mathbf{C}$ $(K \times F)$ with F components (using the same number for each mode). The CP model can be written formally as

$$\mathbf{X}_{::k} = \hat{\mathbf{X}}_{::k} + \mathbf{E}_{::k} = \mathbf{A}\mathbf{D}_k\mathbf{B}^t + \mathbf{E}_{::k} \quad k = 1, \dots, K \tag{1}$$

$\mathbf{X}_{::k}(I \times J)$ is the generic k-th frontal slice of $\underline{\mathbf{X}}$, obtained by fixing the third-mode index. $\mathbf{D}_k$ are the diagonal matrices storing the kth rows of the third-mode for $k = 1, \dots, K$. Lastly $\mathbf{E}_{::k}$ is the corresponding frontal slice of the error array $\underline{\mathbf{E}}$. The model is a rank decomposition if $F = R$ which corresponds to the real dimensionality of the latent solution. Over-factoring implies $F > R$.

To estimate the optimal component matrices the residual sum of squares

$$||\underline{\mathbf{E}}||^2 = \sum_{i=1}^{I}\sum_{j=1}^{J}\sum_{k=1}^{K}(x_{ijk} - \hat{x}_{ijk})^2 = \sum_{i=1}^{I}||\mathbf{x}_i - \hat{\mathbf{x}}_i||^2 = \sum_{i=1}^{I} RD_i^2 \tag{2}$$

is minimized. The residual distance for observation i is thus given by

$$RD_i = ||\mathbf{x}_i - \hat{\mathbf{x}}_i|| = \sqrt{\sum_{j=1}^{J}\sum_{k=1}^{K}(x_{ijk} - \hat{x}_{ijk})^2} \tag{3}$$

and the estimation is equivalent to the minimization of the sum of the squared distances. With ALS the component matrices are estimated one at a time, keeping the estimates of the other component matrices fixed, i.e. we start with initial estimates of $\mathbf{B}$ and $\mathbf{C}$ and find an estimate for $\mathbf{A}$ conditional on $\mathbf{B}$ and $\mathbf{C}$ by minimizing the objective function. Estimates for $\mathbf{B}$ and $\mathbf{C}$ are found analogously. The iteration continues until the relative change in the model fit is smaller than a predefined constant.

3 The ASD and INT-3 Algorithms

The ASD algorithm was introduced in Jiang et al. (2000). It is an iterative procedure based on the minimization of a slice-wise diagonalization (SD) loss criterion. Assuming the two loading matrices $\mathbf{A}$ and $\mathbf{B}$ have full column rank Eq. (1) can be reformulated as

$$\mathbf{P}^t\mathbf{X}_{::k}\mathbf{Q} = \mathbf{D}_k + \mathbf{P}^t\mathbf{E}_{::k}\mathbf{Q} = \mathbf{D}_k + \tilde{\mathbf{E}}_{::k} \tag{4}$$

where $\mathbf{P}$ and $\mathbf{Q}$ are defined by the equations $\mathbf{P}^T\mathbf{A} = \mathbf{I}_F$ and $\mathbf{P}^T\mathbf{A} = \mathbf{I}_F$ and $\mathbf{I}_F$ is the f-dimensional identity matrix. The SD loss criterion is then defined:

$$L_{SD} = \sum_{k=1}^{K}\left\|\mathbf{P}^t\mathbf{X}_{::k}\mathbf{Q} - \mathbf{D}_k\right\| + \lambda(\left\|\mathbf{P}^t\mathbf{A} - \mathbf{I}_f\right\| + \left\|\mathbf{B}^t\mathbf{Q} - \mathbf{I}_f\right\|) \tag{5}$$

Here λ is a predefined positive constant weighting the penalty terms. Sets of parameters are estimated in an alternating fashion. To speed up convergence the loss function is not minimized directly but is first transformed into a reduced problem by the use of compression operators obtained by SVD of $\sum_{k=1}^{K} \mathbf{X}_{::k}\mathbf{X}_{::k}^t$ and $\sum_{k=1}^{K} \mathbf{X}_{::k}^t\mathbf{X}_{::k}$. For more details on the procedure, please see the original contribution.

INT-3 can now be introduced. A simple integrated scheme is implemented in two subsequent phases. In the first phase, ASD estimation is carried out to ensure quick convergence and insensitivity to over-factoring. The procedure stops to an interim convergence criterion expressed as relative loss of fit. In this work, the interim parameter was set to $10^{-}02$. Following this, estimation is resumed with ALS steps. This phase ensures a cleaner solution in terms of noise, increases accuracy, and stabilizes convergence.

4 Simulation Study

The performance of the newly proposed algorithm INT-3 for fast estimation of trilinear CP models will be demonstrated in a brief simulation study comparing classical CP estimation by the ALS algorithm and the integrated INT-3 one.

Artificial data are generated by considering the following combinations of parameters: $I, J, K = 100$; $R = 3, 5$, congruence level $\text{CONG} = 0.2, 0.5, 0.8$ and heteroscedastic and homoscedastic noise $\text{HO} = 15\%$, 20%, 25% and $\text{HE} = 10\%$, 15%, 20%. See Simonacci and Gallo (2020) for details on data creation. All procedures were performed using the R programming language and environment and the R package **rrcov3way** (Todorov et al. 2022). The ASD and INT-3 functions were programmed in R referring to the MATLAB code provided in Faber et al. (2003) and to the original work of Jiang et al. (2000).

First of all, we want to verify that INT-3 works at least as good as ALS in retrieving solutions with good statistical quality on data sets with different configurations of noise and factor congruence. At the same time, we want to confirm that the convergence is improved significantly and thus the computational time is reduced.

The mean squared error of the fit measures how precisely the data can be recovered through the lower dimensional decomposition and is given by

$$MSE = \frac{1}{w} \sum_{i=1}^{I} \sum_{j=1}^{J} \sum_{k=1}^{K} w_i (x_{ijk} - \hat{x}_{ijk})^2 \qquad (6)$$

with $w = \sum_{i=1}^{I} w_i$ and $w_i = 0$ if the i-th observation is outlier or $w_i = 1$ otherwise. Thus the MSE will be computed only for the regular observations. For now we do not add any outlying observations. In Fig. 1 we observe that for both $R = 3$ and $R = 5$ INT-3 works just as well as ALS in the rank decomposition case while it performs better when over-factoring.

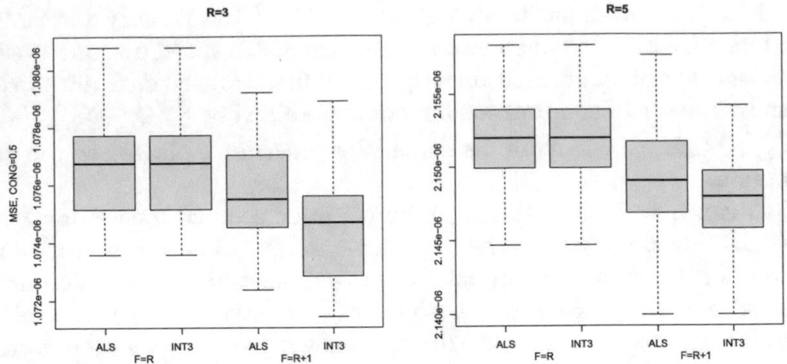

Fig. 1. MSE of the fit by rank in the cases of rank decomposition and over-factoring for $CONG = 0.5$

To compare the methods with respect to how well they recover the underlying structure the Tucker congruence coefficient (also referred to as *uncorrected correlation*) of the columns of the estimated loading matrices with those of the underlying true matrices was used (Lorenzo-Seva and ten Berge 2006). The congruence coefficient ϕ between two components $\mathbf{x}$ and $\mathbf{y}$ is given by the following formula:

$$\phi_{xy} = \frac{\sum \mathbf{xy}}{\sqrt{\sum \mathbf{x}^2 \sum \mathbf{y}^2}}. \tag{7}$$

We say that full recovery occurred if the congruence coefficient between any two columns was 0.99 for any of the loading matrices. Thus, averaging on the three modes, we obtain the threshold $0.99^3 \approx 0.97$ under which the recovery is considered failed (FR). Since the methods do not uniquely determine the order and the sign of the loading vectors, the estimated matrices were reordered and reflected in all possible ways and the one that gave the best recovery was kept and used for deciding if the recovery failed.

Table 1. Percentages of failed recoveries by rank and number of factors (CONG = 0.5, HO = 15%, HE = 10%)

	F = R = 3	F = R + 1 = 4	F = R = 5	F = R + 1 = 6
ALS	0	72	0	38
INT-3	0	10	0	6

This diagnostic, displayed in Table 1, shows that none of the procedures experiences issues in the rank decomposition. On the other hand, ALS records more degeneracies in over-factoring than INT-3, especially for $R = 3$.

Let us now focus on computational efficiency. Figure 2 displays the distribution of computational time and number of iterations, respectively, of ALS and INT-3. The left-side column shows the rank decomposition and the right-side one the over-factoring case. In the first row of the figures the outliers are not shown, to make the comparison easier. The computational efficiency gain is clear in this representation.

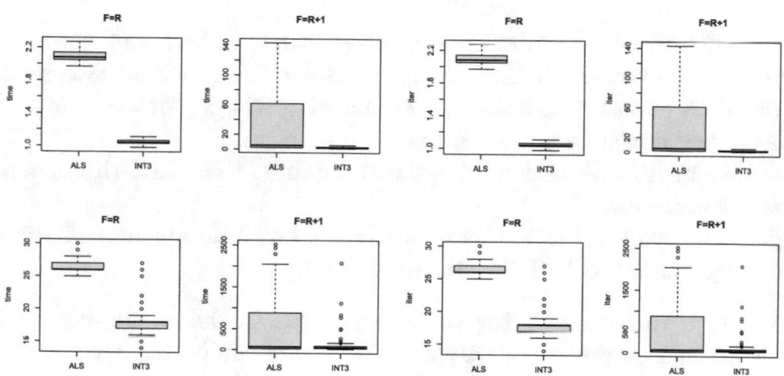

Fig. 2. Distribution of computational time and number of iterations of the standard PARAFAC-ALS procedure and the integrated ASD-ALS procedure (INT-3) in case of rank decomposition and over-factoring

In the rank decomposition, a significant gap is recorded while variability is low for both procedures. When over-factoring, ALS instability is visible in the wide-ranging distributions.

The computational time efficiency is also evaluated by congruence level, to study the effect of factor collinearity for different ranks (3 and 5), in the cases of rank decomposition and over-factoring. Due to space limitation Fig. 3 shows the distribution of CPU time for only one congruence level, namely $CONG = 0.5$.

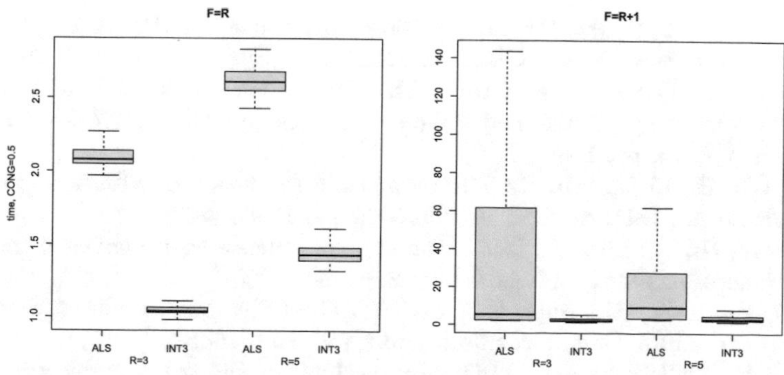

Fig. 3. Distribution of CPU time of the standard PARAFAC-ALS procedure and the integrated ASD-ALS procedure by rank (CONG level is fixed to 0.5)

5 Summary and Conclusions

In this contribution we propose the novel efficient algorithm INT-3 which integrates ASD and ALS steps to ensure fast convergence, robustness to over-factoring, and stability. Performance is assessed by testing INT-3 and the baseline algorithm ALS in a Monte Carlo simulation study, considering varied data conditions. The preliminary findings can be summarized as follows:

- The MSE and FR diagnostic demonstrate that, in terms of quality of the solution, INT-3 outperforms ALS. In the rank-decomposition case, both algorithms work just as well, while in over-factoring INT-3 provides a more precise solution and encounters fewer degeneracies.
- INT-3 is much faster and stable (less variability) than ALS throughout simulating conditions.
- Higher noise contamination and congruence levels do not impact simulation results by much and INT-3 is always reliable.

The algorithm discussed in the paper and used in the simulation study are available in the R package **rrcov3way**. This study can be further developed by implementing a comparison with previous integrated procedures and by simulating difficult data conditions such as outlier contamination and compositional structures.

References

Carroll, J., Chang, J.: Analysis of individual differences in multidimensional scaling via an N-way generalization of Eckart-Young decomposition. Psychometrica **35**(3), 283–319 (1970)

Cattell, R.B.: Parallel proportional profiles and other principles for determining the choice of factors by rotation. Psychometrika **9**(4), 267–283 (1944)

Chen, Z.P., Wu, H.L., Jiang, J.H., Li, Y., Yu, R.Q.: A novel trilinear decomposition algorithm for second-order linear calibration. Chemom. Intell. Lab. Syst. **52**(1), 75–86 (2000)

Faber, N.M., Bro, R., Hopke, P.K.: Recent developments in CANDECOMP/PARAFAC algorithms: a critical review. Chemom. Intell. Lab. Syst. **65**, 119–137 (2003)

Harshman, R.A.: Foundations of the PARAFAC procedure: models and conditions for an "explanatory" multi-modal factor analysis. In: UCLA Working Papers in Phonetics, vol. 16, pp. 1–84 (1970)

Jiang, J., Wu, H., Li, Y., Yu, R.: Three-way data resolution by alternating slice-wise diagonalization (ASD) method. J. Chemom. **14**, 15–36 (2000)

Lorenzo-Seva, U., ten Berge, J.M.F.: Tucker's congruence coefficient as a meaningful index of factor similarity. Methodology **2**, 57–64 (2006)

Simonacci, V., Gallo, M.: Improving PARAFAC-ALS estimates with a double optimization procedure. Chemom. Intell. Lab. Syst. **192**, 103822 (2019)

Simonacci, V., Gallo, M.: An ATLD–ALS method for the trilinear decomposition of large third-order tensors. Soft. Comput. **24**(18), 13535–13546 (2019). https://doi.org/10.1007/s00500-019-04320-9

Tomasi, G., Bro, R.: A comparison of algorithms for fitting the PARAFAC model. Comput. Stat. Data Anal. **50**(7), 1700–1734 (2006)

Wu, H.L., Shibukawa, M., Oguma, K.: An alternating trilinear decomposition algorithm with application to calibration of HPLC-DAD for simultaneous determination of overlapped chlorinated aromatic hydrocarbons. J. Chemom. **12**(1), 1–26 (1998)

Yu, Y.J., et al.: A comparison of several trilinear second-order calibration algorithms. Chemom. Intell. Lab. Syst. **106**(1), 93–107 (2011)

Todorov, V., Di Palma, M.A., Gallo, M.: rrcov3way: Robust Methods for Multiway Data Analysis. R package version 0.2-3 (2022). https://CRAN.R-project.org/package=rrcov3way

On Quantifying and Estimating Directed Dependence

Wolfgang Trutschnig(✉) and Florian Griessenberger

Department for Artificial Intelligence and Human Interfaces (AIHI), University of Salzburg, Hellbrunnerstraße 34, Salzburg, Austria
wolfgang@trutschnig.net, florian.griessenberger@plus.ac.at

Abstract. Considering that a (random) variable X may provide more information about a (random) variable Y than vice versa it is natural that dependence measures, i.e., notions quantifying the extent of dependence, are not necessarily symmetric. Working with Markov kernels (regular conditional distributions) allows to construct the measure which assigns every copula a value in $[0, 1]$, which is 0 exactly in the case of independence, and 1 exclusively for Y being a function of X. More importantly, given samples of X and Y and considering so-called checkerboard estimators it is possible to derive a strongly consistent estimator for the dependence measure which also exhibits a good performance for small and medium sample sizes. After sketching the background on the dependence measure and its checkerboard estimator, and illustrating its performance in terms of a small simulation study we discuss how the studied approach can be generalized to the multivariate question of quantifying the extent of dependence of a random variable Y on an ensemble of random variables.

1 Introduction

Standard dependence/association measures like Pearson correlation, Spearman correlation, and Schweizer and Wolff's σ (see Schweizer and Wolff 1981) are symmetric, i.e. they assign each pair of random variables (X, Y) the same dependence as they assign the pair (Y, X). Independence of two random variables is a symmetric concept (modelling the situation that knowing X does not change our knowledge about Y and vice versa) - dependence, however, is not. As a direct consequence, notions 'measuring' dependence should not necessarily be symmetric since in many situations the dependence structure may be highly asymmetric - think, for instance, of a sample $(x_1, y_1), \ldots, (x_n, y_n)$ in the shape of the letter V, in which case it is without doubt easier to predict the y-value given the x-value than vice versa.

The copula-based, hence scale-invariant dependence measure ζ_1 introduced in Trutschnig (2011) (also see Siburg and Stoimenov (2010), as well as Wiedermann et al. (2020)) was developed in order to overcome this problem. The measure ζ_1 detects asymmetries in the dependence structure and clearly separates independence and so-called complete dependence describing the exact opposite - the

L. A. García-Escudero et al. (Eds.): SMPS 2022, AISC 1433, pp. 382–389, 2023.
https://doi.org/10.1007/978-3-031-15509-3_50

situation where $Y = f \circ X$ for some measurable $f : \mathbb{R} \longrightarrow \mathbb{R}$ (i.e. the situation of maximal information gain about/predictability of Y when knowing X).

Considering that ζ_1 is based on conditional distributions (Markov kernels) and that (in the continuous setting) estimating conditional distribution is a difficult endeavor, it was a-priori unclear if good estimators can be derived at all. It is therefore to a certain extent surprising that so-called empirical checkerboard estimators (ECBEs) can be shown to be strongly consistent in the general setting, i.e., without any smoothness assumptions (see Junker et al. 2021).

The rest of this contribution paper is organized as follows: Sect. 2 gathers some preliminaries and notations, Sect. 3 recalls the definition of empirical copulas and checkerboard aggregations as well as the main result saying that ECBEs are strongly consistent if the resolution (used for the aggregation) is chosen suitably. Finally, Sect. 4 sketches, how the underlying idea can be extended to the general setting for quantifying and estimating the extent of dependence of a (continuous) random variable Y on an ensemble of (continuous) random variables $X_1, ...X_d$ via so-called linkages.

2 Notation and Preliminaries

Throughout the whole contribution $\mathscr{C}$ will denote the family of all two-dimensional copulas (for background on copulas we refer to Durante and Sempi (2016); Nelsen (2006)). For every copula $A \in \mathscr{C}$ the corresponding doubly stochastic measure will be denoted by μ_A. As usual, $d_\infty(A, B)$ will denote the uniform metric on $\mathscr{C}$, i.e. $d_\infty(A, B) := \max_{(x,y)\in[0,1]^2} |A(x, y) - B(x, y)|$, A^t will denote the transpose of $A \in \mathscr{C}$. For every metric space (Ω, d) the Borel σ-field will be denoted by $\mathscr{B}(\Omega)$, λ will denote the Lebesgue measure on $\mathscr{B}([0, 1])$. A mapping $K : \mathbb{R} \times \mathscr{B}(\mathbb{R}) \longrightarrow [0, 1]$ is called a Markov kernel from $\mathbb{R}$ to $\mathscr{B}(\mathbb{R})$ if $x \mapsto K(x, B)$ is measurable for every fixed $B \in \mathscr{B}(\mathbb{R})$ and $B \mapsto K(x, B)$ is a probability measure for every fixed $x \in \mathbb{R}$. Suppose that X, Y are random variables on $(\Omega, \mathscr{A}, \mathbb{P})$. Then a Markov kernel $K : \mathbb{R} \times \mathscr{B}(\mathbb{R}) \longrightarrow [0, 1]$ is called regular conditional distribution of Y given X if for every $B \in \mathscr{B}(\mathbb{R})$

$$K(X(\omega), B) = \mathbb{E}(\mathbf{1}_B \circ Y|X)(\omega)$$

holds $\mathbb{P}$-a.s. It is well known that a regular conditional distribution of Y given X exists and is unique $\mathbb{P}^X$-almost sure. For every $A \in \mathscr{C}$ the corresponding regular conditional distribution (i.e. the regular conditional distribution of Y given X in the case that $(X, Y) \sim A$) will be denoted by $K_A(\cdot, \cdot)$ and considered as function from $[0, 1] \times \mathscr{B}([0, 1])$ to $[0, 1]$. Note that for every $A \in \mathscr{C}$ and Borel sets $E, F \in \mathscr{B}([0, 1])$ we have

$$\int_E K_A(x, F) d\lambda(x) = \mu_A(E \times F).$$

For more details and properties of conditional expectations and regular conditional distributions see Kallenberg (1997); Klenke (2008).

In the current paper we will mainly work with the metrics D_1 and D_∞ introduced in Trutschnig (2011). These metrics are defined by

$$D_1(A, B) := \int_{[0,1]} \Phi_{A,B}(y) d\lambda(y) \tag{1}$$

$$D_\infty(A, B) := \sup_{y \in [0,1]} \Phi_{A,B}(y), \tag{2}$$

whereby $\Phi_{A,B}(y) = \int_{[0,1]} |K_A(x, [0, y]) - K_B(x, [0, y])| \, d\lambda(x)$. It can be shown that both metrics generate the same topology (without being equivalent), that the metric space $(\mathscr{C}, D_1)$ is complete and separable. Furthermore, firstly, $D_1(A, \Pi)$ attains only values in $[0, \frac{1}{3}]$ and that, secondly, $D_1(A, \Pi)$ is maximal if and only if A is completely dependent, i.e. if a λ-preserving transformation $h : [0, 1] \to [0, 1]$ exists such that $K_A(x, \{h(x)\}) = 1$ for λ-a.e. $x \in [0, 1]$. In the sequel we will let $\mathscr{C}_d$ denote the family of all completely dependent copulas, and write A_h and $K_h(\cdot, \cdot)$ for the completely dependent copula and the Markov kernel of the completely dependent copula induced by the λ-preserving transformation h respectively. For equivalent definitions and properties of completely dependent copulas we refer to Trutschnig (2011) and the references therein.

Letting (X, Y) denote a pair of continuous random variables with joint distribution function H and copula A the dependence measure ζ_1 is defined by (see Trutschnig 2011)

$$\zeta_1(X, Y) = \zeta_1(A) := 3D_1(A, \Pi). \tag{3}$$

As a direct consequence of the properties of D_1 it follows that for all continuous random variables X, Y we have $\zeta_1(X, Y) \in [0, 1]$, that $\zeta_1(X, Y) = 0$ if and only if X and Y are independent, and that $\zeta_1(X, Y) = 1$ if and only if the copula A of (X, Y) is completely dependent.

3 Empirical Checkerboard Estimators

Let (X, Y) be a random vector with joint distribution H, margin distributions F, G and copula $A \in \mathscr{C}$. Given a sample $(x_1, y_1), \ldots, (x_n, y_n)$ of (X, Y) we want to estimate $\zeta_1(X, Y) = \zeta_1(A)$. The natural idea of simply calculating the standard empirical (bilinear) copula $\hat{E}_n$ and considering $\zeta_1(\hat{E}_n)$ does not yield a reasonable estimator. In fact, it is straightforward to show that for every copula A we have

$$\lim_{n \to \infty} \zeta_1(\hat{E}_n) = 1$$

with probability one (the reason being that empirical copulas are 'close' to complete dependence since the conditional distributions have an interval of length $\frac{1}{n}$ as support).

A natural and simple way to overcome this problem is to smooth or aggregate the empirical copula. Aggregation leads to so-called checkerboard copulas with which we are going to work in the sequel. To simplify notation, for every $N \in \mathbb{N}$ and $i, j \in \{1, \ldots, N\}$ define the square R_{ij}^N by $R_{ij}^N = \left[\frac{i-1}{N}, \frac{i}{N}\right] \times \left[\frac{j-1}{N}, \frac{j}{N}\right]$.

Definition 1. Suppose that $N \in \mathbb{N}$. A copula $A \in \mathscr{C}$ is called N-checkerboard copula if A is absolutely continuous and (a version of) its density k_A is constant on the interior of each square R_{ij}^N. We call N the resolution of A.

Letting $\mathfrak{CB}_N(A)$ denote the so-called N-checkerboard approximation of A, i.e.,

$$\mathfrak{CB}_N(A)(x,y) := \int_0^x \int_0^y N^2 \sum_{i,j=1}^{N} \mu_A(R_{ij}^N) \mathbf{1}_{R_{ij}^N}(s,t)\, d\lambda(t) d\lambda(s)$$

then according to Li et al. (Li et al. [1998]) (also see Trutschnig 2011) for every copula $A \in \mathscr{C}$ it can be shown that

$$\lim_{N \to \infty} D_1(\mathfrak{CB}_N(A), A) = 0 = \lim_{N \to \infty} D_1(\mathfrak{CB}_N(A)^t, A^t).$$

Having this we can use the fact that, according to Janssen et al. (2012), for continuous random variables with underlying copula A

$$d_\infty(\hat{E}_n, A) = O\left(\sqrt{\frac{\log(\log(n))}{n}}\right) \tag{4}$$

holds with probability one, and choose the resolution N adequately as a function of the sample size n to prove the following consistency result (for a proof based on several lemmata linking d_∞, D_1 and D_∞ we refer to Junker et al. (2021)) (Fig. 1).

Theorem 1. *Suppose that* $(X_1, Y_1), (X_2, Y_2), \ldots$ *is a random sample from* (X, Y) *and assume that* (X, Y) *has continuous joint distribution function* H *and copula* A. *Furthermore suppose that* $s \in (0, \frac{1}{2})$ *and set* $N(n) := \lfloor n^s \rfloor$ *for every* $n \in \mathbb{N}$. *Then:*

$$\lim_{n \to \infty} \zeta_1(\mathfrak{CB}_{N(n)}(\hat{E}_n)) = \zeta_1(A) \ [\mathbb{P}]. \tag{5}$$

In what follows we will refer to $\zeta_1(\mathfrak{CB}_{N(n)}(\hat{E}_n))$ as empirical checkerboard estimator (ECBE) for $\zeta_1(A)$. Summing up, our chosen procedure for estimating $\zeta_1(X,Y) = \zeta_1(A)$ based on a sample $(x_1, y_1), \ldots, (x_n, y_n)$ of (X, Y) with underlying copula A is as follows:

1. Calculate the empirical copula $\hat{E}_n$.
2. Select the resolution $N = N(n)$ and calculate the empirical N-checkerboard $\mathfrak{CB}_N(\hat{E}_n)$.
3. Estimate $\zeta_1(A)$ by $\zeta_1(\mathfrak{CB}_N(\hat{E}_n))$.

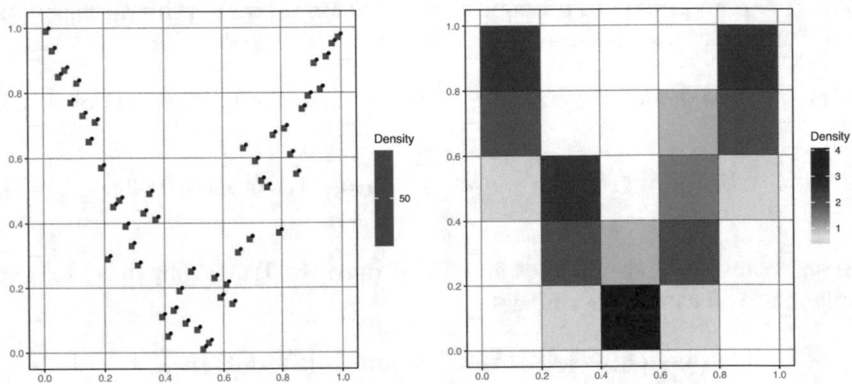

Fig. 1. Left panel: Pseudoobservations of the original sample of size $n = 50$ (black points) and density of the empirical copula $\hat{E}_n$ (magenta squares), Right panel: Density of the empirical checkerboard copula $\mathfrak{CB}_5(\hat{E}_n)$

Remark 1. Notice that Theorem 1 holds without any smoothness assumptions about the copula A, not even continuous first order partial derivatives are required. Numerous simulations insinuate that $\mathfrak{CB}_{N(n)}(\hat{E}_n)$ might also be a strongly consistent estimator for more flexible choices of $N(n)$, particularly for the case $N(n) := \lfloor n^s \rfloor$ and some $s \geq \frac{1}{2}$ - a clarification of this question is future work, in the R-package 'qad' $s = \frac{1}{2}$ was considered.

3.1 Simulations

In order to illustrate the performance of the estimator $\zeta_1(\mathfrak{CB}_{N(n)}(\hat{E}_n))$ we consider the Marshall-Olkin copula $M_{0.3,1}$ and the completely dependent copula A_h, where h is given by $h(x) = 5x\,(mod 1)$. Figures 2 and 3 (left panel) depict a sample of size $n = 5000$ of these two copulas. According to Theorem 1 in both cases the empirical checkerboard estimator converges with probability one to $\zeta_1(M_{0.3,1})$ and $\zeta_1(A_h)$, respectively, the theorem does, however, not provide information on the speed of convergence. Generating samples of sizes $n \in \{10, 50, 100, 500, 1000, 5000, 10000\}$ a total of 1000 times yields the results depicted in Figs. 2 and 3 (right panel), respectively.

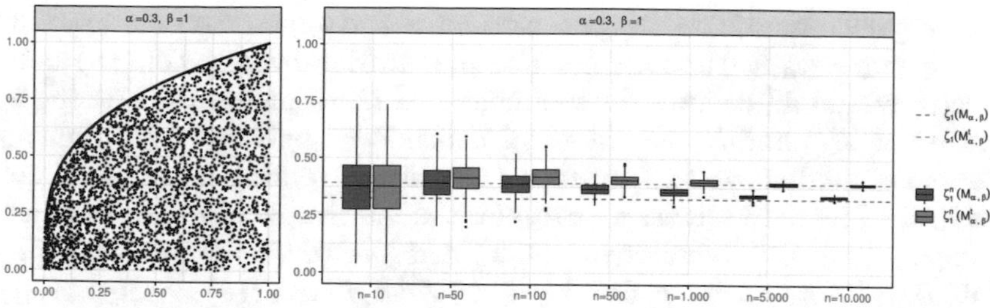

Fig. 2. Boxplots summarizing the 1000 obtained estimates for $\zeta_1(M_{0.3,1})$ (magenta) and $\zeta_1(M^t_{0.3,1})$ (gray). The dashed lines depict $\zeta_1(M_{0.3,1}) \approx 0.324$ and $\zeta_1(M^t_{0.3,1}) \approx 0.391$

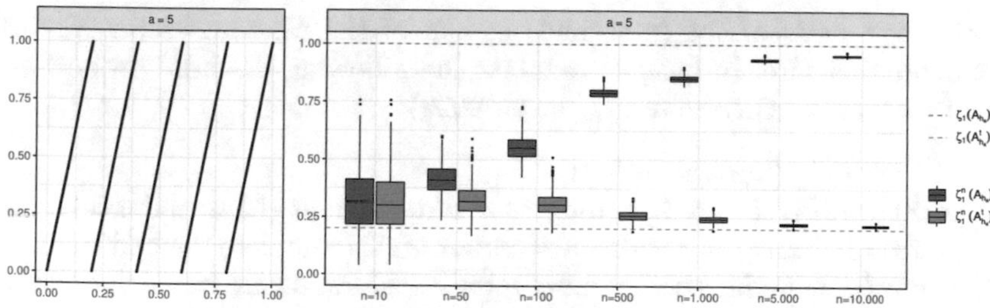

Fig. 3. Boxplots summarizing the 1000 obtained estimates for $\zeta_1(A_h)$ (magenta) and $\zeta_1(A_h^t)$ (gray). The dashed lines depict the true dependence measure $\zeta_1(A_h) = 1$ and $\zeta_1(A_h^t) = \frac{1}{5}$

4 Tackling the Multivariate Setting

It is not straightforward to extend the dependence measure ζ_1 to the multivariate setting in which we want to quantify and estimate the dependence of a (continuous) random variable Y on an ensemble of (continuous explanatory) variables $X_1, \ldots, X_d$. Simply substituting the bivariate kernel in the definition of the metric D_1 by its multivariate version (describing the conditional distribution of Y given $X_1, \ldots, X_d$) does not even yield a metric - the d-dimensional marginal copulas (i.e., the copula of $X_1, \ldots, X_d$) have to be taken into account. Working with so-called linkages (see Griessenberger et al. 2022; Li et al. 1996) allows to transform a general d-dimensional random vector $\boldsymbol{X} := (X_1, \ldots, X_d)$ to a random vector uniformly distributed on $[0,1]^d$ and hence provides a way to overcome the afore-mentioned problem. Given $(d+1)$-dimensional copulas $A, B \in \mathscr{C}^{d+1}$ and letting $L(A)$ and $L(B)$ denote the corresponding linkages of A and B (see Griessenberger et al. 2022), then

$$D_1(A, B) := \int_{[0,1]} \int_{[0,1]^d} \left| K_{L(A)}(\boldsymbol{x}, [0, y]) - K_{L(B)}(\boldsymbol{x}, [0, y]) \right| d\lambda^d(\boldsymbol{x}) d\lambda(y)$$

defines a pseudo-metric on the space of copulas and a metric on the space of linkages, whereby $K_{L(A)}(\cdot,\cdot)$ and $K_{L(B)}(\cdot,\cdot)$ denote the Markov kernel of $L(A)$ and $L(B)$, respectively, and λ^d denotes the d-dimensional Lebesgue measure on $\mathscr{B}([0,1]^d)$.

In order to quantify the extent of dependence between a (continuous) d-dimensional random vector $\boldsymbol{X}$ and a (continuous) univariate random variable Y, as in the bivariate setting we consider the multivariate D_1-distance between the copula $A \in \mathscr{C}^{d+1}$ underlying $(\boldsymbol{X}, Y)$ and the $(d+1)$-dimensional product copula Π_{d+1}. More precisely, defining ζ^1 by $\zeta^1(\boldsymbol{X},Y) := \zeta^1(A) := 3D_1(A, \Pi_{d+1})$ or, equivalently,

$$\zeta^1(A) = 3 \int_{[0,1]} \int_{[0,1]^d} |K_A(\boldsymbol{x},[0,y]) - y| \, d\mu_{A^{1:d}}(\boldsymbol{x}) d\lambda(y),$$

whereby $A^{1:d}$ denotes the (marginal) copula of the random vector $\boldsymbol{X}$ yields a dependence measure. In fact, ζ^1 exhibits the following essential properties (for a proof we refer to Griessenberger et al. 2022):

1. $\zeta^1(\boldsymbol{X},Y) \in [0,1]$ (normalization).
2. $\zeta^1(\boldsymbol{X},Y) = 0$ if and only if Y and $\boldsymbol{X}$ are independent (independence).
3. $\zeta^1(\boldsymbol{X},Y) = 1$ if and only if Y is a function of $\boldsymbol{X}$ (complete dependence).
4. ζ^1 is invariant under strictly monotone transformations (scale-invariance).
5. ζ^1 fulfils the information gain inequality

$$\zeta^1(X_1,Y) \le \zeta^1((X_1,X_2),Y) \le \cdots \le \zeta^1((X_1,\ldots,X_d),Y).$$

Mimicking the bivariate case and working again with empirical checkerboard estimators (ECBEs) yields the following strong result (for a proof see Griessenberger et al. 2022):

Theorem 2. *Let $(\boldsymbol{X}_1,Y_1),(\boldsymbol{X}_2,Y_2),\ldots$ be a random sample from the $(d+1)$-dimensional random vector $(\boldsymbol{X},Y)$ and assume that $(\boldsymbol{X},Y)$ has continuous distribution function H and underlying copula $A \in \mathscr{C}^{d+1}$. Then setting the resolution $N(n) := \lfloor n^s \rfloor$ for some s fulfilling $0 < s < \frac{1}{2d}$*

$$\lim_{n\to\infty} \hat{\zeta}_n^1 := \lim_{n\to\infty} \zeta^1(\mathfrak{CB}_{N(n)}(\hat{E}_n)) = \zeta^1(A)$$

holds with probability 1.

Remark 2. The estimator $\hat{\zeta}_n^1$ works without any regularity assumptions on the copula A. Furthermore, various simulations demonstrated its good performance (see Griessenberger et al. 2022), as a consequence of which the estimator was implemented in the R-package ‘qmd’.

Acknowledgements. The first author gratefully acknowledges the support of the WISS 2025 project ‘IDA-lab Salzburg’ (20204-WISS/225/197-2019 & 20102-F1901166-KZP), the second author gratefully acknowledge the support of the Austrian FWF START project Y1102 ‘Successional Generation of Functional Multidiversity’.

References

Durante, F., Sempi, C.: Principles of Copula Theory. CRC Press, New York (2016)

Griessenberger, F., Junker, R.R., Trutschnig, W.: On a multivariate copula-based dependence measure and its estimation. Electron. J. Stat. **16**(1), 2206–2251 (2022)

Janssen, P., Swanepoel, J., Ververbeke, N.: Large sample behaviour of the Bernstein estimator. J. Stat. Plann. Infer. **142**, 1189–1197 (2012)

Junker, R.R., Griessenberger, F., Trutschnig, W.: Estimating scale-invariant directed dependence of bivariate distributions. Comput. Stat. Data Anal. **153**, 107058 (2021)

Kallenberg, O.: Foundations of Modern Probability. Springer, Heidelberg (1997)

Klenke, A.: Wahrscheinlichkeitstheorie. Springer, Heidelberg (2008). https://doi.org/10.1007/978-3-540-77571-3

Li, H., Scarsini, M., Shaked, M.: Linkages: a tool for the construction of multivariate distributions with given nonoverlapping multivariate marginals. J. Multivar. Anal. **56**(1), 20–41 (1996)

Li, X., Mikusinski, P., Taylor, M.D.: Strong approximation of copulas. J. Math. Anal. Appl. **255**, 608–623 (1998)

Nelsen, R.B.: An Introduction to Copulas. Springer, New York (2006). https://doi.org/10.1007/0-387-28678-0

Schweizer, B., Wolff, E.F.: On nonparametric measures of dependence for random variables. Ann. Stat. **9**(4), 879–885 (1981)

Siburg, K.F., Stoimenov, P.F.: A measure of mutual complete dependence. Metrika **71**(2), 239–251 (2010)

Trutschnig, W.: On a strong metric on the space of copulas and its induced dependence measure. J. Math. Anal. Appl. **384**, 690–705 (2011)

Wiedermann, W., Kim, D., Sungur, E.A., von Eye, A.: Direction Dependence in Statistical Modeling: Methods of Analysis. Wiley, Hoboken (2020)

Explaining Cautious Random Forests via Counterfactuals

Haifei Zhang[1(✉)], Benjamin Quost[1], and Marie-Hélène Masson[2]

[1] UMR CNRS 7253 Heudiasyc, Université de Technologie de Compiègne, 60200 Compiègne, France
{haifei.zhang,benjamin.quost}@hds.utc.fr

[2] UMR CNRS 7253 Heudiasyc, Université de Picardie Jules Verne, IUT de l'Oise, 60000 Beauvais, France
mylene.masson@hds.utc.fr

Abstract. Cautious random forests are designed to make indeterminate decisions when tree outputs are conflicting. Since indeterminacy has a cost, it seems desirable to highlight why a precise decision could not be made for an instance, or which minimal modifications can be made to the instance so that the decision becomes a single class. In this paper, we apply an efficient extractor to generate determinate counterfactual examples of different classes, which are used to explain indeterminacy. We evaluate the efficiency of our strategy on different datasets and we illustrate it on two simple case studies involving both tabular and image data.

1 Introduction

Machine learning models now achieve high performances in many fields such as medical diagnosis, recommendation systems, image and speech recognition. The outputs of these models are traditionally precise: in a classification problem, they consist in a single class for a given instance. However, when training data are scarce, or when mistakes have a very high cost, cautious classifiers can alternatively be used to provide set-valued decisions rather than single classes and thus control the risk. Cautious random forests (CRF) (Zhang et al. 2021) are one of those classifiers. A CRF combines the classical random forest (RF) strategy (Breiman 2001), the Imprecise Dirichlet Model (IDM) (Walley 1996) and the theory of belief functions (Shafer 1976). The major difference with a classical RF is that an indeterminate decision can be reached in presence of both epistemic uncertainty (when the tree outputs are based on scarce information) and aleatoric uncertainty (the conflict between these outputs is high), which typically happens near decision boundaries. Making imprecise predictions has a cost, since indeterminacy must be resolved via further analysis. Therefore, it seems crucial to understand what led to an undetermined decision, and what could be done to change it into a determinate one. Such questions fall under the emerging topic of explainable machine learning (Molnar 2019). In this paper,

L. A. García-Escudero et al. (Eds.): SMPS 2022, AISC 1433, pp. 390–397, 2023.
https://doi.org/10.1007/978-3-031-15509-3_51

we address the second problem using counterfactual explanations (Wachter et al. 2017), which provide clear and intuitive explanations for turning an original instance x into a modified one x' in a minimal way, so that $f(x')$ corresponds to a desired prediction $y' \neq f(x)$. Our approach is inspired by the one proposed by Blanchart (2021), specifically developed for tree ensembles. Our contributions consist in improving the efficiency of the procedure, and in exploiting counterfactuals for explaining indeterminate CRF outputs. Their benefits for explaining indeterminacy are illustrated by experimental results, in particular via two case studies.

The paper is structured as follows. In Sect. 2, we recall general background knowledge on cautious random forests and counterfactual explanations. Their application to explaining indeterminacy are discussed in Sect. 3. Section 4 details the experiments and discusses the results. A short conclusion is drawn in Sect. 5.

2 Background

2.1 Cautious Random Forests

Cautious random forests (CRF) have been proposed as an alternative to precise random forests, so as to make decisions from scarce data. In a binary classification problem, for each test instance x, each tree t in the forest provide pieces of evidence about its actual class $Y \in \{1, 0\}$ in the form of lower and upper bounds $\underline{p}_1^t(x)$ and $\overline{p}_1^t(x)$ over the posterior probability $\Pr(Y = 1|x)$. These bounds are obtained using the Imprecise Dirichlet Model, and reflect the estimation uncertainty due to the lack of training data. These intervals can be pooled using the theory of belief functions, by computing the belief and plausibility $bel(Y = 1|x)$ and $pl(Y = 1|x)$, which can then be used in a cautious decision-making process such as interval dominance, possibly resulting in indeterminate decisions (Zhang et al. 2021).

As can be seen in Fig. 1, imprecision occurs principally around the decision boundaries, where tree leaves are prone to contain few instances and tree outputs are often conflicting with each other.

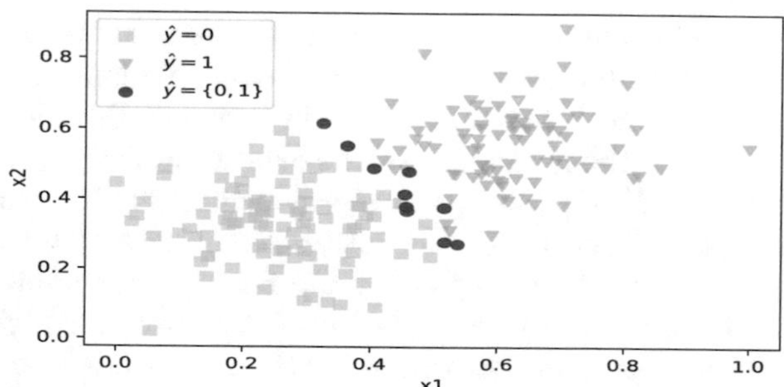

Fig. 1. Predictions of a CRF with $T = 100$ trees

2.2 Counterfactual-Based Explanations

Explainable artificial intelligence is an emerging field of artificial intelligence that helps humans to understand the outputs of machine learning algorithms. Counterfactuals (CF) (Wachter et al. 2017) are local and example-based explanations, which can be seen as minimal alterations of an original query instance x leading to different decisions. Such examples can either be queried for in the training set, or synthesized. Given a classifier f, a query instance $x \in \mathscr{X}$, and a desired prediction label $y' \in \mathscr{Y}$, we aim at efficiently computing x' by solving

$$x' = \arg\min_{z \in \mathscr{X}} \text{dist}(x, z) \text{ s.t. } f(z) = y', \tag{1}$$

where dist is a suitable distance measure (e.g., Euclidean) between instances.

Many methods have been designed to solve (1) exactly or approximately, such as selecting the most similar sample in the training set, or creating a virtual sample by optimizing a loss function (for differentiable models), searching CFs by a heuristic strategy, or approximating the model at hand (e.g. by a decision tree) so as to simplify the search of a CF (Guidotti 2022). Besides the inherent complexity of counterfactual generation algorithms, additional challenges make designing an actionable decision process difficult (Verma et al. 2020), such as protecting some attributes or immutable features (such as gender, ethnicity, etc.), restricting the number of modified features (sparsity), and generating plausible (or realistic) CFs.

3 Explaining Imprecision Using Counterfactuals

This work proposes to apply CF explanations to cautious binary classification: given an instance x with indeterminate prediction $f(x) = \{0, 1\}$, we want to identify its two minimal modifications x^1 and x^0 such that $f(x^1) = \{1\}$ and $f(x^0) = \{0\}$. These two synthetic examples will not only reveal the features which should be modified to remove indeterminacy, but also to which extent they should be modified so as to reach a precise decision.

3.1 Extracting Determinate Counterfactuals

We propose to extract counterfactuals using the geometrical method of Blanchart (2021), which explicitly computes the smallest decision regions of a tree ensemble model, and generates the closest virtual CF in terms of Euclidean distance. A random forest separates the input space $\mathscr{X}$ into decision regions, each of which is itself the intersection of T decision regions provided by the T trees in the forest. Computing the optimal and exact CF with desired class y' for an instance x requires to explore all decision regions of the forest, the complexity of which is exponential. This exhaustive search is thus intractable for high-dimensional data or forests with deep trees. Blanchart (2021) proposed a branch-and-bound strategy to search for CFs only around x, by ignoring decision regions R such that $d(x, R) > d_{\max}$, with $d_{\max}$ (initialized to positive infinity) the distance to the current counterfactual found during the extraction procedure. We refer the reader to this reference for further information.

3.2 Region Filtering and Counterfactual Initialization

Given the complexity of determining a CF x' for a given query instance x with indeterminate decision $f(x) = \{1, 0\}$, we propose two amenities to speed up the procedure. These preliminary steps make it possible to drastically reduce the complexity of the search, as will be shown in Sect. 4.

1. Following a suggestion of Blanchart (2021), in presence of protected features, we filter out the regions that do not correspond to the same protected values as in x.
2. We use an alternative approach to "initialize" the CF search, i.e. to compute the first CF based upon which the initial distance threshold $d_{\max}$ will be determined, rather than positive infinity.

Embedding the filtering step mentioned above in any branch-and-bound procedure is straightforward. The initialization step is critical, since it determines the threshold $d_{\max}$ and therefore the number of regions to be explored.

The Minimum Observable (MO) approach, which selects the nearest instance x' with desired class y' in the training set, is commonly used for this purpose. However, in scarce regions of the input space, the distance between the query point and the closest training CF may be large. Even worse, when several protected features (PF) are considered, the approach may not give an initial CF which meets the requirements. Therefore, we propose a new strategy to find an initial virtual CF, which we call One-dimensional Change CounterFactual (OCCF). In a nutshell, for a given x, OCCF aims at solving Eq. (1) with the additional constraint that x and z differ by one feature only. This problem can be quickly solved using individual conditional expectation (ICE) plots (Goldstein et al. 2015), which estimate how the probability (or the decision) $f(x)$ of a classifier varies according to a modification in x when all other values are fixed.

Note that in a random forest, for a query instance x, we need only consider a finite number of modifications of the value of a mutable feature $\mathscr{X}_d$, defined by the split values for this feature obtained across all trees. However, an OCCF may still not exist when several features are protected, although the experiments suggest that this is much less likely than with the MO approach. In this case, some constraints should be relaxed by "unprotecting" some immutable features.

4 Experimental Results

4.1 Counterfactual Extraction Efficiency

In this experiment, we evaluate the efficiency of the proposed CF extraction procedure on four datasets. The number of trees in the ensemble is 50 for all datasets, and the maximal depth of the trees are respectively 10, 8, 7, and 14. The efficiency is evaluated in three ways: the number of regions to explore after filtering by different initialization approaches, the distance between the query point and the initial CF, and the elapsed time to extract all CFs. Note that

Table 1. Average number of leaves to explore

Dataset	Original	PF	MO	PF+MO	OCCF	PF+OCCF
Compas	7236	2226.86	849.87	732.54	418.36	**305.36**
Heloc	8784	——	5268.12	——	**106.52**	——
Pima	2522	1081.27	1007.43	719.53	133.90	**128.77**
Wine	8949	——	3277.05	——	**761.38**	——

Table 2. Average distance from the query example to the initial counterfactual (left), and average elapsed time for searching the final counterfactual (right)

	Initial CF distance				CF Searching Time (s)			
Dataset	MO	PF+MO	OCCF	PF+OCCF	MO	PF+MO	OCCF	PF+OCCF
Compas	0.078	0.134	**0.040**	0.058	1.091	0.421	0.580	**0.284**
Heloc	0.273	——	**0.011**	——	4.570	——	**1.274**	——
Pima	0.215	0.273	**0.034**	0.041	5.600	4.991	3.589	**3.277**
Wine	0.192	——	**0.060**	——	5.745	——	**4.667**	——

Compas and Pima have one and four protected features, respectively, whereas no protected features were considered for Heloc and Wine.

Tables 1 and 2 indicate that exploiting the protected features can help to reduce the amount of regions to explore, since it restricts searching the CFs to a feature subspace. Our proposed OCCF initialization can generate initial CFs which are much closer to the query point x compared to MO: as a consequence, we may filter out many more regions, and thus reduce the amount of time needed to reach the solution.

4.2 Case Studies

Case 1: Pima

The Pima dataset can be used to predict whether a patient has diabetes or not, based on various measurements: Pregnancies (PGs): number of times pregnant; Glucose; Blood Pressure (BP); Skin Thickness (ST); Insulin: 2-Hour serum insulin (mu U/ml); BMI: body mass index; Diabetes Pedigree Function (DPF); Age. The class is $y = 0$ for a non-diabetic, $y = 1$ for a diabetic. Here, Age, number of pregnancies, DPF values, and Skin Thickness are difficult to change (considered as protected features), while Glucose, Insulin, BMI, and blood pressure are actionable (mutable) features. We chose Pima as an example because, as a medical dataset, explainability may have a great practical interest.

Table 3. Examples of counterfactual explanations from Pima dataset

	PGs	Glucose	BP	ST	Insulin	BMI	DPF	Age
x_1	0	165	90	33	680	52.3	0.427	23
x_1^0	0	**154.5**↓	90	33	680	**47.7**↓	0.427	23
x_1^1	0	**165.5**↑	90	33	680	52.3	0.427	23
x_2	1	122	90	51	220	49.7	0.325	31
x_2^0	1	**121.5**↓	90	51	**128**↓	**49.05**↓	0.325	31
x_2^1	1	**126.5**↑	90	51	220	49.7	0.325	31

In Table 3, two examples are provided for illustration. The query instance x_1 corresponds to a non-diabetic patient. First, note that x_1 is close to being classified as diabetic since the CF x_1^1 of this class is very close. This demonstrates that the cautious random forest can help managing the uncertainty arising from scarce data, by detecting instances for which the decision is uncertain and providing insights about their actual labels. Second, the non-diabetic CF x_1^0 suggests a possible way to maintain a healthy condition, i.e. reducing BMI and the Glucose level. The query instance x_2 corresponds to a diabetic patient. The indeterminacy comes from the Glucose feature, since we can get a correct prediction (diabetic) by only modifying its value. On the other hand, to obtain the non-diabetic CF x_2^0, an important decrease of Insulin is needed, which is coherent with the fact that high 2-Hour serum insulin levels are common for type-II diabetic patients.

Case 2: MNIST

MNIST is a large database of handwritten numbers containing about 60,000 training cases and 10,000 test cases. In our experiment, numbers of 4 and of 9 were selected and 40 principal components of the original data were extracted to train a CRF consisting of 50 trees of depth 10. We generated CFs, which we required to belong to the objective class with a belief of at least 0.75, so as to ensure that the instance is credible after applying the inverse PCA transformation. Generating CFs of a query instance helps understanding which parts of the image are responsible for the indeterminacy of the decision. This point is illustrated using two instances drawn in Fig. 2. We can see how the two indeterminate examples (center) should be modified to be determinately classified as a "4" or as "9", and that these modifications make sense.

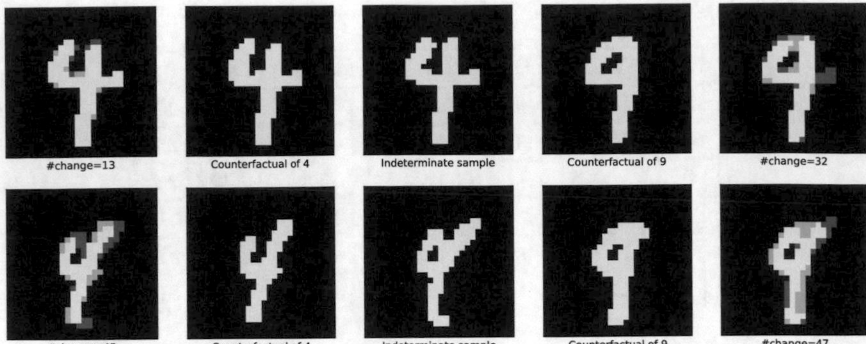

Fig. 2. Examples of indeterminate numbers (center) and corresponding counterfactuals of class 4 (left) and 9 (right). Left- and right-most images display pixels to be added (green) and to be deleted (blue) in order to obtain the counterfactual

5 Conclusion

In this paper, we have proposed a procedure to extract CFs of indeterminate instances, i.e. for which no precise decision could be made, so as to interpret and explain the indeterminacy of the classifier. The algorithm presented in this paper is specific to cautious random forests. It is based on an algorithm proposed in the case of precise CF extraction. Our modifications make it possible to filter the regions of the input space to be explored and to generate CFs closer to the query instance, thus speeding up the extraction process. This increased efficiency, as well as the usefulness of our approach for explaining the indecision of the classifier, has been demonstrated on several experiments. In future works, we will investigate how CFs can be used to estimate the importance of features and to identify regions of significant uncertainty in the feature space. We also plan to use CFs in an active learning process to reduce the indeterminacy of cautious classifiers.

References

Blanchart, P.: An exact counterfactual-example-based approach to tree-ensemble models interpretability. arXiv preprint arXiv:2105.14820 (2021). https://doi.org/10.48550/arXiv.2105.14820

Breiman, L.: Random forests. Mach. Learn. **45**(1), 5–32 (2001)

Goldstein, A., Kapelner, A., Bleich, J., Pitkin, E.: Peeking inside the black box: visualizing statistical learning with plots of individual conditional expectation. J. Comput. Graph. Stat. **24**(1), 44–65 (2015)

Guidotti, R.: Counterfactual explanations and how to find them: literature review and benchmarking. Data Min. Knowl. Disc. (2022, in press). https://doi.org/10.1007/s10618-022-00831-6

Molnar, C.: Interpretable Machine Learning. A Guide for Making Black Box Models Explainable (2019). https://christophm.github.io/interpretable-ml-book/

Shafer, G.: A Mathematical Theory of Evidence. Princeton University Press, Princeton (1976)

Verma, S., Dickerson, J., Hines, K.: Counterfactual explanations for machine learning: a review. In: NeurIPS 2020Workshop. ML Retrospectives, Surveys & Meta-Analyses (ML-RSA). arXiv preprint arXiv:2010.10596 (2021). https://doi.org/10.48550/arXiv.2010.10596

Wachter, S., Mittelstadt, B., Russell, C.: Counterfactual explanations without opening the black box: automated decisions and the GDPR. Harv. J. Law Technol. **31**, 841 (2017)

Walley, P.: Inferences from multinomial data: learning about a bag of marbles. J. Roy. Stat. Soc.: Ser. B (Methodol.) **58**, 3–34 (1996)

Zhang, H., Quost, B., Masson, M.H.: Cautious random forests: a new decision strategy and some experiments. In: Proceedings of the Twelfth International Symposium on Imprecise Probability: Theories and Applications. Proceedings of Machine Learning Research, vol. 147, pp. 369–372 (2021)

Remarks on Martingale Representation Theorem for Set-Valued Martingales

Jinping Zhang[1(✉)] and Kouji Yano[2]

[1] School of Mathematics and Physics, North China Electric Power University, Huilongguan Street, Changping District, Beijing 102206, People's Republic of China
zhangjinping@ncepu.edu.cn

[2] Graduate School of Science, Kyoto University, Kyoto 606-8501, Japan
kyano@math.kyoto-u.ac.jp

Abstract. It is proved that the martingale representation theorem for set-valued martingales proposed by Kisielewicz (2014b) holds only in the degenerate case. A revised representation theorem for a special kind of non-degenerate set-valued martingales is presented.

1 Introduction

Set-valued function has been received much attention. A set is called *degenerate* if it is a singleton and otherwise *non-degenerate*. A difficulty to handle non-degenerate set-valued functions lies in the fact that the power set of a set is not linear. We define the operations of intervals as $\lambda[a, b] := \{\lambda h : a \leq h \leq b\}$ and $[a, b] + [c, d] := [a + c, b + d]$, then $(\lambda + \eta)[a, b] \neq \lambda[a, b] + \eta[a, b]$ if λ and η have different signs. For the expectation, which will be defined by (1), two integrable real random variables f, g with $f \leq g$ a.s. satisfy $E(\lambda[f, g]) = \lambda E([f, g])$, while $E(f[a, b]) = E(f)[a, b]$ may not hold. We have to be very careful to deal with set-valued variables.

Set-valued (super/sub)martingales have been defined by Hiai and Umegaki (1977). After that, there have been many studies on set-valued martingales. For example, Zhang et al. (2009) proved that a set-valued stochastic integrals with respect to Brownian motion given by Jung and Kim (2003) is a set-valued submartingale. For a non-degenerate set-valued stochastic integral w.r.t. the compensated Poisson measure, Zhang et al. (2021) proved that it is not a set-valued martingale but a set-valued submartingle. For one w.r.t. Brownian motion, we shall show, in a very simple way, that it is not a set-valued martingale.

The martingale representation theorem for point-valued martingale plays an important role in classical stochastic analysis, see e.g. Theorem (3.4) of Revuz and Yor (2005). For set-valued case, is there a similar representation theorem? It is our task in this short paper to discuss this question.

When taking as the underlying space the r-dimensional Euclidean space $\mathbb{R}^r$, Kisielewicz (2014b) proposed a representation theorem for set-valued martingale as follows:

L. A. García-Escudero et al. (Eds.): SMPS 2022, AISC 1433, pp. 398–405, 2023.
https://doi.org/10.1007/978-3-031-15509-3_52

Theorem 1. *For every set-valued* $\mathbb{A}$*-martingale* $F = (F_t)_{t\geq 0}$ *with* $F_0 = \{0\}$ *defined on a probability space* $(\Omega, \mathscr{A}, P)$ *equipped with a* d*-dimensional Brownian motion* $B = (B_t)_{t\geq 0}$ *and its augmented natural filtration* $\mathbb{A} = (\mathscr{A}_t)_{t\geq 0}$, *there exists a set* $\mathscr{G} \in \mathscr{P}(\mathscr{L}^2_{\mathbb{A}})$ *such that* $F_t = \int_0^t \mathscr{G}\, dB_\tau$ *a.s. for every* $t \geq 0$.

Here $\mathscr{L}^2_{\mathbb{A}}$ denotes the family of all $\mathbb{A}$-nonanticipative $\mathbb{R}^{r\times d}$-valued integrable processes on $[0, \infty)$ and $\mathscr{P}(\mathscr{L}^2_{\mathbb{A}})$ denotes the power set of $\mathscr{L}^2_{\mathbb{A}}$. For $\mathscr{G} \in \mathscr{P}(\mathscr{L}^2_{\mathbb{A}})$, the generalized stochastic integral $\int_0^t \mathscr{G}\, dB_\tau$ was defined in Kisielewicz (2014a) as a set-valued random variable for each t.

We will see in Example 2 that Theorem 1 imposes so strong assumption $F_0 = \{0\}$ as to exclude non-degenerate set-valued martingales. We will propose in Theorem 2 a revised representation theorem.

This paper is organized as follows. In Sect. 2 we recall several notations and preliminary facts about set-valued processes. In Sect. 3 we develop our main theorem. We shall prove that Theorem 1 does not hold in the non-degenerate case. A revised martingale representation will be given.

2 Notations and Preliminaries

We consider a complete non-atomic probability space $(\Omega, \mathscr{A}, P)$ equipped with a filtration $\mathbb{A} = (\mathscr{A}_t)_{t\in[0,T]}$ for $T > 0$ satisfying the usual condition, that is, $\mathscr{A}_0$ includes all P-null sets in $\mathscr{A}$ and the filtration is non-decreasing and right continuous. We denote by $(\mathfrak{X}, \|\cdot\|)$ a separable Banach space equipped with the Borel sigma-algebra $\mathscr{B}(\mathfrak{X})$. Let $K(\mathfrak{X})$ (resp. $K_c(\mathfrak{X})$, $K_{bc}(\mathfrak{X})$, $K_{kc}(\mathfrak{X})$) denote the family of all nonempty closed (resp. closed convex, bounded closed convex, compact convex) subsets of $\mathfrak{X}$. For a subset C of $\mathfrak{X}$, we write $\|C\| := \sup_{x\in C}\|x\|$. Let $L^p(\Omega, \mathscr{A}, P; \mathfrak{X}) = L^p(\Omega; \mathfrak{X})$ $(1 \leq p < \infty)$ denote the set of all $\mathfrak{X}$-valued Borel functions $f : \Omega \to \mathfrak{X}$ such that $\|f\|_p := \{\int_\Omega \|f(\omega)\|^p dP(\omega)\}^{1/p} < \infty$.

A mapping $F : \Omega \to K(\mathfrak{X})$ is called a *set-valued random variable* (or a *random set*) if $\{\omega \in \Omega : F(\omega) \cap O \neq \emptyset\} \in \mathscr{A}$ for all open set $O \subset \mathfrak{X}$. For $\mathfrak{K} = K(\mathfrak{X})$, $K_c(\mathfrak{X})$, $K_{bc}(\mathfrak{X})$ and $K_{kc}(\mathfrak{X})$, we write $\mathscr{M}(\Omega, \mathscr{A}, P; \mathfrak{K}) = \mathscr{M}(\Omega; \mathfrak{K})$ for the family of all $\mathfrak{K}$-valued random variables. For $F \in \mathscr{M}(\Omega, K(\mathfrak{X}))$ and $1 \leq p < \infty$, the family of all L^p-integrable selections is denoted by

$$S^p_F(\mathscr{A}) = S^p_F := \{f \in L^p(\Omega, \mathscr{A}, P; \mathfrak{X}) : f(\omega) \in F(\omega) \text{ a.s.}\}.$$

A set-valued random variable F is called *integrable* if S^1_F is nonempty. F is called L^p*-integrably bounded* if there exists $h \in L^p(\Omega, \mathscr{A}, P; \mathbb{R})$ such that $\|F(\omega)\| \leq h(\omega)$, i.e. [for all $x \in F(\omega)$ we have $\|x\| \leq h(\omega)$], almost surely. For $\mathfrak{K} = K(\mathfrak{X})$, $K_c(\mathfrak{X})$, $K_{bc}(\mathfrak{X})$ and $K_{kc}(\mathfrak{X})$, we write $L^p(\Omega, \mathscr{A}, P; \mathfrak{K}) = L^p(\Omega; \mathfrak{K})$ for the family of all $\mathfrak{K}$-valued L^p-integrably bounded random variables.

A set Γ of measurable functions $f : \Omega \to \mathfrak{X}$ is called *decomposable* with respect to the σ-algebra $\mathscr{A}$ if for any finite $\mathscr{A}$-measurable partition $A_1, \ldots, A_n$ of Ω and for any $f_1, \ldots, f_n \in \Gamma$ we have $f_1 1_{A_1} + \cdots + f_n 1_{A_n} \in \Gamma$. By Hiai and Umegaki (1977), we know that a nonempty subset Γ of $L^p(\Omega, \mathscr{A}, P; \mathfrak{X})$ is represented as

$\Gamma = S_F^p$ for some p-integrable set-valued random variable F iff Γ is decomposable with respect to $\mathscr{A}$. Note also that $F = G$ a.s. iff $S_F^p = S_G^p$. Therefore, in order to study F, we only have to study S_F^p.

The following *Castaing representation* is due to Hiai and Umegaki (1977):

Lemma 1. *For a p-integrable set-valued random variable $F \in \mathscr{M}(\Omega, \mathscr{A}, P; K(\mathfrak{X}))$, there exists a sequence $\{f^i\}_{i=1}^{\infty} \subset S_F^p$ such that $F(\omega) = \mathrm{cl}\{f^i(\omega) : i \in \mathbb{N}\}$ for all $\omega \in \Omega$, where the closure is taken in $\mathfrak{X}$. In addition, $S_F^p = \overline{\mathrm{de}}\{f^i : i \in \mathbb{N}\}$, where the $\overline{\mathrm{de}}$ denotes the decomposable closure of the sequence $\{f^i : i \in \mathbb{N}\}$ in the space $L^p(\Omega; \mathfrak{X})$.*

The expectation of a set-valued r.v. F was defined by Aumann (1965) as

$$E(F) := \{E(f) : f \in S_F^1\}, \tag{1}$$

where $E(f) = \int_\Omega f dP$ denotes the Bochner integral. As the probability space is non-atomic, we have $\mathrm{cl}\{E(F)\}$ is convex. The expectation $E(F)$ is closed in $\mathfrak{X}$ under some conditions; for example, in the case where the Banach space $\mathfrak{X}$ has the Radon–Nikodym property and $F \in L^1(\Omega, K_{kc}(\mathfrak{X}))$, and in the case where the space $\mathfrak{X}$ is reflexive and $F \in L^1(\Omega, K_c(\mathfrak{X}))$ (Hiai and Umegaki 1977). It is well-known that finite dimensional spaces and $L^p(\Omega; \mathfrak{X})$ for $1 < p < \infty$ are reflexive.

A set-valued process $F = \{F_t : t \in [0, T]\}$ is called *uniformly integrable* if the real-valued process $\|F\| = \{\|F_t\| : t \in [0, T]\}$ is uniformly integrable, and is called $\mathbb{A}$-*adapted* if $F_t \in \mathscr{M}(\Omega, \mathscr{A}_t, P; K(\mathfrak{X}))$ for each t. A point-valued process $f = \{f_t : t \in [0, T]\}$ is called a *martingale selection* of F if f is an $\mathfrak{X}$-valued $\mathbb{A}$-martingale such that $f_t(\omega) \in F_t(\omega)$ a.s. for each t. The family of all martingale selections of F is denoted by $MS(F)$. The set-valued *conditional expectation* $G = E[F_t|\mathscr{A}_s]$ is defined by $S_G^1(\mathscr{A}_s) = \mathrm{cl}\{E[f_t|\mathscr{A}_s] : f_t \in S_{F_t}^1(\mathscr{A}_t)\}$ for $0 \le s \le t$, where the closure is taken in $L^1(\Omega; \mathfrak{X})$.

Definition 1. *An integrable convex set-valued $\mathbb{A}$-adapted process $F = \{F_t : t \in [0, T]\}$ is called a* set-valued $\mathbb{A}$-martingale *if, for any $0 \le s \le t$ it holds that $E(F_t|\mathscr{A}_s) = F_s$ in the sense that $S^1_{E(F_t|\mathscr{A}_s)}(\mathscr{A}_s) = S^1_{F_s}(\mathscr{A}_s)$. It is called a* set-valued submartingale *(resp.* supermartingale*) if, for any $0 \le s \le t$, it holds that $E(F_t|\mathscr{A}_s) \supset F_s$ (resp. $E[F_t|\mathscr{A}_s] \subset F_s$) in the sense that $S^1_{E(F_t|\mathscr{A}_s)}(\mathscr{A}_s) \supset S^1_{F_s}(\mathscr{A}_s)$ (resp. $S^1_{E(F_t|\mathscr{A}_s)}(\mathscr{A}_s) \subset S^1_{F_s}(\mathscr{A}_s)$).*

It is obvious that a set-valued martingale satisfies $\mathrm{cl}\{E(F_t)\} = \mathrm{cl}\{E(F_0)\}$ for all $0 \le t \le T$. In the case $\mathfrak{X} = \mathbb{R}$, it is known (Theorem 3.1.1 of Zhang 2009) that $\{F_t = [f_t, g_t] : t \in [0, T]\}$ is an interval-valued $\mathbb{A}$-martingale iff both of the endpoints are $\mathbb{R}$-valued $\mathbb{A}$-martingales.

3 Main Result

In the following, we write $\mathscr{L}^2_{\mathbb{A}}(\mathfrak{K})$ for the family of all $\mathfrak{K}$-valued square-integrable $\mathbb{A}$-nonanticipative processes.

Theorem 2. *Take $\mathfrak{X} = \mathbb{R}^r$. Let $(B_t)_{0\le t\le T}$ be the d-dimensional Brownian motion and let $\mathbb{A} = (\mathscr{A}_t)_{t\in[0,T]}$ denote its augmented natural filtration. Let $\{M_t : t \in [0,T]\}$ be a $K_c(\mathbb{R}^r)$-valued square-integrable $\mathbb{A}$-martingale. Then the following assertions are equivalent:*

(i) There exists a process $G = (G_t)_{0\le t\le T} \in \mathscr{L}^2_{\mathbb{A}}(K(\mathbb{R}^{r\times d}))$ such that

$$M_t = E(M_0) + \int_0^t G_s dB_s \quad a.s.\ for\ each\ t. \tag{2}$$

(ii) There exist a process $g = \{g_t : t \in [0,T]\} \in \mathscr{L}^2_{\mathbb{A}}(\mathbb{R}^{r\times d})$ and a bounded closed convex subset $C \subset \mathbb{R}^r$ such that

$$M_t = C + \left\{\int_0^t g_s dB_s\right\} \quad a.s.\ for\ each\ t.$$

(iii) There exists a sequence $\{f^n\}_{n=1}^{\infty}$ of $\mathbb{R}^r$-valued square-integrable $\mathbb{A}$-martingales such that $f_t^i - f_t^j$ is non-random and independent of t for any $i, j \ge 1$ and

$$M_t = \mathrm{cl}\left\{f_t^n : n \in \mathbb{N}\right\} \quad a.s.\ for\ each\ t.$$

Remark 1. For the M-type 2 Banach space $\mathfrak{X}$, the set-valued stochastic integral $I_t(G) = \int_0^t G_s dB_s$ w.r.t. 1-dimensional Brownian motion was defined in Definition 4.2 of Zhang et al. (2009), which is a $K(\mathfrak{X})$-valued random variable for each t. Here in (2), the integral is defined in the same way as that of Zhang et al. (2009), i.e., $I_t(G)$ is determined by

$$S^2_{I_t(G)}(\mathscr{A}_t) := \overline{\mathrm{de}}\left\{\int_0^t g_s dB_s : g \text{ is the integrable selection of } G\right\}.$$

Before proving Theorem 2, firstly, we give a result about expectation of set-valued random variable.

Lemma 2. *Let $\mathfrak{X}$ be a separable Banach space. For any set-valued random variable $F \in L^1(\Omega, K(\mathfrak{X}))$ and any deterministic element $a \in \mathfrak{X}$, it holds that $E(F) = \{a\}$ iff F degenerates to a random singleton $\{f\}$ with $E(f) = a$.*

Proof. The sufficiency is obvious. For the necessity, take $f, g \in S^1_F(\mathscr{A})$ and $A \in \mathscr{A}$ arbitrarily. We have $E(f) = E(g) = E(f1_A + g1_{\Omega\setminus A}) = a$ since $E(F) = \{a\}$. Hence we obtain $E((f - g)1_A) = 0$, which yields $f = g$ a.s. Thus $S_F(\mathscr{A})$ contains only one element, i.e. F degenerates to a random singleton. □

Lemma 2 is sometimes convenient for disproving martingale property of set-valued processes. There are some examples.

Example 1. Let $\mathfrak{X}$ be a separable Banach space. Assume $G \in \mathscr{L}^2_{\mathbb{A}}(K_c(\mathfrak{X}))$ is non-degenerate. Then, for the set-valued stochastic integral $I_t(G) = \int_0^t G_s dB_s$, the process $\{I_t(G) : t \in [0,T]\}$ is not a set-valued martingale. In fact, we know $I_0(G) = \{0\}$ a.s. so that $E(I_0(G)) = \{0\}$. Since there exists $t > 0$ such that $I_t(G)$ is non-degenerate, Lemma 2 shows that $E(I_t(G)) \neq \{0\} = E(I_0(G))$, which yields that $\{I_t(G) : t \in [0,T]\}$ is not a set-valued martingale. Note that, by Zhang et al. (2009), we know that it is a set-valued submartingale.

Example 2. Now let us concentrate on the martingale representation Theorem 1. By Lemma 2, we can easily see that there is no non-degenerate set-valued martingale such that its expectation is a singleton. Particularly, there is no non-degenerate set-valued martingale with expectation zero. From this point of view, the martingale representation Theorem 1 does not hold for any non-degenerate set-valued martingale, since $F_0 = \{0\}$ a.s. In fact, if the subset $\mathscr{G}$ in Theorem 1 is not a singleton, then the stochastic integral $I_t(\mathscr{G}) := \int_0^t \mathscr{G} dB_\tau$ is non-degenerate, and hence $\{I_t(\mathscr{G}) : t \in [0,T]\}$ is not a non-degenerate set-valued martingale, since $I_0(\mathscr{G}) = \{0\}$ a.s. Note that, if the set $\mathscr{G}$ only includes one $\mathbb{R}^{r\times d}$-valued integrable process, then Theorem 1 becomes the classical martingale representation theorem.

Example 3. Let us give a concrete example of a stochastic integral which is not a set-valued martingale. Let $\{B_t : t \geq 0\}$ be a 1-dimensional Brownian motion. We denote by $\mathscr{G}$ the set of constant functions taking λ for $\lambda \in [0,1]$. For each $t \in (0,T]$, the stochastic integral $\int_0^t \mathscr{G} dB_\tau$ is determined by

$$S^1_{\int_0^t \mathscr{G} dB_\tau}(\mathscr{A}_t) = \overline{\mathrm{de}}\left\{\int_0^t \lambda dB_\tau : \lambda \in [0,1]\right\} = \overline{\mathrm{de}}\left\{\lambda B_t : \lambda \in [0,1]\right\}.$$

Obviously, $S^1_{\int_0^t \mathscr{G} dB_\tau}(\mathscr{A}_t)$ is convex. For any $f = \sum_{i=1}^n \lambda_i B_t 1_{A_i}$ with some $\lambda_1, \ldots,$ $\lambda_n \in [0,1]$ and some finite $\mathscr{A}_t$-measurable partition $\{A_1, \ldots, A_n\}$ of Ω, we have

$$E\left[\left|\sum_{i=1}^n \lambda_i B_t 1_{A_i}\right|\right] \leq E\left[|B_t| \sum_{i=1}^n 1_{A_i}\right] = E[|B_t|] \leq \sqrt{t} < \infty.$$

Then $S^1_{\int_0^t \mathscr{G} dB_\tau}(\mathscr{A}_t)$ is bounded in L^1, which shows that $\int_0^t \mathscr{G} dB_\tau$ is convex and L^1-integrably bounded. (In fact, the L^2-integrable boundedness can be proved easily). By the Castaing representation, we obtain that

$$\int_0^t \mathscr{G} dB_\tau(\omega) = [\min\{0, B_t(\omega)\}, \max\{0, B_t(\omega)\}] =: I_t$$

for each $t \in (0,T]$. Note that $\{I_t : t \in (0,T]\}$ is not an interval-valued martingale.

Remark 2. For discrete time set-valued martingales, Ezzaki and Tahri (2019) (in Corollary 3.8) proposed that in a separable RNP (Radon–Nikodym Property) Banach space $\mathfrak{X}$, a sequence $\{F_n\}_{n=1}^\infty$ is a uniformly integrable $K_{bc}(\mathfrak{X})$-valued martingale if and only if it admits a Castaing representation of regular martingale selections such that $\liminf \int_\Omega \|F_n\| dP < \infty$. Unfortunately, the 'if' part may not hold. For example, let $\{f_n\}_{n=1}^\infty$ be a uniformly integrable $\mathbb{R}$-valued martingale. For each n, we assume that the value of f_n changes its signs with positive probability. Set $M_n = \mathrm{cl}\{\lambda f_n + 2(1-\lambda) f_n; \lambda \in [0,1] \cap \mathbb{Q}\}$. Then $M_n = [\min\{f_n, 2f_n\}, \max\{f_n, 2f_n\}]$. Note that $\{M_n\}_{n=1}^\infty$ is not an interval-valued martingale. But $\mathrm{cl}\{\lambda f_n + 2(1-\lambda) f_n : \lambda \in [0,1] \cap \mathbb{Q}\}$ is a Castaing representation of regular martingale selections with $\liminf \int_\Omega \|M_n\| dP \leq 2 \liminf \int_\Omega |f_n| dP < \infty$.

Proof of Theorem 2:

Proof. (i) $\Longrightarrow$ (ii): Assume there exists $G \in \mathscr{L}^2_{\mathbb{A}}(K(\mathbb{R}^{r\times d}))$ such that $M_t = E(M_0)+ \int_0^t G_s dB_s$ a.s. for each t. Then $E(M_t) = \text{cl}\left\{E(M_0) + E(\int_0^t G_s dB_s)\right\} = E(M_0)$, which implies $E\left(\int_0^t G_s dB_s\right) = \{0\}$. By Lemma 2, we obtain that $G = \{g\}$ for some $g \in \mathscr{L}^2_{\mathbb{A}}(\mathbb{R}^{r\times d})$. Hence (ii) holds with $C = E(M_0)$.

(ii) $\Longrightarrow$ (iii): Since $E(M_0)$ is a closed subset of $\mathbb{R}^r$, there exists a sequence $\{x_n\}_{n=1}^{\infty} \subset \mathbb{R}^r$ such that $E(M_0) = \text{cl}\{x_n : n \in \mathbb{N}\}$. Then we have

$$M_t = \text{cl}\{x_n : n \in \mathbb{N}\} + \left\{\int_0^t g_s dB_s\right\} = \text{cl}\{f_t^n : n \in \mathbb{N}\}$$

with $f_t^n := x_n + \int_0^t g_s dB_s$, which apparently satisfies the desired requirement.

(iii) $\Longrightarrow$ (i): Assume there exists a sequence $\{f^n\}_{n=1}^{\infty}$ of $\mathbb{R}^r$-valued $\mathbb{A}$-martingales such that $M_t = \text{cl}\{f_t^n : n \in \mathbb{N}\}$ a.s. for all t. In terms of Theorem 3.4 of Revuz and Yor (2005), we have a stochastic integral representation $f_t^1 = E(f_0^1) + \int_0^t g_s dB_s$ for some $g \subset \mathscr{L}^2_{\mathbb{A}}(\mathbb{R}^{r\times d})$. Then, for any n, we have $f_t^n = E(f_0^n) + \int_0^t g_s dB_s$, since the difference $f_t^n - f_t^1$ is non-random and is independent of t. Thus

$$M_t = \text{cl}\{f_t^n : n \in \mathbb{N}\} = \{E(f_0^n) : n \in \mathbb{N}\} + \left\{\int_0^t g_s dB_s\right\} = E(M_0) + \int_0^t G_s dB_s,$$

where $G = \{g\}$. In fact, each f_0^n is constant a.s since $\mathscr{A}_0$ is trivial. □

We now give a representation theorem of interval-valued martingale, which is easy to deal with because it is determined by its endpoints.

Theorem 3. *Let $B = \{B_t : t \in [0,T]\}$ be the one-dimensional Brownian motion. Let $M = \{M_t = [a_t, b_t] : t \in [0,T]\}$ be a $K_c(\mathbb{R})$-valued square-integrable $\mathbb{A}$-martingale. Then there exists an interval-valued process $G \in \mathscr{L}^2_{\mathbb{A}}(K(\mathbb{R}))$ such that*

$$M_t = [E(a_0), E(b_0)] + \int_0^t G_s dB_s, \quad for\ every\ 0 \le t \le T \tag{3}$$

if and only if $\{a_t : t \in [0,T]\}$ is a square-integrable $\mathbb{A}$-martingale and $b_t - a_t$ is non-random and is independent of t. In this case, it holds that G degenerates to a singleton process $\{g\}$ with $g \in \mathscr{L}^2_{\mathbb{A}}(\mathbb{R})$.

Proof. This is a direct consequence of Theorem 2. □

Example 4. Let $B = \{B_t : t \in [0,T]\}$ be the one-dimensional Brownian motion. For $t \in [0,T]$, set $M_t = \left[e^{B_t - \frac{t}{2}}, 1 + e^{B_t - \frac{t}{2}}\right]$. Then $\{M_t : t \in [0,T]\}$ is an interval-valued martingale with the following martingale representation:

$$M_t = [1,2] + \left\{\int_0^t e^{B_s - \frac{s}{2}} dB_s\right\} \quad \text{a.s. for all } t.$$

In fact, $E\left(\int_0^T (e^{B_t-\frac{t}{2}})^2 dt\right) = e^T - 1 < \infty$ and hence $\left\{e^{B_t-\frac{t}{2}} : t \in [0,T]\right\} \in \mathscr{L}^2_{\mathbb{A}}(\mathbb{R})$. By Ito's formula, we have $e^{B_t-\frac{t}{2}} = 1 + \int_0^t e^{B_s-\frac{s}{2}} dB_s > 0$, and so $E\left(e^{B_t-\frac{t}{2}}\right) = 1$. Thus we obtain $E(M_t) = [1,2]$.

Example 5. Let $B = \{B_t : t \in [0,T]\}$ be the one-dimensional Brownian motion. Set $M_t = \left[e^{B_t-\frac{t}{2}}, 2e^{B_t-\frac{t}{2}}\right]$. Then $\{M_t : t \in [0,T]\}$ is an interval-valued martingale since both of the endpoints are martingales and $E(M_t) = [1,2]$. But M does not have a martingale representation since the difference between the endpoints of M_t is $e^{B_t-\frac{t}{2}}$, which is not a constant.

Interval-valued martingale is a concrete example of set-valued martingales with wide applications in statistical modelling and practical fields such as econometrics, mathematical finance etc. For example, Sun et al. (2018) proposed a threshold autoregressive model based on interval-valued data. In the model, $\{u_t : t \in [0,T]\}$ is an interval-valued martingale difference sequence (the difference between two intervals is the Hukuhara difference which guarantees the difference of two identical sets is zero). There is an assumption in the model: $E(u_t|I_{t-1}) = [0,0]$ where I_{t-1} is the information set up to time $t-1$. Unfortunately, the assumption is not appropriate for non-degenerate interval-valued stochastic process.

Acknowledgements. We are deeply indebted to Yoshiaki Okazaki, Kosuke Yamoto and Toru Sera for valuable discussions. We are also would like to thank the anonymous reviewers of the manuscript for valuable comments. Jinping Zhang was supported by Beijing Municipal Natural Science Foundation (No.1192015). The research of Kouji Yano was supported by JSPS KAKENHI grant No.'s JP19H01791, JP19K21834 and JP18K03441 and by JSPS Open Partnership Joint Research Projects grant No. JPJSBP120209921.

References

Aumann, R.: Integrals of set-valued functions. J. Math. Anal. Appl. **12**, 1–12 (1965)

Ezzaki, F., Tahri, K.: Representation theorem of set valued regular martingale: application to the convergence of set valued martingale. Stat. Probab. Lett. **154**, 108–148 (2019)

Hiai, F., Umegaki, H.: Integrals, conditional expectations and martingales of multivalued functions. J. Multivar. Anal. **7**, 149–182 (1977)

Kisielewicz, M.: Properties of generalized set-valued stochastic integrals. Discussiones Mathematicae Differ. Inclusions Control Optim. **34**, 131–147 (2014)

Kisielewicz, M.: Martingale representation theorem for set-valued martingales. J. Math. Anal. Appl. **409**, 111–118 (2014)

Jung, E.J., Kim, J.H.: On set-valued stochastic integrals. Stoch. Anal. Appl. **21**(2), 401–418 (2003)

Revuz, D., Yor, M.: Continuous Martingales and Brownian Motion, 3rd edn. Springer, Heidelberg (2005). https://doi.org/10.1007/978-3-662-06400-9

Sun, Y., Han, A., Hong, Y., Wang, S.: Threshold autoregressive models for interval-valued time series data. J. Econometrics **206**(2), 414–446 (2018)

Zhang, J.: Integrals and stochastic differential equations for set-valued stochastic processes. Ph.D. thesis, Saga University, Japan (2009)

Zhang, J., Li, S., Mitoma, I., Okazaki, Y.: On set-valued stochastic integrals in an M-type 2 Banach space. J. Math. Anal. Appl. **350**, 216–233 (2009)

Zhang, J., Mitoma, I., Okazaki, Y.: Submartingale property of set-valued stochastic integration associated with Poisson process and related integral equations on Banach spaces. J. Nonlinear Convex Anal. **22**(3), 775–799 (2021)

Author Index

L. A. García-Escudero et al. (Eds.): SMPS 2022, AISC 1433, pp. 407–408, 2023.
https://doi.org/10.1007/978-3-031-15509-3